Acoustics—A Textbook for Engineers and Physicists

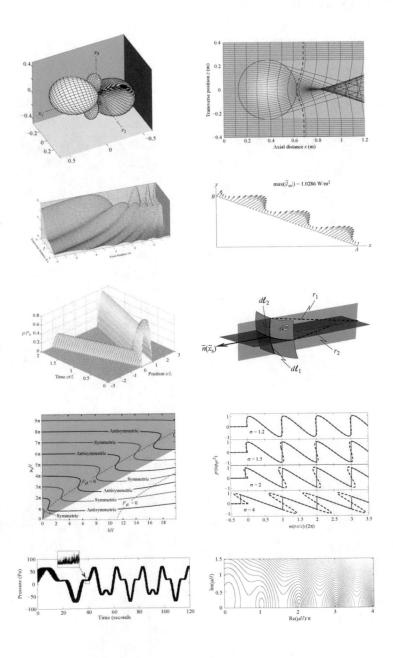

Jerry H. Ginsberg

Acoustics—A Textbook for Engineers and Physicists

Volume I: Fundamentals

Jerry H. Ginsberg
G. W. Woodruff School of Mechanical
    Engineering
Georgia Institute of Technology
Dunwoody, GA
USA

ISBN 978-3-319-86016-9   ISBN 978-3-319-56844-7   (eBook)
DOI 10.1007/978-3-319-56844-7

Library of Congress Control Number: 2017937706

Printed on acid-free paper

This Springer imprint is published by Springer Nature
The registered company is Springer International Publishing AG
The registered company address is: Gewerbestrasse 11, 6330 Cham, Switzerland

The ASA Press

The ASA Press imprint represents a collaboration between the Acoustical Society of America and Springer dedicated to encouraging the publication of important new books in acoustics. Published titles are intended to reflect the full range of research in acoustics. ASA Press books can include all types of books published by Springer and may appear in any appropriate Springer book series.

 ASA Press

The Acoustical Society of America

On December 27, 1928, a group of scientists and engineers met at Bell Telephone Laboratories in New York City to discuss organizing a society dedicated to the field of acoustics. Plans developed rapidly and the Acoustical Society of America (ASA) held its first meeting on May 10–11, 1929, with a charter membership of about 450. Today, ASA has a worldwide membership of 7000.

The scope of this new society incorporated a broad range of technical areas that continues to be reflected in ASA's present-day endeavors. Today, ASA serves the interests of its members and the acoustics community in all branches of acoustics, both theoretical and applied. To achieve this goal, ASA has established technical committees charged with keeping abreast of the developments and needs of membership in specialized fields as well as identifying new ones as they develop.

The technical committees include acoustical oceanography, animal bioacoustics, architectural acoustics, biomedical acoustics, engineering acoustics, musical acoustics, noise, physical acoustics, psychological and physiological acoustics, signal processing in acoustics, speech communication, structural acoustics and vibration, and underwater acoustics. This diversity is one of the Society's unique and strongest assets since it so strongly fosters and encourages cross-disciplinary learning, collaboration, and interactions.

ASA publications and meetings incorporate the diversity of these technical committees. In particular, publications play a major role in the Society. *The Journal of the Acoustical Society of America* (JASA) includes contributed papers and patent reviews. *JASA Express Letters* (JASA-EL) and *Proceedings of Meetings on Acoustics* (POMA) are online, open-access publications, offering rapid publication. *Acoustics Today*, published quarterly, is a popular open-access magazine. Other key features of ASA's publishing program include books, reprints of classic acoustics texts, and videos.

ASA's biannual meetings offer opportunities for attendees to share information, with strong support throughout the career continuum, from students to retirees. Meetings incorporate many opportunities for professional and social interactions, and attendees find the personal contacts a rewarding experience. These experiences result in building a robust network of fellow scientists and engineers, many of whom became lifelong friends and colleagues.

From the Society's inception, members recognized the importance of developing acoustical standards with a focus on terminology, measurement procedures, and criteria for determining the effects of noise and vibration. The ASA Standards Program serves as the Secretariat for four American National Standards Institute Committees and provides administrative support for several international standards committees.

Throughout its history to present day, ASA's strength resides in attracting the interest and commitment of scholars devoted to promoting the knowledge and practical applications of acoustics. The unselfish activity of these individuals in the development of the Society is largely responsible for ASA's growth and present stature.

To Leah Morgan, Elizabeth Rachel,
and Abigail Rose, my grandchildren.
Each is talented, each is beautiful,
each is unique, each is amazing.
I love them.

Preface

The Basic Concept

As is stated by its title, this is a textbook. It is not a treatise that will make the reader an expert in any topic. However, it also is not merely an introduction to acoustics. The intent is to provide students with a foundation of core concepts and tools that will permit in-depth study of any of the physics and engineering acoustics specialties. This book, like my others in engineering dynamics and vibrations, is based on some basic precepts. A textbook must prepare a student to work in an area where phenomena are complicated. These complications must be addressed in a manner that is accessible to students. Rather than merely deriving and applying basic formulas, a textbook should assist the reader to recognize the connection between formulas and physical phenomena and correspondingly to use understanding of each of these aspects to anticipate and explain results from the other. Equally importantly, a well-done text should motivate students for further study by exposing them to the many remarkable phenomena that make the subject of acoustics so interesting. In view of these objectives, the coverage is quite comprehensive. I have not found some of the topics fully addressed in other textbooks. Indeed, some I developed especially for this book. Furthermore, almost every chapter has features that are not commonly encountered in other texts. In my estimation, complete coverage of all topics would require more than a two-semester course sequence. In recognition of the conflict between depth and breadth, each chapter is organized such that instruction of the early sections, which address fundamental concepts, should be adequate to proceed to the next chapter.

I have taken great care to explain analytical steps, the physical interpretation of key results, and especially how the student should proceed to solve problems in each topic. Some of the derivations and explanations I believe are unique to this book. Examples are numerous, and most are more than simple applications of derived formulas. Rather, they are intended as case studies that explain why the example is important, why the solution proceeds as it does, how to perform unfamiliar operations, what can be learned from the results about fundamental behaviors, and why the qualitative aspects of the results are consistent with the underlying fundamental principles. Some examples analyze systems by more than one method. This serves to enhance the student's fundamental understanding of the underlying physical processes, as well as enhancing the ability to make the appropriate line of attack when confronted with a new situation. All of the 122 examples are my own creation. With the exception of established standards, all data and graphs are newly generated. The advent and wide availability of computational software is exploited to lend greater realism to some examples. When the usage of software entails any potentially problematic aspects, especially concerning algorithms and their implementation, those issues are addressed explicitly, sometimes with program fragments. An examination of the List of Examples will help one see the diversity of the examples that receive this type of treatment. In recognition of the importance of computations, and as an aid to students to succeed in their efforts to solve homework exercises, the MATLAB code used to solve the examples is available for download from the Springer server.

A broad range of physical principles and mathematical concepts arise in the various topics of acoustics, but students enter the study of acoustics from a variety of paths. Consequently, it is likely that a student cohort will not have equal preparation to handle the task at hand. The presentation assumes that the reader is familiar with the basic concepts of algebra, vector analysis, calculus, and ordinary differential equations with constant coefficients, but little else. Any aspect of a derivation or problem solution that has a degree of subtlety is fully explained. Mathematics is essential, but this book is not a treatise in applied mathematics. Nevertheless, some topics do require sophisticated operations. If that is the case, the steps are carefully explained.

One of the features that will not be found herein is extensive use of analogies. This decision stems in part from the diverse background of students. An analogy that is meaningful to an electrical engineering student might be merely something else to learn for a physics student. Furthermore, I believe that one should learn a subject at its core. Analogies to me are equivalent to listening or reading a foreign language by translating it mentally. The best way to learn a foreign language is to become immersed in it, and a thorough immersion is the best way to learn acoustics. Recognition of connections between subjects is best attained by understanding each profoundly.

Technical Content

The organization of the topics is my vision of how the realm of acoustics fits together. My education was in civil engineering as an undergraduate and then in engineering mechanics as a graduate student. I had a rather extensive background in applied mathematics, dynamics, vibrations, and solid and fluid mechanics, but no formal education in acoustics. (I did become involved for a short time with a project involving sonic booms prior to embarking on my doctoral work analyzing nonlinear vibrations of shells.) My acoustics efforts began six years after I became an assistant professor. That work was in nonlinear waves, and it caused me to begin to interact with acousticians. Several years later, I had the good fortune to meet Allan Pierce when I moved to Georgia Tech. It was he and his wonderful book, *Acoustics*, that were responsible for most of my early education in the subject. I continued to work in nonlinear acoustics and also pursued research in structural acoustics and vibration. My interests were broadened by teaching a two-semester acoustics sequence several times. With each meeting of the Acoustical Society of America, I came to realize how much was left to learn. Writing this book has served to fill that gap because it has caused me to delve into topics with which I had little prior experience. This is especially so because I wished that each topic be given an extensive treatment, and I could not write without thoroughly understanding what I was writing about.

As I said, the content is organized in a manner that seems to me to be most logical. Much of Chap. 1 is not specific to acoustics. It introduces the basic tools of complex representation of harmonic functions, Fourier series, complex frequency response functions, and fast Fourier transforms, all of which are core concepts for any discipline that is concerned with oscillations. However, the discussion of frequency bands, musical scales, and noise models distinguishes this treatment from those for other applications. All readers should find the last example, which analyzes the spectral properties of noise data measured in the open fuselage of NASA's SOFIA aircraft, to be quite interesting.

Chapter 2 is devoted to time-domain analysis of plane waves. One-dimensional conservation laws are derived as the foundation of the derivation of the wave equation. The derivation of the d'Alembert solution follows a heuristic, but rigorous, approach. This solution, which describes waves in an unbounded medium, is used as the building block for construction of solutions for initial conditions and for boundary excitation of semi-infinite and closed waveguides. This is done by the development of the method of wave images, which converts multiple reflections to an equivalent problem in an unbounded medium. This content is somewhat lengthy to develop, but it offers numerous rewards. Most importantly, it greatly enhances a student's ability to switch between spatial and temporal views of waves. The development introduces some basic phenomena, such as multiple reflections, standing waves, natural frequencies, modal patterns, and resonance, through a set of simple operations. Doing so should enhance a student's desire for further study. The results of Example 2.7, which concerns the transient response of a closed waveguide that is resonantly excited, are likely to surprise many readers.

Chapter 3 examines from a frequency-domain perspective many of the topics in Chap. 2. This arrangement is chosen in order to avoid the tendency of students to forget that acoustics signals evolve in time, as well as propagating spatially. The frequency-domain formulation is essential because it allows for the development of techniques that are required for more general systems, and it allows for introduction of the local impedance model for boundaries. One of the highlights of this chapter is the analysis of synthetic data for an impedance tube, which amply demonstrates how small measurement errors can seriously lessen the quality of experimental

results. The treatment of various dissipation mechanisms and their influence on the
propagation properties is comprehensive, yet it should be accessible to students
who do not have an extensive background in fluid mechanics and thermodynamics.
The treatment of waveguide networks using transfer functions is a concept I refined
for this book. The topic is not essential for later study, but the variety of systems it
can analyze serves to motivate students by allowing them to exercise their cre-
ativity, simultaneously with introducing them to real-world issues.

Chapter 4 focuses on the core principles governing multidimensional waves. The
derivations of the continuity equation and of the momentum-impulse equation differ
from those in Chap. 2 by using the tools of vector calculus, which also are needed
for many later topics. The analysis of plane waves propagating in an arbitrary sense
in a three-dimensional sense introduces the fundamental concept of a wavenumber
vector without simultaneously dealing with the task of solving the wave equation in
several spatial coordinates. The concluding example in Chap. 4 uses a crude model
of the field radiated by a piston to demonstrate the utility of the power balance
principle. Seeing how the combination of a basic principle and some fundamental
reasoning can provide a "back of the envelope" prediction reinforces a student's
confidence that acoustics is an accessible subject.

Various phenomena that arise when a plane wave is obliquely incident at a
planar interface are the subject of Chap. 5. This chapter establishes the concepts of
subsonic and supersonic trace velocities, which are core factors for a wide range of
later topics. Diagrams illustrating the difference between the propagating and
evanescent transmitted fields enhance the mathematical descriptions. To my
knowledge, the algorithm for propagation through multiple layers has not appeared
in prior publications.

The time-domain analysis of radially symmetric spherical waves opens Chap. 6.
The study of transient radiation from a vibrating sphere is followed by a thorough
exploration of the transient waves in a spherical cavity. The example that illustrates
the analysis of the singularity at the center tends to be especially interesting to
students because it leads to the concept underlying a lithotripter. The sequence of
topics for the frequency-domain analysis of spherical waves follows the
time-domain treatment. Intensity and power flow for combinations of sources
receive special attention. A thorough exposition on the meaning of a Green's
function leads to the concept of a point source. Combinations of point sources to
create dipole and quadrupole fields are analyzed by Taylor series expansions. The
reader will find the diagrams illustrating various combinations of lateral and longitudinal quadrupoles to be quite informative. Chapter 6 closes with an examination
of the Doppler effect for various moving sources. The analysis and results for a
supersonically moving, harmonically varying, point source were developed for this
book. Many individuals might be unaware of the effects that are identified.

Radiation from vibrating objects is the subject of Chap. 7. The development
begins with derivation of the general spherical harmonic solution for axisymmetric
signals. The first applications are to vibrating spheres for cases of an imposed
vibration pattern, followed by forced response of an elastic spherical shell. An
important aspect is a detailed examination of the influence of the acoustic surface

impedance on the occurrence of fluid-loaded resonances. Radiation from infinitely long cylinders leads to a representation of the field as a result of helical surface waves. The Kirchhoff-Helmholtz integral theorem for enclosed and open domains is derived. An example uses the analytical solution for the radially symmetric field interior to a vibrating sphere to generate the surface pressure required to formulate KHIT. The pressure along a line running from the center of the sphere out to a distant location is found by evaluating the integral. Examination of the result very close to the surface illustrates the analytical prediction that the pressure computed from the theorem at a field point interior to a radiating body will be the actual pressure, whereas the computed pressure at any interior point will be zero. Moreover, it goes on to presage the limiting form known as the Surface Helmholtz Integral Equation, which is the basis for several boundary element formulations. Finite length effects for radiation from a cylinder are assessed with a pair of examples, one for a line source and the other using the Helmholtz integral theorem. Once again, the objective of these examples is to demonstrate that one can employ their knowledge of fundamental concepts and the behavior of related systems to address issues that would not be accessible otherwise. This chapter closes with treatments of numerical techniques that are frequently used to evaluate radiation from nonstandard bodies, specifically source superposition, boundary elements, and finite elements.

Although the field generated by a piston embedded in a baffle is a type of acoustic radiation, it is treated separately in Chap. 8. This arrangement was chosen because analytical results are available for a wide range of parameters and locations, and also because the phenomena have much importance. Students tend to be motivated by the relevance of the results to their experiences. The standard topics are addressed, but extensive mappings of the entire field, which recall a classical computation, are seldom found in other texts. Seeing these properties helps students understand the rational for a simplified standard model and also serves to place the various specialized analyses on a common foundation. An example applies the radiation impedance formulas in combination with Fourier series analysis to evaluate the instantaneous power required to make a piston execute a square wave oscillation. Students interested in transducers find this example to be quite interesting.

The field within a waveguide is the subject of Chap. 9. First to be studied is the Webster horn equation for one-dimensional waveguides. The closure of this topic is an example that compares the analysis according to the WKB solution to a numerical analysis for the case of an extreme narrowing of the cross section. The discussion of these analyses, which shows that both have faults, is intended to encourage students to be critical of any results with which they are presented. The remainder of Chap. 9 is devoted to two-dimensional and three-dimensional rectangular and cylindrical waveguides. Modal decomposition of the pressure dependence transverse to the direction of propagation is a common thread. Special consideration is devoted to the analysis of waveguides whose walls are locally reactive. An example of the analysis of a two-dimensional waveguide whose walls are elastic plates has numerous interesting features, including a novel technique for identifying dispersion curves.

Chapter 10 is devoted to the sound field within an enclosed region. It begins by using the waveguide representation to explain the alternative descriptions of the field as a set of waves that propagate in multiple directions, or as a set of cavity modes that are standing waves. Both analytical descriptions are developed, with emphasis on the situations where each is best employed. An example uses an infinite series of cavity modes to describe the field within a two-dimensional rectangular enclosure due to a point source. Using the viewpoint of the method of images to assess the modal solution serves multiple purposes, including refreshing students' capabilities with earlier topics, recognizing that the selection of an ana-lytical approach sometimes requires consideration of what one wishes to learn, and understanding of the computational issues that might arise when a modal description is used for highly localized excitations. The closure of this chapter describes two widely employed approximate treatments: the Rayleigh-Ritz method and an analysis based on Dowell's approximation. The former is often included in standard texts, but the latter is relatively recent.

Chapter 11 is devoted to geometrical acoustics. This is a rather thorough development of the subject, except for the exclusion of mean flow effects. The development begins with a discussion of the basic concepts associated with the description of rays and wavefronts. An example of reflection from a spherical mirror introduces the concept of a caustic. The initial development is limited to vertically stratified media, which is typified by a large body of water heated from above. Descriptions of the ray path and time of travel are derived as integrals. Examples consider a channel whose vertical sound speed varies quadratically with depth, and a channel whose sound speed profile was suggested as a prototype for studies of propagation in the ocean. The former leads to a sequence of foci with increasing range, whereas shadow zones and caustics are displayed for the latter sound speed profile. Surface reflections are shown to have a major effect. The treatment of vertically stratified fluids closes with discussions of how an eigenray may be determined and how energy conservation may be employed to analyze the amplitude dependence along a ray. The next part of Chap. 11 is devoted to ray tracing equations, which describe propagation through an arbitrary heterogeneity. The application of numerical methods to the solution of these equations, as well as the transport equation, is given a detailed development. The primary example for these topics analyzes refraction of a plane wave as it passes transversely through a cylindrical region in which the sound speed is a function of radial distance from the axis. The field is shown to contain a single caustic. A fine-scale scan shows that the wavefront is folded after it passes through the caustic, and the numerical evaluation of the transport equation shows the amplitude enhancement in the vicinity of the caustic. This chapter closes with the presentation of Fermat's principle. The cal-culus of variations is developed here in the context of a proof that the Euler-Lagrange equation derived from Fermat's principle is the same as the general ray tracing equations.

Scattering from bodies surrounded by a fluid is the subject of Chap. 12. The Born approximation is derived and employed to evaluate monostatic scattering of a plane wave at arbitrary incidence to a cylindrical region containing a second fluid. Rayleigh scattering and its relation to the Born approximation are the next topic. The metrics commonly sought from a scattering study, such as target strength, are discussed. After that, Kirchhoff scattering theory and its relation to geometrical acoustics are explored. This chapter closes with the application of spherical harmonics to analyze scattering from a rigid sphere and a spherical shell. These studies shed light on the transition from Rayleigh to Kirchhoff scattering, as well as the fundamental importance of fluid loading relative to elasticity.

Chapter 13 closes the textbook with an exploration of nonlinear acoustic analyses and phenomena. The bulk of the chapter is devoted to simple plane waves. The Riemann solution, which is derived by the method of characteristics, is used as the foundation for qualitative and quantitative explorations of the distortion of waveforms and spatial profiles. Graphical interpretations are given much attention. FFT analysis of waveforms in an example of a multiharmonic excitation is used to anticipate the Fubini-Ghiron analysis of frequency content. Identification of criteria that mark the existence of shocks is followed by derivation of the Rankine-Hugoniot relations for weak shocks. These conditions lead to the equal-area rule for fitting shock discontinuities to a waveform. A numerical algo rithm for implementing the equal-area rule is developed and applied to monitor an initially harmonic signal from its inception to the old-age stage. This development goes on to apply the equal-area rule to follow propagation of an initial square wave, which ultimately leads to explanation of the phenomenon of acoustic saturation. A nonlinear wave equation is derived as the basis for study of multidimensional nonlinear waves. Its first usage leads to differential equations governing the position dependence of Fourier series coefficients for a plane wave. The effects of dissipation are incorporated into these equations, which are solved numerically. The solution illustrates how dissipation effects tend to work counter to nonlinear distortion effects. Perturbation analysis techniques for the nonlinear wave equation are developed on the foundation of plane waves and then extended to radially symmetric spherical waves. This chapter closes with a study of the waves that radiate from a periodically supported vibrating plate. One of the interesting results of that analysis is interpretation that rays rotate periodically based on the tangential particle velocity. The result is the formation of caustics that periodically move closer and farther from the plate. These phenomena are not observed in nonlinear plane waves nor in any type of linear acoustic wave.

Although the overriding precept of the text is that all topics must be fully explained as they arise, two appendices are provided for further assistance. One is devoted to derivation of the coordinate transformations and vector differential operators in spherical and cylindrical coordinates. The second describes Fourier transforms and their application. Fourier transforms are invoked in the body of the text only for a few topics. One such situation is the analysis of radiation from a cylinder whose vibration pattern varies arbitrarily in the axial direction. The means for lessening the importance of this mathematical tool is the numerical implementation of FFT concepts. Nevertheless, familiarity with Fourier transforms is an essential skill for advanced studies, as well as for understanding much of the technical literature. This recognition was the primary reason that this mathematical tool was included.

Allan Pierce, in the Preface to *Acoustics*, wrote that "…a deep understanding of acoustical principles is not acquired by superficial efforts." I fully agree with this statement, yet somewhat paradoxically believe that such understanding is accessible to all students. The key is the support afforded to the student by the instructor, textbook, and supporting materials. I cannot do much about the discourse between student and instructor, but I have done my best to address the formal foundation for instruction. It is up to you, the reader, to decide whether I have met my objectives.

Dunwoody, GA, USA   Jerry H. Ginsberg

The original version of the book was revised: The Electronic Supplementary Materials have been included. The correction to the book is available at
https://doi.org/10.1007/978-3-319-56844-7_7

Acknowledgements

I am indebted to many individuals for providing motivation to write these books. Above all, neither would exist if I had not met my good friend, Allan Pierce, when I interviewed in 1980 for a professorship at the Georgia Institute of Technology. Working with him convinced me to extend my knowledge of acoustics beyond the specialized subject of nonlinear acoustics that was part of my early career. Over the years, our discussions were quite revelatory regarding where there were gaps in my knowledge of the subject. Furthermore, I hope he is not offended by this remark, but learning from, and then teaching from, his book convinced me of the necessity that I write *Acoustics—A Textbook for Engineers and Physicist*.

In the six-year interval during which I wrote these books what I needed most was assurance that the effort was worth pursuing. Some of my colleagues at Georgia Tech were quite supportive. Karim Sabra convinced me on several occasions that I would be filling an important need. Students solving the homework exercises should thank him because he suggested that the MATLAB code I used for the examples should be publically available. Pete Rogers provided extremely useful critical remarks for an early draft of my treatment of geometrical acoustics. My former Ph.D. students, especially J. Gregory McDaniel at Boston University and Kuangchung Wu at the NSWC Carderock Division of the Naval Sea Systems Command, were especially enthusiastic. I also am indebted to those attendees at many Acoustical Society of America meetings who I waylaid to discuss my writing efforts. They are too numerous to list, and I am sure that I have forgotten some names, but I greatly appreciate their attention. I owe Mark Hamilton of the University of Texas at Austin a special debt because he convinced me to participate in the ASA Book program under the aegis of Springer Publishing. Sara Kate Heukerott, my Editor at Springer, was quite understanding of my requests. Her expertise was a great aid as we assembled this project. Some might be surprised at the inclusion of my granddaughter, Leah Morgan Ginsberg, in the list of folks deserving recognition. Early in the writing stage, because she was a proficient clarinetist, I sought her assistance for the discussion of music in Chap. 1. Then at the conclusion, as she approached graduation from Georgia Tech in the

G. W. Woodruff School of Mechanical Engineering, from which I had retired, she served as my sounding board and spokeswoman for students when I deliberated how best to disseminate this work.

In addition to Leah's role, my family was essential to the effort. The forbearance of my wife, Rona, while I focused on writing, ignored other responsibilities, and forgot many things that I still cannot remember, astonishes me, even now that my efforts are over. She went through this experience before when I wrote my prior books on statics, dynamics, and vibrations. However, none of those experiences could have prepared her for the intensity and duration of the present effort. My sons, Mitchell and Daniel, had similar experiences when they lived at home. Although they and their wives, Tracie and Jessica, were not as strongly impacted now, I greatly appreciate their forbearance when I was not as communicative as I should have been. My granddaughters, Leah, Beth, and Abby, inspire me by their dedication to their own activities. I hope that recognition of the pleasure their Papa derived from creating these books will inspire them.

Dunwoody, GA, USA   Jerry H. Ginsberg

Contents

About the Author

Jerry H. Ginsberg began his technical education at the Bronx High School of Science, from which he graduated in 1961. This was followed by a B.S.C.E. degree in 1965 from the Cooper Union and an E.Sc.D. degree in engineering mechanics from Columbia University in 1970, where he held Guggenheim and NASA Fellowships. From 1969 to 1973, he was an Assistant Professor in the School of Aeronautics, Astronautics, and Engineering Science at Purdue University. He then transferred to Purdue's School of Mechanical Engineering, where he was promoted to Associate Professor in 1974. In the 1975–1976 academic year, he was a Fulbright Hayes Advanced Research Fellow at the École Nationale Supérieure d'Électricité et de Mécanique in Nancy, France. He came to Georgia Tech in 1980 as a Professor in the School of Mechanical Engineering, which awarded him the George W. Woodruff Chair in 1989. He retired in June 2008. His prior publications include five textbooks in statics, dynamics, and vibrations, most in several editions, as well as more than one hundred and twenty refereed papers covering these subjects. Dr. Ginsberg became a Fellow of the Acoustical Society of America in 1987 and a Fellow of the American Society of Mechanical Engineers in 1989. The awards and recognitions he has received include Georgia Tech Professor of the Year (1994), ASEE Archie Higdon Distinguished Educator in Mechanics (1998), ASA Trent-Crede Medal (2005),

ASME Per Bruel Gold Medal in Noise Control and
Acoustics (2007), and the ASA Rossing Prize in
Acoustics Education (2010). In addition to his tech-
nical activities, he is an exceptional photographer.

List of Examples

Chapter 1
Descriptions of Sound

Acoustics is the science of sound. For most people, the word "sound" is synonymous with "hearing," but the realm of acoustics is far greater than phenomena associated with audible signals. The signals we hear are sound waves, in which pressure fluctuates. Hearing is a complex process of detecting and decoding these signals. Our ability to hear is as marvelous as any of our other senses. Indeed, no mechanical device can match the auditory capability of the average person. Thus, it might be appropriate to say that those of us who have an undiminished ability to hear are already acoustical experts. Many individuals take this sense for granted, but the mere fact that you are reading this book indicates that you have some awareness of sound as a basic experience that needs to be better understood. Our emphasis will be ways that sound is generated and modified by the environment in which it occurs. The topics in this text are grounded in physics and mathematics. However, the way we interpret sound cannot be ignored because it dictates the properties of a sound signal that we must characterize and evaluate. Understanding this aspect of acoustics requires expertise in biology and psychology.

Our hearing mechanism is quite complicated. A simplified version is that a major component of the signals received by the brain originates at the ear drum. The ear drum is a membrane that is stretched across the ear canal. It vibrates as a result of fluctuating forces associated with external pressure oscillations relative to the ambient atmospheric pressure. This vibration is magnified by a chain of tiny ear bones attached at one end to the ear drum and at the other end to nerves that transmit signals to the brain. We also respond to motion of the hair cells in the ear canal, which is induced by the oscillating air flow associated with the pressure fluctuation. Given these mechanisms, it is not surprising that key elements of an acoustical analysis will be determination of the spatial and time dependence of both the pressure fluctuation and the velocity of fluid particles.

Usually, we hear acoustical signals in the air, but swimmers might hear a passing boat or other noises while submerged. However, it actually is too restrictive to limit the term "sound" to pressure signals that are heard, because there are many

Electronic supplementary material The online version of this chapter (DOI 10.1007/978-3-319-56844-7_1) contains supplementary material, which is available to authorized users.

© Springer International Publishing AG 2018, corrected publication 2021
J.H. Ginsberg, *Acoustics—A Textbook for Engineers and Physicists, Volume I*,
DOI 10.1007/978-3-319-56844-7_1

signals that occur at frequencies (a term that will figure prominently in our acoustics study) outside the range of human audibility. In some cases, this frequency might be extremely low, in which case it is referred to as *infrasound*. An elephant's ability to sense infrasound is well documented, but some studies suggest that they actually are sensing ground vibration through their feet. This is comparable to humans who sense very low frequencies as a pounding sensation in the chest. Sounds at frequencies that are higher than humans can hear are said to be *ultrasound*. Dogs have a keen ability to hear ultrasound. Bats use ultrasound to locate their food in a process of echo location, which is a similar process to that used by marine mammals. However, limiting the term "sound" to signals that are within the hearing ability of living beings is still far too restrictive, for there are many applications, especially in the areas of biomedicine and nondestructive testing, where the signals occur at frequencies that are a factor of one hundred or more greater that which can be heard by any animal.

For the purposes of this book, sound will be taken to mean pressure signals that occur in a fluid, either liquid or gas, which in the beginning will be taken to have ideal properties, especially that it is inviscid. Under this restriction, the governing equations, other than those describing the fundamental material properties, apply equally to gases and liquids. Later, we will remove the restriction to ideal properties, but for the most part doing so will only differentiate gases and liquids by the amount and conditions for which the various nonideal effects become significant.

Although propagation of stress waves in solids is an essential aspect of nondestructive testing and some biomedical applications, and research articles on stress waves are frequently published, wave propagation through solid materials is not explored here. This subject is best studied after one has become familiar with the manner in which signals propagate through liquids. This is so because many of the concepts and methods encountered in the study of stress waves also arise for signals in fluids. The governing equations for fluids are more readily formulated and solved because ideal fluids can sustain only a pressure. In contrast, solid materials can sustain a general stress state that consists of direct and shear stresses whose values depend on the plane to which they are referenced.

Many technical societies headquartered around the world are dedicated to various aspects of acoustics. The largest is the Acoustical Society of America, whose members come from many nations. A good way to recognize the breadth of interest is to consider the emphasis of each of its thirteen technical committees.

♦ The ASA Acoustical Oceanography Technical Committee is concerned with the development and use of acoustical techniques to measure and understand the physical, biological, geological, and chemical processes and properties of the sea.

♦ The ASA Architectural Acoustics Technical Committee is interested in phenomena relevant to buildings and their environment, including the physical and psychological acoustical character of rooms, sound enhancement systems, and noise transmission and control for buildings.

♦ The ASA Animal Bioacoustics Technical Committee is concerned with all aspects of sound in nonhuman animals, including communication, sound production, auditory mechanisms, and the effects of human-made noise.

♦ The ASA Biomedical Acoustics Technical Committee is concerned with the interaction of sound with biological materials, with a specific interest in diagnostic and therapeutic applications of ultrasound. The technical concern is with the interactions of acoustic waves with biological materials, including cells, tissues, organ systems, and entire organisms.

♦ The ASA Engineering Acoustics Technical Committee is dedicated to concepts and methods for developing and improving acoustical devices for generating and measuring sound. These devices are used in a broad range of applications, including audio engineering, acoustic holography and acoustic imaging, acoustic instrumentation and monitoring, and underwater sonar.

♦ The ASA Musical Acoustics Technical Committee focuses on the application of science and technology to the field of music. Topics of particular current interest include the physics of musical sound production, music perception and cognition, and the analysis and synthesis of musical sounds and composition.

♦ The ASA Noise Technical Committee is dedicated to the study of noise generation and propagation, passive and active noise control, and the effects of noise. Specific interests include sound source mechanisms, propagation, perception, prediction, measurement, evaluation, analysis, effects, regulation, mitigation, and legal aspects of noise.

♦ The ASA Physical Acoustics Technical Committee is concerned with fundamental acoustic wave propagation phenomena, including transmission, reflection, refraction, interference, diffraction, scattering, absorption, dispersion of sound, and sonoluminescence, and their use to study and alter physical properties of matter.

♦ The ASA Psychological and Physiological Acoustics Technical Committee is dedicated to the investigation and dissemination of information about psychological and physiological responses of humans and animals to acoustic stimuli. Some of the areas of interest are detectability of sound in a noisy environment, description and recognition of sound, and biological hearing mechanisms.

♦ The ASA Signal Processing in Acoustics Technical Committee is concerned with encoding and extracting information from acoustic signals, which encompasses a broad range of applications in telecommunications, sonar, remote sensing, noise control, and acoustical holography.

♦ The ASA Speech Communication Technical Committee focuses on the production, transmission, and perception of speech. Some areas of specific interest are the physiology and mechanics of speech production, speech perception for native and foreign languages, speech perception, speech and hearing disorders, and the neuroscience of speech.

♦ The ASA Structural Acoustics & Vibration Technical Committee is dedicated to the development and improvement of methods for measuring, analyzing, and controlling the dynamic interaction of machines and structures with their surrounding environment. Some topics of interest are acoustical radiation and scattering from structures, shock and vibration response, active control, and methods for harvesting and dissipating energy.

♦ The ASA Underwater Acoustics Technical Committee is concerned with all aspects of underwater sound resulting from both natural and human-made sources. Particular attention is paid to sound generation underwater and the propagation, reflection, and scattering of sound in the underwater environment including the seabed and the sea surface.

These committees provide a home for society members sharing interest in that specialization, but many members are active in more than one committee. (Incidentally, ASA fosters interactions with all individuals, especially students, who are interested in acoustics. The publications and activities it provides its members make it a great value to join, especially because its dues for students are maintained at a low level.)

Although this book is restricted to sounds that propagate through fluids, and it does not delve into the biological and psychological aspects of hearing, the topics that are explored greatly facilitate learning about all aspects of acoustics. At some time, every member of each ASA technical committee has made use of some of the developments to be found herein. Furthermore, most acousticians find that the emphasis of their work shifts as their career advances. Thus, every topic covered in this text might some day be a vital resource for you.

1.1 Harmonic Signals

What is sensed as sound is the fluctuation of pressure relative to ambient conditions. Ambient pressure might be the atmospheric pressure at sea level, or the enormous pressure at a great ocean depth, or the much reduced pressure in the upper atmosphere. This ambient pressure has little direct effect on an acoustic analysis. The quantity that is significant to an acoustics investigation is the amount by which the pressure differs from the ambient value. This quantity is called the *pressure perturbation* or acoustic pressure, but it often is the case that it is merely called the pressure.

The basic building block in acoustics is a sinusoidal pressure fluctuation. There are multiple reasons for the importance of this type of signal. It represents a pure tone from which the music scale is constructed. It is for this reason that a sinusoidal signal is synonymously referred to as a *harmonic signal*, even if it is outside the audible range. Sinusoidal signals have the property that they are the simplest to analyze because of the nature of the equations that govern acoustics. Equally important, we will see that we can determine how a system responds to any excitation if we can determine how it responds to harmonic excitation.

1.1.1 Basic Properties

The pressure signals plotted in Fig. 1.1 are two harmonics p_a and p_b that only differ by the instants at which they are zero. Both signals have the same *amplitude A*, which is their largest absolute value. The time interval in which each pattern repeats is the *period T*. Each repetition is a cycle. The number of cycles in a unit of time is the reciprocal of T; it is the *cyclical frequency f*,

$$\boxed{f = \frac{1}{T}} \tag{1.1.1}$$

If T is measured in seconds, then f has units of *hertz* (Hz).

The only difference between the two signals in Fig. 1.1 is that one is shifted in time relative to the other. Which is shifted, and whether the shift is earlier or later, is ambiguous because of the repetitive nature of their pattern. A reference signal

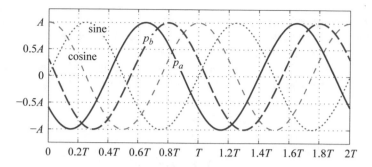

Fig. 1.1 Harmonic signals having the same amplitude and frequency

is needed to define time shifts. (The best way to visually identify time shifts is to compare the portion of the signal where it crosses the zero axis from negative to positive values.) This reference might be a true sine function (the dotted line) or a true cosine function (the dashed line), or it could be a known signal, such as an excitation that is applied to a system. Specification of the reference signal does not remove the ambiguity, as exemplified by the fact that we can take p_a to occur prior to the sine by $0.6T$ or after the sine by $0.4T$. Similarly, p_b can be considered to occur $0.2T$ before, or $0.8T$ after, the cosine. What is needed is a convention, that is, an agreement to use a common standard. The most common one is to restrict the amount of advance or retardation of one signal relative to another to be less than or equal to one half of a period. (When the time shift is exactly $T/2$, it does not matter if it is taken to be an advancement or retardation.)

The mathematical description of a harmonic signal can be posed in terms of a sine or cosine function as

$$p = A \sin(\omega t + \phi_s) \text{ or } p = A \cos(\omega t + \phi_c) \qquad (1.1.2)$$

The only parameter in these expressions that actually appears in Fig. 1.1 is the *amplitude A*. The term used to refer to ω is the *circular frequency*. Novices often confuse ω and f because both are referred to as frequencies. They are indeed related by a simple proportionality, which is obtained by recognizing that a sine function's value repeats when its argument is increased by 2π, and the time increment for such a repetition must be T, so it must be that $\omega T = 2\pi$, from which we find that

$$\boxed{\omega = \frac{2\pi}{T} = 2\pi f} \qquad (1.1.3)$$

The dimensions of ω are radians per unit time, which usually is rad/s. **Mathematical analyses must be carried out using the circular frequency**, but we usually will describe a signal in terms of its cyclical frequency, often with a suitable prefix, such as kilohertz or megahertz.

The remaining parameters appearing in Eq. (1.1.2) are ϕ_s and ϕ_c, both of which are *phase angles*. Both are proportional to the time by which the signal is shifted from the reference function. Let τ_s be this time delay relative to a sine, which is synonymous with saying that the signal looks like a sine except that the elapsed time is $t - \tau_s$. Thus, the signal in Eq. (1.1.2) could be written equivalently as

$$p = A \sin \left(\omega \left(t - \tau_s \right) \right) \tag{1.1.4}$$

from which it follows that

$$\phi_s = -\omega \tau_s \tag{1.1.5}$$

Similarly, if the reference signal is a cosine and the time shift from a cosine is τ_c, then

$$\phi_c = -\omega \tau_c \tag{1.1.6}$$

The dimension of the phase angle derived from above is radians, which is required for mathematical formulas. However, just as we use the circular frequency to describe a harmonic signal, even though it would be incorrect to use such units mathematically, a phase angle usually is described in units of degrees.

A standard terminology for both acoustics and alternating electrical currents is to say that the sign of a harmonic function is its phase, and its argument is its phase variable, which could be $\omega t + \phi_s$ or $\omega t + \phi_c$. For example, if $\omega t + \phi_s$ is between -2π and $-\pi$, zero and π, 2π and 3π, etc., a sine is in its positive phase. Correspondingly, ϕ_s and ϕ_c are said to be phase angles. (In the next section, we will see a graphical representation of harmonic functions that will make evident the reason for calling them angles of phase.) These two angles must be related, because they represent the same signal. To identify this relation, we observe that a cosine looks like a sine that begins at time $-T/4$. The elapsed time in that sine function would be $t + T/4$, so it must be that $\cos \left(\omega t + \theta_c \right) = \sin \left(\omega \left(t + T/4 \right) + \theta_c \right) = \sin \left(\omega t + \theta_s \right)$. Because $\omega T = 2\pi$, it follows that

$$\theta_s = \theta_c + \pi/2 \tag{1.1.7}$$

This relation is expressed in words by saying that a sine function *lags* a cosine function by 90°, or by one-quarter of a period.

The similarity of sines and cosines has many implications for acoustics, starting from the fact that they combine to form a single harmonic if both are present at the same frequency. One way of evaluating this combination is to rely on trigonometric identities, but these identities are difficult for many folks to remember. In the following, we will see that there are ways in which operations can be performed algebraically without reliance on auxiliary formulas.

EXAMPLE 1.1 The graph is a pressure signal taken as measurement of a tuning fork. What are the amplitude, period, circular frequency, and phase angle relative to a pure cosine? Also, determine the minimum time shift that should be applied to the signal if it is to be converted to a pure sine.

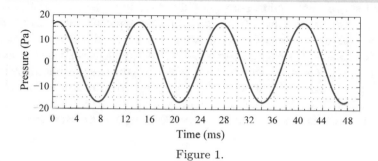

Figure 1.

Significance

The process of extracting basic parameters from a measured signal serves to reinforce understanding of the meaning of those parameters.

Solution

The easiest quantity to identify is the amplitude, which is the largest absolute value of the pressure. An estimate of 17 Pa is consistent with the size of graph and the fineness of its subdivisions, which prevent a more precise estimate. Precision also is an issue for the estimation of frequency. One cycle is the time to repeat the pattern, so the period could be estimated as the time between adjacent maxima, or adjacent minima. However, the relative flatness of the plotted curve at maxima and minima lessens the precision with which the corresponding time can be identified. It is better to use the fact that the zeros are separated by $T/2$. The slope of the curve is greatest at those instants, which increases the precision with which we can locate the intersection of the curve with the zero axis. Also, the error in locating such an intersection will be the same regardless of which zero is used, so using the longest interval between zeros will reduce measurement error. We estimate the earliest zero to occur at 0.0048 s, and the last to occur at 0.0448 s. Three cycles occur in that interval, so the period is estimated to be $T = (0.0448 - 0.0048)/3 = 0.0133$ s. The cyclical frequency is $f = 1/T = 75$ Hz, and the circular frequency is $\omega = 2\pi f = 471$ rad/s.

It is desired to find the phase angle of the response relative to a pure cosine, so we set the measured response to $17 \cos(\omega t + \phi)$. Here too, using instants when the pressure is zero gives the greatest precision. The mathematical form has a zero value when its argument is an odd multiple of $\pi/2$. Instants when $p = 0$ and the slope dp/dt is positive correspond to $\omega t_0 + \phi = -\pi/2, \ 3\pi/2, \ ...$ while $p = 0$ and $dp/dt < 0$ when $\omega t_0 + \phi = -3\pi/2, \ \pi/2, \ ....$ Thus, we can obtain multiple estimates for ϕ by using any of the several instants t_0 at which the graph crosses the zero axis. The alternative fraction of π to use for each case is the one that places ϕ in the range from $-\pi$ to π. For example, the zero at $t_0 = 0.0307$ corresponds to $dp/dt < 0$, so $\phi = -471(0.0307) + (-3\pi/2, \ \pi/2, \ ...)$. The appropriate choice may be found by trial and error, or else by rounding $\omega t_0/\pi$ to the nearest integer, which gives int(471 (0.0307)) = 5. The alternative closest to 5 is $\phi = -471\,(0.0307) + 4.5\pi = -0.323$. This value is negative, so the pressure lags a pure cosine. Other values of ϕ may be found by a similar calculation for each instant when $p = 0$. Averaging those values would greatly increase the precision of the estimate.

The last question requires comparing the curve to a sine function described in terms of a time delay τ as $A\sin(\omega\,(t - \tau))$. One approach compares the desired mathematical form to the cosine form that has been determined, but a simpler alternative is available. If the given curve is advanced by the earliest time t_0 at which $p = 0$ and $dp/dt > 0$, the shifted curve will be a pure sine. It follows that this t_0 is the requested value of τ, so we examine the graph to estimate $\tau = 0.0108$ s. This value is positive, so it represents a lag. However, this value of τ is greater than $T/2$, so it is preferable to subtract T to find $\tau = -0.0025$ s, which corresponds to a phase lead.

1.1.2 Vectorial Representation

A pictorial representation of a harmonic variation stems from the conversion from polar to rectangular coordinates, in which the projections of a radial line of length R at angle θ from the x axis are $R\cos\theta$ and $R\sin\theta$ onto the x and y axes, respectively. If we let $\theta = \omega t$, we obtain pure sine and cosine functions. If we add an angle ϕ to θ, we obtain an arbitrary harmonic function, as shown in Fig. 1.2. In other words, a harmonic function may be considered to be the projection onto either axis of a vector that rotates counterclockwise with an angular speed that is the circular frequency of the signal. Either axis can be used for the projection, but the same axis should be used for all functions to avoid confusion. For example, suppose we wish to use projections onto the x axis to depict pure sine and cosine functions having the same frequency. The sine is the y component when ωt is measured from the x axis, but we wish to depict it as an x component. It is necessary that all vectors rotate in the same sense, (counterclockwise for us). Both conditions can be met if ωt for the sine function is measured counterclockwise relative to the negative y axis. Such a construction appears in Fig. 1.3. It is apparent there that the vector for a cosine at any instant has rotated by $90° = \pi/2$ rad more than the vector for the sine, which is a different way of demonstrating that a cosine leads a sine by $90°$.

Fig. 1.2 Representation of a harmonic function as rotating vector

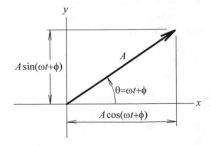

Fig. 1.3 Vectorial diagram
showing addition of two
harmonic signals whose
phase differs by 90°

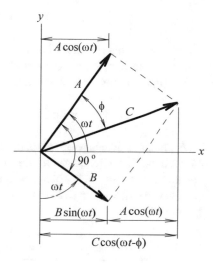

Suppose that we wish to add the functions depicted by the vectors in Fig. 1.3, which means that we wish to sum the x projection of each vector. Vectors are added graphically according to the parallelogram rule. Thus, Fig. 1.3 indicates that $B \sin(\omega t) + A \cos(\omega t)$ also is the x projection of the vector $\bar{C}$ that is the sum of the $\bar{A}$ and $\bar{B}$ vectors. The amplitude of $\bar{C}$ and its angle ϕ relative to $\bar{A}$ are readily found from trigonometry. The angle from the x axis to $\bar{C}$ is $\omega t - \phi$, which leads to

$$A \cos(\omega t) + B \sin(\omega t) = C \cos(\omega t - \phi)$$
$$C = \left(A^2 + B^2\right), \quad \phi = \tan^{-1}\left(\frac{A}{B}\right) \tag{1.1.8}$$

The concept of using a vector resultant to describe the sum of harmonic functions may be generalized to handle functions whose phase angle is arbitrary. In such situations, the parallelogram that forms the resultant will not be a rectangle. The consequence is that it would be necessary to use the trigonometry of scalene triangles to determine the amplitude and phase angle of the resultant. Such an approach can be cumbersome, so it seldom is used in practice.

1.1.3 Complex Exponential Representation

The most common and convenient way to handle harmonic functions results from changing the axis labels in Figs. 1.2 and 1.3 from x and y to "real" and "imaginary." We shall use the convention of letting $i = \sqrt{-1}$, in which case the construction in Fig. 1.2 corresponds to writing

$$Ae^{i\omega t} = A\cos(\omega t) + iA\sin(\omega t) \tag{1.1.9}$$

When $A = 1$, this is *Euler's identity*. The projections of a complex number onto the real and imaginary axes are referred to as its real and imaginary parts, so pure sine and cosine functions may be obtained by taking the respective parts,

$$A\cos(\omega t) = \mathrm{Re}\left(Ae^{i\omega t}\right), \quad A\sin(\omega t) = \mathrm{Im}\left(Ae^{i\omega t}\right) \tag{1.1.10}$$

A primary advantage of complex exponentials is the greater ease with which they are manipulated. One reason for this ease is that the time dependence can be factored out and the phase angle folded into the amplitude factor. Because $\exp(a + b) \equiv \exp(a)\exp(b)$, we can define the *complex amplitude* to be the factor of $\exp(i\omega t)$, specifically

$$\boxed{p = A\cos(\omega t + \phi) = \mathrm{Re}\left(Pe^{i\omega t}\right), \quad P = A\exp(i\phi)} \tag{1.1.11}$$

It is appropriate at this juncture to discuss notation. To the greatest possible extent, lower case symbols will be used to denote response variables that depend on time, while the corresponding capitalized letter will denote the complex amplitude of that time-dependent variable. The (real) amplitude of the signal is the magnitude of its complex amplitude, and the phase angle is the polar angle of that complex number. If the phase angle is in the first or second quadrants of the complex plane ($0 < \phi < \pi$), it is a phase lead, whereas it is a phase lag if it is in the third or fourth quadrants ($0 < \phi < -\pi$).

It often happens that an analysis yields the complex amplitude of a signal in rectangular coordinates, that is, $P = b + ci$, from which we wish to extract the amplitude and phase angle. This is done by invoking the polar to rectangular conversion, which states that

$$A\cos\phi = b, \quad A\sin\phi = c \tag{1.1.12}$$

The value of A is readily found, but one must be careful to select the proper quadrant in which the phase angle is placed. An error often occurs if the arctangent function, which states that $\phi = \tan^{-1}(c/b)$, is employed, because that function only depends on the sign of c relative to b.

For the vectorial representation, it was stated that it would be best to avoid mixing projections onto different axes. Similarly, it is best to avoid mixing real and imaginary parts. In this text **we will use real parts**, which leads to the question of how to extract a sine function. The answer comes from observing that dividing Eq. (1.1.9) by i yields a real part that is a sine function. More generally, we have

$$\boxed{A\cos(\omega t + \phi) = \mathrm{Re}\left(Ae^{i\phi}e^{i\omega t}\right), \quad A\sin(\omega t + \phi) = \mathrm{Re}\left(\frac{Ae^{i\phi}}{i}e^{i\omega t}\right)} \tag{1.1.13}$$

A different way to view Eq. (1.1.13) is to let C and S be the complex amplitudes of the respective functions. They are related by $S = C/i$, but $i \equiv \exp(i\pi/2)$, so $S = C \exp(-i\pi/2)$. This is equivalent to stating that the phase lag of the sine function relative to the cosine function is $\pi/2$, which we found earlier. An extension of this notion is a (complex) transfer function. Suppose the complex amplitude of some input to a system is X and the response is complex amplitude Y. The ratio of these quantities is the *transfer function*, whose value usually depends on the frequency, so that

$$G(\omega) = \frac{Y}{X}$$

(1.1.14)

The magnitude of G is often referred to as the magnification factor, while the phase angle of G indicates whether the response leads or lags the input.

EXAMPLE 1.2 A purely harmonic pressure signal has an amplitude of 60 Pa, and its frequency is 8 kHz. The earliest time at which the signal enters its positive phase is $t = 0.01$ ms. Describe this signal as a complex exponential function.

Significance

Proficiency in representing a signal as a complex exponential is essential to future work.

Solution

The earliest time at which a pure sine enters its positive phase is $t = 0$, so the given time represents the time delay relative to a pure sine. Thus, our strategy is to represent the given signal as a sine function, and then convert it to complex exponential form. The amplitude is 60 Pa, and the circular frequency is $\omega = 2\pi (8000)$ rad/s. Thus, we know that

$$p = 60 \sin(16000\pi (t - 0.00001))$$

The complex exponential form of this function is

$$p = \mathrm{Re}\left[\frac{60}{i} e^{(i(16000\pi(t-0.00001)))}\right]$$

$$= \mathrm{Re}\left[\frac{60}{i} e^{i(16000\pi)(-0.00001)} e^{(i16000\pi t)}\right]$$

The complex amplitude is the coefficient of the time exponential, so

$$p = \mathrm{Re}\left(A e^{(i16000\pi t)}\right)$$

$$A = \frac{60}{i} e^{i(16000\pi)(-0.00001)} = 60 e^{-i0.34\pi}$$

As an aside, note that the exponential function will be displayed as $\exp{(i\omega t)}$ when it appears in a line of text, but it will be $e^{i\omega t}$ when it appears in a displayed equation.

1.1.4 Operations Using Complex Exponentials

Operations on harmonic signals, such as comparison, addition, multiplication, differentiation, and integration, are fundamental tasks. This section is devoted to situations where all signals share the same frequency. In such cases, the complex exponential representation is advantageous because $\exp{(i\omega t)}$ is a common factor that may be cancelled in an equation, so we may focus on the complex amplitude of quantities. A common operation entails establishing the conditions for which harmonic functions will satisfy an equation. A simple case involves stating that two harmonic functions are equal. As an example, suppose we know a pressure signal as a cosine, say $p = 5\cos{(200t - \pi/3)}$, and we wish to express it as a sine function with a phase angle. We may write each function as a complex exponential and equate the two representations. Although this equation only governs the real parts, it must be satisfied at every instant. This can only be true if the respective complex exponentials are equal. Furthermore, the part of the exponential containing the frequency is common to both representations, so it may be canceled. Thus, we arrive at the basic fact that

> *If an equation is satisfied by harmonic functions at the same frequency, then it must be that the complex amplitudes on either side of the equality sign are equal.*

For the example posed above this leads to

$$p = 5\cos{(200t - \pi/3)} = A\sin{(200t + \phi)}$$
$$5e^{-i\pi/3} = \frac{A}{i}e^{i\phi} \tag{1.1.15}$$
$$Ae^{i\phi} = 5ie^{-i\pi/3} \equiv 5e^{i(\pi/2 - \pi/3)}$$

Two complex numbers are equal, if and only if, their magnitude and polar angle are equal, so we have

$$A = 5, \quad \phi = \frac{\pi}{6} = 30° \tag{1.1.16}$$

The notion that complex amplitudes should match substantially simplifies working with harmonic functions. Consider a linear combination of p_1 and p_2, both of which are oscillations at frequency ω. Let P_1 and P_2 denote the complex amplitude of each. A linear combination multiplies each signal by constant factors, then adds them. The same linear combination yields the complex amplitude of the resulting signal,

$$p = \mathrm{Re}\left(Pe^{i\omega t}\right) = ap_1 + bp_2 \quad \Longleftrightarrow \quad P = aP_1 + bP_2 \tag{1.1.17}$$

It is worth noting that numerical evaluation of these operations only requires conversions between the rectangular and polar representations of a complex number, thereby fulfilling our desire to avoid trigonometric identities. Furthermore, these operations are readily implemented in calculators and computer software.

Multiplying two harmonic functions, which is an operation that arises in a computation of acoustic power, requires that one exercise care. Consider any two harmonic functions p_1 and p_2, not necessarily at the same frequency. Let each function be the real part of complex variables z_1 and z_2, respectively. A common source of error for beginning students is to ignore the fact that *the product $p_1 p_2$ is not the real part of the product $z_1 z_2$.* This can be recognized by comparing the two products,

$$p_1 p_2 = \mathrm{Re}\,(z_1)\,\mathrm{Re}\,(z_2)$$
$$z_1 z_2 = [\mathrm{Re}\,(z_1)\,\mathrm{Re}\,(z_2) - \mathrm{Im}\,(z_1)\,\mathrm{Im}\,(z_2)]$$
$$+ i\,[\mathrm{Re}\,(z_1)\,\mathrm{Im}\,(z_2) + \mathrm{Re}\,(z_2)\,\mathrm{Im}\,(z_1)]$$

To evaluate $p_1 p_2$ without foregoing the use of complex exponentials, we follow a procedure that entails using the complex conjugate, which is denoted by an asterisk superscript,

$$z = \mathrm{Re}\,(z) + i\,\mathrm{Im}\,(z) \implies z^* = \mathrm{Re}\,(z) - i\,\mathrm{Im}\,(z) \qquad (1.1.18)$$

The proper mathematical operation for recovering the real part of z is to add it to its complex conjugate, which cancels the imaginary part, specifically

$$\mathrm{Re}\,(z) \equiv \frac{1}{2}\left(z + z^*\right) \qquad (1.1.19)$$

When this identity is applied to the complex exponential representation, the result is

$$
\boxed{
\begin{aligned}
A \cos\,(\omega t + \phi) &\equiv \frac{P}{2} e^{i\omega t} + \frac{P^*}{2} e^{-i\omega t} \\
A \sin\,(\omega t + \phi) &\equiv \frac{P}{2i} e^{i\omega t} - \frac{P^*}{2i} e^{-i\omega t}
\end{aligned}
}
\qquad (1.1.20)
$$

If the complex amplitude $P = 1$, these relations reduce to the complex variable definitions of the sine and cosine functions.

As an aside, one should be cognizant that many works in the acoustics literature use an exponential with a negative argument to describe a harmonic function, that is, $p = \mathrm{Re}\,(B \exp\,(-i\omega t))$. This corresponds to the second part of the above representations. A comparison of the alternative representation to $p = \mathrm{Re}\,(A \exp\,(i\omega t))$ makes it clear that B must equal A^*. Whether to use the positive or negative sign in the exponent is the author's choice. Neither is actually correct from a formal mathematical viewpoint because it requires that one perform a nonmathematical operation (dropping the imaginary part). The important aspect to realize is that if two works employ different sign conventions for their complex representations, all complex amplitudes in one work will be the complex conjugate of the corresponding amplitudes in the second work.

Any product of real variables q_1 and q_2 that are the real parts of z_1 and z_2 may be found by applying the rules of algebra in conjunction with Eq. (1.1.19), so that

$$q_1 q_2 = \left[\frac{1}{2}\left(z_1 + z_1^*\right)\right]\left[\frac{1}{2}\left(z_2 + z_2^*\right)\right] = \frac{1}{4}\left(z_1 z_2 + z_1 z_2^* + z_1^* z_2 + z_1^* z_2^*\right)$$

$$= \frac{1}{4}\left(z_1 z_2 + z_1 z_2^* + \text{c.c.}\right) = \frac{1}{2}\,\text{Re}\left(z_1 z_2 + z_1 z_2^*\right) = \frac{1}{2}\,\text{Re}\left(z_1^* z_2 + z_1^* z_2^*\right)$$

$$(1.1.21)$$

In the first resulting form, the term "c.c." is used as shorthand to eliminate writing terms that are the complex conjugate of those that precede it. Its use is a corollary of the fact that $q_1 q_2$ is a real quantity, and the only way to obtain a real result from a complex number is to add its complex conjugate. Application of Eq. (1.1.21) to evaluate the product of two pressure signals having complex amplitudes $\hat{P}_1$ and $\hat{P}_2$ at frequency ω yields

$$p_1 p_2 = \frac{1}{2}\,\text{Re}\left(P_1 P_2 e^{i2\omega t} + P_1 P_2^*\right) \equiv \frac{1}{2}\,\text{Re}\left(P_1^* P_2^* e^{-i2\omega t} + P_1^* P_2\right) \qquad (1.1.22)$$

This result indicates that the product of two harmonic functions at the same frequency will contain a constant part (unless $P_1 P_2^*$ is purely imaginary), in addition to a part that oscillates at twice the original frequency.

The advantage of representing harmonic functions as complex exponentials becomes even more significant for calculus operations. Real harmonic functions switch between sine and cosine in these operations, while the complex exponential form is always maintained. Both differentiation and integration are linear operations, so it is permissible to use real parts to represent them. Thus,

$$p = \text{Re}\left(Pe^{i\omega t}\right) \implies \frac{dp}{dt} = \text{Re}\left(i\omega Pe^{i\omega t}\right) \implies \frac{d^2 p}{dt^2} = -\,\text{Re}\left(\omega^2 Pe^{i\omega t}\right)$$

$$(1.1.23)$$

In other words, the nth derivative of a harmonic function is also harmonic, with a complex amplitude that is $(i\omega)^n$ times the complex amplitude of the function. Conceptually, this means that differentiation advances the function by 90°, or $T/4$ in time, and multiplies the amplitude by ω. Conversely, integration divides the complex amplitude by $i\omega$,

$$\int p\,dt = \text{Re}\left(\frac{\hat{A}}{i\omega}e^{i\omega t}\right) \qquad (1.1.24)$$

Thus, integration retards a harmonic by 90° and divides its amplitude by ω. An interesting observation arises when a particle velocity $\bar{v}$ is differentiated and integrated. Acceleration $\bar{a}$ is the derivative of velocity, and displacement $\bar{u}$ is its integral, so that

$$\bar{v} = \text{Re}\left(\bar{V}e^{i\omega t}\right) \implies \bar{a} = \text{Re}\left(i\omega\bar{V}e^{i\omega t}\right), \quad \bar{u} = \text{Re}\left(\frac{\bar{V}}{i\omega}e^{i\omega t}\right) \qquad (1.1.25)$$

(When a complex amplitude is a vector quantity, each of its components may be a complex number.) The preceding may be interpreted to say that the acceleration's amplitude is proportional to the velocity's, with a phase angle that leads the velocity by 90°. Similarly, the displacement amplitude is proportional to the velocity's, with a phase angle that lags the velocity by 90°. The phase shifts make sense physically, because acceleration must be present for some time before the velocity becomes appreciable. In the same way, displacement is the consequence of the presence of velocity over an interval.

EXAMPLE 1.3 Determine the amplitude and phase angle relative to a pure sine of the sum of two harmonic signals given by $20 \sin (600t + 0.7)$ and $50 \cos (600t - 4)$. Also, determine the earliest instant $t > 0$ at which the sum signal enters its positive phase.

Significance

Individuals who are comfortable with the complex arithmetic operations required to describe harmonic signals as complex exponentials eventually cease to think in terms of real functions. In this respect, complex exponentials are like a foreign language— you know you are proficient when you think in that language, rather than mentally translating it to your native tongue.

Solution

Our strategy is to convert the two harmonic functions to complex exponential form, then add their complex amplitudes. The most direct identification of the earliest instant at which the resultant p becomes positive comes from transforming the complex form back to a real sine function, which is the form requested for the phase angle. Thus, we have

$$p_1 = \text{Re}(A_1 \exp(i600t)), \quad p_2 = \text{Re}(A_2 \exp(i600t)),$$

$$p = p_1 + p_2 = \text{Re}\left(\frac{A}{i} \exp(i600t)\right)$$

Matching p_1 and p_2 to the given functions yields A_1 and A_2, and equating their sum to A/i yields

$$A_1 = \frac{20}{i} e^{i0.7}, \quad A_2 = 50e^{-i4}, \quad A = i(A_1 + A_2)$$

The next step depends on the tools being used for calculations. Most software packages and many calculators can add complex numbers in any form, but if arithmetic can only be performed for real numbers, then polar to rectangular conversion is the best route to adding the coefficients. Specifically,

$$A_1 = \frac{20}{i} [\cos (0.7) + i \sin (0.7)] = 12.8844 - 15.2968i$$
$$A_2 = 50 [\cos (4) - i \sin (4)] = -32.6822 + 37.8401i$$
$$A = -22.5433 - 19.7978i$$

Manual conversion of A to polar form proceeds by equating $|A|$ first, which leads to

$$|A| = \left(22.5433^2 + 19.7978^2\right)^{1/2} = 30.0026$$
$$|A|\cos\theta = -22.5433 \quad \& \quad |A|\sin\theta = -19.7978 \Longrightarrow \theta = -2.4209 \text{ rad} = -138.71°$$

To find the earliest instant at which the pressure vanishes, we convert it back to a real form that displays the time delay,

$$p = 30.0026 \sin\left(600t - 2.4209\right) = 30.0026 \sin\left(600\left(t - \tau\right)\right)$$

The smallest value of t at which $p = 0$ is $t = \tau$, so

$$\tau = \frac{2.4209}{600} = 4.035 \text{ ms}$$

EXAMPLE 1.4 Two harmonics are $p_1 = 0.05\sin\left(50t - \pi/3\right)$ and $p_2 = 0.04\cos\left(10t + 5\pi/6\right)$. Describe the product $p_1 p_2$ as a sum of harmonics. Is this variable periodic, and if so, what is its period?

Significance

Carrying out a product of harmonics is best done on a case by case basis as will be done here, rather than by reference to standard formulas. Consideration of the question of periodicity will be enlightening.

Solution

The first step is to represent each signal as a complex exponential and its complex conjugate, then form the product taking care to account for all terms. Conversion back to real form will allow us to examine the properties of the product. Thus, we begin by writing

$$p_1 p_2 = (0.05) \left[\frac{\exp\left(i\left(50t - \pi/3\right)\right)}{2i} - \frac{\exp\left(-i\left(50t - \pi/3\right)\right)}{2i} \right]$$
$$\times (0.04) \left(\frac{\exp\left(i\left(10t + 5\pi/6\right)\right)}{2} + \text{c.c.} \right)$$
$$= \frac{20\left(10^{-4}\right)}{4} \left[\frac{\exp\left(i\left(60t + \pi/2\right)\right)}{i} - \frac{\exp\left(i\left(-40t + 7\pi/6\right)\right)}{i} + \text{c.c.} \right]$$
$$= 20\left(10^{-4}\right)\left(\frac{1}{2}\right)\text{Re}\left[\exp\left(i\left(60t\right)\right) - \exp\left(-i\left(40t - 2\pi/3\right)\right)\right]$$
$$= 0.001\left[\cos\left(60t\right) - \cos\left(40t - 2\pi/3\right)\right]$$

We see that the product consists of two harmonics whose frequencies are the sum and difference of the frequencies of the signals forming this product. Each harmonic

in the product obviously is periodic, but periodicity of the result requires that each
factor repeat in an interval T that might contain several cycles of the individual terms.
The time required for m cycles of a signal at ω_1 is $m\,(2\pi/\omega_1)$, and n cycles at ω_2
corresponds to an elapsed time of $n\,(2\pi/\omega_2)$. Thus, we conclude that the product is
periodic if there are a pair of integers for which

$$T = \frac{2\pi m}{\omega_1} = \frac{2\pi n}{\omega_2} \implies \frac{m}{n} = \frac{\omega_1}{\omega_2}$$

Hypothetically, this condition could not be attained if ω_1/ω_2 were irrational. Integers for any rational fraction can be found by suitable multiplication by a power
of ten to clear any decimals followed by factoring out common factors in the
numerator and denominator. For example, $\omega_1 = 3.5$ and $\omega_2 = 7.50$ gives m/n
$= 3.5/7.5 = 35/75 = 7/15$, so $m = 7$ and $n = 15$. In the present case $\omega_1 = 60$ and
$\omega_2 = 40$, which corresponds to $m = 3$ and $n = 2$. Thus, the period of the product is
$T = 3\,(2\pi/\omega_1) = 2\,(2\pi/\omega_2) = \pi/10\,\text{s}$, whereas the periods of the individual signals
are $\pi/30$ and $\pi/20\,\text{s}$.

1.2   Averages

Most sounds we hear, including music, are not harmonic. Even when we hear a
harmonic sound, our awareness of its properties does not arise from a conscious
monitoring of its instantaneous oscillation relative to the ambient pressure. Rather,
our hearing system apparently performs averages to estimate a signal's characteristics. Various time-averaged quantities also arise in other aspects of acoustics, such
as issues regarding acoustical power. If s is any function of time, its average over an
interval $t_1 < t < t_2$ is defined as

$$(s)_{\text{av}} = \frac{1}{t_2 - t_1} \int_{t_1}^{t_2} s\,dt \tag{1.2.1}$$

If s is a harmonic function, the interval of interest is a period, $t_2 - t_1 = 2\pi/\omega$. Over
such an interval, all harmonic functions have a zero average because every interval in
which the function is positive is matched by a negative interval. This is true regardless
of the value of t_1.

Clearly, the mean value of a pressure signal p is not indicative of its average
magnitude. A measure that is useful for any oscillating pressure, not just a harmonic
sound, is the *mean-squared pressure*, which is the term used to describe the average
value of the square of the signal,

$$\boxed{(p^2)_{\text{av}} = \frac{1}{t_2 - t_1} \int_{t_1}^{t_2} p^2\,dt} \tag{1.2.2}$$

In some situations, the acoustic signal will be a *pulse,* which means that it exists only for a finite interval. If the interval for the average extends beyond the duration of the pulse, then the integrand in the extended interval will be zero identically. The consequence is that the average decreases monotonically as the averaging interval increases. The mean-squared pressure is not a good metric for a pulse, but a metric called sound exposure level described in the next section compensates for this shortcoming.

When p is harmonic, the time interval for the integration will be the period, $T = 2\pi/\omega$, so we set $t_2 = t_1 + T$, and apply the method in Eq. (1.1.21) to write

$$\left(p^2\right)_{\text{av}} = \frac{\omega}{2\pi} \int_{t_1}^{t_1+2\pi/\omega} \left(\frac{1}{4}P^2 e^{2i\omega t} + \frac{1}{4}PP^* + \text{c.c.}\right) dt \qquad (1.2.3)$$

The first term in the integrand gives no contribution. This leads to the important result that the mean-squared pressure for a harmonic signal is one half the square of its amplitude,

$$\boxed{\left(p^2\right)_{\text{av}} = \frac{1}{4}PP^* + \text{c.c.} \equiv \frac{1}{2}PP^* \equiv \frac{1}{2}|P|^2} \qquad (1.2.4)$$

Closely related to this is the *root-mean-squared value* (rms), which we abbreviate as p_{rms} and define as

$$\boxed{p_{\text{rms}} = \sqrt{\left(p^2\right)_{\text{av}}}} \qquad (1.2.5)$$

For a harmonic signal, the rms value is the square root of the amplitude, $|P|/\sqrt{2}$. This relation should be used with caution because *it is valid only for harmonic functions.*

Another important acoustical quantity is the *intensity,* which is defined as the product of the pressure and particle velocity. Because velocity is a vector, it follows that intensity also is a vector,

$$\boxed{\bar{I} = p\bar{v}} \qquad (1.2.6)$$

If p and $\bar{v}$ are harmonic at frequency ω with complex amplitudes $\hat{P}$ and $\hat{\bar{V}}$, evaluation of the average intensity gives

$$\bar{I}_{\text{av}} = \frac{1}{T} \int_{t_1}^{t_1+T} \left(\frac{P\bar{V}}{4} e^{i2\omega t} + \frac{P\bar{V}^*}{4} + \text{c.c.}\right) dt$$

$$\qquad (1.2.7)$$

$$\boxed{\bar{I}_{\text{av}} = \frac{1}{4}\left(P\bar{V}^* + \text{c.c.}\right) \equiv \frac{1}{2}\text{Re}\left(P\bar{V}^*\right) \equiv \frac{1}{2}\text{Re}\left(P^*\bar{V}\right)}$$

A more general situation involves a signal that is a sum of harmonics at a variety of arbitrary frequencies ω_n. In that case, there is not a specific period over which to average, so the time base is taken as the limit for a very long interval. Thus, the general definition of the average for any quantity s that has no identifiable period is

$$(s)_{\text{av}} = \lim_{T \to \infty} \frac{1}{T} \int_{-T/2}^{T/2} s \, dt \tag{1.2.8}$$

Let us evaluate the mean-squared pressure for a signal that consists of only two harmonics,

$$p = \frac{1}{2} \left(P_1 e^{i\omega_1 t} + P_2 e^{i\omega_2 t} + \text{c.c.} \right) \tag{1.2.9}$$

Sixteen terms arise in the product p^2, but eight are complex conjugates of the others, so we have

$$(p^2)_{\text{av}} = \lim_{T \to \infty} \frac{1}{T} \int_{-T/2}^{T/2} \frac{1}{4} \left[(P_1)^2 e^{i2\omega_1 t} + 2P_1 P_2 e^{i(\omega_1+\omega_2)t} + P_1 P_1^* + P_1 P_2^* e^{i(\omega_1-\omega_2)t} \right.$$
$$\left. + (P_2)^2 e^{i2\omega_2 t} + P_1^* P_2 e^{i(-\omega_1+\omega_2)t} + P_2 P_2^* + \text{c.c.} \right] dt \tag{1.2.10}$$

We could carry out this integral, but it is simpler to observe that the complex exponentials oscillate with positive lobes that exactly match negative lobes. Thus, the largest value that these terms can contribute to the integral is the area of one lobe, depending on the value of T. These terms therefore give bounded contributions to the integral, so dividing them by T, then letting $T \to \infty$, causes their contribution to vanish. Consequently, all that remains is

$$(p^2)_{\text{av}} = \frac{1}{4} \left(P_1 P_1^* + P_2 P_2^* + \text{c.c.} \right) = \frac{1}{2} \left(|P_1|^2 + |P_2|^2 \right) \tag{1.2.11}$$

In other words, the contributions to the average value are a superposition of the contributions of each harmonic acting independently. The preceding may be generalized to the case of many harmonics, specifically

$$\boxed{(p^2)_{\text{av}} = \frac{1}{2} \sum_{n=1}^{N} |P_n|^2} \tag{1.2.12}$$

This is *Parseval's theorem*. (Some individuals reserve this name to the result when $p(t)$ is periodic, which is addressed later in this chapter.) The average intensity for a multiharmonic signal is derived in a similar manner. The result is

$$\boxed{(\bar{I})_{\text{av}} = \frac{1}{2} \text{Re} \left(\sum_{n=1}^{N} P_n \bar{V}_n^* \right) \equiv \frac{1}{2} \text{Re} \left(\sum_{n=1}^{N} P_n^* \bar{V}_n \right)} \tag{1.2.13}$$

Here too, each harmonic's contribution to the average value is uninfluenced by the presence of other harmonics. An important aspect of p^2 and $\bar{I}$ is that either at a specific instant might differ greatly from its average.

The preceding relation for $\left(p^2\right)_{\text{av}}$ is applicable when the individual harmonics $\hat{P}_n$ are mutually *incoherent*, meaning that there is no correlation between the fluctuations of each term. Mathematically, two signals are defined to be incoherent if the mean value of their product is zero,

$$\text{mutually incoherent} \iff (p_1 p_2)_{\text{av}} = \lim_{T \to \infty} \frac{1}{T} \int_{-T/2}^{T/2} p_1 p_2 \, dt = 0 \qquad (1.2.14)$$

The question of coherence crucially affects how different signals combine. Two harmonic signals at the same frequency are coherent. Their relative phase angles dictate whether the combination is weaker or stronger than the individual terms. In contrast, the manner in which harmonics at different frequency combine does not depend on their relative phase. The issue of coherence does not affect measurements of mean-squared pressure and average intensity with instrumentation that actually averages signals according to Eq. (1.2.8). It is an issue when we invoke Eq. (1.2.12) or (1.2.13) because the formulas require that all coherent terms be collected into a single complex amplitude.

A corollary of these developments is that averaging over a long interval yields the correct mean-squared pressure, even if the signal consists of a single harmonic. This frees us from the need to determine the period prior to evaluating an average. However, for any signal, the evaluation of the required averages must address the fact that it is impossible to actually use an infinite time interval. It is essential that T be sufficiently large, but how large depends on the mix of amplitudes and frequencies. In the case of two harmonics having equal amplitude, the error of $\left(p^2\right)_{\text{av}}$ will be approximately 10% if $T = 0.05/|\omega_2 - \omega_1|$. (Determination of this estimate is the subject of Exercise 1.15.)

In some situations, averaging is not appropriate because it fails to capture the basic nature of the sound. For example, suppose a signal consists of a pair of harmonics at the same frequency ω whose phase angles drift slowly in time, so that the change of the phase angles only seems to be significant when viewed over an interval for which T is much greater than a period. Over part of this interval, the two harmonics might reinforce each other, while they might cancel in another part. Another example is a pair of frequency modulated signals if their frequency changes slowly. These signals might seem to have the same frequency and a constant relative phase shift for a short interval, but over a sufficiently long interval, they will be recognized as being incoherent.

Even if the signal consists of true harmonics it might not be useful to consider averages. Specifically, if two frequencies are very close, an individual will not consider them to be incoherent. To see why let $\omega_1 < \omega_2$ be two frequencies and define average and difference frequencies such that

$$\omega_{\text{av}} = \frac{1}{2}(\omega_2 + \omega_1), \quad \sigma = \frac{1}{2}(\omega_2 - \omega_1) \qquad (1.2.15)$$

which leads to

$$\omega_1 = \omega_{av} - \sigma, \quad \omega_2 = \omega_{av} + \sigma \tag{1.2.16}$$

The sum of the two associated signals is then

$$
\begin{aligned}
p &= \mathrm{Re}\left(P_1 e^{i\omega_1 t}\right) + \mathrm{Re}\left(P_2 e^{i\omega_2 t}\right) \\
&= \mathrm{Re}\left(P_1 e^{i\omega_{av} t} e^{-i\sigma t} + P_2 e^{i\omega_{av} t} e^{i\sigma t}\right) \\
&= \mathrm{Re}\left[e^{i\omega_{av} t}\left(P_1 e^{-i\sigma t} + P_2 e^{i\sigma t}\right)\right] \\
&= \mathrm{Re}\left\{e^{i\omega_{av} t}\left[(P_1 + P_2)\cos(\sigma t) - i(P_1 - P_2)\sin(\sigma t)\right]\right\} \tag{1.2.17}
\end{aligned}
$$

To interpret this expression, note that $\sigma \ll \omega_{av}$ because $\omega_2 \approx \omega_1$. This means that the harmonic terms whose frequency is σ change much more slowly than the complex exponential whose frequency is ω_{av}. In essence, the bracketed term acts as a slowly varying complex amplitude $P(t)$ of a signal whose frequency is ω_{av}. The mathematical representation is

$$
\begin{aligned}
P(t) &= \left[(P_1 + P_2)\cos(\sigma t) - i(P_1 - P_2)\sin(\sigma t)\right] \\
p &= \mathrm{Re}\left[P(t) e^{i\omega_{av} t}\right]
\end{aligned}
\tag{1.2.18}
$$

The ratio of the real and imaginary parts of $P(t)$ is not constant, which means that both its magnitude and phase angle vary slowly. When the typical person hears a sound such as this, their perception of loudness is not based on $\left(p^2\right)_{av}$. Rather, it is perceived as a pure harmonic at frequency ω_{av} whose instantaneous amplitude is defined by the magnitude of $P(t)$. Figure 1.4 shows a typical situation. The time scale

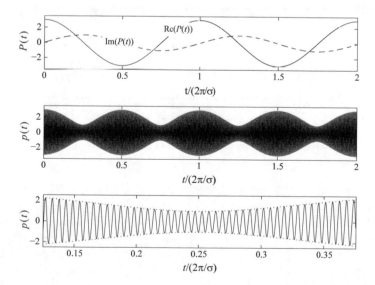

Fig. 1.4 The sum of two harmonics whose frequencies are close, $\omega_1 = 0.99\omega_{av}$, $P_1 = 2$, $\omega_2 = 1.01\omega_{av}$, $P_2 = 1$

there is nondimensionalized by the period of $P(t)$, which is $2\pi/\sigma$. Two cycles of $P(t)$ are shown in the upper graph. In the lower graph, there are so many oscillations of p that its individual cycles are indistinguishable. Over a short interval, $\exp(i\omega_{av}t)$ will vary in magnitude between ± 1, while $P(t)$ is essentially constant. Thus, $\pm|P(t)|$ represents an envelope bounding p. The interval in which the pattern repeats is one half the period of $P(t)$ because in each half cycle of $P(t)$ there are many instants when the magnitude of $\exp(i\omega_{av}t)$ is either $+1$ or -1. The third graph shows the signal and envelope function over a time span that is sufficiently small to see the oscillation.

A special case arises when the $|P_1| = |P_2|$ for two signals whose frequencies are close. The simplest such case is $P_2 = P_1$, in which case $P(t) = 2P_1\cos(\sigma t)$. Thus, the combined signal seems to have frequency ω_{av} with an amplitude that oscillates slowly at the low-frequency σ. At intervals of π/σ, this amplitude is zero so the pressure vanishes. This is the *beating signal,* or more simply the *beat,* displayed in Fig. 1.5. The envelope of the signal is $P(t)$. Usage of mean-squared pressure to describe a beating signal would miss the fundamental nature of what would be heard.

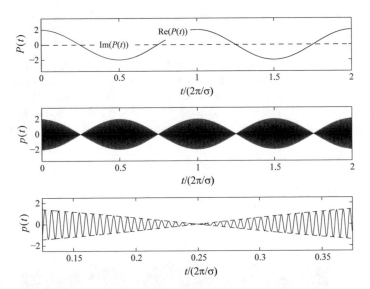

Fig. 1.5 A signal exhibiting beats, $\omega_1 = 0.99\omega_{av}$, $\omega_2 = 1.01\omega_{av}$, $P_1 = P_2 = 1$

EXAMPLE 1.5 Measurement of the mean-squared value of a harmonic function can be done with data taken over one period, or alternatively, over a very long interval. If the measurement window does not match either specification, the result will be erroneous. Estimate the maximum per cent error if the measurement is taken with a time window that is longer than one period but less than two. Then, compare that error to the maximum error if the window is between ten and eleven periods in duration.

Significance

The relationship between time averages and instantaneous values is an essential feature for many topics. This example highlights the importance of measuring signals over a sufficiently long interval.

Solution

We begin by evaluating the average value of p^2 as an integral over a time window T_w. That operation is straightforward, but identifying the maximum error in that expression will require careful consideration. For an arbitrary harmonic function $p = \mathrm{Re}\,(P\exp(i\omega t))$, the average is

$$
\begin{aligned}
\left(p^2\right)_{\text{av}} &= \frac{1}{T_w}\int_0^{T_w} \left(\frac{1}{4}\right)\left(Pe^{i\omega t} + P^*e^{-i\omega t}\right)\left(Pe^{i\omega t} + \text{c.c.}\right)dt \\
&= \frac{1}{4T_w}\int_0^{T_w}\left(P^2 e^{2i\omega t} + P^*P + \text{c.c.}\right)dt \\
&= \frac{1}{4T_w}\left[\frac{P^2}{i2\omega}\left(e^{2i\omega T_w} - 1\right) + P^*P\,T_w + \text{c.c.}\right] \\
&\quad -\frac{1}{2}|P|^2 + \frac{1}{4\omega T_w}\,\mathrm{Re}\left[\frac{P^2}{i}\left(\exp\left(i2\omega T_w\right) - 1\right)\right]
\end{aligned}
$$

The correct mean-squared pressure is the first term, so the second term is the error. The percent error is the absolute error ratioed to the true value, then multiplied by 100, so

$$
\varepsilon_\% = \frac{100}{2\omega T_w}\left|\mathrm{Re}\left[\left(\frac{P}{|P|}\right)^2\left(\frac{\exp\left(i2\omega T_w\right) - 1}{i}\right)\right]\right|
$$

We could try to identify the maximum value by plotting $\varepsilon_\%$ as a function of ωT_w, but the phase angle of P also is important. We will consider the factors forming $\varepsilon_\%$ individually. Let $B = \exp\left(i2\omega T_w\right) - 1$. In Fig. 1, B is a number in the second quadrant. Because the magnitude of the exponential is one, the only other possibility is that B is in the third quadrant. In any case, we can say that $B = |B|\exp(i\beta)$. Both $|B|$ and β depending on the value of ωT_w. Their range is $0 \le |B| \le 2$ and $\pi/2 \le \beta \le 3\pi/2$.

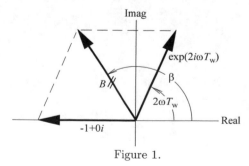

Figure 1.

The polar representation of the signal is $P = |P| \exp(i\theta)$, so

$$\varepsilon_\% = \frac{100}{2\omega T_w} |B| \operatorname{Re}\left[\exp\left(i\left(\beta + 2\theta - \pi/2\right)\right)\right]$$

We seek the maximum possible value of $\varepsilon_\%$. The maximum of $|B|$ is 2 corresponding to $\beta = \pi$ and $2\omega T_w = \pi,\ 3\pi$. The extremes of the real part of the complex exponential are ± 1, which can result for any β because θ can have any value.

It follows that the extreme values of $|B| \operatorname{Re}\left[\exp\left(i\left(\beta + 2\theta - \pi/2\right)\right)\right]$ are ± 2, and that these extremes can occur when $\omega T_w = \pi/2,\ 3\pi/2,\ ...$ This does not exactly maximize $\varepsilon_\%$ because ωT_w also appears in the denominator of $\varepsilon_\%$. However, that denominator changes much less rapidly than the exponential as ωT_w varies. Hence, the condition leading to maximum $|\varepsilon_\%|$ is well approximated by setting ωT_w to an odd multiple of $\pi/2$. For the first case, it is specified that $2\pi/\omega < T_w < 2\,(2\pi/\omega)$, so we set $\omega T_w = 5\,(\pi/2)$, which gives

$$|\varepsilon_\%| \approx \frac{100}{2.5\pi} = 12.7\%$$

In the second case, $10\,(2\pi/\omega) < T_w < 11\,(2\pi/\omega)$, so the worst case corresponds to $\omega T_w = 41\,(\pi/2)$, which gives

$$|\varepsilon_\%| = \frac{100}{20.5\pi} = 1.6\%$$

The error for the short window might be excessive in some situations, whereas the result obtained with the longer window would be acceptable in almost any application.

This guideline for selection of an averaging interval should be used judiciously because it only applies to harmonic signals. As was noted previously, the averaging interval required to attain a desired accuracy for a multiharmonic signal is likely to be much larger.

EXAMPLE 1.6 The pressure and fluid particle velocity at a certain location are known to be

$$p = 80\cos(500t) + 120\sin(1000t + 5) \text{ Pa}$$
$$\bar{v} = 0.0012\sin(500t - 3)\,\bar{e}_x + [0.002\cos(500t) + 0.001\sin(1000t)]\,\bar{e}_y \text{ m/s}$$

where the value of t has units of seconds and $\bar{e}_x$ and $\bar{e}_y$ are unit vectors in the horizontal and vertical directions, respectively. Use graphs of the components of the instantaneous intensity $\bar{I}$ to estimate the extreme values of each component. Then, compare those values to the time-averaged intensity.

Significance

This example highlights the fact that intensity is a vectorial quantity. This property is important because developments in Chap. 4 will show that the acoustic power flow can be determined from the intensity vector. The results emphasize the distinctly different nature of instantaneous and average values.

Solution

The intensity is $\bar{I} = p\bar{v}$, but it is convenient to work individually with the vector components, which are found from dot products with the respective unit vectors. Thus, we will evaluate

$$I_x = \bar{I} \cdot \bar{e}_x = pv_x, \quad I_y = \bar{I} \cdot \bar{e}_y = pv_y$$

Our plan is to use complex exponentials to describe each component. The operations are somewhat simpler if the complex amplitudes are described algebraically. Let $\omega_1 = 500$ rad/s, $\omega_2 = 1000$ rad/s, and

$$P_1 = 80, \quad P_2 = \frac{120}{i}e^{5i}$$
$$V_{x1} = \frac{0.0012}{i}e^{-3i}, \quad V_{x2} = 0$$
$$V_{y1} = 0.002, \quad V_{y2} = \frac{0.001}{i}$$

Then, the pressure and particle velocity components are

$$p = \frac{1}{2}\left(P_1 e^{i\omega_1 t} + P_2 e^{i\omega_2 t} + P_1^* e^{-i\omega_1 t} + P_2^* e^{-i\omega_2 t}\right)$$
$$v_x = \frac{1}{2}\left(V_{x1} e^{i\omega_1 t} + V_{x2} e^{i\omega_2 t} + \text{c.c.}\right)$$
$$v_y = \frac{1}{2}\left(V_{y1} e^{i\omega_1 t} + V_{y2} e^{i\omega_2 t} + \text{c.c.}\right)$$

The products forming the intensity components are

$$I_x = \frac{1}{4}\left(P_1 V_{x1} e^{2i\omega_1 t} + P_1 V_{x2} e^{i(\omega_1+\omega_2)t} + P_2 V_{x1} e^{i(\omega_2+\omega_1)t} + P_2 V_{x2} e^{2i\omega_2 t}\right.$$
$$\left. + P_1^* V_{x1} + P_1^* V_{x2} e^{i(-\omega_1+\omega_2)t} + P_2^* V_{x1} e^{i(-\omega_2+\omega_1)t} + P_2^* V_{x2}\right) + \text{c.c.}$$
$$I_y = \frac{1}{4}\left(P_1 V_{y1} e^{2i\omega_1 t} + P_1 V_{y2} e^{i(\omega_1+\omega_2)t} + P_2 V_{y1} e^{i(\omega_2+\omega_1)t} + P_2 V_{y2} e^{2i\omega_2 t}\right.$$
$$\left. + P_1^* V_{y1} + P_1^* V_{y2} e^{i(-\omega_1+\omega_2)t} + P_2^* V_{y1} e^{i(-\omega_2+\omega_1)t} + P_2^* V_{y2}\right) + \text{c.c.}$$

In light of the values of the frequencies, the result is that

$$I_x = \frac{1}{2}\left[(I_x)_0 + (I_x)_1 e^{500it} + (I_x)_2 e^{1000it} + (I_x)_3 e^{1500it} + (I_x)_4 e^{2000it}\right] + \text{c.c.}$$

$$I_y = \frac{1}{2} \left[(I_y)_0 + (I_y)_1 \, e^{500it} + (I_y)_2 \, e^{1000it} + (I_y)_3 \, e^{1500it} + (I_y)_4 \, e^{2000it} \right] + \text{c.c.}$$

where the complex amplitudes are

$$(I_x)_0 = \frac{1}{2} \left(P_1^* V_{x1} + P_2^* V_{x2} \right), \quad (I_y)_0 = \frac{1}{2} \left(P_1^* V_{y1} + P_2^* V_{y2} \right)$$

$$(I_x)_1 = \frac{1}{2} \left(P_1^* V_{x2} + P_2 V_{x1}^* \right), \quad (I_y)_1 = \frac{1}{2} \left(P_1^* V_{y2} + P_2 V_{y1}^* \right)$$

$$(I_x)_2 = \frac{1}{2} P_1 V_{x1}, \quad (I_y)_2 = \frac{1}{2} P_1 V_{y1}$$

$$(I_x)_3 = \frac{1}{2} \left(P_1 V_{x2} + P_2 V_{x1} \right), \quad (I_y)_3 = \frac{1}{2} \left(P_1 V_{y2} + P_2 V_{y1} \right)$$

$$(I_x)_4 = \frac{1}{2} P_2 V_{x2}, \quad (I_y)_4 = \frac{1}{2} P_2 V_{y2}$$

The values that are obtained are

$$(I_x)_0 = 0.048e^{1.7124i}, \quad (I_y)_0 = 0.1128e^{0.53528i}$$

$$(I_x)_1 = 0.072e^{1.71268i}, \quad (I_y)_1 = 0.13683e^{-2.5699i}$$

$$(I_x)_2 = 0.048e^{1.7124i}; \quad (I_y)_2 = 0.08$$

$$(I_x)_3 = 0.072e^{-1.1416i}, \quad (I_y)_3 = 0.13683e^{-2.5699i}$$

$$(I_x)_4 = 0, \quad (I_y)_4 = 0.06e^{1.8584i}$$

The mean values of the x and y intensity components are

$$(I_x)_{\text{av}} = \text{Re}\left((I_x)_0 \right) = -0.00677$$

$$(I_y)_{\text{av}} = \text{Re}\left((I_x)_0 \right) = 0.09702$$

The units of the intensity amplitudes are Pa-m/s, which is the same as J/m^2.

The easiest way to identify the extreme values of the intensity is to inspect sampled data. The intensity consists of harmonics at 500, 1000, 1500, and 2000 rad/s. One period T of this sum consists of n_{500}, n_{1000}, n_{1500}, and n_{20000} cycles of each, such that $T = n_{500} \, (2\pi/\omega_1) = n_{1000} \, (2\pi/\omega_2) = n_{1500} \, (2\pi/\omega_3) = n_{2000} \, (2\pi/\omega_4)$. These ratios are $n_{500} : n_{1000} : n_{1500} : n_{2000} = \omega_1 : \omega_2 : \omega_3 : \omega_4$. Thus the period is $T = 1 \, (2\pi/500)$ $= 2 \, (2\pi/1000) = 3 \, (2\pi/1500) = 4 \, (2\pi/2000)$. (Later in this chapter, we will see that these intensity components constitute a Fourier series whose fundamental frequency is 500 rad/s. The period of such a series is 2π divided by the fundamental frequency.)

Sampling this interval with 200 instants should be sufficient resolution to assure that one of the data points is close to each extreme value. Figure 1 graphs I_x and I_y versus t for one period of the product.

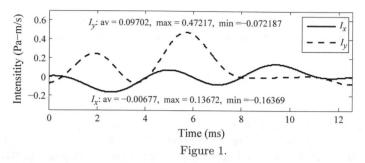

Figure 1.

Both averages listed in the graph are substantially smaller than the corresponding extreme values. Furthermore, the ratio I_y/I_x is not constant, which means that the instantaneous intensity vector is not parallel to the time-averaged quantity. To understand these trends, we could consider the individual harmonics of p and $\bar{v}$. Over some interval pressure and velocity, harmonics at the same frequency will have similar phase, and therefore reinforce an intensity component. In other intervals like harmonics will cancel each other because they have opposing phases.

Another interesting observation is derived from considering the direction of $\bar{I}$, according to which $I_y(t)/I_x(t) = v_y(t)/v_x(t)$. This means that the instantaneous intensity is always parallel to the instantaneous particle velocity. This is not true for the mean-squared values. For the velocity components, they are

$$\left(v_x^2\right)_{av} = \frac{1}{2}\left(|V_{x1}|^2 + |V_{x2}|^2\right) = 0.72\left(10^{-6}\right) \text{ m}^2/\text{s}^2$$

$$\left(v_y^2\right)_{av} = \frac{1}{2}\left(|V_{y1}|^2 + |V_{Y2}|^2\right) = 2.56\left(10^{-6}\right) \text{ m}^2/\text{s}^2$$

It is evident that $\left(v_y^2\right)_{av} / \left(v_x^2\right)_{av} \neq \left(I_y\right)_{av} / \left(I_x\right)_{av}$, from which we conclude that the time-averaged intensity is not parallel to the mean-squared velocity.

1.3 Metrics of Sound

1.3.1 Sound Pressure Level

A common measure of the loudness of an acoustic signal is sound pressure level (SPL). It is based on psychoacoustic experiments in the 1920s at Bell Labs[1] aimed at identifying how people react to changes in the level of a signal. These investigations led to recognition that the average person does not perceive an increase in the amplitude of a sound as a proportional increase in loudness. Rather, most individuals react logarithmically when comparing two signals. Sound pressure level, denoted as $\mathcal{L}_p$

---

[1] H. Fletcher, "Physical Measurements and Their Bearing on the Theory of Hearing," Bell Systems Technical Journal, **2–4**, 145–173 (1923).

and abbreviated as "SPL," is a measure of loudness that accounts for this perception. It uses the mean-squared pressure as the measure of the magnitude of the sound. The definition is

$$\mathcal{L}_p = 10 \log_{10} \left(\frac{\left(p^2\right)_{\text{av}}}{p_{\text{ref}}^2} \right) \text{dB} \qquad (1.3.1)$$

It is implicit to this definition that $\left(p^2\right)_{\text{av}}$ is a meaningful quantity. This excludes the case of a pulse that lasts for a finite interval. In the typical situation, the signal consists of several harmonics. Based on the assumption that they are incoherent, the mean-squared pressure is given by Eq. (1.2.12), from which it follows that

$$\boxed{\mathcal{L}_p = 10 \log_{10} \left(\sum_{n=1}^{N} \frac{|P_n|^2}{2p_{\text{ref}}^2} \right) dB} \qquad (1.3.2)$$

As indicated, the units of sound pressure level are decibels (dB). In keeping with the somewhat qualitative nature of what it indicates, sound pressure levels are usually rounded to the nearest integer. In this definition of sound pressure level, p_{ref} is a reference pressure that sets the 0 dB state, that is, $\mathcal{L}_p = 0$ dB when $\left(p^2\right)_{\text{av}} = \left(p_{\text{ref}}\right)^2$. The overall nature of the dependence, including the factor of 10, is based on the experiments at Bell Labs. The parameters are set to match observations regarding how individuals with average hearing acuity perceive changes in a signal's loudness. An increase in 1 dB would be barely noticeable to an individual who is listening carefully, an increase in 5 dB would be noticeable under any circumstance, 10 dB would be perceived as making the sound twice as loud, and a 20 dB increase would be thought to make the signal four times louder. Solution of the definition of $\mathcal{L}_p$ yields the mean-squared pressure corresponding to a given sound pressure level,

$$\left(p^2\right)_{\text{av}} = p_{\text{ref}}^2 \left(10^{\mathcal{L}_p/10} \right) \qquad (1.3.3)$$

Two values of p_{ref} are in current use. For sound in air, $p_{\text{ref}} = 20 \left(10^{-6}\right)$ Pa, while $p_{\text{ref}} = \left(10^{-6}\right)$ Pa for signals in water. The first time a sound pressure level is reported, one should also state the reference, for example, $\mathcal{L}_p = 100$ dB//20 μPa. The reference value for sound in air was selected such that for most people 0 dB is the minimum audible level of a harmonic signal in the 2–5 kHz frequency range. At the other end, the largest possible amplitude for a harmonic signal in air at the earth's surface is 1 atm ≈ 100 kPa, because the minimum total pressure (signal + ambient) in a cycle would be zero. Such a signal corresponds to $\mathcal{L}_p = 191$ dB. Signals in air above 160 dB usually are more properly considered to be explosions. Various levels have been reported as typical in a variety of references, but in each case, there is much variability depending on the details of the device and the circumstances in which the measurement was taken. For example, it has often been stated that a sound of 30 m from a jet engine is 140 dB, but modern engines have been quieted significantly. It has been stated that the sound level at a loud rock concert is 120 dB, but that does not account for the genre of the music or the dependence on location within an auditorium or stadium. Similarly, a statement that the sound in the interior of an automobile in traffic is 80 dB does not recognize the actual traffic conditions or the enormous dependence on design features of the vehicle. Normal conversation is said to occur between 80 and 90 dB, but that level depends on many aspects regarding environmental conditions, as well as the hearing acuity and personality of the conversants. It is known that exposure to loud sounds at specific levels can damage human hearing. Levels below 90 dB are considered to be safe regardless of the duration of exposure, while 140 dB will cause permanent damage, even if the exposure time is very short. The latter corresponds to an rms value of 200 Pa.

The reference value of 10^{-6} Pa for water is merely selected for convenience. To get an idea of the corresponding scale, consider a harmonic signal whose amplitude is 1 atm $\approx$ 100 kPa. If it occurs at the surface of a body of water, the minimum absolute pressure (ambient plus acoustic) will be zero. Further increase in the acoustic amplitude would lead to cavitation (formation of microbubbles). For such a signal, $(p^2)_{\text{av}} = (10^5)^2 / 2$, which leads to $\mathcal{L}_p = 217$ dB$//(10^{-6})$ Pa. Much higher sound pressure levels may occur without cavitation well below the surface because the ambient pressure increases linearly with depth.

Confusion sometimes arises because sound pressure level may be computed in a slightly different way if a signal consists of a single harmonic. In that case $(p^2)_{\text{av}} = \left| \hat{P} \right|^2 / 2$, which leads to

$$\boxed{\mathcal{L}_p = 20 \log_{10} \left(\frac{|P|}{\sqrt{2} p_{\text{ref}}} \right) \, dB} \tag{1.3.4}$$

The decibel scale also may be used to compare signals. The ratio of $(p^2)_{\text{av}}$ for two signals is given by the difference of their sound pressure levels

$$\mathcal{L}_2 - \mathcal{L}_1 = 10 \log_{10} \left(\frac{(p_2^2)_{\text{av}}}{(p_1^2)_{\text{av}}} \right) \tag{1.3.5}$$

Thus, if the level of a signal is multiplied by ten at every instant, thereby raising $(p^2)_{\text{av}}$ by a factor of one hundred, the resulting sound pressure level is 20 dB higher. The table describes some useful ratios.

$|p_2/p_1|$	10	2	$\sqrt{2}$	1.2	1.1
$\mathcal{L}_2 - \mathcal{L}_1 (dB)$	20	6	3	2	1

The logarithmic nature of sound pressure level leads to some aspects that are difficult for the lay person to grasp. Suppose $(p_1^2)_{\text{av}}$ and $(p_2^2)_{\text{av}}$ are two signals that are mutually incoherent. Under this assumption, the individual mean-squared pressures may be added without concern for the relative phase of harmonics. The sound pressure level for the combination of two signals therefore is

$$\mathcal{L}_p = 10 \log_{10} \left(\frac{\left(p_1^2\right)_{av} + \left(p_2^2\right)_{av}}{p_{ref}^2} \right) = 10 \log_{10} \left[\left(10^{\mathcal{L}_1/10}\right) + \left(10^{\mathcal{L}_2/10}\right) \right] \quad (1.3.6)$$

Clearly, the result is not simple addition of decibel levels. Without loss of generality, let $\mathcal{L}_1$ denote the larger of the values, so that

$$\mathcal{L}_p = 10 \log \left\{ \left(10^{\mathcal{L}_1/10}\right) \left[1 + \left(10^{-(\mathcal{L}_1-\mathcal{L}_2)/10}\right) \right] \right\}$$
$$= \mathcal{L}_1 + 10 \log \left[1 + \left(10^{-(\mathcal{L}_1-\mathcal{L}_2)/10}\right) \right] \quad (1.3.7)$$

It follows from this relation that if $\mathcal{L}_2$ is much smaller than $\mathcal{L}_1$, say $\mathcal{L}_1 - \mathcal{L}_2 > 10\,\mathrm{dB}$, then the result will be that $\mathcal{L}_p \approx \mathcal{L}_1$. At the other end, for $\mathcal{L}_1 - \mathcal{L}_2 = 0$ or 1 dB, the combined signal is $\mathcal{L}_p = \mathcal{L}_1 + 3$ dB. Between these limits are $\mathcal{L}_p = \mathcal{L}_1 + 2$ dB for $\mathcal{L}_1 - \mathcal{L}_2 = 2$ or 3 dB, and $\mathcal{L}_p = \mathcal{L}_1 + 1$ dB for $\mathcal{L}_1 - \mathcal{L}_2$ from 4 to 8 dB. Similar reasoning applies if one wishes to consider what the result would be if a portion of a signal is removed $\left(p^2\right)_{av} = \left(p_1^2\right)_{av} - \left(p_2^2\right)_{av}$. The remaining signal's level will be

$$\mathcal{L}_p = \mathcal{L}_1 + 10 \log \left[1 - \left(10^{-(\mathcal{L}_1-\mathcal{L}_2)/10}\right) \right] \quad (1.3.8)$$

An important consequence of this property is that if one wishes to reduce the sound pressure level when two sources are present, removal of the weaker one will be relatively ineffective unless its magnitude is close to the stronger one's.

We are left with the question of how we can quantify the loudness of a pulse or other type of transient. The concept of a mean-squared pressure is meaningful only for sounds whose duration exceeds the time window for which the average is created. Even for continuous signals, SPL might not fit our needs. For example, suppose it is necessary to identify suitable flight paths around an airport based on their noise impact on surrounding residential neighborhoods. Two alternative paths will lead to different levels, but the path along which the sound is loudest might also be the one for which the duration is shortest. Which path is preferable? One way to decide this is to evaluate the *sound exposure level* $\mathcal{L}_E$, which is a measure of the cumulative value of p^2. The definition is that *sound exposure* is the integral of the mean-squared pressure over the duration of the signal,

$$\boxed{E = \int_0^{t_{max}} p^2 d\tau} \quad (1.3.9)$$

The pulse should begin no sooner than $t = 0$ and it should end no later than t_{max}. The units of E are Pa^2-s, but we need to account for the logarithmic nature of human sensitivity to sound level. Thus, we divide E by the same $(p_{ref})^2$ used for sound pressure level, and also divide it by a reference time. It is standard practice to use $t_{ref} = 1$ s, but any other value would merely add a constant to the sound exposure level. Thus,

$$\mathcal{L}_E = 10 \log_{10} \left(\frac{E}{p_{ref}^2 \, t_{ref}} \right) \mathrm{dB} \quad (1.3.10)$$

This quantity can be measured with a sound level meter. Multiplication of the measured value of $(p^2)_{av}$ by the time interval $t - t_1$ for the measurement yields E, so it follows that

$$\mathcal{L}_E = \mathcal{L}_p + 10 \log_{10} \left(\frac{t_2 - t_1}{t_{ref}} \right) \tag{1.3.11}$$

EXAMPLE 1.7 The primary noise-producing components within the computer workstation that Leah is building consists of four hard disks that rotate at a high speed and five cooling fans. When the computer case is open, the sound pressure level at a certain location is 40 dB when only one hard disk is operational and no fans are attached, while running one fan and no hard disks gives a sound pressure level of 30 dB. Also, it is known that installing the cover of the computer case will lower the sound pressure level by 6 dB. What is the sound pressure level when the computer is fully functional?

Significance

The nature of the decibel scale is such that a quantitative evaluation is needed to truly appreciate how several noise sources combine.

Solution

We must convert the individual sound pressure levels to mean-squared pressure values in order to add them. It is reasonable to assume that the sound emanating from all sources are mutually incoherent, because each source is an independently operating mechanical device. (This might not be true if the rotational speeds are extremely close, which is especially likely for hard disks.) We also may assume that no sound emanates from the computer if the hard disks and fans do not rotate. The analysis will add $(p^2)_{av}$ for each source, convert that value to dB to obtain the open-case sound pressure level, then apply the attenuation of the case to get the requested level.

The open-case value of $\mathcal{L}_p$ for a single hard disk is 40 dB, so

$$\text{disk: } (p^2)_{av} = p_{ref}^2 \left(10^{40/10} \right)$$

The comparable level for a single fan is

$$\text{fan: } (p^2)_{av} = p_{ref}^2 \left(10^{30/10} \right)$$

There are 4 hard disks and 5 fans. Adding the value of $(p^2)_{av}$ for each source gives

$$\text{total: } (p^2)_{av} = 4 \left[p_{ref}^2 \left(10^{40/10} \right) \right] + 5 \left[p_{ref}^2 \left(10^{30/10} \right) \right] = p_{ref}^2 (4.5) \left(10^4 \right)$$

The corresponding open-case sound pressure level is

$$\mathcal{L}_p = 10 \log_{10} \left((4.5) \left(10^4 \right) \right) = 47 \text{ dB}$$

The computer case reduces this level by 6 dB, so the noise level in the closed configuration is 41 dB. For comparison, if the fans were silent, the result would be 40 dB. In general, if one wishes to quiet a machine, it is best to first try to reduce the level of the loudest sources. If that cannot be achieved, there will be little benefit in reducing the other sources.

1.3.2 Human Factors

The audible frequency range for humans is often said to be 20 Hz–20 kHz, but few individuals respond to signals at the upper limit. Frequencies below 20 Hz are infrasonic and are sensed for the most part as a thumping sensation in the chest, while those above 20 kHz are ultrasonic. Within the audible range, human perception of loudness is highly dependent on its frequency, so any metric of loudness must account for this attribute. Pioneering experiments at Fletcher[2] at Bell Labs measured the lowest level that an average person can hear as a function of frequency. This minimum level is referred to as the *absolute threshold of hearing*. Fig. 1.6 is fairly recent data describing its frequency dependence. The curves represent averages derived from a sample population. It is interesting to note that the zero dB level is comparable to the threshold of hearing only in a narrow mid-frequency band.

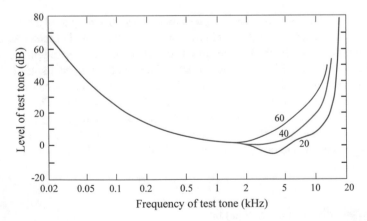

Fig. 1.6 Threshold of hearing for individuals at the indicated age. (Data extracted from a graph in *Psychoacoustics: Facts and Models,* E. Zwicker and H. Fastl, Springer (1999)

Other experiments at Bell Labs by Fletcher and Munson[3] measured a related attribute of human hearing, specifically how the perception of loudness of a pure

---

[2]H. Fletcher, "Frequency-dependent thresholds of hearing and for people with acute hearing," Rev. Mod. Phys, **12**, 47–65 (1940).

[3]H. Fletcher and W.A. Munson, "Loudness, its definition, measurement and calculation," J. Acoust. Soc. Am. **5**, 82–108 (1933).

tone varies as the frequency is changed. The specific question was addressed by starting with a 1 kHz tone at a certain sound pressure level as the reference. The amplitude of signals at other frequencies was adjusted until they were perceived by a listener to be equal in loudness to the reference tone. This level of loudness was given the unit of a *phon*. In other words, the phon level of a signal is defined to be the SPL of a 1 kHz pure tone that seems to be equally loud. The result of the Bell Labs experiments were a set of constant phon contours that plot sound pressure level (dB) against a frequency. These curves are extremely important for psychoacoustics, as indicated by the fact that they are specified in an international standard, ISO 226:2003 revision. This data, which is plotted in Fig. 1.7, differs somewhat from the data reported by Fletcher and Munson. Nevertheless, a graph of equal phon contours is referred to generically as a *Fletcher–Munson curve*. An example extracted from the graph is that a 92 dB tone at 100 Hz seems to be equally loud as a 1 kHz sound at 80 dB. The phon level of both is 80.

The threshold of hearing and the equal loudness curves have a similar appearance. They both tell us that we perceive signals in the 500 Hz–5 kHz range to be loudest, and that the falloff is fairly rapid as the frequency shifts out of this range. Any metric of loudness that is concerned with the human experience must account for this phenomenon. If the signal is a pure tone at frequency f, we could use Fig. 1.7. To do so, we would compute $\mathcal{L}_p$ according to Eq. (1.3.1) or (1.3.4), then look for the constant phon contour that passes through the point $(f, \mathcal{L}_p)$, which might require interpolation. However, this does not address the need for a metric describing a multiharmonic signal. The concept of a weighted sound pressure level was introduced to handle such a signal. Weighting involves emphasizing or de-emphasizing the contribution of signals at different frequencies by introducing a factor $W(f)$, where f

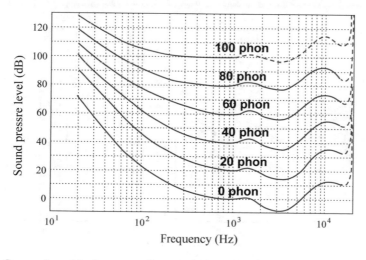

Fig. 1.7 Curves of equal loudness according to ISO 226:2003 revision. The phon level of a curve is its sound pressure value at 1 kHz

is the cyclical frequency (Hz). This factor multiplies the mean-squared pressure. If f_n, $n = 1, 2, ..., N$ are the frequencies contained in a signal, and P_n are the corresponding complex amplitudes, then the weighted mean-squared pressure and corresponding sound pressure level are defined as

$$\left(p^2\right)_{\text{weighted}} = \frac{1}{2} \sum_{n=1}^{N} W(f_n) \left|\hat{P}_n\right|^2$$

$$\left(\mathcal{L}\right)_{\text{weighted}} = 10 \log_{10} \left(\frac{\left(p^2\right)_{\text{weighted}}}{p_{\text{ref}}^2} \right) \text{dB}$$

(1.3.12)

A value of $W(f_n)$ greater than one emphasizes that frequency's contribution, and a value less than one de-emphasizes it. It should noted that the prescriptions for combining sound pressure levels described by Eqs. (1.3.6) and (1.3.8) are equally valid for weighted sound pressure levels.

Given the nature of the threshold of hearing and the Fletcher–Munson curves, the weighting function should be less than one outside a frequency band where our sensitivity is highest, whereas harmonics within this band should have a weighting function value of one or greater. Figure 1.8 displays A, B, and C weighting functions. A-weighting, which is the most common, is used to describe industrial and environmental noise, as well as health effects. B-weighting is an old standard, and C-weighting is sometimes used in entertainment venues to account for dominant bass tones that often are present. All three are described in an ANSI standard: S1.4-1983. Other weighting functions sometimes arise, such as D-weighting, which strongly emphasizes contributions in the band from 1 to 10 kHz where older jet engines had a lot of content.

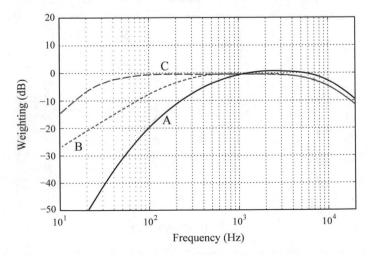

Fig. 1.8 Standard weighting functions used to compute the weighted sound pressure level

The value that is plotted in Fig. 1.8 is $\Delta\mathcal{L}$ in dB, where the symbol $\Delta\mathcal{L}$ indicate that it is the incremental adjustment to a sound pressure level. Plotting $\Delta\mathcal{L}$ rather than W facilitates reading the graph at the upper and lower frequency limits where W is very small. The weighting factor corresponding to a value of $\Delta\mathcal{L}$ is given by

$$W = 10^{\Delta\mathcal{L}/10} \tag{1.3.13}$$

The A-weighted correction to the sound pressure level shall be denoted as $(\mathcal{L})_A$. Despite its wide use, it is not entirely satisfactory as a metric of our perception. One weakness is that it is descended from the threshold of hearing, which describes how we respond to low-level pure tones. As the Fletcher–Munson curves show, our frequency response depends on the amplitude of a signal. It also has been found that our response to multiharmonic signals is somewhat different from how we perceive pure tones. Indeed, researchers continue to seek metrics that describe our perception of the loudness of noise, as well as metrics that are strongly correlated to the amount of physiological hearing damage caused by noise.

1.3.3 Frequency Bands

There are several reasons to decompose a frequency range into contiguous subintervals called *bands*. Let $b = 1, 2, \ldots$ denote the band number, and let $(f_b)_L$ and $(f_b)_U$, respectively, denote the lower and upper cyclical frequency limits (Hz) of band b. It follows that

$$\text{Band } b: \quad (f_b)_L \leq f \leq (f_b)_U; \quad b = 1, 2, \ldots, B \tag{1.3.14}$$

where B is the total number of bands spanning the interval of interest. In order that there be no gap, it must be that $(f_b)_U = (f_{b+1})_L$.

Standard Bands

Suppose a signal contains many harmonics, with some, but not necessarily all, frequencies lying in the interval $(f_1)_L \leq f \leq (f_B)_U$. Let N_b denote the number of frequencies that lie within band b, and let $f_{b,n}$, $n = 1, 2, \ldots, N_b$ denote the specific frequencies that lie within a band. The mean-squared pressure for band b is denoted as $(p_b^2)_{av}$. Recall that the total mean-squared pressure is evaluated by summing the individual contributions $|P_n|^2/2$ for all frequency components. This sum can be decomposed into the contributions from each band. To do so, one can list the frequencies in ascending magnitude and collect those that fall between the lower and upper limit of each band. Let $P_{b,n}$ denote the set of harmonics whose frequency lies in band b, so that

$$\boxed{(p_b^2)_{av} = \frac{1}{2}\sum_{n=1}^{N_b}|P_{b,n}|^2} \tag{1.3.15}$$

The band sound pressure level $\mathcal{L}_b$ for band b is found by the same definition as Eq. (1.3.1), except that the mean-squared pressure for that band is used,

$$\mathcal{L}_b = 10\log_{10}\left(\frac{\left(p_b^2\right)_{\text{av}}}{p_{\text{ref}}^2}\right) \qquad (1.3.16)$$

The band-limited mean-squared pressure describes the contributions of only those frequency components that fall in the frequency range of interest, that is, that lie on one of the bands. In view of the additive nature of the mean-squared value at each frequency, this quantity is

$$\left(p^2\right)_{\text{band-limited}} = \sum_{b=1}^{B}\left(p_b^2\right)_{\text{av}} \qquad (1.3.17)$$

Clearly, the band-limited value cannot exceed the total obtained from adding all frequency components. Any of the weighting functions may be applied to the banded quantities by using a mean value of $W(f)$ for the respective bands, that is,

$$\left(p^2\right)_{\text{weighted}} \approx \frac{1}{2}\sum_{b=1}^{B} W\left(f_b\right)_{\text{av}}\left(p_b^2\right)_{\text{av}} \qquad (1.3.18)$$

If the band intervals are not too large, the result should be quite close to the value obtained from Eq. (1.3.12).

Proportional frequency bands are the most widely used. In such a scheme, the ratio of the upper to lower limits of each band, $(f_b)_U / (f_b)_L$, is constant. A fundamental one for acoustics is *octave bands,* in which the upper limit of a band is twice the lower limit, $(f_b)_U / (f_b)_L = 2$. This practice apparently is a consequence of the significance of octaves for the music of western civilizations. If an individual hears a purely harmonic signal that is twice the frequency of another harmonic, the pitch (that is, the frequency) of the first signal will be recognized as being higher than the second, but both signals will also be recognized as being somewhat alike. Both signals would be assigned the same musical note, but in different octaves. Fine musical instruments generate a tone that consists of the same musical note in many octaves. The note in the lowest octave is the *fundamental harmonic,* while the notes in the higher octaves are said to be *higher harmonics.* It is the mix of these harmonics that gives each instrument its unique character.

Finer divisions may be obtained by geometrically subdividing the octave bands. For example, a 1/3-octave scale uses $(f_b)_U = 2^{1/3}(f_b)_L$. Similarly, a $1/N$-octave scale uses

$$(f_b)_U = 2^{1/N}(f_b)_L \qquad (1.3.19)$$

Increasing the value of N decreases bandwidths. The corollary is that fewer of a signal's harmonic constituents will lie within each band.

It often is more useful to describe $1/N$-octave bands in terms of their center frequencies, rather than their upper and lower limits. A center frequency is defined as the *geometric mean* of the upper and lower limits,

$$(f_b)_C = [(f_b)_U \, (f_b)_L]^{1/2} = 2^{1/(2N)} \, (f_b)_L = 2^{-1/(2N)} \, (f_b)_U \qquad (1.3.20)$$

These relations are readily solved for the limit frequencies corresponding to a set of center frequencies. The proportionality factor for the center frequencies is the same value as that for the upper and lower limits of the bands,

$$(f_{b+1})_C = 2^{1/N} \, (f_b)_C \qquad (1.3.21)$$

Note that the center frequencies, being geometric means, are always lower than the arithmetic mean of the lower and upper limits.

A standard 1/3 octave band scheme is in wide use, especially for noise control applications. It is based on the fact that $2^{1/3} \approx 1.260$ is close to a 25% progression, $(f_b)_U = 1.25 \, (f_b)_L$, while $2^{10/3} \approx 10.080$ leads to a progression in decades, $(f_b)_U = 10 \, (f_b)_L$. Thus, by adjusting the center frequencies from precise values, one can define a standard set of 1/3-octave-band center frequencies that also contain octave and decade scales. The center frequency of the first band is defined to be $(f_1)_C = 1$ Hz. For convenience, Fig. 1.9 lists the center and limit frequencies for standard 1/3-octave bands in a part of the sonic range. It can be seen there that increasing a band number by three shifts to the next octave to an accuracy that is adequate for engineering applications, while increasing a band number by 10 gives the next decade exactly. Thus, values for bands outside the tabulation may be found by shifting the band numbers by ten and multiplying or dividing the corresponding frequencies by ten.

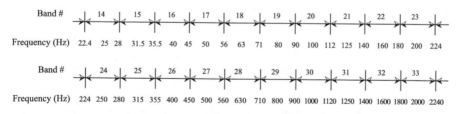

Fig. 1.9 Center and limit frequencies for standard 1/3-octave bands

EXAMPLE 1.8 A signal is generated by a number of harmonic sources. Sound pressure level measurements have been taken at a certain location in which each of the sources is turned on individually. These measurements are listed in the tabulation. For the case where all sources are active, evaluate the overall sound pressure level, pressure levels for octave bands whose lowest limit is 90 Hz, and pressure levels for standard 1/3-octave bands. Compare the unweighted and A-weighted values for each property.

Source #	1	2	3	4	5	6	7	8	9	10
Frequency (Hz)	110	170	230	300	340	400	490	570	650	700
$\mathcal{L}_p$	80	68	86	76	70	92	78	64	81	88

Significance

The best way to understand frequency band averages and the influence of weighting on sound pressure levels is to use actual data.

Solution

The solution will convert $\mathcal{L}_p$ for each source to the corresponding mean-squared pressure. The values in the various bands will be collected, then converted to band sound pressure levels. A useful procedure is to append rows to the given table in order to track the properties for each source. Equations (1.3.12) and (1.3.18) are posed in terms of the weighting factors because the mean-squared pressures are known. In the present situation, the sound pressure level for each harmonic is known, which enables a shortcut that avoids computing W corresponding to the value of $\Delta \mathcal{L}$ extracted from Fig. 1.8. Instead, we can use $\mathcal{L}_A = \mathcal{L} + \Delta \mathcal{L}$ for each harmonic signal, from which we can evaluate the individual A-weighted mean-squared pressures. The values of $\Delta \mathcal{L}$ appearing in the following tabulation are imprecise because they were read from the graph, but the values are acceptable in view of the qualitative nature of sound pressure level.

Source # n	1	2	3	4	5	6	7	8	9	10
$f(Hz)$	110	170	230	300	340	400	490	570	650	700
$\mathcal{L}_p$	80	68	86	76	70	92	78	64	81	88
$\Delta \mathcal{L}$	-18	-14	-9	-7	-6	-4	-3	-2	-1	-1
$\dfrac{(p^2)_{\text{av}}}{(10^8)\, p_{\text{ref}}^2}$	1.000	0.063	3.981	0.398	0.100	15.849	0.631	0.025	1.259	6.310
$\dfrac{(p^2)_{\text{av-A}}}{(10^8)\, p_{\text{ref}}^2}$	0.016	0.0025	0.501	0.080	0.025	6.310	0.316	0.016	1.000	5.012

Figure 1.9 describes the 1/3-octave bands in the frequency range of the given signal. To evaluate the sound pressure level for each band, we create another table that lists the band number, the harmonics whose frequency lies in that band, the sum of the mean-squared pressures of these harmonics, first unweighted then weighted, and the corresponding sound pressure levels for each band.

1/3-octave band #b	20	22	24	25	26	27	28
In-band source #	1	2	3	4, 5	6	7	8, 9, 10
$\dfrac{\left(p_b^2\right)_{av}}{(10^8)\,p_{ref}^2}$	1.000	0.063	3.981	0.498	15.849	0.631	7.594
$\dfrac{\left(p_b^2\right)_{av\text{-}A}}{(10^8)\,p_{ref}^2}$	0.0159	0.0025	0.501	0.105	6.310	0.316	6.028
$\mathcal{L}_b$	80	68	86	77	92	78	89
$(\mathcal{L}_b)_A$	62	54	77	70	88	75	88

Octave sound pressure levels may be found by combining the 1/3-octave band values of mean-squared pressure, for which another tabulation is useful. The octaves of interest are stipulated in the problem statement to begin at 90 Hz. The following tabulation collects the 1/3-octave band values that belong in each octave.

Octave frequency range (Hz)	$90 - 180$	$180 - 355$	$355 - 710$
Bands in octave	$20 - 22$	$23 - 25$	$26 - 28$
$\dfrac{\left(p_{octave}^2\right)_{av}}{(10^8)\,p_{ref}^2}$	1.0631	4.4792	24.0735
$\dfrac{\left(p_{octave}^2\right)_{av\text{-}A}}{(10^8)\,p_{ref}^2}$	0.018366	0.6057	12.6535
$\mathcal{L}_{octave}$	80	87	94
$(\mathcal{L}_{octave})_A$	63	78	91

The total sound pressure level is obtained by adding the mean-squared pressures for either the individual harmonics or the 1/3-octave values or the octave values. Using the latter gives

$$\mathcal{L} = 10\log\left(\sum_{octaves}\frac{\left(p_{octave}^2\right)_{av}}{p_{ref}^2}\right) = 95 \text{ dB},$$

$$\mathcal{L}_A = 10\log\left(\sum_{octaves}\frac{\left(p_{octave}^2\right)_{av\text{-weighted}}}{p_{ref}^2}\right) = 91 \text{ dB}$$

An overview shows that A-weighting has the greatest effect for band 20, thereby reducing the level of a strong source to relative insignificance. The A-weighting factor above 500 Hz is close to one, and this is the frequency range in which most of the strong sources lie. Consequently, the upper band levels are not reduced greatly, with the result that $\mathcal{L}_A$ is only 4 dB less than $\mathcal{L}$.

Music

The music scale is another scheme by which octave bands are subdivided. Most music of the Western World divides octave bands into twelve notes, which are pure harmonics tones. A music scale begins with any of these notes, then ascends sequentially to the highest of the twelve notes. The next octave consists of the same scale with the frequency of each note multiplied by two. A variety of tuning schemes, called *intonations,* assigns frequencies to the notes in an octave. A fundamental aspect is that most people accustomed to hearing music find that two harmonic tones played either consecutively or simultaneously are pleasant sounding if the ratio of their frequencies is a fraction composed of small integers. If this criterion is not met, the music is sensed as being discordant. Ratios of 2:1, 3:2, 4:3, and 5:4 are considered to be especially harmonious. The layout of piano keys provides a visual aid to understand how notes are defined. The basic pattern of a piano keyboard consists of repeated groups of 12 keys. Each group constitutes an octave. The layout is depicted in Fig. 1.10.

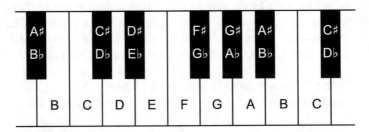

Fig. 1.10 The pattern of keys of a modern piano. An octave scale begins with C

Seven white keys and five interspersed black keys form an octave. The pattern begins with two black keys between three white keys, followed by three black keys between four white keys. The white keys are said to be *natural* notes, while the black keys are *accidentals.* Each note played on a piano consists of a fundamental frequency generated by striking a key, as well as harmonics in other octaves. The relative amplitudes of the harmonics are intrinsic to each instrument. The combination of notes at any instant might be the result of simultaneously striking more than one key, or they might be the ongoing sound resulting from keys that were previously struck.

The naturals are designated with letters, with the first key in an octave denoted as "C". In a certain octave band, C is assigned to 264 Hz. The six other naturals within an octave are "D", "E", "F", "G", "A", "B". (Obviously, any eight naturals in sequence span an octave, but the sequence described here constitutes the musical scale.) The accidental notes of the black keys are set to frequencies that are intermediate to those of the naturals on either side. A black key may be referenced to the natural to the left, in which case it is said to be a *sharp* meaning higher, of that natural. The symbol ♮ is used to designate a sharp note. Confusingly, the same black key also may be referenced to the adjacent natural to the right, in which case it is said to be a *flat,* meaning lower, of that natural. A flat is designated by the symbol ♭. Thus, the black key between F and G may be said to be F♮ or G♭. The full keyboard begins with A, A♮/B♭, and B in octave number 0, followed by seven full octaves C to B numbered 1 to 7, then terminated by C in octave number 8, for a total of 88 keys.

Just intonation dominated the music of Western civilizations from Ptolemy in the second century AD until the early nineteenth century. It arose from a desire to have frequency ratios between notes that are consistent with the small integer fraction quality. In just intonation the frequency ratios of the naturals starting with C are 1/1,

9/8, 5/4, 4/3, 3/2, 5/3, and 15/8. The next higher C, which begins the next group of keys to the right, has a ratio of 2/1, while the ratio for the next lower C, which begins the group to the left, is 1/2. For the octave beginning with C at 264 Hz, the other naturals are 297, 330, 352, 396, 440, and 495 Hz.

Fixing the frequency of notes according to just intonation leads to some difficulties. To see why suppose a piece of music begins with C. Playing D, E, F, or G either next or concurrently would meet the desire for small integer ratio fractions. The difficulty is that this criterion would not be attained if the music piece were to begin with a different note. For example, E followed by F gives a ratio of 16/15. One alternative is to adjust the scale by shifting frequencies for the notes, depending on where the music begins, in other words, varying the intonation. This would not be a problem for some instruments, like a violin, because the musician has complete freedom to select the frequencies. However, other instruments, notably the piano, can only play the frequencies to which they have been tuned. To circumvent the difficulty of where music begins, several just intonation schemes used different definitions for the frequency ratios of the accidental notes. In each case, the scheme led to some difficulties. Composers found just intonation to be too confining and sought alternatives. Some shifted the tunings of the naturals, and some adopted scales consisting of more than twelve notes per octave.

Around the time of Beethoven, *equal intonation* took hold in the Western world. Equal intonation uses notes whose frequencies are the centers of 1/12-octave bands, so each key on a piano is tuned to a frequency that is a factor $2^{1/12} = 1.0595$ higher than the key to the left. Thus, the ratios for the naturals in an octave are (C) 1, (D) $2^{2/12} = 1.1225$, (E) $2^{4/12} = 1.2599$, (F) $2^{5/12} = 1.3348$, (G) $2^{7/12} = 1.4983$, (A) $2^{9/12} = 1.6818$, (B) $2^{11/12} = 1.8877$. For comparison, the ratios for just intonations are (C) 1, (D) 1.125, (E) 1.25, (F) 1.3333, (G) 1.5, (A) 1.6667, and (B) 1.875. None of the ratios for equal intonation exactly matches the small integer ratio specification. The deviations are small and not generally perceptible, but some individuals find equal temperament to be discordant. (One website says it produces a "tinny" sound.) For this reason, alternative intonations continue to have their advocates, and music based on those intonations continues to be written. One can also hear other intonation schemes in some non-Western music.

1.4 Transfer Between Time and Frequency Domains

Many systems in acoustics are difficult, if not impossible, to analyze when their excitation is an arbitrary function of time, even though it is possible to find their response to an excitation that varies sinusoidally in time. One reason for this greater ease is the ability to remove time as an independent variable by factoring out the complex exponential factor. It might seem that being able to determine only a response to harmonic excitations is a severe limitation, but that is not so. Rather, such solutions may be used to derive transfer functions. When we consider an excitation or response to be an arbitrary function of time, we are working in the *time domain*, whereas *frequency domain* refers to analyses that begin from the premise of a purely harmonic behavior. Suppose a system's response to harmonic excitation is known. A strategy that is used in such situations converts the time-dependent excitation to an equivalent frequency-domain excitation. The known frequency-domain transfer function is applied to determine the system's response in that domain. The last step converts the frequency-domain response to the time domain. This process is illustrated in Fig. 1.11.

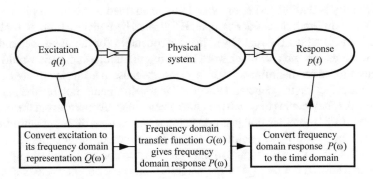

Fig. 1.11 Alternate analysis paths leading to a time-domain solution of the equations governing acoustical phenomena

Strictly speaking, transformation of response variables between the time and frequency domains is a topic in applied mathematics, rather than an essential part of the study of acoustics. Nevertheless, techniques for performing such transformations shall be explored here because a complete acoustical analysis might require them. The development begins with a restriction to situations in which the excitation and response of a system may be viewed as a superposition of harmonics. This is the essential aspect of periodic excitations, which may be described by Fourier series. We will use the complex exponential representation to formulate Fourier series for excitations and responses, then see how all operations may be carried out with great efficiency by application of Fast Fourier Transform (FFT) algorithms.

Transient situations are any that exist for some time interval, then vanish. We will see that the procedures for using FFTs to study periodic functions can be extended to handle transients. Fourier transforms are another technique for transferring between the time and frequency domains for arbitrary nonperiodic excitations. The development in Appendix B[4] of Volume 2 derives a Fourier transform by considering the limit as the period becomes infinite. This analytical tool has been employed to handle the time dependence in many important research problems in acoustics. A primary feature of that tool is that it can lead to algebraic solutions. However, its usage entails a level of mathematical sophistication that is beyond the intent of the present development. Their use to determine time responses can be avoided if numerical results, rather than mathematical formulas, are adequate.

---

[4]J.H. Ginsberg, *Acoustics-A Textbook for Scientists and Engineers, Volume II- Applications*, Appendix B, Springer (2017).

1.4.1 *Fourier Series*

Any periodic function may be represented as a Fourier series. Periodicity means that the function's graph repeats in fixed intervals. Here, we are concerned with functions of time, so we let T be the period. Analysis of a steady-state situation entails taking the periodic excitation to have existed for all time. However, sometimes, we are interested in a periodic process that begins at time t_0, in which case periodicity only applies for for $t > t_0$, so that

$$p(t + T) = p(t), \quad t_0 < t < \infty \tag{1.4.1}$$

In such situations, the series may be used in place of the actual function for any period subsequent to the initial time t_0.

A Fourier series is a sum of harmonics. Each term has a frequency that is an integer multiple of the *fundamental frequency*, ω_1, which is the frequency of the harmonic whose period is T,

$$\boxed{\omega_1 = \frac{2\pi}{T}} \tag{1.4.2}$$

Introductory texts typically use a sum of cosine and sine terms to describe a Fourier series, but summing complex exponentials is consistent with the developments thus far. Let integer n denote the harmonic number, whose corresponding frequency is

$$\omega_n = n\omega_1 = \frac{2\pi n}{T} \tag{1.4.3}$$

One way of forming the Fourier series is to use the real part representation of each harmonic. It is convenient to denote the mean value of $p(t)$, that is, the zero frequency term, as $\frac{1}{2} P_0$, so

$$\boxed{p = \frac{1}{2} P_0 + \sum_{n=1}^{\infty} \text{Re}\left(P_n e^{in\omega_1 t}\right)} \tag{1.4.4}$$

The usage of real parts causes difficulty for some mathematical operations, which are best done when the complex conjugate is listed explicitly,

$$p = \frac{1}{2} P_0 + \frac{1}{2} \sum_{n=1}^{\infty} \left(P_n e^{in\omega_1 t} + P_n^* e^{-in\omega_1 t}\right) \tag{1.4.5}$$

To consolidate this expression, complex amplitudes for negative harmonic indices, $n < 0$, are defined to be the complex conjugate of the positive index harmonic amplitudes,

$$P_{(-n)} = P_n^*, \quad n = 1, 2, \dots \tag{1.4.6}$$

The result is the complex Fourier series,

$$p = \frac{1}{2} \sum_{n=-\infty}^{\infty} P_n e^{in\omega_1 t} \equiv \frac{1}{2} \sum_{n=-\infty}^{\infty} P_n e^{i2\pi nt/T} \tag{1.4.7}$$

This is the form we shall use to perform analysis. The process of actually evaluating a Fourier series is synthesis, which is best carried out according to Eq. (1.4.4). Great care should be taken to assure that the number of terms at which the series terminates is sufficient to attain convergence to the desired accuracy.

An obvious question is how does one obtain the complex Fourier series amplitudes P_n corresponding to a specified time function $p(t)$? The answer to this question originates in an important property of the complex exponentials terms, which is their orthogonality. Let j be the index of a specific harmonic. Multiply Eq. (1.4.7) by $\exp(-i2\pi jt/T)$, then integrate over a period. It does not matter when the period begins, so the integration interval is taken to be $t_1 \leq t \leq t_1 + T$. Thus, we form

$$\int_{t_1}^{t_1+T} p\, e^{-i2\pi jt/T} dt = \int_{t_1}^{t_1+T} \left(\frac{1}{2} \sum_{n=-\infty}^{\infty} P_n e^{i2\pi nt/T} \right) e^{-i2\pi jt/T} dt$$
$$= \frac{1}{2} \sum_{n=-\infty}^{\infty} P_n \int_{t_1}^{t_1+T} e^{i2\pi(n-j)t/T} dt \tag{1.4.8}$$

Because $n - j$ is an integer, the period T is an integer multiple of the period of the complex exponential, so the integral is zero. The only exception occurs in the single term for which the running index n for the summation equals the selected index j, because then the exponential reduces to a unit value. Hence, the integration process filters out of the summation the single term associated with j. The result is that

$$P_j = \frac{2}{T} \int_{t_1}^{t_1+T} p\, e^{-i2\pi jt/T} dt \tag{1.4.9}$$

As was noted previously, this expression is valid for any t_1. However, if the function is analytical, we usually take the period to be centered on $t = 0$, which corresponds to setting $t_1 = -T/2$. If the function has a discontinuity at time t_d, $t_d + T$, ..., then it is advisable to set $t_1 = t_d$. Doing so allows for an analytical integration over a period. However, if more than one discontinuity exists within a period, then analytical evaluation of the integral must be performed piecewise over each subinterval in which the function is analytical.

If $p(t)$ is not too complicated, the integral may be evaluated manually, perhaps with the aid of a modest table of integrals. Alternatively, it might be more convenient to use symbolic mathematical software. If neither approach seems to promising, it is not advisable to use computational software like MATLAB to evaluate the integrals numerically. Clearly, computational software cannot yield formulas for the

Fourier coefficients. But the reason for the caution is that the same operations can be performed with a numerical algorithm called the Fast Fourier Transform (FFT). The usage of FFTs will be covered in the next section, where it will be seen that they are far more efficient than conventional numerical procedures for deriving and synthesizing Fourier series, and equally important, easy to implement.

The notion of shifting the integration limits brings to the fore a useful property. Consider two periodic functions $p(t)$ and $q(t)$ that are the same, except that one is shifted in time relative to the other, that is,

$$q(t) = p(t - \tau), \quad p(t + T) = p(t) \tag{1.4.10}$$

The Fourier coefficients of $q(t)$ are found from Eq. (1.4.9), and the integration variable is changed to $t' = t - \tau$. This leads to be

$$Q_j = \frac{2}{T} \int_{\tau - T/2}^{\tau + T/2} p(t') \, e^{-i2\pi j\tau/T} e^{-i2\pi jt/T} dt' = P_j e^{-i2\pi j\tau/T} \tag{1.4.11}$$

In other words, the Fourier coefficient of each harmonic contains a phase lag that is the product of the time delay and the harmonic's frequency.

The delay property has an interesting implication when one seeks to interpret Fourier coefficients that are extracted from measured data. If that data describe a variable that is delayed, the phase angles will not seem to be proportional to the harmonic number because the phase angles will be assigned values in a 2π range. The identified phase angles may be plotted against harmonic number, after which one can increment or decrement individual values by multiples of 2π in an attempt to obtain a smoother curve from which the value of τ could be estimated. This operation is said to be *phase unwrapping*. It can be performed with an unwrap function in some computational software.

The mean-squared pressure of a Fourier series is found by adding the contribution of each harmonic. A strict interpretation of the average gives

$$\left(p^2\right)_{\mathrm{av}} = \left(p_{\mathrm{mean}}\right)^2 + \frac{1}{2} \sum_{n=1}^{\infty} |P_n|^2, \quad p_{\mathrm{mean}} = \frac{1}{2} P_0 \tag{1.4.12}$$

This relation is *Parseval's theorem*; it is a special case of Eq. (1.2.12). Although the preceding is correct mathematically, the mean pressure term should be omitted for a computation of sound pressure level, because it represents a nonoscillatory, and therefore inaudible, effect.

An important type of excitation is a periodic train of impulses. An impulse is a signal that is very large over a very small interval. A Dirac delta function captures the primary effect of an impulse. There are many ways of introducing a Dirac delta function; the simplest uses a function that is constant over a small interval ε starting at time τ, with the value of the function being $1/\varepsilon$, as shown in Fig. 1.12

Let us denote this representation as $f(t, \tau, \varepsilon)$, where

Fig. 1.12 Representation of
a Dirac delta function

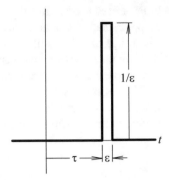

$$f(t, \tau, \varepsilon) = \begin{cases} 1/\varepsilon, & \tau < t < \tau + \varepsilon \\ 0, & t < \tau \text{ and } t > \tau + \varepsilon \end{cases} \tag{1.4.13}$$

The Dirac delta function, denoted with the symbol δ, is the limit of $f(t, \tau, \varepsilon)$ as $\varepsilon \to 0$. Its value is undefined when $t = \tau$, and it is zero at every other instant, so we may consider it to depend solely on the difference $t - \tau$, which leads us to the first basic property,

$$\boxed{\delta(t - \tau) = 0 \text{ if } t \neq \tau} \tag{1.4.14}$$

The fact that a Dirac delta is not defined at the instant it occurs is not an issue because we are solely concerned with how it affects integrals.

The second basic property comes from integrating a Dirac delta over an interval that includes $t = \tau$. We evaluate the integral by representing the function as $f(t, \tau, \varepsilon)$ for finite, but very small, ε, after which we take the limit as $\varepsilon \to 0$. Thus, we have

$$\int_a^b \delta(t - \tau)\, dt = \lim_{\varepsilon \to 0} \int_a^b f(t, \tau, \varepsilon)\, dt = \lim_{\varepsilon \to 0} \int_\tau^{\tau + \varepsilon} \frac{1}{\varepsilon}\, dt$$

$$\boxed{\int_a^b \delta(t - \tau)\, dt = 1 \text{ if } a < \tau < b} \tag{1.4.15}$$

The third property pertains to the integral of the product of an arbitrary analytical function $z(t)$ and a Dirac delta. This is obtained by using $f(t, \tau, \varepsilon)$ to represent $\delta(t - \tau)$, and noting that f is nonzero only over a short interval, so that

$$\int_a^b z(t)\, \delta(t - \tau)\, dt = \lim_{\varepsilon \to 0} \int_a^b z(t)\, f(t, \tau, \varepsilon)\, dt$$

$$= \lim_{\varepsilon \to 0} \int_\tau^{\tau + \varepsilon} z(t) \left(\frac{1}{\varepsilon}\right) dt \text{ if } a < \tau < b \tag{1.4.16}$$

The central limit theorem states that if ε is sufficiently small, then $z(t)$ may be replaced with its value at any instant in the interval of the integration, which leads to

$$\boxed{\int_a^b z(t)\,\delta(t-\tau)\,dt = z(\tau) \quad \text{if } a < \tau < b}$$ (1.4.17)

Equations (1.4.14), (1.4.15), and (1.4.17) will be invoked in several topics. In the present context, they lead to the Fourier series of a periodic sequence of impulses. Suppose an impulse having a magnitude of A occurs at multiples of T,

$$p(t) = \begin{cases} A\delta(t), & -T/2 \le t \le T/2 \\ p(t \pm T), & |t| > T/2 \end{cases}$$ (1.4.18)

The corresponding Fourier coefficients are given by Eq. (1.4.9), which is greatly simplified by Eq. (1.4.17). The result is

$$P_j = \frac{2}{T}\int_{-T/2}^{T/2} A\delta(t)\, e^{-i2\pi jt/T}\,dt = \frac{2}{T}A$$ (1.4.19)

In some situations, it might be preferable to consider the sequence of impulses to occur at instants that are shifted from multiples of T. If the impulse is such that it occurs at $t = \tau$, then it is $A\delta(t - \tau)$. Equation (1.4.11) states that the Fourier coefficients have the same magnitude as the preceding with an additional phase delay that is product of τ and the corresponding frequency,

$$p(t) = A\delta(t-\tau) \iff P_j = \frac{2}{T}Ae^{-i2\pi j/T}$$

This result shows that the Fourier coefficients for a periodic sequence of impulses have the same magnitude regardless of the impulse's timing relative to $t = 0$. This property leads to a failure of the Fourier series to converge. Figure 1.13 shows the synthesis of the series for a periodic sequence of impulses when the series is truncated at $N = 10, 20,$ and 30. As N increases, the region where the values are large becomes increasingly confined to the vicinity of $t = 0, \pm T, ...$, which is where the impulses occur, and the peak value increases with increasing N. However, the values oscillate around zero between the impulses, rather than actually being zero, although the magnitude of this fluctuation decreases with increasing N.

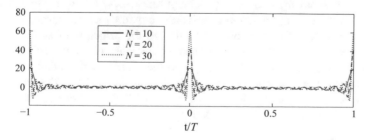

Fig. 1.13 Synthesis of the Fourier series describing an periodic sequence of impulses, $p(t) = \delta(t)$, $p(t + T) = p(t)$

A periodic sequence of impulses represents an extreme case. If $p(t)$ is a periodic function that has a finite value at every instant, then its Fourier coefficients will decrease with increasing harmonic number. Synthesis of the summation for such series will approach the actual function as the series length is increased. This applies to all instants except at discontinuities. Let t_d be the instant at which some function $p(t)$ has a discontinuity and let p^- and p^+ denote the function's value immediately before and after $t = t_d$. If the series is evaluated at $t = t_d$, it will converge to the mean of p^- and p^+, so that

$$\lim_{\varepsilon \to 0} p(t_d - \varepsilon) = p^-, \qquad \lim_{\varepsilon \to 0} p(t_d + \varepsilon) = p^+$$
$$p(t_d) = \frac{1}{2}\left(p^- + p^+\right) \tag{1.4.20}$$

Fourier series techniques provide the foundation for the conversions between the time and frequency domains described generally in Fig. 1.11. An essential requirement for doing so is linearity of the governing equations, which allows us to add fundamental solutions. (Most of our study of acoustics will meet this criterion because acoustical signals typically are quite small physically, even when they seem to be quite loud.) The principle of superposition is a corollary of linearity. If an excitation is periodic, each of its Fourier series terms represents a harmonic excitation. Analysis of the response of the system at hand to a generic one of these excitations yields the complex amplitude of the harmonic response. These are the Fourier series coefficients of the time-domain response.

To make the procedure more explicit, consider any system that is governed by linear differential equations and is subjected to a harmonic excitation. The response will consist of homogeneous and particular solutions of the differential equations. The former does not require the existence of an excitation. It can be generated by initial conditions, which would be the response at the instant when an excitation is removed. The particular solution is that response that is directly induced. The portion of the response that is the homogeneous solution eventually becomes negligible due to dissipation effects like viscosity, even if those effects are not contained in the theory. The particular solution persists as long as the harmonic excitation is present. It is the *steady-state response*.

The transfer function $G(\omega)$ is the steady-state response corresponding to a harmonic excitation having a unit complex amplitude. According to the linearity property, if $Q(\omega)$ is the complex amplitude of the excitation, then the harmonic response will be $G(\omega)Q(\omega)$. Furthermore, according to the principle of superposition, if the excitation is a sum of harmonics, then the response is a sum of the associated harmonic responses. Specifically, let Q_n be the Fourier series coefficients of the excitation $q(t)$. The frequencies of the excitation are $n\omega_1 = 2\pi n/T$, so the response coefficient for the nth harmonic is $G(2\pi n/T)Q_n$. Then, the Fourier series for the time-domain response $p(t)$ is described at

$$q(t) = \frac{1}{2}\sum_{n=-\infty}^{\infty} Q_n e^{i2\pi nt/T} \implies p(t) = \frac{1}{2}\sum_{n=-\infty}^{\infty} G\left(\frac{2\pi n}{T}\right) Q_n e^{i2\pi nt/T} \tag{1.4.21}$$

This description of the steady-state response seems to be quite straightforward to evaluate, but there are two aspects to bear in mind. The first concerns convergence. Recall that $G(\omega)$ is the response to an excitation at frequency ω. All systems have some mass, and acceleration is proportional to ω^2. It follows that the force required to make any system move approaches infinity as the frequency is increased beyond limit, in other words, all responses must approach zero in this limit, so that $G(\omega) \to 0$ as $\omega \to \infty$. This means that the response series will converge to a given accuracy more quickly than the excitation series. An extreme example is that of an infinite sequence of impulses, whose Fourier series never converges. Nevertheless, its Fourier coefficients lead to a convergent Fourier series of the time-domain response.

Another aspect requiring care is the selection of instants at which a Fourier series is synthesized because doing so requires that the series be truncated. The time instants at which such a series is evaluated should be selected in accord with the Nyquist sampling criterion developed in the next section. Evaluation of the series at an excessive number of instants is shown there to be an attempt to extract more information from the series than it contains. If this criterion is ignored, the consequence will be evident in the vicinity of discontinuities, where there will be fluctuations that are not found in the actual function. In other words, the evaluation will not satisfy Eq. (1.4.20), although the mean of the values on either side probably will be computed. This feature will be seen in the next example.

EXAMPLE 1.9 The graph depicts a sawtooth pressure signal. Determine the Fourier coefficients P_n for this signal. Identify the harmonic number M beyond which all higher coefficients are smaller than one hundredth of the maximum value, that is, $|P_n| < 0.01 \max |P_n|$ for $n > M$. Let the harmonic number N at which the series is truncated equal M. Evaluate the truncated series at instants separated by the constant interval T/J, for $J = N/2$, N, and $4N$. Compare the values of $p(t)$ obtained from each evaluation to the actual sawtooth curve. Then, repeat the analysis when the series is truncated prematurely by setting $N = M/4$.

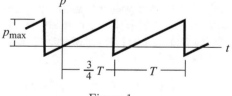

Figure 1.

Significance

That complex Fourier series are no more difficult to evaluate than the real version is demonstrated here. Some results will exhibit Gibb's phenomenon, which is associated with Fourier series of discontinuous functions. This will motivate the study in the next section of the Nyquist criterion.

Solution

Direct application of the formulas will suffice for the solution. The integration interval for Eq. (1.4.9) is best taken to be $-T/4 < t < 3T/4$, because the function is a continuous straight line in this interval, specifically, $p = p_{max}t/(3T/4)$. The coefficients are

$$P_n = \frac{2}{T} \int_{-T/4}^{3T/4} \left(p_{max} \frac{4t}{3T} \right) \exp\left(-i\frac{2\pi nt}{T} \right) dt$$

The integral may be found with the aid of a standard table integrals or with symbolic mathematical software. The integral for the mean value term, $n = 0$, is different from the one for $n \neq 0$. The results are

$$P_n = \begin{cases} \dfrac{4i^{n+1}}{3\pi n} p_{max}, & n \neq 0 \\[2mm] \dfrac{2}{3} p_{max}, & n = 0 \end{cases}$$

The harmonic number M beyond which the coefficients are negligible may be identified directly from this result because $|P_n|$ is inversely proportional to n for $n > 0$. The largest coefficient is P_0, so we wish that $|P_M| < 0.01 |P_0|$, which gives

$$\frac{4}{3\pi M} < 0.01 \left(\frac{2}{3} \right)$$

The smallest integer satisfying this condition is $M = 63$, but we will use $M = 64$ in order to avoid rounding fractions to set N.

Truncation of the Fourier series at N terms means that the result will depend on N as well as the value of t. Thus, we write Eq. (1.4.4) as

$$p_{fs}\left(t_j, N \right) = \frac{1}{2} P_0 + \sum_{n=1}^{N} \text{Re} \left(P_n e^{in\omega_1 t_j} \right)$$

The sum of products of P_n and the exponential factor may be computed efficiently as a row-column matrix product. Let $\{P\}$ be a column vector of the Fourier series coefficients from P_1 to P_N and let $[E_j]$ be a row vector holding the values of the exponentials from $n = 1$ to $n = N$ at t_j, that is,

$$\left[E_j \right] = [e^{i\omega_1 t_j} \; e^{i2\omega_1 t_j} \; \cdots \; e^{iN\omega_1 t_j}]$$

Evaluation of the series is described by

$$p_{fs}\left(t_j, N \right) = \frac{1}{2} P_0 + \text{Re} \left(\left[E_j \right] \{P\} \right)$$

Evaluation of the actual sawtooth function at the discontinuity requires special attention. The function jumps from p_{max} to $-p_{max}/3$ at $t = 3T/4$, but it is not defined at the discontinuity. The Fourier series at $t = 3T/4$ will evaluate to the average of the function values on either side, which gives $p = p_{max}/3$. Let us assign this value to the given function. Thus, we evaluate

$$p_{exact}(t_j) = \begin{array}{l} (4p_{max}/3)\, t_j/T \text{ if } t_j < 3T/4 \\ P_{max}/3 \text{ if } t_j = 3T/4 \\ (4p_{max}/3)\,(t_j/T - 1)\, - \text{ if } t_j > 3T/4 \end{array}$$

It is easier to interpret a graph if it displays more than one cycle, which can be done without calculation by replicating the data for another period according to

$$t_{j+N} = t_j + T, \quad p(t_{j+N}) = p(t_j)$$

Values of T and P_{max} are not given, so we set each to a unit value. This corresponds to computing p/p_{max} as a function of t/T, which is how the data will be plotted.

Figure 2 describes the results for $N = M = 64$. Evaluation of $p_{fs}(t_j, N)/p_{max}$ for $J = N/2$ and $J = N$ yield values that are very close to the actual sawtooth function. However, $J = 32$ might be considered to undersample the function because the slope across the discontinuity is not extremely high for both the series synthesis and the actual function. The results for $J = 4N$ adequately captures the discontinuity. However, it exhibits an artifact in the vicinity of the discontinuity in which there is a small rapid fluctuation about the true value. This fluctuation decreases with increasing time from the discontinuity.

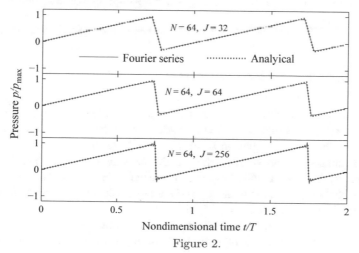

Figure 2.

Figure 3 shows the results obtained when the Fourier series is truncated at $N = 16$. The trends are similar to those for $N = 64$. Here too, using many more points than the series length leads to a small fluctuation of the synthesized values in the vicinity of the discontinuity. The magnitude of this fluctuation seems to be comparable to that obtained for $N = 64$, but the frequency of the fluctuation seems to be lower.

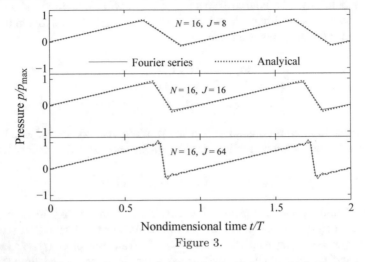

Figure 3.

The important conclusion to be derived from this example is that it is a good policy to estimate the number of terms required for convergence by monitoring the magnitude of the series coefficients. Once the series length is set, the time increment with which the series is evaluated should be set at the largest interval that captures discontinuities and other large changes over a small interval. The deviation of the series values for a large number of data points is known as Gibbs' phenomenon. The next section will show that it is a consequence of sampling a discontinuous function too finely.

1.4.2   Discrete Fourier Transforms

Determination of Fourier series coefficients by analytically evaluating the integral in Eq. (1.4.9) might be extremely difficult, or not possible, because the function is complicated. Furthermore, integration is not possible if the variable is only known at a set of discrete instants, as would be the case for measurements taken by a digital data acquisition system. One could fit a curve to such data to obtain an approximate representation. However, full numerical approximations can handle complicated functions and sampled data with equal ease.

The first task is to decide how integrals will be evaluated numerically. Let $g(t)$ be a function that we wish to integrate over the interval $0 \leq t < T$. This interval is subdivided into equal intervals Δt, and the function is sampled at the instants t_n. Let N be the number of subintervals, so that the sampling instants are

$$t_n = n\Delta t, \quad n = 0, 1, ..., N, \quad \Delta t = \frac{T}{N} \tag{1.4.22}$$

A variety of numerical algorithms, such as the trapezoid rule or a Simpson's rule, are available. However, the best one (in terms of numerical accuracy, as well as efficiency) in the case of the integrals for Fourier coefficients is a crude approximation known as the strip rule. It is depicted in Fig. 1.14.

Fig. 1.14 Strip rule for numerical evaluation of Fourier coefficients

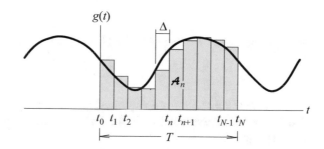

The construction approximates $g(t)$ by strips whose height is the function value at the start of each subinterval. A stepped approximation might not seem to be good as an interpolation method. However, periodicity of the function causes overestimation of strip areas in intervals where g is decreasing to be compensated by underestimation in intervals where g is increasing. The area of the nth strip is $A_n = g(t_n) \Delta t$. There are N intervals. There is no strip for t_N, which is appropriate because $g(t_N) \equiv g(t_0 + T) \equiv g(t_0)$ as a consequence of periodicity, and the contribution of $g(t_0)$ is contained in the first strip. The approximate integral of $g(t)$ over a period is obtained by adding the area of each strip, so that

$$\int_0^T g(t)\, dt \approx \Delta t \sum_{n=0}^{N-1} g(n\Delta) \tag{1.4.23}$$

Let us apply this approximation to an evaluation of the Fourier series coefficients Q_j of an excitation $q(t)$ whose period is T. We set the integration interval in Eq. (1.4.9) to $0 \le t < T$ in order to employ directly the preceding numerical rule. The fundamental frequency is $2\pi/T$ and the integrand is evaluated at instants $t_n = nT/N$, so the argument of the exponential is $-ij(2\pi/T)(nT/N)$. Hence, application of the strip rule to evaluate the Fourier coefficients of $q(t)$ yields

$$Q_j \approx \left(\frac{2}{T}\right)(\Delta t) \sum_{n=0}^{N-1} q_n e^{-i2\pi jn/N} \tag{1.4.24}$$

The unadorned summations are denoted as $\hat{Q}_j$ values,

$$\hat{Q}_j = \sum_{n=0}^{N-1} q_n e^{-i2\pi jn/N}, \quad q_n \equiv q(t_n) \tag{1.4.25}$$

These quantities are the *discrete Fourier transform* of the q_n data. They are referred to as *DFT* coefficients. It follows from the equation preceding the above that a Fourier series coefficient is related approximately to the corresponding DFT coefficient by

$$Q_j \approx \frac{2}{N}\hat{Q}_j \qquad (1.4.26)$$

Note that replacing j with $-j$ in Eq. (1.4.25) gives

$$\hat{Q}_{(-j)} = \hat{Q}_j^* \qquad (1.4.27)$$

which is consistent with the complex conjugate property of the Fourier coefficients. The frequencies corresponding to the DFT data are integer multiples of the fundamental frequency,

$$\omega_j = j\frac{2\pi}{T} \qquad (1.4.28)$$

An important question is how many coefficients $\hat{Q}_j$ should be computed? The answer lies in an examination of the number of independent numbers that are contained in a set of Fourier series coefficients. A period has been sampled to give N (real) data values, which means that the DFT set should consist of no more than N real numbers. A Fourier coefficient generally is complex, so it consists of two real numbers, which would seem to indicate that the independent coefficients are the $N/2$ coefficients $\hat{Q}_0$ to $\hat{Q}_{N/2-1}$. However, $\hat{Q}_0$ is the real mean value of $p(t)$, so it contains only one real number. Also, evaluating Eq. (1.4.25) for $j = N/2$ shows that $\hat{Q}_{N/2}$ is real. It follows that the coefficients in the range from $\hat{Q}_0$ to $\hat{Q}_{N/2}$ are composed of N real numbers. Thus, the DFT values $\hat{Q}_j$ for $j = 0, 1, ..., N/2$ represent an independent data set that may be computed from N samples of a real function $q(t)$. The complex conjugate property is applied to determine the coefficients $\hat{Q}_j$ for $j = -1, -2, ..., -N/2 + 1$, and the coefficient $\hat{Q}_{-N/2}$ is discarded because it repeats $\hat{Q}_{N/2}$.

Let us now consider the evaluation of pressure $p(t)$ when a set of DFT coefficients $\hat{P}_j$ have been evaluated, as would be the case if we applied a transfer function to the excitation coefficients $\hat{Q}_j$. We have represented the excitation as a set of sampled values at the discrete time instants. Therefore, it is reasonable to restrict evaluation of the response to the same instants. The summation in Eq. (1.4.7) adds the contribution of each Fourier coefficient, which is now replaced with $(2/N)\hat{P}_j$ in accord with Eq. (1.4.26). The range of indices for these coefficients are the same as those for the excitation, that is, $j = -N/2 + 1, ..., 0, ..., N/2$. Because the discrete instants are $t_n = nT/N$, and $\omega_1 = 2\pi/T$, the value of T disappears from the summation in Eq. (1.4.7), which becomes

$$p_n = \frac{1}{N} \sum_{j=-N/2+1}^{N/2} \hat{P}_j e^{i2\pi nj/N} \tag{1.4.29}$$

The process of constructing the time values of a variable from its DFT coefficients is said to be an *inverse discrete Fourier transform* (IDFT).

In summary, the DFT of a periodic function $q(t)$ that has been sampled at N intervals covering a full period is given by

$$\boxed{\hat{Q}_j = \sum_{n=0}^{N-1} q_n e^{-i2\pi jn/N}, \quad , \hat{Q}_{(-j)} = \hat{Q}_j^*, \quad j = 0, 1, ..., N/2} \tag{1.4.30}$$

The IDFT yields the discrete time values of a variable p whose DFT coefficients are known to be $\hat{P}_j$. These values are obtained by evaluating

$$\boxed{p_n = \frac{1}{N} \sum_{j=-(N/2-1)}^{N/2} \hat{P}_j e^{i2\pi nj/N}, \quad n = 0, 1, ...N - 1} \tag{1.4.31}$$

The p_n values correspond to the period $0 \le t < T$, but they may be periodically replicated.

The IDFT can be proven to have a remarkable property. Although the $\hat{P}_j$ values are approximately proportional to the Fourier series coefficients, as described by Eq. (1.4.27), the inverse DFT is exact. That is, if a set of DFT coefficients has been evaluated according to Eq. (1.4.30), then the IDFT of those DFT coefficients will yield the original time data to the numerical precision of the computing device. For this reason, a DFT could be considered to be a data transformation that exists independently of Fourier series concepts.

1.4.3   Nyquist Sampling Criterion

An important aspect of digitizing a time signal is selection of the sample size N. Suppose we have evaluated the $N/2 + 1$ DFT coefficients $\hat{P}_j$. We could graph the magnitude of these coefficients as a function of their index j in order to identify the largest index J_{max} for which the DFT coefficients are significant in comparison with the largest values. If J_{max} is substantially smaller than $N/2$, we would conclude that the associated Fourier series would converge. In contrast, if $J_{max} \approx N/2$, or if there is no identifiable index at which the DFT are insignificant, then it is quite likely that enlarging N would yield additional coefficients whose contribution would be significant. Thus, we deduce that a convergent DFT is marked by its coefficients being relatively small above index J_{max}, with J_{max} being substantially smaller than $N/2$.

This is the fundamental requirement, but we will not leave it in its present form because J_{max} does not appear to have any relation to basic properties of the signal. The fundamental frequency for the DFT is $\Delta f \equiv 1/T$, so the maximum frequency that must be included is $J_{max}\Delta f = f_{max}$. It is important that f_{max} is a property of the time signal, but not how that signal was sampled. In essence, f_{max} is the cutoff frequency beyond which the DFT coefficients are negligible.

It is necessary that the DFT coefficients at least cover the range $0 \leq f \leq f_{max}$, but the highest cyclical frequency for the coefficients is $(N/2)(1/T)$. Thus, it is necessary that $N/(2T)$ be greater than or equal to f_{max}, that is

$$N \geq 2f_{max}T \tag{1.4.32}$$

This is the *Nyquist sampling criterion*, and the frequency $2f_{max}$ is called the *Nyquist frequency*,

$$f_{Nyquist} = 2f_{max} \tag{1.4.33}$$

The Nyquist criterion usually is posed in terms of the sampling interval $\Delta t = T/N$, because that is usually the primary parameter for data acquisition equipment. The *sampling rate* is $1/\Delta t$, that is, the number of samples contained in a unit of time. (The usual unit is hertz.) Replacing N/T by $1/\Delta t$ in Eq. (1.4.32) yields the form in which the Nyquist criterion usually is stated,

$$\frac{1}{\Delta t} \geq 2f_{max} = f_{Nyquist} \tag{1.4.34}$$

In words, the Nyquist sampling criterion requires that **Data should be sampled at a rate that is at least twice the highest frequency at which the DFT coefficients are significant**.

Now that we have identified a criterion, the logical question is how can we select the sampling rate to satisfy the Nyquist criterion prior to acquisition of the data, when the nature of the signal is not yet known? One strategy is to examine the time data over a short interval in order to identify the shortest time scale over which the variable fluctuates. The Nyquist sampling criterion is equivalent to saying that the period of the most rapid fluctuation should contain at least two samples. Thus, Δt should be no larger than the time between the maximum and minimum of a wiggle.

The difficulty with this approach is the identification of the most rapid fluctuation, especially if the record is long. An alternative is to perform a trial run. The value of Δt is selected on the basis of past experiences, intuition, and knowledge of the process being measured, although sometimes Δt is dictated by the measurement equipment that is available. The time data is collected and processed to find its DFT coefficients, whose magnitude $|P_n|$ is plotted against either n or $f_n = n/T$. The sampling is acceptable if $|P_n|$ is much less than max $|P_n|$ for a range of n ending at $N/2$. If the sampling is adequate, the full set of data may be acquired.

Failure to select a sufficiently high sampling rate has a much more severe consequence than merely missing some significant DFT coefficients. Rather, such a failure introduces an error called *aliasing* that contaminates the DFT values over a broad frequency range. The frequency increment for the coefficients is $1/T$, so the DFT values represent the range $-N/(2T) < f \leq N/(2T)$, where the negative frequency range is associated with the DFT coefficients having negative index. By definition, the significant DFT coefficients lie in the frequency band $-f_{max} < f < f_{max}$. The corresponding Nyquist frequency is $f_{Nyquist} = 2f_{max}$. Aliasing occurs when $f_{max} > N/(2T)$.

Such a situation is depicted in Fig. 1.15, where $N/(2T)$ and f_{max} are replaced by $1/(2\Delta t)$ and $f_{Nyquist}/2$, respectively. The coefficients that are outside the frequency interval of the computation, which is $-1/(2\Delta t) < f < -1/(2\Delta t)$, are not negligible. The coefficients associated with frequencies above $1/(2\Delta t)$ are downshifted by the period, $1/\Delta t$, whereas coefficients associated with frequencies below $-1/(2\Delta t)$ are upshifted by $1/\Delta t$. Both shifts assign the coefficients to a frequency inside the band of the computation. These shifts are said to be *aliasing*, because the shifted coefficients have been given a false "name," that is, frequency. The coefficients that are aliased combine with the coefficients whose frequency is properly assigned.

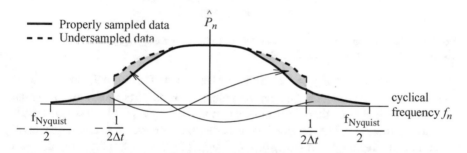

Fig. 1.15 How aliasing introduces error to DFT coefficients

There is no way to identify the amount by which the aliased coefficients contaminate the DFT values. Hence, if a plot of the DFT coefficients in the manner of Fig. 1.15 shows that the values in the vicinity of $\pm 1/(2\Delta t)$ are not small, the only remedy is to reacquire the time data using a smaller Δt. If the aliasing is not too severe, as would be the case if the highest coefficients are reasonably small, the alternative is to proceed with the flawed data.

The Nyquist sampling criterion also has relevance in regard to the synthesis of time values from a set of Fourier coefficients. In Example 1.9, we encountered the Gibbs phenomenon, which is associated with the usage of Fourier series to describe functions that have discontinuities. Let us examine what happens if we fix the number of DFT coefficients at N and use those coefficients to evaluate the function at a sequence of instants that are selected independently. To do this, the square wave in Fig. 1.16 was sampled at $N = 32$ instants, $t_n = n(T/32)$. The sampled values were set to give the mean $p = 0.5$ at the discontinuities. The DFT coefficients for this data were evaluated according to Eq. (1.4.30). (Actually, the complex conjugate property required computation of only $\hat{P}_0$ to $\hat{P}_{16}$.) Then, the time data was synthesized at

N_{synth} instants $t'_n = n \left(T/N_{\text{synth}} \right)$ using these $\hat{P}_j$ values. The results of $N_{\text{synth}} = 32$, 64, and 256 appear in the figure.

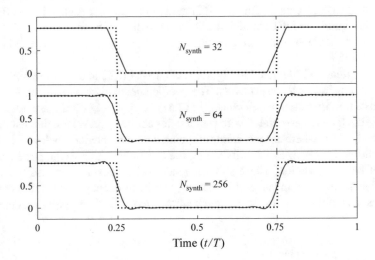

Fig. 1.16 Comparison of the results of synthesizing a square wave at N_{synth} instants using 16 DFT coefficients acquired from data at 32 instants

The first result, for $N_{\text{synth}} = 32$, verifies that application of the DFT, then the IDFT, to a data set will reproduce the data. Examination of the data shows that $\hat{P}_1$ is the largest coefficient other than the mean value, and that $|\hat{P}'_{N/2-1}/\hat{P}_1|$ is less than 0.001. This means that $N = 32$ meets the Nyquist criterion. However, the sampling interval is too large to capture the rapidity of the change of pressure. Synthesizing the data at more points cannot overcome this fundamental shortcoming. This is evidenced by the fact that neither $N_{\text{synth}} = 64$ nor $N_{\text{synth}} = 256$ improve the description of the discontinuities. The consequence of setting the number of instants for the IDFT greater than the number of instants for the time data is the appearance of a small fluctuation in the vicinity of the discontinuities. This feature is a different manifestation of the Gibbs phenomenon. Recall that an IDFT exactly reproduces the data that led to the DFT coefficients. Raising N_{synth} beyond N merely increases the resolution of that Fourier series, which is why the plots for $N_{\text{synth}} = 64$ and $N_{\text{synth}} = 256$ have similar appearance. The aphorism "Garbage in, garbage out" is not quite true here, but there is no merit in letting N_{synth} exceed N.

Mathematicians often imply that Gibbs phenomenon is an error that originates from truncating the Fourier series. It would be more correct to say that it is an error in the way the series is evaluated. The conventional evaluation of a Fourier series leaves us free to select the instants at which to synthesize the time data. The proper usage would be to evaluate a Fourier series containing M harmonics at no more than $2M$ equally spaced time instants within the period. This is the maximum amount of information that is contained in the data. (An interesting observation here is that this is a situation where numerical methods tell us how to perform mathematical analysis.)

1.4.4 Fast Fourier Transforms

Digital signal processing is at the heart of a wide range of electronic devices. The core concept is the DFT. However, these advances would not have been possible if it were necessary to evaluate DFTs and IDFTs by following the definitions in the previous section. It would take an extremely long time to carry out the operations, and a high-powered computer would be required, whereas DFT computations are now carried out on small devices in close to real-time. The advances were made possible by the derivations of fast Fourier transform (FFT) algorithms for evaluating DFTs, of which there are many, going back to Gauss, whose derivation in 1805 did not appear in published form until 1866.[5]

To set up the scheme, it is useful to adjust the definitions in Eqs. (1.4.30) and (1.4.31) to have similar forms. This is done by rearranging the DFT coefficients. To avoid confusion, let us denote the rearranged set of coefficients as P'_j. They are sequenced such that the DFT values whose subscript is negative are shifted to come after the DFT coefficients whose subscript is positive. This operation is achieved by defining

$$P'_j = \begin{cases} \hat{P}_j, & j = 0, 1, ..., N/2 \\ \hat{P}_{j-N}, & j = N/2+1, ..., N-1 \end{cases} \tag{1.4.35}$$

In terms of the rearranged coefficients, the inverse DFT is

$$\begin{aligned} p_n &= \frac{1}{N} \sum_{j=-N/2+1}^{-1} P'_{j+N} e^{i2\pi nj/N} + \frac{1}{N} \sum_{j=0}^{N/2} P'_j e^{i2\pi nj/N} \\ &= \frac{1}{N} \sum_{j=N/2+1}^{N-1} P'_j e^{i2\pi n(j-N)/N} + \frac{1}{N} \sum_{j=0}^{N/2} P'_j e^{i2\pi nj/N} \end{aligned} \tag{1.4.36}$$

Because $\exp(i2\pi n (j - N)/N) \equiv \exp(i2\pi nj)$, the inverse transform is converted to

$$p_n = \frac{1}{N} \sum_{j=0}^{N-1} P'_j e^{i2\pi nj/N} \tag{1.4.37}$$

Let $E_{j,n}$ denote the complex exponential that is the coefficient of q_n in the DFT definition, Eq. (1.4.30). The upward shift of the negative subscripts changes the evaluation of DFT coefficients to

$$Q'_j = \sum_{n=0}^{N-1} E_{j,n} q_n, \quad E_{j,n} = e^{-i2\pi jn/N}, \quad j = 0, 1, .., N-1 \tag{1.4.38}$$

---

[5]M.T. Heidemann, D.H. Johnson, & C.S. Burrus, "Gauss and the history of the Fast Fourier Transform," IEEE ASSP Mag., October 1984, 14–21.

The above description of the inverse DFT has a similar form,

$$p_n = \frac{1}{N} \sum_{j=0}^{N-1} (E_{n,j})^* P'_j$$

The equivalent matrix representation of these operations defines $[E]$ to be an $N \times N$ array whose elements are $E_{j,n}$. Placing the time-domain data and DFT coefficients into column vectors yields

$$\left\{ \hat{Q} \right\} = [E]\{q\}, \quad \{p\} = \frac{1}{N} [E]^{\mathrm{H}} \left\{ \hat{P} \right\} \tag{1.4.39}$$

The superscript "H" denotes the Hermitian of a matrix, which means that one should take the transpose of $[E]$, then replace each element of the transpose with its complex conjugate value. (The fact that $[E]$ is symmetric makes the transpose operation unnecessary, but that is unimportant because the FFT algorithm will not entail matrix operations.)

Before we consider how Eq. (1.4.39) are implemented efficiently, it is appropriate to examine the implications of rearranging the DFT coefficients. According to Eq. (1.4.27), the original coefficients with negative subscripts are the complex conjugate of the coefficients with positive subscript. (This property applies only of the data that is transformed is real. There are applications that require processing of complex data.) If Eq. (1.4.27) applies, then the DFT data will have the property that

$$P'_j = \hat{P}_{j-N} = \hat{P}_{N-j}, \quad j = N/2 + 1, ..., N - 1 \tag{1.4.40}$$

Hence, the first $N/2 + 1$ coefficients resulting from FFT processing will be the same as the original $N/2 + 1$ DFT coefficients. The remaining $N/2 - 1$ FFT values will be the complex conjugate of the first set, sequenced in the backward order. In other words, the sequence of DFT coefficients will be $\hat{P}_0, \hat{P}_1, ..., \hat{P}_{N/2-2}, \hat{P}_{N/2-1}, \hat{P}_{N/2}$, $\hat{P}^*_{N/2-1}, \hat{P}^*_{N/2-2}, ..., \hat{P}^*_1$. This arrangement may be said to be a "complex conjugate mirror image" relative to the highest positive subscript, $N/2$. "Mirror image" comes from visualization of the arrangement relative to the midvalue $j = N/2$. It will be necessary to account for this arrangement when we employ transfer functions.

The basic nature of the $E_{j,n}$ coefficients in Eq. (1.4.38) makes it evident that shortcuts are possible. An element of $[E]$ at row j and column n only depends on the row-column index product jn. All elements for which this product is $N/2$ greater are the negative of those whose index product is less by $N/2$ because

$$e^{-i2\pi(jn+N/2)/N} = -e^{-i2jn/N} \tag{1.4.41}$$

Thus, only the $E_{j,n}$ values for which $jn < N/2$ are independent; the others may be filled in by repetition with appropriate sign changes. Further simplifications stem from the fact that the $E_{j,n}$ values may all be computed from a single exponential because $E_{j,n} = \exp\left(-i2\pi/N\right)^{jn}$. Although these properties reduce the number of function evaluations, many mathematical operations would be required to multiply matrices. The greatest improvement is obtained by decomposing the summations appearing in Eq. (1.4.39). This entails subdividing a sum into two groups according to whether jn is odd or even. Then, each group is further subdivided into odd–odd, odd–even, even–odd, and even–even. The subdivision continues until the smallest groups are formed. Suppose $N = 2^M$, where M is a positive integer. Then, the smallest group is a single term, which leads to the most efficient computation. In addition, the most efficient algorithms manipulate the binary addresses of the data without actually evaluating complex exponentials and performing multiplications. The Cooley–Tukey algorithm is widely employed.

An excellent text for further study is Numerical Recipes.[6] It is shown there that the number of operations entailed in a direct evaluation of a set of DFT coefficients is of the order of N^2, whereas the number of operations for an FFT algorithm is of the order of $N \ln(N)$. This is an enormous reduction if N is large. For example, if $N = 2^{10} = 1024$, the number of FFT operations is approximately 0.7% of the number of DFT operations. It is for this reason that the these algorithms are said to be Fast Fourier Transforms (FFTs). Each FFT algorithm has a companion inverse Fast Fourier Transform (IFFT).

Every popular mathematical software suite, including MATLAB, Mathcad, and mathematical libraries for Fortran or C+, contains at least one subroutine that yields an FFT and a companion for the corresponding IFFT. Alternative FFT implementations might be encountered in the software, with the differences being the attributable to several factors, including the definition of a DFT. How one can identify differences between the present definitions and the FFT implementation in a specific package is the focus of the next example. Once these differences are understood, performing the operations will require little programming effort.

EXAMPLE 1.10 Consider the sawtooth pressure function in Example 1.9. Use an FFT computer routine to determine the DFT coefficients corresponding to sampling a period with $2^4 = 16$, $2^6 = 64$, and $2^8 = 256$ points. Compare the approximate values obtained in each case to the actual series coefficients.

Significance

Calling an FFT routine within a software package is straightforward, but setting up the data for usage in a specific package and correctly extracting results from the processed data require that one be cognizant of several aspects that will be demonstrated here. The example also will disclose another aspect of discontinuous data that requires careful handling.

---

[6]W.H. Press, S.A. Teukolsky, W.T. Vetterling, & B.P. Flannery. *Numerical Recipes,* 3rd Ed., Chap. 12, Cambridge University Press (2007).

Solution

The process of sampling data and providing it as input for an FFT routine would require little explanation were it not for differences between how various software package implement FFTs relative to the basic definitions. We will elucidate each issue, then describe how to address it in the 2009b version of MATLAB. The first consideration is subscripting. Data sets here were allowed to start from subscript zero, but many packages require that the subscript for data arrays begin at one.

Another aspect to consider stems from adoption of the $\exp(i\omega t)$ convention, whereas some prefer $\exp(-i\omega t)$. If mathematical software is based on the latter convention, then its FFT coefficients will be the complex conjugates of those derived from the present definition. For MATLAB, the definition of the DFT coefficients is described in the Help system. The description of the $fft(x)$ function stated there is

$$ X(k) = \sum_{j=1}^{N} x(j) \omega_N^{(j-1)(k-1)} \text{ where } \omega_N = e^{(-2\pi i)/N} $$

The subscripts j and k in MATLAB begin at one. To compare this definition to the present one, we replace j and k with $j+1$ and $k+1$, respectively, with both indices now beginning from zero. Then, the coefficients defined in Eq. (1.4.38) are $E_{j,n} = (\omega_N)^{jn}$, from which it follows that the preceding is the same as the DFT definition in Eq. (1.4.38). Thus, MATLAB will yield values that are consistent with the convention adopted here.

A third possible difference between the DFT definition in a software package and the one here is the scaling factor, but identification of that factor is important only if one wishes to compare the software's output to results derived elsewhere. No factor appears in Eq. (1.4.38), nor is there a factor in MATLAB's definition. However, some packages divide the summation by N or $\sqrt{N}$. This factor is compensated in the package's inverse FFT.

Other differences between FFT routines pertain to restrictions that might be imposed on the data set. Some are limited to real data, so that only the first $N/2 + 1$ DFT coefficients are output. In such situations, the full set of DFT coefficients may be obtained from Eq. (1.4.40). Some routines require that the length of the time data fit $N = 2^m$ in order to maximize its numerical efficiency. Further confusion is found in the fact that one software package might offer several alternatives. MATLAB offers only one function, but its internal programming defaults to the algorithm that is most efficient for the data set it has received. It is crucial that each of these considerations be resolved, which means that one needs to carefully read the software's documentation. Once we understand the ways in which its definition and implementation differ from Eq. (1.4.38), the actual operations are quite simple.

To use MATLAB to solve the present example, we begin by creating a column array of time instants covering an interval by writing

```
n = [0:N-1]'; t = n*(1/N);
```

This definition of the time variable may be considered to set $T = 1$, or to let the time variable be the nondimensional parameter t/T. The next step is to evaluate the sawtooth function at these time instants. It is a straight line having slope 4/3 until $t = 3/4$, at which time the straight line is reduced by a constant amount. The function to the left of $t = 3/4$ is P_{max}, whereas it is $-P_{max}/3$ to the right, so the decrease across the discontinuity is $(4/3) P_{max}$. We set $P_{max} = 1$, so that the time data is p/P_{max}. The sampled values may be evaluated in a single program line by using a logical operator, specifically

```
p = (4/3)*(t-(t > 3/4));
```

This data is a column vector, but that is not important because MATLAB's fft routine allows the data to be arranged as a row or column. MATLAB uses the same DFT definition as Eq. (1.4.25), so we leave the output in its original form,

```
P = fft(p);
```

The DFT coefficients we seek are the first $N/2 + 1$ values of P. Because MATLAB's implementation is consistent with the present DFT definition, we apply Eq. (1.4.26) to the P(j) output to compare them to the Fourier series coefficients found in Example 1.9.

The Fourier coefficients P_0 to P_8 and the DFT values scaled by $2/N$ are listed in the tabulation. The Fourier series coefficients alternate between being real or imaginary, whereas the DFT values are generally complex. The part of a DFT value that matches the Fourier series coefficient becomes increasingly accurate as N increases. The out-of-phase parts of the DFT values are much smaller than the magnitude of the largest coefficients, but their magnitude does not vary much with increasing N.

	Fourier Coefficient	$N = 16$	$N = 64$	$N = 256$
P_0	0.6667	0.75	0.6875	0.6719
P_1	−0.4244	−0.4189 + 0.0833i	−0.4241 + 0.0208i	−0.4244 + 0.0052i
P_2	−0.2122i	−0.0833 − 0.2012i	−0.0208 − 0.2115i	−0.0052 − 0.2122i
P_3	0.1415	0.1247 − 0.0833i	0.1405 − 0.0208i	0.1414 − 0.0052i
P_4	0.1061i	0.0833 + 0.0833i	0.0208 + 0.1047i	0.0052 + 0.1060i
P_5	−0.0849	−0.0557 + 0.0833i	−0.0832 + 0.0208i	−0.0848 + 0.0052i
P_6	−0.0707i	−0.0833 − 0.034518i	−0.0208 − 0.0687i	−0.0052 − 0.0706i
P_7	0.0606	0.0166 − 0.0833i	0.0582 − 0.0208i	0.0605 − 0.0052i
P_8	0.0531i	0.0833	0.0208 + 0.0503i	0.0052 + 0.0529i

The erroneous appearance of a substantial out-of-phase part in the coefficients is not a fault of the DFT concept and implementation. Rather, it is a consequence of dealing with discontinuous time data. The sampling described above yields $p = P_{max}$ at $t = 3T/4$, which is the value to the left of the discontinuity. However, evaluation of a Fourier series at an instant where the time function is discontinuous will yield a value that is the average of the function on either side of the discontinuity. The second tabulation is the result of setting the sampled time value at the discontinuity to the average, $p = P_{\max}/3$, as was done in Example 1.9. Although only one element of the time data has been altered, the approximate DFT coefficients are much more accurate.

	Fourier Coefficient	$N = 256$ Adjusted data	$N = 256$ Original data
P_0	0.6667	0.66667	0.6719
P_1	−0.4244	$-0.4244 + 1.78(10^{-17})i$	$-0.4244 + 0.0052i$
P_2	−0.2122i	$3.04(10^{-18}) - 0.21216i$	$-0.0052 - 0.2122i$
P_3	0.1415	$0.1414 + 5.20(10^{-18})i$	$0.1414 - 0.0052i$
P_4	0.1061i	$4.77(10^{-18}) + 0.1060i$	$0.0052 + 0.1060i$
P_5	−0.0849	$-0.0848 - 1.43(10^{-17})i$	$-0.0848 - 0.0052i$
P_6	−0.0707i	$3.03(10^{-18}) - 0.0706i$	$-0.0052 - 0.0706i$
P_7	0.0606	$0.0605 - 3.04(10^{-18})i$	$0.0605 - 0.0052i$
P_8	0.0531i	$-8.67(10^{-18}) + 0.0529i$	$0.0052 + 0.0529i$

An important conclusion to be drawn from this example is that care must be taken to sample time data if the quantity changes greatly over a very small fraction of a period. In that case, convergence of the DFT values should be judged by an assessment of the real and imaginary parts, rather than only the magnitudes. In fact, it is good idea to make such an assessment standard practice.

1.4.5   Evaluation of Time Responses

The first task for our consideration is that in which we seek the steady-state response to a periodic excitation. Before we employ FFT technology to evaluate a time-domain response, we must have already determined the frequency-domain transfer function $G(\omega)$. As defined in Eq. (1.1.14), this quantity is the complex amplitude of the response variable when the excitation is harmonic with unit amplitude. (Many of the acoustics studies carried out in the following chapters determine a transfer function.)

The jth DFT coefficient $\hat{Q}_j$ of the excitation corresponds to the jth harmonic, whose frequency is $\omega_j = 2\pi j/T$. Thus, the jth DFT of the response must be $\hat{Q}_j$ multiplied by the transfer function at that frequency, that is,

$$\hat{P}_j = G(2\pi j/T)\,\hat{Q}_j \tag{1.4.42}$$

The procedure for determining the steady-state response to a periodic excitation is rather straightforward. The time-domain excitation over period T is sampled at N instants spaced at T/N. (Of course, N must be consistent with the Nyquist criterion. Also, if possible, N should be a power of two for maximum numerical efficiency in FFT operations.) This data is placed in a vector that is input to an FFT subroutine, whose output will be the DFT coefficients of the excitation. Each DFT coefficient is multiplied by the transfer function for its frequency, as specified in Eq. (1.4.42). This operation requires that one be cognizant of how the DFT coefficients are arranged in the output, especially because the coefficients for negative indices are placed at the end of the column vector. At this juncture, one should verify that the DFT coefficients of the response are not aliased. If they are not, these coefficients are input to the IFFT routine that is the companion of the routine that performed the FFT. The output of the IFFT routine will be a vector of time values of the response at the instants when the excitation was sampled. If one wishes to plot more than one period, the values for a period may be replicated in the vector of values, with the corresponding time instants incremented by the period.

This procedure may be extended to handle transient excitations, that is, an excitation that exists for a finite time interval, but then is negligibly small. The concept entails creating a periodic excitation by artificially extending the excitation data.

It is assumed that the excitation is zero before $t = 0$. Let T' be the latest time for which we seek to determine the response. The minimum value of T' is the earliest time at which the excitation may be considered to be negligibly small. However, T' is usually taken to be larger than this minimum because a response can last well after the excitation goes away. The sampling interval for this data is Δt, which presumably meets the Nyquist sampling criterion. Thus, we have excitation time values q_n at instants $n \Delta t$ for $n = 0, 1, ..., N'$, where N' is the integer closest to $T'/\Delta t$. We create a periodic excitation by extending the q_n values with zeros. This operation is called zero padding, but how many zeros should be inserted?

Criteria for zero padding originate from the recognition that multiplying the excitation DFT by a transfer function, then performing the inverse FFT, is the frequency-domain version of convolution. The equivalent operation in the time domain is an integral. The reader probably has encountered this version in the context of Laplace transforms, and a similar operation arises for Fourier transforms, see Appendix B of Volume 2. Analysis of convolution in the time domain[7] reveals that *wraparound error* will occur if the number of zeros is inadequate. This phenomenon is such that the response at the beginning of the period is contaminated by the response at the end, and vice versa. Wraparound error is avoided if the number of zeros that is added covers a time interval that is greater than or equal to T', which means that the excitation data should be padded by at least N' zeros. However, as long as we are adding zeros, we might as well make the total length of the time data be a power of two in order to use the most efficient FFT algorithms. Thus, we set $N = 2^m$, subject to the requirement that $N \geq 2N'$. Padding the data with $N - N'$ zeros creates a data set that consists of N values, for which the time duration is $T = N \Delta t = (N/N') T'$.

After the excitation has been zero-padded, the evaluation of the transient response proceeds as though the excitation actually is periodic. That is, the DFT coefficients of the excitation are multiplied by the transfer function at the corresponding frequency. The coefficients that result are the DFT coefficients of the response. An IFFT gives the time values of the response. The only adjustment to the process occurs at the last step. The IFFT routine will return an array whose length is N. However, the response that is obtained is only uncontaminated by wraparound error in the first half of the

---

[7]J.H. Ginsberg, *Mechanical and Structural Vibrations*, John Wiley and Sons (2000) pp. 184–191.

artificial period, $0 < t < T/2$, which corresponds to response values whose index is $n < N/2$. It follows that the response data obtained from the IFFT routine should be truncated at a length that does not exceed $N/2$.

EXAMPLE 1.11 A simple model describing the response of a microphone is a one-degree-of-freedom oscillator, which consists of a piston of mass M supported by a spring having stiffness K and a dashpot, whose viscous resistance is C. Application of a harmonic force $F(t) = \text{Re}\left(\hat{F}\exp(i\omega t)\right)$ results in a steady-state harmonic displacement of the piston, $x = \text{Re}(X\exp(i\omega t))$, that is proportional to the voltage drop across the microphone's terminals. The force $F(t)$ is the incident pressure signal multiplied by the area of the piston that is exposed to the atmosphere. The ratio $X/\hat{F}$ is governed by the oscillator's complex frequency response, which is

$$\frac{X}{\hat{F}} = \frac{1}{M}\left(\frac{1}{\Omega^2 + 2i\zeta\Omega\omega - \omega^2}\right)$$

where $\Omega = (K/M)^{1/2}$ is the oscillator's natural frequency and $\zeta = C/2(KM)^{1/2}$ is the damping ratio. If F_0 is the maximum value of the force, then evaluation of Kx/F_0 only requires specification of the natural frequency, damping ratio, and period of the excitation, which are $\Omega = 480$ rad/s, $\zeta = 0.1$, and $T = 40$ ms. Use this knowledge to determine the steady state response $x(t)$ when $F(t)$ is a square wave excitation like that in Fig. 1.16.

Significance

This example will demonstrate how convergence may be verified when transfer functions are used in conjunction with FFTs. It also will lead to recognition of some design criteria for a microphone.

Solution

The transfer function is the value of X obtained from the given formula when $\hat{F} = 1$. The next step is to obtain the DFT coefficients of the square wave. We sample a period of the excitation, then take the FFT of that data to obtain the DFT coefficients F_j of the excitation. Each coefficient is multiplied by $G(\omega_j)$ to obtain DFT response coefficients. These are input to an IFFT routine to obtain the time response. The analysis will describe the operational steps for MATLAB, but the procedures for other software would be similar.

We are free to select the number of samples N. In view of the experience in Example 1.10, it is preferable that sample instants fall at the discontinuities, $t = T/4$ and $T = 3T/4$, which will occur if N is a multiple of 4. Such a selection also fits the suggestion that N be a power of two, so we set $N = 2^m$. The excitation in the interval $0 \le t < T$ is F_0 for $t < T/4$ and $t > 3T/4$, $F_0/2$ for $t = T/4$ and $t = 3T/4$, and zero otherwise. The following uses logical variables to sample the excitation function into a row vector.

```
           n = [0:N-1]; t = n*T/N;
  force = 1*((n < N/4)+(n > 3*N/4))+0.5*((n==N/4)+(n==3*N/4));
```

The force data constitutes the values of $F(t)/F_0$.

Application of an FFT yields the DFT force coefficients F_fft of the force. MATLAB's algorithm produces the values we seek without adjustment, and the first $N/2 + 1$ elements of F_fft correspond to the zero and positive frequencies. We save this part of the output. The discrete circular frequencies, which are multiples of $2\pi/T$, are evaluated at this stage in anticipation of calculating the transfer function values.

```
  Q=fft(force);F_fft=Q(1:N/2+1); w = (0:N/2)*2*pi/T;
```

Because force is a row vector, so too is F_fft.

We wish to determine the nondimensional response given by xK/F_0. The corresponding DFT coefficients are X_jK/F_0, but the complex frequency response has the form $X_j = G\left(\omega_j\right)F_j$, so that

$$\frac{X_jK}{F_0} = KG\left(\omega_j\right)\frac{F_j}{F_0}$$

Recall that $K/M = \Omega^2$, so setting $\hat{F} = 1$ in the given complex frequency response leads to

$$KG\left(\omega_j\right) = \frac{\Omega^2}{\Omega^2 + 2i\zeta\Omega\omega_j - \omega_j^2}$$

These value are stored in a vector KG(j). In our MATLABah program F_j/F_0 is F_fft(j), so the nondimensional DFT coefficients of the response are KG(j) * F_fft(j). Evaluation of these products may be vectorized in MATLAB by using a period before an operator, which denotes that the operation is to be performed individually on each element. The corresponding program steps are

```
  KG = W^2./(W^2+2*i*zeta*W*w-w.^2); X_nondim=KG.*Q_fft
```

where W is the natural frequency.

The last step is to return to the time domain by taking the IFFT of X_nondim. At this juncture, these coefficients have been evaluated for the nonnegative Fourier frequencies, but Matlab's ifft routine requires the negative frequencies be positioned after the positive frequency elements. There is no need to calculate these additional values, because they must be consistent with the complex conjugate mirror image property. Because X_nondim has been defined to be row vector, the fliplr function can rearrange the elements in the required manner,

```
X_neg = conj(fliplr(X_nondim(2:N/2))); X_fft = [X_nondim   X_neg];
```

The last set of calculations performs the IFFT, which produces the nondimensional response over one period. This data is replicated to the next period in order to clarify a plot of the response,

```
x = ifft(X_fft), x_nondim = [x  x];  t = [t  t+T]
```

Implementation of this program requires selection of N. The minimum permissible value is set by the Nyquist criterion, which governs the maximum allowable time interval $\Delta t = T/N$ between samples. However, the discontinuous nature of the step function obscures identification of a suitable Δt. Thus, we shall select a value of N that seems sufficiently large, then verify that the DFT coefficients are not aliased. An important aspect here is that it is only necessary that the DFT of the response, which is X_fft in the Matlab program, not be aliased. To see why, suppose that the DFT of the excitation is mildly aliased, but the transfer function is very small in the range of the higher DFT frequencies. Then the DFT coefficients of the response also would be small in the upper frequency range, which would mean that the response data is not aliased. This is the reason why a Fourier series representation of a periodic sequence of impulses, which never converges, can be used to obtain a convergent response. (This discussion assumes that aliasing of the excitation DFT has not con-taminated the entire spectrum.) Figure 1 displays the DFT coefficients of the force, the transfer function, and the response corresponding to $N = 64$. It can be seen that each becomes extremely small well before the highest harmonic number, $N/2$. This confirms that $N = 64$ meets the Nyquist criterion. The graphs suggest that N could be even smaller, but making N too small will lead to jagged curves describing the time response because the time increment T/N would be course.

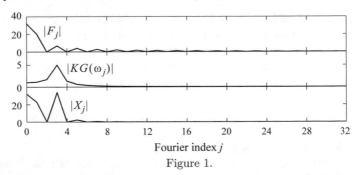

Figure 1.

The response x_nondim is plotted against t in Fig. 2. The most notable feature of the response is that it bears no resemblance to the excitation. Such a microphone would be useless as a measuring device. Although the response is periodic in interval T, a higher frequency component is quite strong. The fact that there are three peaks in a period tells us that there is a strong third harmonic in the response.

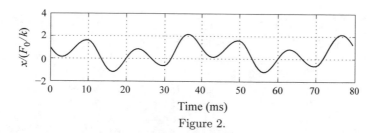

Figure 2.

One of the primary virtues of frequency-domain analysis is that it can be used to understand fundamental processes. The microphone will faithfully reproduce the incident pressure if the response DFT coefficients are proportional to the excitation DFT coefficients. There is no mechanical device for which $G(\omega)$ is truly constant, but the objective would be met if $G(\omega)$ were nearly constant over the frequency band where the DFT coefficients of the excitation are most significant. Inspection of Fig. 1 shows that this is not the property of $|G(\omega)|$. Rather $|G(\omega)|$ shows a strong peak at the third harmonic. (This is a manifestation of the phenomenon of resonance, in which a system's response is strongly magnified if the frequency of a harmonic excitation is close to a natural frequency.) The graph of $|X|$ versus ω shows that the response's third harmonic is comparable to its first, even though the third harmonic of the excitation is quite small. These graphs enable a designer to improve the microphone. Doing so entails shifting the peak of $|G(\omega)|$ to a much higher frequency, because $K|G(w)|$ is very close to one for $\omega < \Omega/2$. This can be achieved by raising K and/or lowering M.

The moving parts of an actual microphones have very low mass and are very stiffly supported. To demonstrate that such a design is effective let us increase Ω by a factor of ten to 4800 rad/s, which is between the 36th and 37th harmonic. The DFT coefficients in this range are extremely small. Furthermore, $\mathrm{Re}\,(KG(f))$ is very close to one and $\mathrm{Im}\,(KG)$ is very small for the lower harmonics, with the deviation of each part remaining below 1% up to the ninth harmonic. As can be seen in Fig. 3, this design functions exceptionally well. The square wave variation is reproduced, aside from some fluctuating discrepancies. This seems to be a Gibb's phenomenon, but we have seen that DFTs do not exhibit such artifacts. Rather, it is the result of large contributions at the 36th and 37th harmonics, where the resonance now lies.

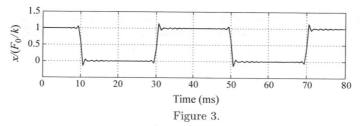

Figure 3.

EXAMPLE 1.12   When a harmonic pressure wave is reflected from a compliant surface, the complex amplitude of the reflected wave P_{ref} is proportional to
the complex amplitude of the incident wave P_{inc}. The factor of proportionality
is the reflection coefficient R, which depends on a mechanical property of the
surface called the specific impedance ζ. The relations are

$$P_{ref} = R P_{inc}, \quad R = \frac{\zeta - 1}{\zeta + 1}$$

If ζ is infinite, $R = 1$, which corresponds to a rigid surface, whereas $R = -1$
when $\zeta = 0$. Such a surface is said to be pressure-release because the sum of
incident and reflected pressures is zero. If $\zeta = 1$, then $R = 0$, which means
that there is no reflected wave.

Polymers, which are commonly used as acoustical coatings, have impedances
that are highly dependent on frequency. Beth, who is a material scientist, claims that
she can manufacture a polymer that gives $\zeta = 1$ for waves above cyclical frequency
f_{max}, with the specific frequency dependence given by

$$\zeta = \begin{cases} 0.2 + 2 \sin\left(\dfrac{\pi f}{f_{max}}\right) + 0.5i \sin\left(\dfrac{\pi f}{f_{max}}\right), & f \leq f_{max} \\ 0.2, \quad f > f_{max} \end{cases}$$

She would like to know if this would be good material for reducing the reflec-
tion of an incident wave that is a sine pulse of duration T given by $p_{inc} = p_0 \sin(\pi t/T) [h(t - T) - h(t)]$, where $h(\tau)$ denotes a step function whose value
is zero when $\tau < 0$ and one when $\tau > 0$. Determine the corresponding wave that is
reflected from the polymer coated surface in the interval $0 \leq t < 2T$, and from that
result determine whether this material meets the design objective. Parameters for the
system are $T = 0.1$ s and $f_{max} = 70$ Hz.

Significance

The focus here is the measures that should be taken to avoid aliasing and wraparound
error when FFT technology is used to evaluate the response to a time-domain excitation. In addition, the solution will disclose a fundamental issue that might arise
when a frequency-domain response is converted to the time domain.

Solution

The only part of the analysis that did not arise in the previous example is adaptation
of the procedure to handle a transient excitation. The duration of interest is $2T$, so the
FFT window is set at $T_{FFT} = 4T$ in order to avoid wraparound error. Our experience
with a square wave in Example 1.11 indicated that approximately 16 harmonics were
significant. The sine pulse is more gradual than the step function, but it lasts only for

1/4 of the duration of the extended data set, as opposed to 1/2 of the period for the square wave. This suggests that 32 harmonics of the incident wave are significant. We will set $N = 64$. This gives a fundamental frequency of $1/T_{FFT} = 2.5$ Hz and a maximum frequency of $(N/2)(1/T_{FFT}) = 80$ Hz.

The analysis begins by sampling the incident wave at N instants $t_n = n(T_{FFT}/N)$ and taking its FFT. The first $N/2 + 1$ coefficients output by the FFT function are stored in a vector $(P_{inc})_j$. The real and imaginary parts of $(P_{inc})_j$ are plotted in the Fig. 1. All values above 20 Hz, which is the eighth harmonic, are essentially zero. We conclude that these DFT values are not aliased, and therefore proceed to the next step.

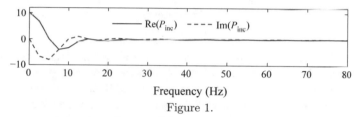

Figure 1.

The reflection coefficient is a transfer function that gives the complex amplitude of the reflected wave when the incident wave is harmonic with a unit complex amplitude. It is plotted in Fig. 2. We see that $|R|$ is not small at any frequency except small intervals around 10 and 60 Hz. Consequently, we anticipate that the reflected wave will not be significantly reduced.

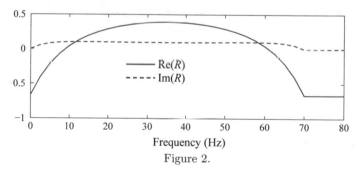

Figure 2.

To see how far from the desired behavior this material is, we proceed to the next step, which is to determine the DFT coefficients of the reflected signal. The DFT coefficients $(P_{ref})_j$ of the reflected wave are found by multiplying a $(P_{inc})_j$ value by the corresponding $R(f_j)$. This gives

$$(P_{ref})_j = R\left(f_j\right)(P_{inc})_j, \quad j = 0, 1, ..., N/2$$

The last step is to take the IFFT of the $(P_{ref})_j$ coefficients. (If the IFFT routine requires the DFT coefficients for negative frequencies, as MATLAB does, the computed values of $(P_{ref})_j$ must be appended with the complex conjugate mirror image of that data, see Eq. (1.4.40)) Only the first half of this time response output, which covers the interval $0 \le t \le T_{\text{fft}}/2$, is retained because the later data is contaminated by wraparound error. The resulting reflected wave, as well as the incident wave, is shown in Fig. 3.

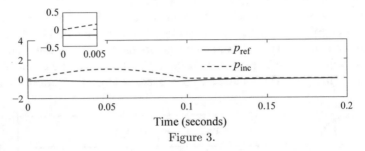

Figure 3.

The question that was posed was whether this is a good material for reducing the incident wave. The graph indicates that the reflected wave's amplitude is reduced almost to zero. Thus, it seems that the answer is yes. However, a closer inspection causes concern for the correctness of the analysis. The zoomed view shows that p_{ref} is less than zero at $t = 0$. In other words, the response is predicted to begin before the excitation that causes it. This is a violation of the fundamental *principle of causality*. The error is a manifestation of the general imperative that a transfer function be derived from a physical model. The given function for ζ does not have this basis. Even if a transfer function is extracted from experiments on an actual system, causality errors will occur if the data is not acquired over a sufficiently long time window or the sampling rate does not meet the Nyquist criterion. Thus, the answer to the question that was posed is that one should look for a physically based system that has an impedance close to the function that was given.

1.5 Spectral Density

1.5.1 Definition

There are many ways that one can define noise, but a working definition is a signal composed of many frequencies having the property that the individual complex amplitudes are random in amplitude and phase angle. This is the case for the sound generated by most machines and by turbulence. If one were to measure such a signal repeatedly over the same time window, the DFT of each measurement would differ from the others by a possibly large amount. It is more meaningful in such situations to consider averages, which is an issue we will address after the concept has been developed.

The phase angle of a complex amplitude is related to the time at which the harmonic begins. Thus, any value between $-180°$ and $180°$ is equally likely to occur. In contrast, $|P_n|$ is likely to to be statistical, such as the bell-shaped curve of a normal distribution. Hence, we focus on the magnitude. Equation (1.2.12) states that the mean-squared pressure for a multiharmonic signal is one half of the sum of the magnitude squared for each harmonic. Consider a plot of each value of $|P_n|^2/2$ against its frequency. However, rather than directly plotting the mean-squared amplitudes, let us plot the values divided by the frequency increment between adjacent values. It is conventional to use the cyclical frequency. Thus, the data points that are plotted are at $f_n = n/T$ along the abscissa, and $|P_n|^2/(2f_1) = |P_n|^2 T/2$ along the ordinate. In a strict sense, the data should be depicted as a bar graph with strips centered on the plotted points, but the strips would be quite narrow, so it is more usual to connect plotted points with straight lines.

The graph that is created in this manner is said to be a *pressure spectral density*. The term "spectral" conveys the fact that the graph describes a distribution over the frequency spectrum, while "density" denotes that the quantity has dimensions of pressure squared per unit frequency, typically Pa^2/Hz. "Pressure" is prepended to the name because that is the quantity being described, but it may be omitted unless there is ambiguity. We will denote the dependence of the spectral density on the cyclical frequency as $P_f^2(f)$. The values of f are the discrete frequencies for the DFT, so that

$$P_f^2(f_n) = \frac{T}{2}|P_n|^2$$
(1.5.1)

Note that a P_n value is the true series coefficient associated with frequency f_n. They are obtained by scaling DFT coefficients, as in Eq. (1.4.26). The reason for using the Fourier coefficients will soon be apparent.

Consider a vertical strip whose width is the frequency increment Δf and whose height is $P_f^2(f_n)$. The area of this strip is $\Delta f P_f^2(f_n) = \Delta f\left(|P_n|^2 T/2\right) \equiv |P_n|^2/2$. In other words it is the contribution of that harmonic to the mean-squared pressure. Integrating $P_f^2(f_n)$ over a frequency interval is equivalent to adding the strip areas. But the sum of $|P_n|^2/2$ values is the mean-squared pressure of the signal in the associated frequency interval. Thus, for a frequency band that is $f_{N_1} \leq f \leq f_{N_2}$, the mean-squared pressure contained in that band is

$$\Delta\left(p_b^2\right)_{av} = \int_{f_{N_1}}^{f_{N_2}} P_f^2(f)\,df \approx \frac{1}{2}\sum_{n=N_1}^{N_2-1}|P_n|^2 = \frac{1}{T}\sum_{n=N_1}^{N_2-1}P_f^2(f_n)$$
(1.5.2)

The upper limit of the sum is set at $N_2 - 1$ in recognition of the fact that the interval contains $N_2 - N_1 - 1$ frequency strips. The reason we employ Eq. (1.4.26) to scale the DFT coefficients up to Fourier series coefficients is that doing so allows us to extract mean-squared pressure values directly.

The spectral density values usually are plotted on a logarithmic scale for the positive frequencies only, because the negative frequencies values are the same. (Often both the frequency and spectral density axes are logarithmic.) The shape of the spectral density conveys a picture of where significant contributions to $\left(p^2\right)_{av}$ lie. The DFT of a Dirac delta impulse is a set of constant values, which would give a flat spectral density. More generally, a signal whose spectral density has significant values distributed over a range of frequencies is said to be *broadband*. Impacts are a common generator of a broad response, and turbulence is an incoherent source. In contrast, a *narrow-band* signal has a spectral density that looks like a tall peak. If the response is a pure harmonic, only the strip at the harmonic's frequency should be nonzero. However, this result requires that the time window for the FFT either exactly match a multiple of the period, or be extremely long. Otherwise, *leakage error* occurs, in which the DFT coefficients are nonzero for a range of frequencies around the actual. Even if leakage error arises, the spectral density will feature a spike that is quite noticeable. A narrow-band signal was generated by a repeating process. The source often is the rotating parts of a motor or wheel; in which case, the frequency of the peak will match the rotational speed and/or a multiple of that speed.

As was mentioned earlier, the random nature of the time data is likely to lead to considerable variation in the spectral density covering different time intervals. Averaging can be done by dividing a long time interval into individual data sets. Alternatively, data sets taken individually may be processed. Let T be the duration of each data set. The value of T should be sufficiently long that $\Delta f \equiv 1/T$ adequately resolves the DFT values. The sampling rate should be consistent with the Nyquist sampling criterion, which requires that $\Delta t = T/N < 1/(2 f_{\text{Nyquist}})$.

Averaging the time data sets will tend to smooth out the data, thereby removing the statistical properties. Furthermore, because a DFT is a linear transformation from time data to frequency data, averaging the Fourier series coefficients will have the same effect. The proper procedure is to obtain the spectral density for each data set, then average the individual spectra.

If a frequency-domain transfer function is known, it may be used to predict the spectral density of the response associated with the excitation's spectral density. Any scaling factor for a DFT applies equally to the response and excitation. Thus, the transfer function in Eq. (1.4.42) also describes the proportionality of Fourier series coefficients. Substitution of this expression into the definition of spectral density yields

$$P_f^2 (f_n) = \frac{T}{2} |G_n Q_n|^2 \equiv |G_n|^2 \, Q_f^2 (f_n) \,, \quad G_n = G \, (2\pi f_n) \tag{1.5.3}$$

where Q_f^2 is the spectral density of the excitation, and G_n is the transfer function evaluated at the nth DFT frequency, $f_n = n \Delta f$. It is evident from this relation that the amount by which physical processes shift the phase of the response relative to the excitation does not affect the response's spectral density. The banded mean-squared response, which is the area under a plot of the response's spectral density, is given by Eq. (1.5.2). Application of the preceding relation between spectral densities leads to

$$\boxed{\left(p_b^2\right)_{\text{av}} = \frac{1}{T} \sum_{n=N_1}^{N_2-1} |G_n|^2 \, Q_f^2 (f_n)} \tag{1.5.4}$$

The total mean-squared pressure in a signal is obtained if the sum extends over the full frequency range. A common metric for a response is its RMS value, which is

$$p_{\text{RMS}} = \left[\frac{1}{T} \sum_{n=1}^{N} |G_n|^2 \, Q_f^2 \, (f_n) \right]^{1/2} \tag{1.5.5}$$

EXAMPLE 1.13   The microphone in Example 1.11 is used to measure a pressure signal that has harmonic form except that its frequency at any instant is a random variable. Its statistics fit a uniform probability density over a frequency band, which means that any value within that frequency range is equally likely. The corresponding force driving the microphone is

$$F = F_0 \sin \left(2\pi f \, (t) \, t \right)$$
$$f \, (t) = f_{\min} + (f_{\max} - f_{\min}) \, \text{rand} \, (t)$$

where rand (t) is a function that has an equal likelihood of having a value between 0 and $+1$. The first task is to sample $F \, (t)$ over an interval that is 100 times the microphone's natural period, which is $2\pi/\Omega$. Then, evaluate the spectral density of the force and use that data to predict the rms amplitude of the microphone's response $x_{\text{rms}} K / F_0$. The frequency range of the excitation is centered on the microphone's natural frequency, $2\pi f_{\min} = 0.9\Omega$, $2\pi f_{\max} = 1.1\Omega$. Other parameters required to evaluate the nondimensional response as a function of time are $\Omega = 5000$ rad/s and $\zeta = 0.01$.

Significance

Many situations require that a system's response be estimated when the excitation has random properties. Here, we will see what can be learned if a long record is processed without averaging to account for the random nature of the data. The next example will examine the benefits of averaging.

Solution

The force $F \, (t)$ is the probabilistic version of frequency modulation of radio waves. The randomness of the frequency only influences the evaluation of the force as a function of time. The duration of the data window is specified to be $T = 100(2\pi/\Omega)$. The highest possible cyclical frequency of the force is $f_{\max} = 1.1\Omega/(2\pi)$. The Nyquist sampling criterion requires that T/N be less than $1/(2 f_{\max})$, so that at least two samples fall in the shortest period of the force. However, this is an uncertain situation because F is not truly harmonic. In anticipation that the minimum requirement might not suffice, let us instead use eight samples for the smallest possible excitation period, $\Delta t = T/N = 1/(8 f_{\max})$, which gives $N = 8 f_{\max} T$. Setting $f_{\max} = 1.1\Omega/(2\pi)$, and $T = 200\pi/\Omega$ gives $N = 1100$. To use the optimum FFT algorithm, we increase

N to the next largest power of two, which is $N = 2048$. The sampling interval is $\Delta t = 0.5113$ ms.

MATLAB's random number function is rand(n,m), which generates an $n \times m$ array of random numbers between 0 and one. These random numbers are used to evaluate the frequency at each instant. The phase of the force function is the product of a time value and the frequency at that instant. Both sets of values are in row vectors of the same length, so the operations may be performed in a vectorized manner. The following is a MATLAB fragment to perform these operations.

```
f_force = (f_min+(f_max-f_min)*rand(1,N));
t = [0:N-1]*T/N;   force = sin(2*pi*f_force.*t);
```

Note the period prior to t, which causes the row vectors to be multiplied element by element. Also, the force values are nondimensionalized relative to F_0. A typical force-time history for an interval of five natural periods, $0 \leq t \leq 5\,(2\pi/\Omega)$ is plotted in the Fig. 1. Although it is oscillatory, this waveform certainly is not sinusoidal.

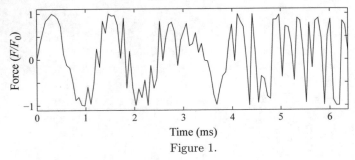

Figure 1.

The next operations evaluate the FFT of the force, scaled to give Fourier series coefficients. The spectral density of the excitation is constructed directly from these coefficients,

```
        Q = (2/N)*fft(force);
SD_Q = (T/2)*abs(Q(1:N/2+1)).^2;
```

Note that only the first $N/2 + 1$ elements of the output of the FFT routine are used to construct the spectral density of the force, because the higher elements belong to the negative frequency range.

The appropriate mathematical notation for the spectral density would be F_f^2 because it describes the properties of a force. Similarly, the spectral density of the displacement x would be denoted as X_f^2. To evaluate this quantity, a row vector of transfer functions at the DFT frequencies is calculated. Note that G is given as a function of circular frequency, so the frequencies for this evaluation are w_dft=2*pi*[0:N/2]*T/N. Then, X_f^2 is found by multiplying each element of the F_f^2 row vector by the corresponding element of $|G|^2$. The transfer function values that are computed are $K\,G$, as was done in Example 1.11. In combination with the fact that we have evaluated the Fourier coefficients for $F\,(t)\,/F_0$, it follows that the quantity that is computed is the spectral density of the nondimensional displacement, Kx/F_0. The MATLAB implementation of these operations is

```
KG=W^2./(W^ 2+2*im*zeta*W*w_dft-w_dft.^2);
       SD_X = abs(KG).^2*SD_Q;
```

where W is the natural frequency. Finally, the rms value of x is found from Eq. (1.5.4), which is implemented as X_rms = sqrt(sum(SD_X)/T);.

Before we consider the results of these procedures, it is appropriate to consider how they may be checked. All steps may be verified by setting $f_{min} = f_{max} = \Omega/(2\pi)$. In that case, the frequency of the excitation is stationary at Ω. The waveform of the force should display this property, and the spectral density of the force should be close to a spike at the natural frequency. Furthermore, because $G(\omega)$ has a very large value at the natural frequency, a plot of X_f^2 should show a distinct peak at that frequency.

The upper plot in Fig. 2 is the force spectral density. It is quite irregular with several broad intervals in which it is significantly larger. This data appears to be broadband noise. The lower graph is the spectral density of the response. There is a pronounced peak in the vicinity of the natural frequency, $\Omega/(2\pi) = 5000/(2\pi) = 795.8$ Hz. This is to be expected because the natural frequency is the mean value of the excitation's frequency. However, the response's spectral density everywhere changes greatly over small frequency intervals, as a consequence of the extreme variability of the force spectral density.

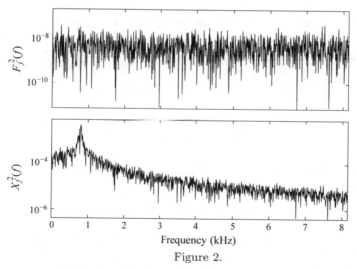

Figure 2.

The rms response obtained from Eq. (1.5.5) for the graphed data was $x_{rms}K/F_0 = 1.9474$. This value is not constant because it depends on the randomly varying excitation frequency as a function of time. Subsequent runs without changing the seed for the random number generator gave $x_{rms}K/F_0 = 2.9221$, 1.8220, and 2.0644. This variability suggests that is advisable to perform averaging.

1.5.2 Noise Models

An idealized noise model is *white noise*. The association of "white" with this type of noise originates from white light, which is composed of equal contributions from all colors. Similarly, the spectral density of white noise is independent of frequency,

$$\text{White noise: } P_f^2 (f) = \tilde{P}_f^2 \tag{1.5.6}$$

The analogy with light leads to another standard noise model. Red light is at the low end of the visible spectrum. A pink color contains the full color spectrum with a dominant red contribution. Correspondingly, the term *pink noise* is used to describe a broadband signal in which the low-frequency contributions are largest. Its more precise definition is a signal in which the spectral density is inversely proportional to the frequency,

$$P_f^2 (f) = \frac{\alpha}{f} \tag{1.5.7}$$

where α is a proportionality factor.

White and pink noise have distinctive manifestations in a log–log plot. In such a plot, $x = \log_{10} (f)$ and $y = \log_{10} \left(P_f^2 (f) \right)$. Then, white noise corresponds to constant y, so the spectrum appears flat. For pink noise, $y = \log_{10} (\alpha) - \log_{10} (x)$, which means that pink noise will be a straight descending line whose slope is minus one.

Actual noise sources might exhibit significant variations of their spectral density, especially if the frequency scale is divided very finely. Averaging over frequency bands might enable us to recognize underlying trends. Consider a band $(f_b)_L \leq f \leq (f_b)_U$. The average spectral density for that band is obtained by integrating $P_f^2 (f)$, then dividing by the width of the frequency band, so that

$$\left(P_f^2 \right)_{\text{av}} = \frac{1}{(f_b)_U - (f_b)_L} \int_{(f_b)_L}^{(f_b)_U} \left(P_f^2 \right) (f) \, df \equiv \tag{1.5.8}$$

According to Eq. (1.5.2) the mean-square pressure in a band is the above integral, so it must be that

$$\left(p_b^2 \right)_{\text{av}} = \left(P_f^2 \right)_{\text{av}} \left[(f_b)_U - (f_b)_L \right] \tag{1.5.9}$$

The usual situation is that the spectral density is computed only at the DFT frequencies f_n. In that case, the integral may be replaced by a sum of strip areas. For a frequency band defined by $f_{bL} \leq f \leq f_{bU}$, for which bL and bU are the indices of the corresponding DFT data, the mean-squared pressure is given by

$$\left(p_b^2 \right)_{\text{av}} = \sum_{n=bL}^{bU-1} P_f^2 (f_n) \, \Delta f \tag{1.5.10}$$

If one considers a set of proportional bands, for which the interval increases with increasing band number, the effect is to give increasing weight to the higher frequencies.

As they do for a spectral density plot, the two standard noise models have a recognizable signature in a plot of banded mean-squared pressure. For white noise, whose constant spectral value is $\tilde{P}_f^2$, the mean-squared pressure is the area of the rectangle in a PSD graph, $\left[(f_b)_U - (f_b)_L \right] \tilde{P}_f^2$. In the case of $1/N$-octave proportional bands, the result is

$$\left(p_b^2 \right)_{av} = \left(2^{1/2N} - 2^{-1/2N} \right) (f_b)_C \ \tilde{P}_f^2 \tag{1.5.11}$$

where $(f_b)_C$ is the center frequency of band b. In other words, the band averages increase linearly with the center frequency.

In the case of pink noise, Eq. (1.5.7), evaluation of the integral for the banded mean-squared pressure gives

$$\left(p_b^2 \right)_{av} = \alpha \ln \left(\frac{(f_b)_U}{(f_b)_L} \right) \tag{1.5.12}$$

The definition of proportional frequency bands is that the ratio of the upper to lower frequency is constant. Thus, a plot of mean-squared pressure within $1/N$-octave bands will be flat if the signal is pink noise,

$$\left(p_b^2 \right)_{av} = \alpha \ln \left(2^{1/N} \right) \tag{1.5.13}$$

Like spectral densities, banded mean-squared pressure is often displayed in a log–log plot. White noise will fit a rising straight line, whereas pink noise will be flat. Specifically, in a log–log plot, the ordinate and abscissa are

$$x = \log_{10} \left((f_b)_C \right), \quad y = \log_{10} \left((p_b^2)_{av} \right) \tag{1.5.14}$$

The plots will fit

$$\text{White noise: } y = \log_{10} \left((2^{1/2N} - 2^{-1/2N}) \right) + \log_{10} \left(\tilde{P}_f^2 \right) + x$$
$$\text{Pink noise: } \quad y = \log_{10} \left(\alpha \ln \left(2^{1/N} \right) \right) \tag{1.5.15}$$

EXAMPLE 1.14 SOFIA is the acronym for the Stratospheric Observatory for Infrared Astronomy, which is a collaboration of the National Aeronautics and Space Administration and the German Aerospace Center to design, construct, and operate a modified 747 aircraft.[8] A 10,000 kg actively controlled telescope is mounted on a specially constructed bulkhead. A large door in the fuselage rotates about the longitudinal axis, thereby creating an opening of approximately one-quarter of the fuselage's circumference, through which the telescope tracks stellar objects. One of the concerns when the aircraft was modified was a phenomenon known as a Rossiter resonance, which is a low-frequency buffeting sound. This phenomenon often is heard inside an automobile when its sun roof or one window is opened. The concern was that this phenomenon might create a tuned acoustical resonance in which the pressure fluctuations within the open compartment of the fuselage are sufficiently large to damage the telescope or the aircraft's structure. Thus, as part of the flight testing program, pressure-time measurements were taken inside the open fuselage. The figure below is one such record. The sampling rate for the 120 s interval was 500 Hz. Here, we will follow the procedure by which a spectral density was obtained from this data. Then, we will examine the result to determine whether it describes a resonance process.

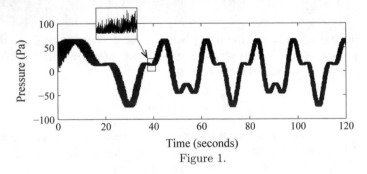

Figure 1.

Significance

The utility of a spectral density as a diagnostic tool for real-world systems will be evident. One of the key points will be the necessity of averaging.

Solution

The data that was measured is a stream of pressure values. The first processing step is reading the data into active memory from wherever it is stored. We begin with the assumption that the data has been placed into a column vector $\{p_{\text{all}}\}$. Note that the large scale graph seems to be nearly periodic beyond $t = 15$ s. However, what appears to be a heavy single curve is indicated in the zoomed view to be a high-frequency oscillation superposed on the slower nearly periodic fluctuation.

Before we examine how the data was actually processed, let us consider what would be obtained if $\{p_{\text{all}}\}$ was treated as a single record. Its length is the sampling rate multiplied by the time duration, which gives a length of 60,000. This is not a power of two. We could truncate the data at the largest lesser length that is a power of two, which is 32,768, but doing so would remove almost half the data. Thus, we take the FFT of the full data. The loss of computational efficiency resulting from data whose length is not a power of two is not important because these operations are only done once here. The frequency increment for these DFT values is $\Delta f = 1/120$ Hz, and the maximum DFT frequency is $f_{N/2} = 250$ Hz. The DFT data is scaled up to Fourier series coefficients by multiplying the values by the appropriate factor, which according to Eq. (1.4.26) is $2/6000$ for the DFT definition used here. The power spectral density is obtained by substituting the Fourier coefficients into Eq. (1.5.1). The results appear in Fig. 2.

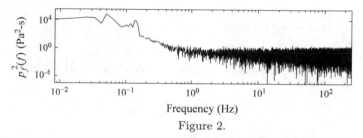

Figure 2.

We see that the frequency range above 1 Hz has extreme and rapid fluctuations. This is the frequency range of the rapid oscillation exhibited in the zoomed view of the first graph, What appears there to be a slower nearly periodic fluctuation occurs over an interval of approximately 25 s, which corresponds to a frequency of approximately 0.04 Hz. Both aspects are quite problematic. Most importantly, if the resonance occurred, it was expected from studies of related acoustics problems to be above 1 Hz. In addition, the instrumentation used to measure this data was not calibrated for frequencies below 1 Hz. Thus, this spectral density provides no insight as to whether a resonance occurred.

The remedy to these difficulties was to divide the full set of 60000 pressure values into subintervals, each of which is processed individually. The selection of a window length was based on the fact that the frequency range below 1 Hz is not a concern. If it is desired that 1 Hz be the fundamental frequency, then successive windows should be no less than one second. The data length in one second is the sampling rate, that is, 500. It is preferable to work with a data length that is a power of two, so the window length was set at $N_{\text{window}} = 512$. The corresponding time duration is $T_{\text{window}} = 1.024$ s. For this value of N_{window}, the original data set consists of 117 windows. Note that windowing the data in this manner does not alter the Nyquist frequency, which is set by the sampling rate. However, doing so does raise the fundamental frequency to $\Delta f = 1/T_{\text{window}} = 0.977$ Hz, which means that the frequency range is resolved less finely.

There is a convenient algorithm for slicing the data into columns whose length is N_{window}. For window n, there are $n-1$ preceding windows, each of which contains N_{window} numbers. Thus, the nth window consists of the data from $p_{1+(n-1)N_{\text{window}}}$ to $P_{(n)N_{\text{window}}}$. It is possible to work with each column vector as it is generated, and thereby avoid using computer memory. However, that was not done because there was some thought that it might be necessary to examine the windowed data at a later time. Thus, the windowed pressure values were stored columnwise in a rectangular array $[p_{\text{window}}]$ whose size is N_{window} rows and 117 columns. Figure 3 shows the pressure values for some typical windows. The abscissa is labeled as relative time because it is elapsed time relative to the start of the window.

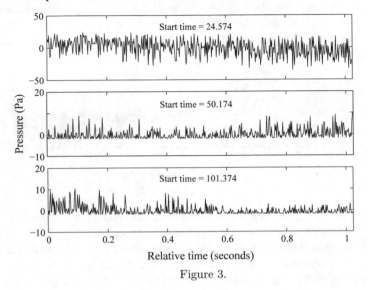

Figure 3.

To obtain pressure spectral densities, each column of $[p_{\text{window}}]$ was input to the FFT routine. The output of that routine was truncated at element $N_{\text{window}}/2 + 1$, corresponding to the nonnegative FFT frequencies. The spectral density for each window was obtained by evaluating Eq. (1.5.1). The result of performing this operation for each time window was a column vector whose length is $N_{\text{window}}/2 + 1$. The spectral values for window n form the nth column of $[PSD_{\text{window}}]$. The last step was averaging of the spectral density for each window. These values are the average of each row of $[PSD_{\text{window}}]$. Figure 4 shows the spectral densities for two selected windows, and the average value obtained from the 117 windows.

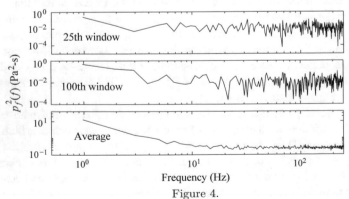

Figure 4.

There is a great deal of variability between the spectral density for different windows. In contrast, the averaged spectral density is unambiguous. Beyond 10 Hz the value of $p_f^2(f)$ is nearly constant, which fits the white noise model. From one to ten Hz, it decreases almost linearly. The data has been plotted on a log–log scale. The slope for a fitted straight line from $f = 1$ to $f = 10$ Hz is approximately

$$\frac{\Delta y}{\Delta x} = \frac{\log_{10}(0.2) - \log_{10}(10)}{\log_{10}(10) - \log_{10}(1)} \approx -1.7$$

This slope is greater than minus one for pink noise, so there is a possibility that it corresponds to some coherent process. As was mentioned earlier, resonances are characterized by relatively narrow and tall peaks in a spectral density. The averaged spectral density does not display this attribute, although the region around 1 Hz might be the tail of one at an extremely low frequency. This uncertainty was not an issue because a load below 1 Hz is essentially static for SOFIA's telescope and fuselage structures, both of which had been designed to withstand much larger loads generated by aircraft maneuvers.

In the course of a flight testing program at a wide range of altitudes and speeds, pressure measurements were taken at many locations within the open fuselage compartment of SOFIA. No significant resonances were observed. The result is that SOFIA is now operational, flying around the world to get access to different parts of the sky. (In July 2013 it was in New Zealand.)

1.6 Closure

This chapter is devoted to the verbal and mathematical language that is used to describe and analyze acoustical phenomena. Representation of a signal as a complex exponential is the most fundamental concept. It is essential for most developments in this text, so one should not proceed until this formalism has been mastered. Fourier series techniques are important to the study of acoustics because they make it possible to convert some very difficult problems to a form that is much easier to study. (A precursor of this capability was encountered when we used a Fourier series to represent a complicated periodic excitation, and then went on to use that representation to predict the corresponding system response.) The importance of FFT technology extends far beyond its use to simplify working with Fourier series. They are essential for processing experimental data taken by a digital data acquisition system. Consequently, it is important that one be aware of how to avoid aliasing and other errors. Other developments, such as the decibel scale, frequency bands, and spectral density, are more focused on the description and interpretation of sound.

1.7 Homework Exercises

Exercise 1.1 It is given that $z = 30\cos(400t + 0.1) - 20\sin(400t - 2.5)$. Use complex exponentials to find the amplitude and phase angle relative to a pure sine. Then, verify the result by depicting harmonics as vectors that rotate in the complex plane.

Exercise 1.2 For each of the following pressure signals, determine the amplitude A, phase angle θ, and the earliest time $t_0 > 0$ at which $p = 0$. (a) $p = A\cos(\omega t + \theta) = 4\sin(\omega t) - 10\cos(\omega t - 2.8)$ Pa. (b) $p = 2\sin(400t - 3\pi/4) = A\sin(400t - \theta) + 5\cos(400t + \pi/6)$ kPa

Exercise 1.3 The pressure at a certain location is known to be the sum of three harmonics, $p = 120\sin(1200t) - 240\cos(1200t - \pi/3) + B\cos(1200t + \theta)$ Pa. What is the smallest positive value of B and the corresponding value of θ for which p is a pure sine?

Exercise 1.4 The velocity at the surface of a vibrating body is measured to be $v = 0.1\sin(\omega t + \pi/3) + 0.15\cos(\omega t - 3\pi/4)$ mm/s. This body is said to be locally reacting because the surface pressure is defined solely by the particle velocity at that location. Specifically, the pressure is $p = \alpha v + \beta dv/dt$. Parameters are $\alpha = 400$ Pa-s/m, $\beta = 0.1$ Pa-s²/m, $\omega = 3000$ rad/s. What are the maximum values of v and p, and what is the time duration between the occurrence of maximum v and maximum p?

Exercise 1.5 Given that the signal plotted below is the sum of two harmonics, estimate the frequency and amplitude of each.

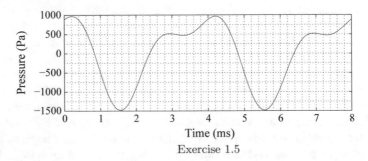

Exercise 1.5

Exercise 1.6 The signal $p = 36\cos(2000t) + C\sin(2500t)$ Pa is known to be the product of two harmonic functions p_1 and p_2, one of which has an amplitude of 9 Pa. Deduce as much as possible about the properties of p_1 and p_2.

Exercise 1.7 When a certain assembly is subjected to a 1 kN harmonic excitation at 30 Hz, the displacement has an amplitude of 40 mm, and the peak displacement occurs 0.25 s after the peak value of the force. Determine the amplitude of the velocity and acceleration, and the corresponding time lag of each quantity relative to the force.

Exercise 1.8 The time-domain response of a system is governed by $\dot{y} + 20y = 0.2 \sin(240t + 1.8)$ where y is the displacement in meters. At sufficiently large t, the steady-state condition, which is the particular solution y_p of this equation, is attained. Express y_p as a complex exponential. From that result, determine the amplitude, frequency, and time delay relative to a pure sine of y_p and $\dot{y}_p$.

Exercise 1.9 Two pressure signals having equal amplitude are a pure harmonic and a quasiharmonic whose frequency slowly varies harmonically relative to the frequency of the first signal. They are described nondimensionally in terms of time variable τ, which is the ratio of elapsed time to the period of the harmonic signal. The parameter ε is the ratio of the fluctuating frequency to the steady frequency. The specific signals are

$$p_1 = \sin(2\pi\tau), \quad p_2 = \sin(2\pi\tau(1 + 0.1\sin(2\pi\varepsilon\tau)))$$

Use numerical software to determine whether these signals are coherent according to Eq. (1.2.14) in the case where $\varepsilon = 0.1$. Then, defend your decision based on the properties of the two signals.

Exercise 1.10 Consider two harmonic signals that have the same amplitude and close frequencies. (a) Prove that they combine to form a beat regardless of their phase angles. (b) With the complex amplitude of one harmonic held fixed at a purely imaginary value, describe the effect of changing the phase angle of the other harmonic.

Exercise 1.11 When a certain system is subjected to harmonic excitation at frequency ω, the pressure at a specified location is known to be

$$p = A\frac{\sin(\omega t) - \sin(\pi c t/L)}{\omega^2 - (\pi c/L)^2} \tag{1.7.16}$$

where all parameters other than t are constants. The quantity $\pi c/L$ is a natural frequency. Resonance occurs if ω is close to the natural frequency. To explore this behavior, follow the analysis of a beat with $\omega_1 = \pi c/L$ and $\omega_2 = \omega$. Determine the representation of $p(t)$ when $|\omega_2 - \omega_1|$ is extremely small compared to ω_{av}. Then, evaluate the limit as ω_2 approaches ω_1. Is the limit finite?

Exercise 1.12 Five automobiles are arranged collinearly as shown. The sound pressure level at 20 m from a single automobile is 80 dB // 20 μPa. It may be assumed that each automobile radiates as a point source, so the acoustic pressure at radial distance r from the center of source n is $p_n(r, t) = (S_n/r)\sin(\omega_n(t - r/c))$. where S_n is the source strength and c is the speed of sound, 340 m/s in air. Determine the sound pressure level due to all automobiles at position $\bar{x}$ when the frequencies are $\omega_1 = 2000$, $\omega_2 = 2500$, $\omega_3 = 3500$, $\omega_4 = 1000$, $\omega_5 = 1500$ rad/s. (b) Consider the case where all automobiles radiate at the same frequency. Evaluate and graph the sound pressure level at position $\bar{x}$ as a function of frequency between 20 Hz and 10 kHz.

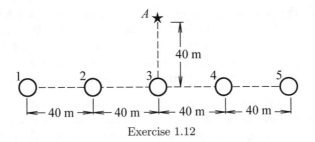

Exercise 1.12

Exercise 1.13 The pressure is $p = 0.1 \sin(\omega_1 t) + 0.2 \cos(\omega_2 t)$ mPa. (a) If the first frequency is 350 Hz and the second is 500 Hz, what is the sound pressure level in air? (b) If both frequencies are 350 Hz, what is the sound pressure level in air? (c) Determine the amplitude of the ω_2 harmonic that will produce a 10 dB increase in the sound pressure level relative to the values found in Parts (a) and (b).

Exercise 1.14 The pressure and particle velocity in the x direction at a certain location within a duct are

$$p = 40 \sin(300\pi t) + 60 \cos(500\pi t - \pi/6) \text{ Pa}$$
$$v = 0.001 \cos(300\pi t - \pi/3) - 0.002 \sin(500\pi t + \pi/8) \text{ m/s}$$

(a) Derive expressions for the instantaneous values of p^2 and intensity $I = pv$ as a function of t. (b) Determine the corresponding values of $(p^2)_{av}$ and $(I)_{av}$.

Exercise 1.15 The pressure at a certain location is $p = 2 \cos(\omega_1 t) + 5 \sin(\omega_2 t)$ kPa. If $\omega_1 = 2$ kHz and $\omega_2 = 5$ kHz, what is the minimum time duration for an averaging meter for which the measured mean-squared pressure differs by no more than 10 % from the true value? What is this duration if $\omega_1 = 2$ kHz and $\omega_2 = 2.1$ kHz?

Exercise 1.16 As was noted previously, mean-squared pressure does not convey the nature of a beating signal. Nevertheless, it is instructive to examine the effect of the time window on this metric. The envelope of a beating signal has a periodic fluctuation at the beat period, π/σ. Such repetition suggests that the beat period could be used as the averaging interval for evaluation of the mean-squared pressure. Evaluate the mean-squared pressure for a beating signal using time windows ranging from $T \ll \pi/\sigma$ to $T = 10\pi/\sigma$. Perform these evaluations for the case where $\omega_{av} = 200$ rad/s and $\sigma = 1$ rad/s, and compare the result to the value obtained when $T \to \infty$.

Exercise 1.17 The time-averaged intensity in a signal whose frequency is 1500 rad/s is observed at a certain location to be $\bar{I}_{av} = 0.360\bar{e}_x + 0.480\bar{e}_y$ N/m/s. The pressure at that location is $p = 800 \cos(1500t)$ Pa. Determine to the maximum possible extent the particle velocity.

Exercise 1.18 The sawtooth curve models the signal associated with a sonic boom from a supersonic aircraft. The other curve, which is a single cycle of a sine function, is a reference signal. (a) Determine the sound exposure level of the reference signal.

(b) Determine the maximum pressure p_{max} of the sawtooth for which both signals have the same sound exposure level. (c) If the sound exposure levels of both waveforms are equal, which one would you expect would be perceived as being more annoying? Justify your answer.

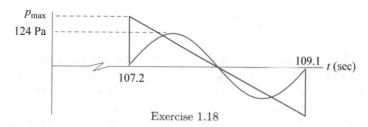

Exercise 1.18

Exercise 1.19 Consider harmonic tones at 50 Hz, 500 Hz, and 5 kHz, with the amplitude at each frequency set at 28 mPa. What is the A-weighted SPL of each tone? What is the phon level of each tone?

Exercise 1.20 The A-weighted sound pressure level is 44 dB for the four octave bands whose center frequencies are 100, 200, 400, and 800 Hz. It also is known that the unweighted sound pressure levels for the second, third, and fourth of these bands are 46, 38, and 39 dB. Determine the unweighted sound pressure level for the octave band whose center frequency is 100 Hz.

Exercise 1.21 The sawtooth pressure signal in Example 1.9 is heard by a person with average hearing. Its fundamental frequency is 200 Hz and $P_{max} = 3$ Pa. Estimate the A-weighted sound pressure level in the range from 25 Hz to 2.5 kHz. Then, repeat the analysis using C-weighting.

Exercise 1.22 The period of the sawtooth wave in Example 1.9 is 4 ms. The signal propagates in air with a maximum value $P_{max} = 0.4$ Pa. (a) Determine the mean-squared pressure and the sound pressure level (dB//20 μPa) in octave bands ranging from 200 Hz to 12.8 kHz.

Exercise 1.23 A periodic sound consists of the positive lobe of a sine for the first half-period, followed by silence for the second half-period. Determine its Fourier coefficients and its mean-squared pressure.

Exercise 1.24 A periodic train of exponentials is described as $p = (B/\beta)$ $\exp(-\beta t/T)$ for $0 < t < T$, $p(t \pm T) = p(t)$. Determine the Fourier series coefficients as a function of T and β. Evaluate the result for $\beta = 1, 5$, and 25. What does the trend of increasing β suggest regarding an alternative representation of p?

Exercise 1.25 A periodic Gaussian waveform is described by $p(t) = p(t + T) =$ $A \exp(-\beta t^2/T^2)$ if $-T/2 < t < T/2$. The peak pressure is 400 mPa and $T = 2$ ms. Use the software of your choice to determine the first eight DFT coefficients of this function for $\beta = 1, 10$, and 100. How does changing β affect the spectral properties of the DFT coefficients?

Exercise 1.26 The pressure at $x = 0$ in a tube is $\varepsilon c \sin (\omega t)$, where c is the speed of sound and ε is the acoustic Mach number. The Fubini solution for the nonlinear pressure wave at $x > 0$ produced by this excitation is

$$p = \rho_0 c^2 \varepsilon \sum_{n=1}^{\infty} \frac{2}{n\sigma} J_n (n\sigma) \sin (n\omega\tau), \quad \sigma = \varepsilon\beta\frac{\omega}{c}x, \quad \tau = t - \frac{x}{c}$$

The symbol $J_n (z)$ denotes the Bessel function of order n and argument z. It is computable in most mathematical software. The variable σ is the propagation distance x ratioed to the distance at which the signal forms a shock, that is, a discontinuity. The variable $\tau = t - x/c$ is the retarded time, where c is the speed of sound. It accounts for the time x/c required for the signal to propagate from the origin to position x where it is observed. Other parameters in this expression are the ambient density ρ_0 and the coefficient of nonlinearity β. (a) Synthesize and plot the nondimensional pressure waveform $p/ (\varepsilon\rho_0 c^2)$ as a function of τ at $\sigma = 1$. (b) Evaluate and plot an estimate for the mean-squared pressure at $\sigma = 1$ as a function of the number of terms at which the series is truncated.

Exercise 1.27 Consider the reflection process described in Example 1.12. Rather than using the impedance ζ described there, use the model for a spring-mass-dashpot system, according to which

$$\zeta = \Gamma \left[2\eta + i \left(\frac{\omega}{\omega_N} - \frac{\omega_N}{\omega} \right) \right], \quad \Gamma = \frac{M\omega_N}{\rho_0 c \mathcal{A}}$$

where $M/\mathcal{A}$ is the mass per unit surface area, ω_N is the natural frequency, and η is the ratio of critical damping. Parameters are $\Gamma = 0.5$. $\eta = 0.1$, and the natural frequency is $\omega_N/(2\pi) = 50$ Hz. The incident wave is the pulse described in the Example. Determine the corresponding wave that is reflected from the polymer coated surface in the interval $0 \leq t < 2T$. Does this wave seem to be consistent with the principle of causality?

Exercise 1.28 A pressure oscillation is a quasiharmonic fluctuation given by $p = 40 (\sin (500t (1 + 0.2 \sin(1000t))$ Pa, where the units of t are seconds. It is desired to determine the spectral density of this function. To do so, a very long time window is required to capture the mix of frequencies. Carry out the computation. Then, examine the result to identify any frequencies at which there are large contributions.

Exercise 1.29 The input velocity to a system varies harmonically, but the amplitude varies randomly relative to a mean value. The function is $v = A (t) \sin (\omega t)$. where $A (t) = v_1 + v_2 (2 \operatorname{rand} (t) - 1)$. Parameters are $\omega = 500$ rad/s, $v_1 = 10$ mm/s and $v_2 = 40$ mm/s. Determine the spectral density of the velocity and the corresponding RMS value based on observing the velocity for a single time window of $16\pi/\omega$. Then, repeat the analysis by averaging the spectral density for eight sequential windows whose duration is $2\pi/\omega$. Use the same data set for all analyses.

Exercise 1.30 Measurement of the pressure radiated by a washing machine indicates that at a certain location, the (unweighted) sound pressure level in the band from 20 to 200 Hz is 38 dB//20 μPa. It is believed that this noise fits a white noise model across its entire frequency range. If so, what it the sound pressure level in the band from 20 Hz to 10 kHz?

Exercise 1.31 Repeat the analysis in Exercise 1.30 for the situation where the noise fits a pink noise model.

Exercise 1.32 The frequency of a loudspeaker is observed to fluctuate randomly about a mean value f_0 (Hz). The consequence is that the pressure measured at a certain location is $p = 8 \sin (\omega(t) t)$ Pa, where $\omega(t) = 2\pi f_0 (1 + \varepsilon \Phi)$. The quantity Φ is a random number having a normal (or Gaussian) distribution. If time is sampled at instants t_n with $n = 1, ..., N$, then a set of Φ values may be generated in MATLAB by the command Phi=random('Normal',0, 1,1,N), where the first argument is the name of the distribution, the second argument is the mean value, the third argument is the standard deviation, the fourth argument is the number of rows for the random numbers, and the fifth argument is the number of columns. The mean frequency is $f_0 = 60$ Hz, and the fluctuation amplitude is $\varepsilon = 0.2$. Consider a time window covering eight periods at the mean frequency, that is $T = 8/f_0$. Evaluate the spectral density of the pressure by sampling the pressure function at 4096 instants. Then, decompose the signal into octave bands centered on f_0. Plot the mean-squared pressure in each band against the corresponding set of center frequencies. Does this plot indicate the presence of a signal at the average frequency? Does the plot indicate that the fluctuation fits the pink noise or white noise models?

Chapter 2
Plane Waves: Time Domain Solutions

Now that we have identified the significant properties of an acoustic signal we may address two fundamental questions: How was the signal generated at its source? How was the signal modified as it traveled from its source to the location where it was observed? These questions often are closely related. Generation of sound is usually associated with oscillation of another medium that shares a surface with the fluid in which the sound is observed. Usually the other body is a solid object like a loudspeaker or an automobile or our vocal cords, but it also could be a region of fluid, such as the gases exhausted by a jet engine. Determination of how one can induce an object to vibrate in the manner required to generate a specific sound is the realm of acoustic specialties such as engineering acoustics, musical acoustics, and structural acoustics and vibrations.

Our study in this chapter is devoted to the simplest kind of acoustic signal, in which the pressure depends only on time and position along a straight line. We say that it is a *plane wave* because all points on a plane perpendicular to the straight line experience the same pressure at every instant. Despite their relative simplicity, plane waves are extremely important to acoustics. They are frequently observed; many other types of signals can be represented as superpositions of plane waves; and some other types of waves tend to become planar at long propagation distances. In addition, the relative simplicity of plane waves makes it easier to identify underlying physical phenomena, and to perform analyses in both the time and frequency domains.

Our first task is to derive the equations whose solution will tell us what pressure will be observed at a specified position and time. We will do so by adapting the basic equations of fluid mechanics to account for the relative smallness of acoustic signals. The result will be the wave equation. A progression of time-domain solutions eventually will demonstrate several acoustical phenomena that also occur for multidimensional signals. These time domain solutions will have certain limitations whose removal will require that we shift in the next chapter to frequency domain

Electronic supplementary material The online version of this chapter (DOI 10.1007/978-3-319-56844-7_2) contains supplementary material, which is available to authorized users.

analysis. In this respect the treatment of planar waves is like our approach for later topics: Find out what we can learn in the time domain, then try to learn more with frequency domain formulations.

2.1 Continuum Equations in One Dimension

Field equations are partial differential equations that govern how variables describing the state of a fluid or solid evolve in time and are distributed spatially. Some assumptions, which usually are quite valid, are made for our initial studies. One is that energy dissipation due to viscosity and other processes is ignored. Another is a restriction to situations where the fluid is homogeneous, which means that its properties in the undisturbed state are the same everywhere. We also consider the fluid to be at rest in the absence of an acoustic disturbance.

These assumptions parallel those that often are invoked in the study of the mechanics of a compressible, inviscid fluid. The transition to the specialized equations of acoustics comes when the small-signal restriction is invoked. This entails simplifying the equations to account for the fact that acoustical signals have an amplitude that is much less than some reference value, such as the ambient pressure. (We have already seen that a very loud sound corresponds to a rather small pressure amplitude.) When these idealizations apply, the field equation will be the linear wave equation for a single state variable, usually pressure. Each of the idealizations is examined in later chapters, but failure to invoke them at the outset of our exploration would obscure learning the core concepts, and it would make any investigation much more difficult.

2.1.1 Conservation of Mass

The particles of a solid have an identifiable position when the solid is undisturbed by outside forces. In contrast, if we look at a fluid in motion, and then remove the causes of that motion, the fluid particles will not return to some reference position. Thus derivation of basic equations requires the concept of a *control volume*, which is merely a region that we artificially isolate. This control volume may be fixed in space, or it may move with the fluid in order to follow a specific group of particles.

We begin with a derivation of the basic equation that governs how a fluid may move without creating or destroying mass. The equation that results is usually referred to as the *continuity equation*, because it enforces the requirement that there may be no voids in which there are no fluid particles. One way to derive the continuity equation is to examine the mass that enters and leaves a *fixed control volume* over some time interval.

A common system in which one might encounter a planar wave is a long straight tube having constant cross-sectional area $\mathcal{A}$. In acoustics such a system usually is said to be a *duct*, or a *waveguide*, both of which convey the fact that the propagation is confined to a specific direction. Position along the tube is specified by the spatial coordinate x. All state variables at a specified x and t in the one-dimensional model have the same value, regardless of where they are situated on a cross section. Figure 2.1 depicts a control volume V that is formed by isolating the region inside the tube between positions x_1 and x_2. The illustration to the left describes the state at some instant t. It shows V and an adjacent region in $x < x_1$ that is the portion of fluid that will enter V in an infinitesimal time interval dt. The illustration to the right depicts the state at time $t + dt$. Here we see V and a region in $x > x_2$ that contains the fluid that has emerged from the control volume.

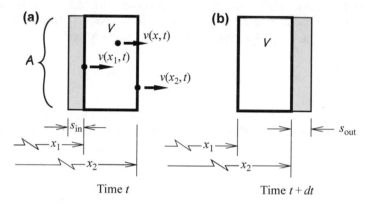

Fig. 2.1 Control volume for formulation of the continuity equation

All particles on a cross section advance in unison at velocity $v(x, t)$. Consequently, the domains that enter and leave are prismatic, as is the control volume. The volume of each region is the product of A and its length in the x direction. For V this length is $x_2 - x_1$, but what are the lengths s_{in} and s_{out} for the domains that enter and leave? To answer this we observe that the distance traveled by a cross-section at position x in an infinitesimal time interval is $v(x, t) \, dt$. It might seem problematic that the movement changes the x and t values on which v depends, but these changes are infinitesimal. For this reason we may use the velocity at x_1 at time t to evaluate s_{in} and the velocity at x_2 at time t to evaluate s_{out}. Thus, we have

$$s_{in} = v(x_1, t) \, dt, \quad s_{out} = v(x_2, t) \, dt \tag{2.1.1}$$

The mass of the two domains depicted in Fig. 2.1a for time t must be the same as the mass of the two domains in Fig. 2.1b for $t + dt$. The density ρ between x_1 and x_2 is not necessarily constant at any instant, so the mass contained in V must be evaluated as an integral over the length of V. The fact that s_{in} and s_{out} are infinitesimal allows us to consider the density of the entering domain to be $\rho(x_1, t)$, and the density of the exiting domain to be $\rho(x_2, t)$, so conservation of mass reduces to

$$\int_{x_1}^{x_2} \rho(x, t+dt)\, \mathcal{A}dx + \rho(x_2, t)\, \mathcal{A}v(x_2, t)\, dt = \int_{x_1}^{x_2} \rho(x, t)\, \mathcal{A}dx$$
$$+ \rho(x_1, t)\, \mathcal{A}v(x_1, t)\, dt$$
$$(2.1.2)$$

Division of this conservation equation by $\mathcal{A}dt$ leads to

$$\int_{x_1}^{x_2} \left[\frac{\rho(x, t+dt) - \rho(x, t)}{dt} \right] dx + \rho(x_2, t)\, v(x_2, t) - \rho(x_1, t)\, v(x_1, t) = 0$$
$$(2.1.3)$$

The integrand is the definition of the partial derivative of density with respect to time. To progress further we consider the terms at x_1 and x_2 to be the limits of an integral over $x_1 < x < x_2$. Specifically, we use the property that

$$f(x_2) - f(x_1) \equiv \int_{x_1}^{x_2} \frac{\partial f}{\partial x} dx \qquad (2.1.4)$$

To some extent it might seem like a trick to introduce this identity, but it actually is the one-dimensional version of the divergence theorem. When Eq. (2.1.4) is used to replace the limit terms in the continuity equation, the result is

$$\int_{x_1}^{x_2} \left[\frac{\partial \rho}{\partial t} dx + \frac{\partial}{\partial x} (\rho v) \right] dx = 0 \qquad (2.1.5)$$

This integral relation must be satisfied for any choice of x_1 and x_2, which can only happen if the integrand vanishes. The result is the *one-dimensional continuity equation*

$$\boxed{\frac{\partial \rho}{\partial t} + \frac{\partial}{\partial x} (\rho v) = 0} \qquad (2.1.6)$$

The first term represents an accumulation of fluid particles within $\mathcal{V}$, while ρv is the mass flux, that is, the net rate per unit surface area at which mass flows through a cross section.

2.1.2 Momentum Equation

The next question is how do fluid particles react to the forces acting on them? The forces are those associated with the pressure. (Gravity is not included here because it has a static role for a homogeneous fluid.) The basic laws that relate motion and forces are those of Newton. We have adopted a continuum model in which an aggregate of fluid particles constitute a differential element. Correspondingly we enforce Newton's second law for a system of particles. It states that the time rate at which the total momentum of all particles in a system changes equals the resultant force acting on that system. This statement is often said to be the momentum-impulse principle, even though we use it in its time derivative form.

Formulation of this principle requires that we track a specific set of particles, so the control volume for the derivation *moves* in unison with the along with the particles. Thus, the control volume is taken to be the prismatic domain extending over $x_1(t) \leq x \leq x_2(t)$, with the stipulation that the lower and upper limits move in unison with the fluid particles at the respective cross-sections. This control volume is depicted as a free body diagram in Fig. 2.2.

Fig. 2.2 Control volume for formulation of the momentum equation

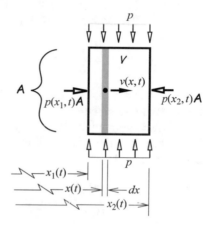

Along the prismatic surface (the side), there is a pressure distribution that is the reaction to the fluid pressing on the wall of the tube. This effect acts perpendicularly to the axis of the tube, and the pressure is the same for all particles at a specified x. It can be proven that the resultant of this force system at each x is zero, which is consistent with the restriction of the fluid to movement in the x direction. Consequently, the pressure exerted by the tube wall can be ignored for the formulation of the momentum equation. There is a resultant force acting in the axial direction. The pressure pushes on the exposed cross-sections faces, and its value is constant along each face. Thus, the force at $x_1(t)$ is $p(x_1, t)\,A$ acting in the direction of increasing x, while the force at $x_2(t)$ is $p(x_2, t)\,A$ in the direction of decreasing x.

All fluid particles situated at the same x have the same velocity, so the momentum of a differential element of mass is $v(x, t)\,dm$. The control volume has been defined to move with the fluid. Correspondingly, we identify $dm = \rho A\,dx$ as the mass contained between cross-sections at x and $x + dx$ that move in unison with the particles. Because mass is conserved, dm is invariant in time, so only $v(x, t)$ needs to be differentiated. Thus, the impulse-momentum principle for a system of particles states that

$$\int_{x_1(t)}^{x_2(t)} \left[\frac{d}{dt} v(x, t) \right] dm = p(x_1, t)\,A - p(x_2, t)\,A \qquad (2.1.7)$$

A subtlety arises at this stage, at the heart of which is what $v(x, t)$ means. Suppose the velocity field is independent time, but varies spatially. This means that a particle's velocity is $v(x)$. Consider a particle that is at $x = \xi$ at time t. In an infinitesimal time interval dt, the particle would displace by distance $v(\xi)\,dt$, so its new position would be $x = \xi + v(\xi)\,dt$. Thus the velocity of the particle would change from $v(\xi)$ at time t to $v(\xi + v(\xi)\,dt)$ at time $t + dt$. In other words, even though the velocity field does not depend on time, the particle accelerates because it has moved to a location where the velocity is different. This effect is incorporated in the evaluation of acceleration by recognizing that x is the position of the particle at time t, and that $dx/dt = v$, so that the general functional dependence of the velocity should be written as $v(x(t), t)$. The total time derivative of velocity, which is the acceleration, is found by the chain rule, which gives

$$\frac{d}{dt}v(x,t) = \frac{\partial}{\partial t}v(x,t) + \frac{\partial}{\partial x}v(x,t)\frac{\partial}{\partial t}x(t) = \frac{\partial v}{\partial t} + v\frac{\partial v}{\partial x} \qquad (2.1.8)$$

where the functional dependence has been omitted because there no longer is a question of interpretation.

The last step applies Eq. (2.1.4) to the boundary terms in Eq. (2.1.7), in order to bring them into the integral. When all terms are collected, and dm is replaced by $\rho d\mathcal{V}$, the result is

$$\int_{x_1(t)}^{x_2(t)} \left[\rho\mathcal{A}\left(\frac{\partial v}{\partial t} + v\frac{\partial v}{\partial x}\right) + \frac{\partial}{\partial x}(\mathcal{A}p) \right] dx = 0 \qquad (2.1.9)$$

As was true for the derivation of the continuity equation, we argue that this integral must vanish for any x_1 and x_2, which will only be true if the integrand is zero. The cross section area $\mathcal{A}$ is constant, so factoring it out leads to the final form of the momentum equation for one-dimensional motion,

$$\boxed{\rho\left(\frac{\partial v}{\partial t} + v\frac{\partial v}{\partial x}\right) = -\frac{\partial p}{\partial x}} \qquad (2.1.10)$$

Although this equation was derived by Euler in 1752, we will reserve the name "Euler's equation" for a simpler version that results from linearization on the next section.

2.2 Linearization and the One-Dimensional Wave Equation

Three variables describing the instantaneous state of the fluid appear in the continuity equation, Eq. (2.1.6), and the momentum equation (2.1.10): pressure p, velocity v, and density ρ, each of which is a function of x and t. Only two equations are available at this juncture, so we are not yet ready to consider solving for the state variables. The missing element is the *equation of state*, which relates pressure and density. For gases, the equation of state is derived from thermodynamic considerations, which introduces temperature T into the mix of variables. Thermodynamic assumptions that have been substantiated by measurements and difficult analyses will lead in the next section to removal of T from the set of unknowns. The consequence is that the equation of state gives a unique function relating pressure and density,

$$p = \tilde{p}(\rho) \tag{2.2.1}$$

Liquids have the same functional representation for their equation of state, but the actual form usually is determined experimentally.

Both the continuity and momentum equations are nonlinear. This means that some of their terms are a product of the state variables, specifically ρv for continuity, and $\rho(\partial v / \partial t + v \partial v / \partial x)$ for momentum. Nonlinear partial differential equations are quite difficult to solve. Fortunately, most phenomena of everyday interest feature acoustic pressure levels that are very small relative to the ambient pressure, as we saw in the discussion of sound pressure level.

The notion that the effects associated with the acoustical response are small is quantified by replacing each state variable by its value in a reference state and a *perturbation*, which is the amount by which the presence of the signal disturbs the reference state. For our initial studies the fluid is homogeneous in the reference state, which means that there is no spatial variation of the density and temperature. Also, we shall restrict our attention to situations where the fluid is at rest in the absence of the signal. The reference state in these conditions consists of ambient pressure p_0, ambient density ρ_0, and ambient temperature T_0, each of which is the same at all locations.

Broad classes of systems and phenomena fit these specifications, but several applications are excluded. For example, a study of propagation to the ground of the sound generated by aircraft must account for the variation of the temperature and density with altitude, and the effect of wind velocity sometimes needs to be addressed. The situation is even more complicated in many aeroacoustic applications where the acoustic signal represents a small perturbation of a very complicated fluid flow that features viscous effects, often including turbulence.

A prime will be used temporarily to indicate the amount by which a variable is perturbed from its value at the reference state. Thus, the absolute pressure, density, and particle velocity are represented as

$$p = p_0 + p', \quad \rho = \rho_0 + \rho', \quad v = v' \tag{2.2.2}$$

The continuity equation that results from this change of variables is

$$\frac{\partial}{\partial t}(\rho_0 + \rho') + \frac{\partial}{\partial x}\left[(\rho_0 + \rho')v'\right] = 0 \tag{2.2.3}$$

Some individuals say the approximation we shall invoke considers the signal to be infinitesimal, because it takes the variables representing the acoustic disturbance to be sufficiently small that their products are negligible. A better description is that we are linearizing the governing equations, because only terms that are linear in the state variables are retained. Accordingly, we drop $\rho' v'$ from the preceding continuity equations. Furthermore, the reference state corresponds to p_0 and ρ_0 being constants. The remaining terms are the *linearized continuity equation,*

$$\boxed{\frac{\partial \rho'}{\partial t} + \rho_0 \frac{\partial v'}{\partial x} = 0} \tag{2.2.4}$$

A linearized momentum equation is derived in a similar manner. Substitution of Eq. (2.2.2) into (2.1.7) yields

$$\left(\rho_0 + \rho'\right)\left(\frac{\partial v'}{\partial t} + v'\frac{\partial v'}{\partial x}\right) = -\frac{\partial}{\partial x}\left(p_0 + p'\right) \tag{2.2.5}$$

Products of primed variables are negligible, and p_0 is taken to be constant. Thus, the linearized *one-dimensional momentum equation* is

$$\boxed{\rho_0 \frac{\partial v'}{\partial t} = -\frac{\partial p'}{\partial x}} \tag{2.2.6}$$

We will employ this relation in many situations, on which occasions we will refer to it as the *Euler equation* in recognition of his contribution.

The variables in the continuity equation are ρ' and v', while those in the momentum equation are v' and p'. We need another equation, which is the equation of state, Eq. (2.2.1). Replacing the absolute pressure and density with their perturbation representations produces

$$p_0 + p' = \tilde{p}\left(\rho_0 + \rho'\right) \tag{2.2.7}$$

Because $|\rho'|$ is restricted to being small relative to ρ_0, it is permissible to linearize the equation of state with the aid of a Taylor series. In the following, a subscript "0" indicates that the quantity is evaluated at the reference state.

$$p_0 + p' = \tilde{p}\left(\rho_0\right) + \left(\frac{d\tilde{p}}{d\rho}\right)_0 \rho' + \frac{1}{2}\left(\frac{d^2\tilde{p}}{d\rho^2}\right)_0 \left(\rho'\right)^2 + \cdots \tag{2.2.8}$$

The term $\tilde{p}\left(\rho_0\right)$ is the pressure corresponding to density ρ_0, which is p_0, so the leading terms cancel. The first derivative of $\tilde{p}\left(\rho\right)$ evaluated at the reference state is a constant. We shall use c^2 to represent it,

$$\boxed{c = \sqrt{\left(\frac{d\tilde{p}}{d\rho}\right)_0}} \tag{2.2.9}$$

The dimensionality of c is $(\text{pressure/density})^{1/2} = ((\text{force/length}^2)/(\text{mass/length}^3))^{1/2} = \text{length/time}$, in other words, speed. It will be apparent in the general solution of the wave equation two sections forward from here that c is the *speed of sound*.

The linear equation of state that results when the quadratic term in Eq. (2.2.8) is dropped is

$$\boxed{p' = c^2 \rho'} \tag{2.2.10}$$

Substitution of this expression into the linearized continuity equation, Eq. (2.2.4), converts it to

$$\boxed{\frac{\partial p'}{\partial t} + \rho_0 c^2 \frac{\partial v'}{\partial x} = 0} \tag{2.2.11}$$

In combination with the linearized momentum equation, Eq. (2.2.6), we now have two differential equations for two state variables, p and v. It is easier to solve a single field equation, and pressure is the primary quantity for an acoustical analysis. To eliminate v' we observe that the term containing this variable in the continuity equation is $\rho_0 c^2 \partial v'/\partial x$, while it shows up as $\rho_0 \partial v'/\partial t$ in the momentum equation. To make v' appear similarly in both equations, we differentiate the momentum equation with respect to x, then divide the continuity equation by c^2 and differentiate it with respect to t. This gives

$$\rho_0 \frac{\partial^2 v'}{\partial x \partial t} = -\frac{\partial^2 p'}{\partial x^2}$$

$$\frac{1}{c^2} \frac{\partial^2 p'}{\partial t^2} + \rho_0 \frac{\partial^2 v'}{\partial t \partial x} = 0 \tag{2.2.12}$$

The result of subtracting the second equation into the first is the *one-dimensional wave equation*,

$$\boxed{\frac{\partial^2 p'}{\partial x^2} - \frac{1}{c^2} \frac{\partial^2 p'}{\partial t^2} = 0} \tag{2.2.13}$$

When we investigate multidimensional systems, we will generalize this field equation, but the result will still be called the wave equation. In addition to its great importance for acoustics, the one-dimensional version is a key element for other subjects. For example, in structural vibration it governs the transverse displacement of a tensioned string, as well as the axial displacement of an elastic bar. For this reason the one-dimensional wave equation has been the subject of much attention for applied mathematicians. It is fortunate for us that we will not require a great deal of mathematical expertise to begin our studies.

EXAMPLE 2.1 A modification that has been suggested as a way to incorporate the effects of viscous friction at the wall of a narrow tube is to add a longitudinal shear force that acts in the direction that is opposite the particle velocity. The model we consider here is restricted to situations where the size of the cross section is sufficiently large that the velocity profile may be considered to be a uniform (planar) distribution that transitions to a narrow boundary layer at the wall. If this is true, the shear stress exerted between the wall and the fluid will be proportional to the mainstream velocity. In addition, this stress is proportional to the shear strain rate, with the dynamic viscosity μ being the factor of proportionality. Multiplying the shear stress by the circumference C of the cross section gives f_v, the viscous force per unit length along the waveguide. Thus, it must be that $f_v = \chi \mu C v$. The factor χ, whose dimensionality is reciprocal length, is described in Sect. 3.3. Derive the analogs of the one-dimensional linear wave equation and Euler's equation corresponding to this model.

Significance

The ability to adapt field equations to address special situations is a vital skill. Doing so for the effect of wall friction reveals that the basic wave equation is ever present, although it is modified.

Solution

The linearized continuity equation remains valid, but the momentum equation requires modification. The wall friction force acting on the section of length dx is $-\chi \mu C v \, dx$, where the negative sign applies because the force acts oppositely to the sense of positive v. The total friction force exerted on the control volume is the accumulation of these differential contributions, which leads to addition of another integral term to the force equation for the control volume, Eq. (2.1.7),

$$\int_{x_1(t)}^{x_2(t)} \frac{d}{dt} [v(x,t)\, dm] = p(x_1,t)\, \mathcal{A} - p(x_2,t)\, \mathcal{A} - \int_{x_1(t)}^{x_2(t)} \chi \mu C v \, dx$$

We convert the pressure terms to an integral representation by invoking the divergence theorem, Eq. (2.1.4), and bring all terms to the left side of the equality, which gives

$$\int_{x_1(t)}^{x_2(t)} \left[\rho \mathcal{A} \left(\frac{\partial v}{\partial t} + v \frac{\partial v}{\partial x} \right) + \frac{\partial}{\partial x} (\mathcal{A} p) + \chi \mu C v \right] dx = 0$$

The integral must vanish for any pair of limits x_1 and x_2, so we set the integrand to zero.

Our interest is in a linear set of equations, which can be obtained directly without introducing primes to represent the acoustical disturbance. Thus, we replace ρ with

ρ_0, and drop the nonlinear acceleration terms. The result is a modified version of the linearized momentum equation,

$$\rho_0 \frac{\partial v}{\partial t} + \frac{\partial p}{\partial x} + \chi\mu\frac{C}{A}v = 0 \tag{1}$$

The linearized continuity equation is unaltered,

$$\frac{\partial p}{\partial t} + \rho_0 c^2 \frac{\partial v}{\partial x} = 0 \tag{2}$$

The task now is to eliminate v from Eqs. (1) and (2). This can be done in a heuristic manner by observing that v appears in Eq. (2) as $\partial v/\partial x$, and differentiating Eq. (1) with respect to x would produce an equation in which particle velocity only occurs as $\partial v/\partial x$ and $\partial^2 v/\partial x \partial t$, that is,

$$\rho_0 \frac{\partial^2 v}{\partial x \partial t} + \frac{\partial^2 p}{\partial x^2} + \chi\mu\frac{C}{A}\frac{\partial v}{\partial x} = 0$$

Substitution of $\partial v/\partial x$ from Eq. (2) into this equation yields

$$c^2 \frac{\partial^2 p}{\partial x^2} - \frac{\partial^2 p}{\partial t^2} - \chi\frac{\mu}{\rho_0}\frac{C}{A}\frac{\partial p}{\partial t} = 0 \tag{3}$$

Hence, the model of wall friction adds a first derivative term to the wave equation.

This heuristic approach to eliminating v might be difficult to implement in more complicated situations. A more formal approach is to write Eqs. (1) and (2) in differential operator form as

$$\mathcal{L}_{v1}(v) + \mathcal{L}_{p1}(p) = 0$$
$$\mathcal{L}_{v2}(v) + \mathcal{L}_{p2}(p) = 0$$

where

$$\mathcal{L}_{v1} = \rho_0 \frac{\partial}{\partial t} + \mu\frac{C}{A}, \quad \mathcal{L}_{p1} = \frac{\partial}{\partial x}$$
$$\mathcal{L}_{v2} = \rho_0 c^2 \frac{\partial}{\partial x}, \quad \mathcal{L}_{p2} = \frac{\partial}{\partial t}$$

We treat the operators like coefficients in a pair of simultaneous equations, which we solve for p,

$$\left(\mathcal{L}_{p1}\mathcal{L}_{v2} - \mathcal{L}_{p2}\mathcal{L}_{v1}\right)(p) \equiv \left(\rho_0 c^2 \frac{\partial^2}{\partial x^2} - \rho_0 \frac{\partial^2}{\partial t^2} - \chi\mu\frac{C}{A}\frac{\partial}{\partial t}\right)(p) = 0$$

Division of this equation by ρ_0 yields Eq. (3).

We will see in Chap. 3 that the term containing $\partial p / \partial t$ is associated with dissipation, which is not surprising because it describe a friction effect. It will be noticed that the magnitude of this effect depends on the kinematic viscosity μ / ρ_0, which is a property of the fluid. It also is proportional to $C/\mathcal{A}$, which is a property of the cross-section. This ratio becomes smaller as the cross-section's size increases (with the shape held fixed). For example for a circle $C/\mathcal{A} = 2/r$. Thus, wall friction decreases in importance as the width of the waveguide is increased.

2.3 Equation of State and the Speed of Sound

The only property of the fluid that appears in the wave equation is the speed of sound c. One way of determining it is to measure it experimentally based on a variety of acoustical phenomena. The value of c does not vary by a great deal for liquids in the usual range of ambient conditions, so experimental measurement usually is sufficient, possibly supplemented by empirical observations regarding the dependence of c on ambient conditions, notably temperature. It is less satisfactory to rely on experiments in the case of gases because c is strongly influenced by the ambient conditions.

It is possible to derive an equation of state for an ideal gas solely from thermodynamic considerations. Even then, some knowledge derived from experimental observations will be required. Pressure p, density ρ, and absolute temperature T (degrees K) for an infinitesimal volume of gas are related by the ideal gas law,

$$p = R\rho T \tag{2.3.1}$$

The parameter R is the ratio of the universal gas constant $R_0 = 8314$ J/(kg-K) to the average molecular weight of the gas. The value for air is $R = 287.1$ J/(kg-K).

Early researchers, starting with Newton in the *Principia* (1687), invoked a plausible assumption to apply the ideal gas law to acoustics. It was argued that the pressure changes extremely rapidly, even at the lowest audible frequencies. If that is so, there should be little time for the fluid to heat up due to an increase in the pressure, or to cool when the pressure decreases. In other words they assumed that acoustical processes are isothermal, so that T is always the ambient value T_0. Because p_0, ρ_0, and T_0 must satisfy the ideal gas law, values of R and T_0 cancel when Eq. (2.3.1) is used to form p/p_0. The result is that

$$p = p_0 \frac{\rho}{\rho_0} \tag{2.3.2}$$

The speed of sound corresponding to this equation of state is found from Eq. (2.2.9) to be

$$c_{\text{isothermal}} = \left(\frac{p_0}{\rho_0} \right)^{1/2} \tag{2.3.3}$$

Despite the plausibility of the isothermal argument, it could not be confirmed experimentally. Measurements generally indicated that the predicted sound speed was 16% lower than the actual value. Many individuals tried to resolve this conflict between theory and experiment, but the fault was not identified until 1816, when Laplace introduced an equally plausible assumption. Based on the same notion that audible signals constitute a rapid change of state, he said that the change is so rapid that there is insufficient time for heat to flow into or out of a volume of gas before the state reverses. A process in which there is no heat flow is said to be **adiabatic**. In such a process the temperature within the volume must change to maintain a fixed heat content when the pressure and density change.

To obtain the adiabatic equation of state consider the differential increase in temperature obtained by application of the chain rule to the ideal gas law,

$$T = \frac{p}{R\rho} \implies dT = dp \frac{\partial}{\partial p}\left(\frac{p}{R\rho}\right) + d\rho \frac{\partial}{\partial \rho}\left(\frac{p}{R\rho}\right) \tag{2.3.4}$$

By definition a partial derivative holds all other variables constant. Hence, the above differential may be considered to describe a pair of processes: Process A increases the pressure with the density held constant, while process B increases the density with the pressure held constant. The associated contributions to the temperature increment are

$$dT_A = dp \left.\frac{d}{dp}\left(\frac{p}{R\rho}\right)\right|_{\text{constant }\rho} = \frac{dp}{R\rho}$$

$$dT_B = d\rho \left.\frac{d}{d\rho}\left(\frac{p}{R\rho}\right)\right|_{\text{constant }p} = -\frac{p}{R\rho^2} d\rho \tag{2.3.5}$$

Specific heat is the heat gain per unit mass resulting from a one degree increase in temperature. For the mass we use a differential mass dm at an arbitrary location. Two kinds of specific heat are used. One is C_v, which applies if density is held constant. (The subscript v in C_v is the specific volume, which is the reciprocal of density.) Process A fits this specification, so we have

$$dq_A = (C_v dT_A)\, dm = \left(C_v \frac{p}{R\rho} dp\right) dm \tag{2.3.6}$$

The other specific heat is C_p, which describes the behavior if the pressure is held constant. This is process B, so the corresponding heat gain is

$$dq_B = (C_p dT_B)\, dm = \left(-C_p \frac{p}{R\rho^2} d\rho\right) dm \tag{2.3.7}$$

where the negative sign merely says that the heat content decreases.

The adiabatic condition requires that the combination of processes A and B not change the heat content of dm. In other words it must be that $dq_A + dq_B = 0$. We use the above relations to describe each increment, and cancel the common factor *dm*, which leads to the requirement that

$$C_v \frac{1}{R\rho} dp - C_p \frac{p}{R\rho^2} d\rho = 0 \tag{2.3.8}$$

This relation may be solved for $dp/d\rho$, The speed of sound, which was defined in Eq. (2.2.9), features this derivative at the ambient state, so we replace p and ρ with p_0 and ρ_0, respectively. The *ratio of specific heats* is typically denoted as γ, so we find that

$$c = \sqrt{\frac{\gamma p_0}{\rho_0}}, \quad \gamma = \frac{C_p}{C_v} \tag{2.3.9}$$

Thus the isothermal sound speed in Eq. (2.3.3) is too small by a factor of $1/\gamma^{1/2}$. For air $\gamma = 1.4$, so $1/\gamma^{1/2} = 0.845$, which matches the 16% error that had been observed. A significant aspect is that γ typically depends solely on the molecular composition of the gas, but there are gases for which the specific heats are temperature dependent. Even if γ is temperature dependent, Eq. (2.3.9) remains valid because the relation was derived by considering only infinitesimal changes in temperature.

For linearized acoustics analyses we only need to know c, but studies of nonlinear effects requires knowing the actual equation of state. Separating variables converts Eq. (2.3.8) to a perfect differential,

$$\frac{dp}{p} = \gamma \frac{d\rho}{\rho} \tag{2.3.10}$$

The indefinite integral is

$$\ln (p) = \gamma \ln (\rho) + \ln (\alpha) \tag{2.3.11}$$

The constant of integration $\ln (\alpha)$ must be such that $p = p_0$ when $\rho = \rho_0$. We impose this condition, then take the antilogarithm of each side, with the result that

$$\frac{p}{p_o} = \left(\frac{\rho}{\rho_0}\right)^\gamma, \quad \gamma = \frac{C_p}{C_v}$$

An alternative form in terms of the acoustic perturbation quantities, $\rho = \rho_0 + \rho'$, $p = p_0 + p'$, is obtained from the preceding by applying the identity that $p_0 = \rho_0 c^2/\gamma$. This leads to

$$\frac{p'}{\rho_0 c^2} = \frac{1}{\gamma}\left[\left(1 + \frac{\rho'}{\rho_0}\right)^\gamma - 1\right] \tag{2.3.12}$$

The quantity $\rho_0 c^2$ is the *bulk modulus K*,

$$K = \rho_0 c^2 \tag{2.3.13}$$

The reciprocal of K is the *adiabatic compressibility*. The bulk modulus is analogous to Young's modulus for an elastic solid. To see why consider an elastic bar having constant cross section $\mathcal{A}$ and length L_0 in the stress-free state. A standard static test measures the increase in length $\Delta L = L - L_0$, when an axial tensile force F is applied slowly. The axial stress is $\sigma = F/\mathcal{A}$ and the axial strain is $\varepsilon = \Delta L/L$. These two quantities are related by $\sigma = E\varepsilon$, where E is Young's modulus. In other words

$$\sigma = E\frac{\Delta L}{L} \tag{2.3.14}$$

To construct the analogy for a fluid replace E with K and σ with the pressure perturbation $-p'$, where the minus sign accounts for the definition of positive stress is tensile, whereas positive p' is compressive. The analog to strain is the ratio of the volume change to the original volume. This is the *condensation* $\Delta \mathcal{V}/\mathcal{V}_0$, where $\mathcal{V}_0$ is the volume in the ambient state. Thus, the analogy is that

$$p' \equiv -K\frac{\Delta \mathcal{V}}{\mathcal{V}_0} \tag{2.3.15}$$

where the negative sign indicates that a positive pressure decreases the volume. Conservation of mass relates the condensation to density. The mass in the ambient state is $\rho_0 \mathcal{V}_0$, while it is $(\rho_0 + \rho')\,\mathcal{V}$ when p' is applied. When we set $\mathcal{V} = \mathcal{V}_0 + \Delta \mathcal{V}$, and equate the mass for each state, we obtain

$$\left(\rho_0 + \rho'\right)\left(\mathcal{V}_0+\Delta \mathcal{V}\right) = C_0 \mathcal{V}_0 \quad\Longrightarrow\quad \frac{\Delta \mathcal{V}}{\mathcal{V}_0} \approx -\frac{\rho'}{\rho_0} \tag{2.3.16}$$

where the approximate equality results from dropping $\rho'\Delta \mathcal{V}$ as a nonlinear effect. The quantity ρ'/ρ_0 is referred to as the dilatation in fluid mechanics. In the linearized approximation it is the negative of the condensation. We have established that $p' = K\left(\rho'/\rho_0\right)$. However, the linearized equation of state, Eq. (2.2.10), indicates that $p' = c^2\rho'$. Equating these alternate description of p' leads to Eq. (2.3.13), which confirms the analogy.

The equation of state for a liquid may be described as a Taylor series expansion of Eq. (2.2.1),

$$p_0 + p' = \tilde{p}\left(\rho_0 + \rho'\right) = \tilde{p}\left(\rho_0'\right) + \left(\frac{d\tilde{p}}{d\rho}\right)_0 \rho' + \frac{1}{2}\left(\frac{d^2\tilde{p}}{d\rho^2}\right)_0 \left(\rho'\right)^2 + \cdots \tag{2.3.17}$$

The ambient pressure corresponding to ρ_0 is p_0, so the leading terms cancel. Furthermore, we do not know the actual functional dependence, so the derivatives may be replaced by coefficients. The result is that the equation of state for a liquid is a simple power series given by

$$\boxed{p' = A\frac{\rho'}{\rho_0} + \frac{1}{2}B\left(\frac{\rho'}{\rho_0}\right)^2 + \cdots} \tag{2.3.18}$$

We know that $c^2 = dp'/d\rho'$ at ambient conditions, so a comparison of this representation to the Taylor series shows that

$$A = \rho_0 \left(\frac{d\tilde{p}}{d\rho} \right)_0 = \rho_0 c^2 = K, \quad B = \rho_0^2 \left(\frac{d^2 \tilde{p}}{d\rho^2} \right)_0 \qquad (2.3.19)$$

The values of c and B/A for liquids usually come from measurements.

The equations of state for a liquid and an ideal gas may be placed in a common form. To do so Eq. (2.3.12) is expanded in a binomial series that is truncated at terms that are quadratic in ρ',

$$\begin{aligned}
\frac{p'}{\rho_0 c^2} &= \frac{1}{\gamma} \left[1 + \gamma \frac{\rho'}{\rho_0} + \frac{1}{2}\gamma (\gamma - 1) \left(\frac{\rho'}{\rho_0} \right)^2 + \cdots - 1 \right] \\
&= \frac{\rho'}{\rho_0} + \frac{1}{2}(\gamma - 1) \left(\frac{\rho'}{\rho_0} \right)^2 + \cdots
\end{aligned} \qquad (2.3.20)$$

Because $A = \rho_0 c^2$, this representation matches that in Eq. (2.3.18) if $B/A = \gamma - 1$. Values of B/A above 10 have seldom been reported for liquids or gases, with $B/A = 0.4$ for air, and B/A ranges from 4.2 to 6.1 for water.[1] (Some research papers have said that B/A is of the order of 1000 for rock, but rock is not a liquid.)

It might seem that halting the equation of state for a liquid at quadratic terms is a limitation. For example, doing so leads to the anomaly that p' has a minimum of $-0.5 \left(\rho_0 c^2 \right) / (B/A)$ at $\rho'/\rho_0 = -1/(B/A)$. This does not limit the utility of Eq. (2.3.18). The bulk modulus is very large for a gas, $\rho_0 c^2 \approx 140$ kPa for air, and it is enormous for liquids, $\rho_0 c^2 \approx 2$ GPa for water. (For comparison, Young's modulus of steel is approximately 200 GPa.). This means that this minimum value is well beyond the range of what is considered to be an acoustic signal. Only in explosions might there be a pressure pulse that is comparable to the bulk modulus.

A further consequence of the largeness of the bulk modulus is that not much error in the density-pressure relation will arise if the equation of state is linearized. Dropping the quadratic term in justifiable if

$$\frac{B}{2A} \left(\frac{\rho'}{\rho_0} \right)_{\text{max}} \ll 1 \qquad (2.3.21)$$

A reasonable upper bound for common liquids and gases is $B/A = 10$, which implies that the linearized equation of state is accurate if $5 \left(\rho'/\rho_0 \right)_{\text{max}} \ll 1$. Let us conservatively say that $\rho/\rho_0 = 0.002$ is the maximum that would be considered to satisfy this condition. The linear equation corresponding to this value of the condensation gives $p = 0.002\rho_0 c^2$. This corresponds to 140 dB//20μPa in air, which is a level that causes permanent auditory damage. In water it is 250 dB//1μPa, which is far

---

[1] An extensive tabulation of B/A for several fluids are provided by R.T. Beyer, "The parameter B/A," in *Nonlinear Acoustics,* M.F. Hamilton and D.T. Blackstock, eds., Academic Press, pp. 34–37 (1998).

above the level at which water near the surface will cavitate because the total pressure $p_0 + p'$ is negative. In other words, it is justifiable to linearize the equation of state for the vast majority of applications.

The primary properties for an acoustical analysis are c and ρ_0, supplemented by B/A if nonlinear effects are to be addressed. Even if the fluid medium has been set, each property depends to some extent on environmental conditions. For air the key factor affecting c is the absolute temperature. This becomes evident when the ideal gas law is used to replace p_0/ρ_0 in Eq. (2.3.9), which leads to

$$\boxed{c = \sqrt{\gamma R T_0}} \tag{2.3.22}$$

The ambient temperature, as well as the ambient pressure, are the primary factors affecting the density of air. In addition, the vapor content (that is, humidity) plays a role by altering the molecular composition on which R depends. (The next example examines standard atmospheric conditions.) For water the important parameters affecting c are temperature, salinity, and ambient pressure, but the density varies little with depth. A comprehensive discussion is provided by Pierce.[2] Unless there is need for precision, one may use as representative values $\rho_0 = 1.2$ kg/m^3 and $c = 340$ m/s for air, and $\rho_0 = 1000$ kg/m^3 and $c = 1480$ m/s for water.

The developments thus far have shown that the only variables requiring determination in a typical situation are the acoustic perturbations of pressure and particle velocity. The ambient state is taken to be known, which means that we know the density ρ_0, pressure p_0, and sound speed c. For this reason, we may discontinue the use of primes to denote acoustic perturbations, which will serve to simplify the notation. Thus, the general relation between the acoustic pressure and particle velocity an any field is the Euler (momentum) equation,

$$\boxed{\frac{\partial p}{\partial x} = -\rho_0 \frac{\partial v}{\partial t}} \tag{2.3.23}$$

The one-dimensional wave equation correspondingly is now written as

$$\boxed{\frac{\partial^2 p}{\partial x^2} - \frac{1}{c^2} \frac{\partial^2 p}{\partial t^2} = 0} \tag{2.3.24}$$

It might be puzzling that the linearized Euler momentum equation is the primary relation, rather that the continuity equation. This choice is made because a common task entails determining the pressure associated with a known motion of a surface. In such situations we know the history of velocity at a spatial location, so may evaluate the acceleration. The continuity equation requires knowledge of the spatial dependence of particle velocity within the fluid, which is not known a priori.

---

[2]A.D. Pierce, *Acoustics,* pp. 28–34, McGraw-Hill (1981), reprinted by ASA Press (1989).

EXAMPLE 2.2 The U. S. Standard Atmosphere is a government recognized model for the dependence of various thermodynamic properties on altitude. The ambient properties relevant to acoustics are tabulated below. (a) Determine whether the entries for pressure, density, and temperature at each altitude are consistent with the ideal gas law. (b) Compare the speed of sound obtained from Eqs. (2.3.9) and (2.3.22) (c) Graph the adiabatic bulk modulus as a function of altitude. (d) When the surface of a sphere of radius a is made to vibrate harmonically as $v_s \sin(\omega t)$, the steady-state pressure at radial distance r from the center of the sphere is

$$p(r, t) = \rho_0 c v_s \frac{a}{r} \sin\left(\omega\left(t - \frac{r-a}{c}\right)\right)$$

Consider a case where $v_s = 20$ m/s, $a = 40$ mm, and $\omega = 1$ kHz. Evaluate the sound pressure level at $r = 10$ m for locations that are at each of the tabulated altitudes. It may be assumed that reflections from the ground are unimportant at any altitude.

Altitude above Sea Level (m)	Temperature (°C)	Absolute Pressure (kPa)	Density (kg/m^3)
−1000	21.50	113.9	1.347
0	15.00	101.3	1.225
1000	8.50	89.88	1.112
2000	2.00	79.50	1.007
3000	−4.49	70.12	0.9093
4000	−10.98	61.66	0.8194
5000	−17.47	54.05	0.7364
6000	−23.96	47.22	0.6601
7000	−30.45	41.11	0.5900
8000	−36.94	35.65	0.5258
9000	−43.42	30.80	0.4671
10000	−49.90	26.50	0.4135
15000	−56.50	12.11	0.1948
20000	−56.50	5.529	0.08891
25000	−51.60	2.549	0.04008
30000	−46.64	1.197	0.01841
40000	−22.80	0.287	0.003996
50000	−2.5	0.07978	0.001027
60000	−26.13	0.02196	0.0003097
70000	−53.57	0.0052	0.00008283
80000	−74.51	0.0011	0.00001846

Significance

In addition to illustrating how to evaluate basic acoustical properties, this example also will lead to some interesting observations regarding acoustics in outer space.

Solution

Because the ambient pressure P_0 and density ρ_0 cover an enormous range, they are plotted on a logarithmic scale in the plot in Fig. 1. The plots of both variables are linear, which corresponds to a power law dependence on altitude H. A fit to the data that uses the values at $H = 0$ and $H = 80$ km to construct a straight line leads to

$$\rho_0 = 1.225 \, (10)^{-0.0603H/1000} \text{ kg/m}^3$$
$$P_0 = 101.03 \, (10)^{-0.0621H/1000} \text{ kPa}$$

As shown in Fig. 1, each empirical fit closely matches the given data.

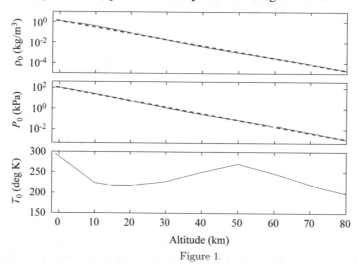

Figure 1.

If the tabulated entries are consistent with the ideal gas law, then each computed value of $p_0/(\rho_0 T_0)$ will equal R, which is 287.1 J/(kg-K) for air. Another consistence check is obtained by comparing the alternative computations of the sound speed, which are

$$c_{\text{density}} = \left(\frac{\gamma p_0}{\rho_0}\right)^{1/2}, \quad c_{\text{temperature}} = (\gamma R T_0)^{1/2}$$

Both would be the same if the ideal gas law applies. Errors for each quantity are

$$\varepsilon_R = \frac{R - p_0/(\rho_0 T_0)}{R}, \quad \varepsilon_c = \frac{\sqrt{\gamma R T_0} - \sqrt{\gamma p_0/\rho_0}}{\sqrt{\gamma R T_0}}$$

Note that the temperature is measured in degrees Kelvin, so 273.13 must be added to the tabulated temperatures. As implied in the definition of ε_c, the speed of sound as a function of temperature is taken as the reference value because that form only depends on one state variable.

The first graph in Fig. 2 compares the altitude dependence of c_T and c_ρ. Also plotted there is a cubic polynomial approximation of the given data obtained by the method of linear least squares. The approximate relation is

$$c = 338.6 - 4.525 \left(\frac{H}{1000} \right) + 0.14421 \left(\frac{H}{1000} \right)^2 - 0.001212 \left(\frac{H}{1000} \right)^3 \text{ m/s}$$

The values of c derived from the tabulated properties are barely different up to an altitude of 60 km, and only differ significantly at 80 km (and above). The discrepancy of the computed value of R, and between the two evaluations of sound speed, are seen in the second graph in Fig. 2 to be small, except at the highest altitude.

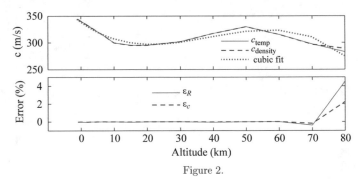

Figure 2.

The variability of the sound speed profile is important because it informs us that we should not consider the atmosphere to be homogeneous if we are interested in propagation over a wide range of altitudes. The fact that c has a maximum and a minimum affects how sound is channeled as it propagates. Chapter 11 introduces the study of propagation through heterogeneous fluids.

Figure 3 describes the bulk modulus. The fact that it too appears to be a straight line in a log-linear scale stems from the fact $K = \rho_0 c^2$, and c is essentially constant relative to the enormous range of ρ_0.

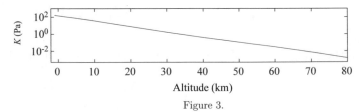

Figure 3.

The amplitude of the sound emitted by the vibrating sphere is $|p| = \rho_0 c v_s (a/r)$. Because the signal is a pure harmonic, we may compute the sound pressure level as

$$\mathcal{L}_p = 20 \log_{10} \left(\frac{|p|/\sqrt{2}}{20 \left(10^{-6} \right)} \right) = 20 \log_{10} (v_s) + 20 \log_{10} \left(\frac{a}{r} \right) + 93.1 + 20 \log_{10} (\rho_0 c)$$

The altitude dependence of this quantity is describe in Fig. 4. The sound pressure level covers an enormous range, from a maximum of 125 dB//20 μPa to a minimum of 26 dB. The part of $\mathcal{L}_p$ that depends on the altitude is $\rho_0 c$, which is called the *characteristic impedance* of the fluid. We will encounter this property in the next section.

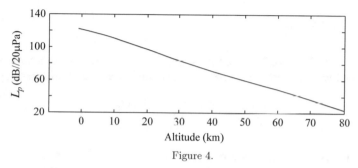

Figure 4.

The linear decrease of $\mathcal{L}_p$ is a consequence of ρ_0 being proportional to a power of H, whereas c changes comparatively little. Continued increase in altitude would yield even smaller values of ρ_0. It is commonly stated that sound cannot be produced in outer space, and that is true. The reason is usually stated that it is a vacuum, which is not true. The density there is variable, but one hydrogen atom per cubic centimeter is a representative value. The scale of an acoustic signal is its wavelength. There are not enough particles in that distance to justify application of the kinetic theory of gases. Hence, it would not be appropriate to consider there to be an acoustic signal in which a disturbance is passed from particle to particle. However, if we consider signals whose wavelength is comparable to the size of planets and stars, there might be a sufficient number of particles to justify a continuum view. If such is the case, c, which depends only on the ambient temperature, would be finite. Nevertheless, unless c is enormous, the nearly infinitesimal density would correspond to an extremely small characteristic impedance. Generating a detectable sound in such an environment would be impossible.

This brings to the fore an interesting question: What is the behavior near a stellar object? If we assume that the ideal gas law is correct there, at least for order of magnitude evaluations, then c will be much larger than it is in the Earth's atmosphere because the temperature is high. Also, the density of the gases will not be small. This would suggest that the sound pressure level near a star will be very high. Should we send a probe containing a microphone to the Sun in order to measure the sound field?

2.4   The d'Alembert Solution

Several paths leading to the general solution of the (linear) wave equation are available. At one end of the spectrum, we could verify that the known solution satisfies the equation. At the other end we could pursue a formal line of mathematical analysis known as the method of characteristics, which also can be used to study signals in inhomogeneous media, as well as nonlinear effects. The approach we follow here is intermediate to those extremes.

2.4.1   Derivation

Combinations of derivatives may be grouped as operators. In this view the one-dimensional wave equation may be factorized into a product of first order operators,

$$\left(\frac{\partial}{\partial x} + \frac{1}{c}\frac{\partial}{\partial t}\right)\left(\frac{\partial}{\partial x} - \frac{1}{c}\frac{\partial}{\partial t}\right) p = 0 \tag{2.4.1}$$

Each of the operator factors is reminiscent of a chain rule derivative associated with changing variables. Specifically, suppose instead of describing the wave equation in terms of derivatives with respect to x and t, we wish to change to a pair of variables, ξ and η. Assuming that we know the transformations, $x = x\,(\xi, \eta)$ and $t = t\,(\xi, \eta)$, the chain rule would state that

$$\frac{\partial}{\partial \xi} = \frac{\partial x}{\partial \xi}\frac{\partial}{\partial x} + \frac{\partial t}{\partial \xi}\frac{\partial}{\partial t}$$
$$\frac{\partial}{\partial \eta} = \frac{\partial x}{\partial \eta}\frac{\partial}{\partial x} + \frac{\partial t}{\partial \eta}\frac{\partial}{\partial t} \tag{2.4.2}$$

We want these derivatives to match the operators in the factorized wave equation, Eq. (2.4.1). At the same time, it is useful for later developments to multiply that equation by 1/4. Then the two operator factors reduce to derivatives with respect to a single variable if

$$\frac{\partial x}{\partial \xi} = \frac{1}{2}, \quad \frac{\partial x}{\partial \eta} = \frac{1}{2}$$
$$\frac{\partial t}{\partial \xi} = -\frac{1}{2c}, \quad \frac{\partial t}{\partial \eta} = \frac{1}{2c} \tag{2.4.3}$$

In a partial differentiation with respect to a variable, other independent variables are held constant. Thus, if we integrate the first equation we obtain $x = \xi/2 + r_x\,(\eta)$, whereas integrating the second equation gives $x = \eta/2 + s_x\,(\xi)$. If each equation is considered in isolation from the other, $r_x\,(\eta)$ and $s_x\,(\xi)$ are arbitrary. However, the alternative equations for x must be consistent, which requires that $r_x\,(\eta) = \eta/2$ and $s_x\,(\xi) = \xi/2$. Similar treatment of the second pair of equations leads to $t = -\xi/\,(2c) + r_t\,(\eta)$ and $t = \eta/\,(2c) + s_t\,(\eta)$, from which it follows that $r_t\,(\eta) = \eta/\,(2c)$ and $s_t\,(\xi) = -\xi/\,(2c)$. It follows that a change of variables given by

$$x = \frac{1}{2}\,(\xi + \eta), \quad t = \frac{1}{2c}\,(\eta - \xi) \tag{2.4.4}$$

will transform the wave equation to

$$\frac{\partial^2 p}{\partial \xi \partial \eta} = 0 \qquad (2.4.5)$$

Because other variables are held constant in a partial differentiation, there are only two possibilities: either $\partial p/\partial \xi$ is an arbitrary function of ξ or $\partial p/\partial \eta$ is an arbitrary function of η. Integrating each possibility leads to

$$p = F(\xi) + G(\eta) \qquad (2.4.6)$$

where F and G are arbitrary functions.

Our interest is in how p depends in x and t, which requires that we determine the inverse of the transformation in Eq. (2.4.4). It is

$$\boxed{\xi = x - ct, \quad \eta = x + ct} \qquad (2.4.7)$$

The variables ξ and η are the *characteristics*. (The reason for this terminology will soon be apparent.) When the characteristic variables in Eq. (2.4.6) are replaced by the physical variables, the result is the *d'Alembert solution* (1747) of the wave equation,

$$\boxed{p = F(x - ct) + G(x + ct)} \qquad (2.4.8)$$

This constitutes the most general solution possible.

No restrictions are placed on the arbitrary functions, other than finiteness and continuity. The latter condition applies because a discontinuity of either function would correspond to dp/dx being infinite. According to the Euler equation, this would require an infinite acceleration, which clearly is impossible. (Occasionally, it will be convenient to consider a discontinuous function, such as a step or square wave. It is justifiable to do so provided that we recognize that the discontinuity is a simplified description of a continuous function that changes greatly over a very small interval).

An important feature of many excitations is that they impose a particle velocity. Hence, we need to know what the particle velocity is according to the d'Alembert solution. Throughout our studies, we will turn to Euler's equation, Eq. (2.2.6), when we need to relate pressure and velocity. Here it requires that

$$\rho_0 \frac{\partial v}{\partial t} = -\frac{\partial}{\partial x} [F(x - ct) + G(x + ct)] \qquad (2.4.9)$$

The quantities inside parentheses for F and G are the respective functions arguments, or equivalently, the characteristics ξ and η, respectively. The formally correct way of carrying out the derivative with respect to x is to differentiate each function with respect to its characteristic, then multiply the result by the derivative of the characteristic with respect to x, which gives

$$\rho_0 \frac{\partial v}{\partial t} = -\left(\frac{\partial \xi}{\partial x} \frac{\partial F}{\partial \xi}\right)\Bigg|_{\xi=x-ct} - \left(\frac{\partial \eta}{\partial x} \frac{\partial F}{\partial \eta}\right)\Bigg|_{\eta=x+ct} \tag{2.4.10}$$
$$= -(1) F'(x - ct) - (1) G'(x + ct)$$

where a prime denotes the derivative of a function with respect to its argument. For example, if $F = \sin(kx - \omega t)$, then $F' = \cos(kx - \omega t)$. A simpler way of arriving at the same result without introducing the characteristics is to differentiate each function with respect to its argument, then multiply that derivative by the derivative of the argument with respect to x.

The next step is to integrate Eq. (2.4.10) with respect to t. To do so, we observe that the time derivatives of the F and G functions are described by

$$\frac{\partial}{\partial t} F(x - ct) = F'(x - ct)\left[\frac{\partial}{\partial t}(x - ct)\right] = -c F'(x - ct)$$
$$\frac{\partial}{\partial t} G(x + ct) = G'(x + ct)\left[\frac{\partial}{\partial t}(x + ct)\right] = +c G'(x + ct) \tag{2.4.11}$$

We solve these for F' and G', and substitute the result into the above form of Euler's equation, Eq. (2.4.10), which gives

$$\rho_0 \frac{\partial v}{\partial t} = \frac{1}{c}\left[\frac{\partial}{\partial t} F(x - ct) - \frac{\partial}{\partial t} G(x + ct)\right] \tag{2.4.12}$$

This relation may be integrated directly to find that

$$\boxed{v = \frac{1}{\rho_0 c} [F(x - ct) - G(x + ct)]} \tag{2.4.13}$$

2.4.2 Interpretation

The portion of the d'Alembert solution that is $p = F(x - ct)$ constitutes a *wave* that propagates in the direction of increasing x. The term "wave" suggests that it is analogous to an ocean waves, in which the elevated water crest seems to be transported along the surface. Here the information that is transported is the pressure value. Recognition that this wave propagates in the direction of increasing x follows from the property that the same F is found at all space-time locations for which $\xi = x - ct$ is constant value, so increasing t must be compensated by an increase in x. Suppose we observe that F has a certain value at location x_1 and time t_1, and we wish to know where that same value of F will be observed at time $t_2 > t_1$. We know that the constant value of the characteristic at x_1, t_1 is $\xi = x_1 - ct_1$, which must also be the value of the characteristic associated with x_2, t_2. It follows that

$$\xi = x_1 - ct_1 = x_2 - ct_2 \implies x_2 = x_1 + c\,(t_2 - t_1) \qquad (2.4.14)$$

In other words the information contained in a value of F is observed at locations whose coordinate x increases proportionally to the elapsed time. The factor of proportionality in such a relation is velocity, and the value of ξ represents the phase of the pressure signal, so $+c$ is the *phase velocity* of the wave described by $p = F\,(x - ct)$. Standard terminology is to say that ξ is the *forward characteristic* because the associated wave transmits the pressure value in the positive x direction with increasing time.

The other part of the d'Alembert solution has a similar interpretation. In this case a specific value of G is observed at two pairs x_1, t_1 and x_2, t_2 if

$$\eta = x_1 + ct_1 = x_2 + ct_2 \implies x_2 = x_1 - c\,(t_2 - t_1) \qquad (2.4.15)$$

In other words the location at which a specific G is observed decreases proportionally to elapsed time, with the factor of proportionality being $-c$. Thus, this solution represents a wave that propagates in the direction of decreasing x, and its phase velocity is $-c$. Correspondingly $\eta = x + ct$ is the *backward characteristic*. The term *phase speed* omits the directionality of the phase velocity. For both waves this quantity is the speed of sound, c.

Sometimes, $p = F\,(x - ct)$ and $p = G\,(x + ct)$ are said to be *simple waves*, because any other wave is more complicated. A more descriptive name is that each is a *progressive plane wave*. This term conveys the fact that they propagate in a single direction, and that there is no variation perpendicular to the direction of propagation. The most specific terminology refers to the two solutions as forward and backward propagating plane waves.

If we were to plot $F\,(x - ct)$ and $G\,(x + ct)$ as functions of x at a specific instant t, each plot would have a shape whose position along the x axis depends on the selected value of t. Such a plot represents a *spatial profile*. In the case of F this shape would translate in the direction of increasing x as time elapses (forward propagation), while the shape of the G plot would translate in the direction of decreasing x (backward propagation). Waves that propagate without changing shape are said to be *nondispersive*. The d'Alembert solution states that plane waves are a superposition (that is, addition) of two nondispersive waves, but the resulting sum does not maintain its shape as time evolves.

One way to visualize the manner in which the waves propagate is to plot lines in an x, t graph that are the locus of all points along which ξ and η are constant. The profiles $F\,(x)$ and $G\,(x)$ of each nondispersive wave at the initial instant, usually $t = 0$, are plotted along the x axis. A three-dimensional plot could use p as the third axis, but we will overlay the profiles onto the x, t plane. Figure 2.3 shows the result when F is parabolic and G is a triangle.

The construction of this picture began by plotting $F\,(x)$ and $G\,(x)$ along the x axis in order to describe the simple waves at $t = 0$. Six values of x are identified as being important. The extent of the forward propagating wave initially is $x_1 \le x \le x_3$ and the maximum value of $F\,(x)$ occurs at x_2. The backward propagating wave is nonzero

Fig. 2.3 Propagation of
spatial pressure profiles
along forward and backward
characteristics in x, t space

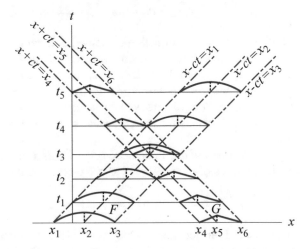
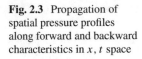

for $x_4 \leq x \leq x_6$, with the maximum $G(x)$ occurring at x_5. (In the sketch, only x_1 is negative, but the whole picture could be shifted farther to the left because the physical domain is considered to be unbounded.) The next step was to construct the characteristics that pass through each of these points. The characteristics are straight lines defined by Eq. (2.4.7), so their slope is $dt/dx = \pm 1/c$, and the constant values of ξ or η for each is the x value at the intersection with the x axis. As shown in the figure, the signals at any subsequent instant are depicted by constructing a line parallel to the x axis at that t, then translating the plot of $F(x)$ and $G(x)$ up to that line along the associated characteristic.

Much insight can be gained from a diagram like Fig. 2.3. At time t_1 the forward propagating wave has passed $x = 0$; after this instant the region $x < 0$ will not be disturbed by this wave. At time t_2 the leading edges (that is, the phase of each wave that is farthest in the direction the wave is propagating) have met. Because the total pressure is the sum of the contribution of each simple wave, we can identify t_2 as the instant when the pressure will be enhanced by their overlap. This is said to be *constructive interference* of the waves. (If F and G have opposite signs, the result would be a *destructive interference*.) We also can see that the waves will no longer interfere when $t > t_4$, because their tails intersect at $t = t_4$. At instant t_3 the characteristics marking the maximum of each simple wave intersect. This is the instant when the maximum total pressure will occur. The last instant indicated in the figure is t_5, which marks the time when the backward propagating wave enters $x < 0$. Thus, $x < 0$ is quiet only in the time interval $t_1 \leq t \leq t_5$.

Figure 2.3 describes the solution to an *initial value problem* for an infinite (one-dimensional) domain. This term stems from the observation that the functions $F(x)$ and $G(x)$ plotted along the x axis represent the simple waves at the initial instant. This is $t = 0$ for the Figure, but it could be any t at which the simple waves are known.

Boundary excitation is another class of problems that can be described by a diagram like Fig. 2.3. This is the situation in which the acoustical response is specified at some location as a function of t. A physical system that would give rise to this condition is a tube in which one end is terminated by a vibrating piston. The medium extends from the boundary to infinity in the x direction. We may define the location of the boundary as $x = 0$, so the fluid's domain is $0 \le x < \infty$. This domain is said to be semi-infinite because it only extends to infinity in one direction. Suppose we know that $p = f(t)$ at $x = 0$. We plot this function along the t axis in the manner of Fig. 2.4. The pressure at each instant propagates into the region $x > 0$ along the forward characteristics that emanates from that point the t axis. There are no backward characteristics emanating from the t axis because they represent propagation into the region $x < 0$, but the fluid does not exist there.

The instants t_1 and t_3 in Fig. 2.4 mark the start and termination of the excitation, and t_2 marks the instant when it is maximum. The diagram shows that the characteristic through t_1 is the front of the forward propagating simple wave, while t_3 marks the tail. The equation of a forward characteristics is $\xi = x - ct$, where the value of ξ is now set by its intersection with the t axis, that is $\xi = -ct_1$, $-ct_2$, and $-ct_3$. It is convenient to rewrite the equation of these characteristics as $t - x/c = t_1, t_2,$ or t_3. Correspondingly, the one-dimensional signal resulting from boundary excitation of a semi-infinite system is

$$ p = f\left(t - \frac{x}{c}\right) \tag{2.4.16} $$

A plot of pressure as a function of time at a specific location is said to be a *waveform*. Figure 2.4 shows that the waveform at any positive x has the same shape as the excitation $f(t)$, but it is delayed by x/c, which is the time required for a specific pressure to propagate there for $x = 0$. This is another way of stating that the waves are nondispersive. But how can we extract from this information a *spatial*

Fig. 2.4 Pressure waveform propagating along the forward characteristic in x, t space

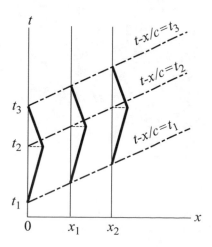

profile, which is p as a function of x at fixed t? A related question for an initial value problem is how can we identify a waveform directly from Fig. 2.3? These questions could be answered graphically by considering the characteristics in those figures to be equal pressure contours in a map. Profiles would be slices parallel to the x axis at any t, and waveforms would be slices parallel to the t axis at any x. Another approach would be to simply evaluate analytical solutions. Section 2.5 will introduce an alternative graphical representation of the signal that simplifies the evaluation of profiles and waveforms.

The best way to understand the description of the particle velocity, Eq. (2.4.13), as well as the overall d'Alembert solution, is to use subscript 1 to designate properties associated with the forward propagating wave (increasing x) and subscript 2 for the backward propagating wave. Then the total acoustic pressure and particle velocity are described by

$$\boxed{\begin{aligned} p &= p_1 + p_2, \quad v = v_1 + v_2 \\ p_1 &= F\,(x - ct)\,, \quad p_2 = G\,(x + ct) \\ v_1 &= \frac{p_1}{\rho_0 c}, \quad v_2 = -\frac{p_2}{\rho_0 c} \end{aligned}} \tag{2.4.17}$$

Recall that v is the particle velocity component in the x direction, so negative v means that a particle is moving in the sense of decreasing x. Thus, the particle velocity corresponding to a positive pressure in each simple wave is in the same sense as the direction in which that wave propagates. The magnitude of that particle velocity is proportional to the magnitude of the pressure. The factor of proportionality, $\rho_0 c$, is the *characteristic impedance* of the fluid. (The ratio of a force variable to a velocity variable is generally referred to as an impedance.) The dimensionality of the characteristic impedance are $F\text{-}T/L^3 = M/(L^2 T)$. In SI units 1 kg/(m^{2}s) is defined to be 1 Rayl in recognition of Rayleigh's many contributions to the science of acoustics.[3]

EXAMPLE 2.3 The pressure field in an infinite domain for $t < 0$ consists of a forward propagating sinusoidal wave in $x < 0$ and a backward propagating sinusoidal wave in $x > 0$. They are timed such that their leading edges both arrive at $x = 0$ when $t = 0$. They are described as

$$p = \begin{cases} -B \sin\left(\dfrac{2\pi\xi}{L}\right) h\,(-\xi)\,, \ \ \xi = x - ct, \ \text{if } x < 0 \,\&\, t < 0 \\[2ex] B \sin\left(\dfrac{2\pi\eta}{L}\right) h\,(\eta)\,, \ \ \eta = x + ct, \ \text{if } x > 0 \,\&\, t < 0 \end{cases}$$

---

[3] His name is John William Strutt, third Baron Lord Rayleigh. The numerous references herein are a small sample of his significant contributions to acoustics. He also was a pioneer in optics, and won the Nobel Prize in 1924 as the codiscoverer of argon.

where $h(z)$ is the step: one if $z > 0$ and zero if $z < 0$. The first task is to construct spatial profiles of pressure at $t = 0$, $1.5L/c$, and $2.25L/c$. Then use those profiles to determine the maximum possible pressure, as well as the instants and locations where that maximum will occur.

Significance

Using characteristic lines in x, t space as the framework will be seen here to be quite useful for tracking how waves evolve.

Solution

The field in $x < 0$ for $t < 0$ is p_1 in Eq. (2.4.17), and the field in $x > 0$ for $t < 0$ is p_2. For $t > 0$ these waves overlap, so we must evaluate their interference pattern. The first step is to plot in an x, t diagram p_1 and p_2 at $t = 0$. At that instant $\xi = \eta = x$, and $h(-x) = 1$ if $x < 0$. Thus, the pressure profile at $t = 0$ for $x > 0$ is a sine curve with amplitude B and wavelength L that starts at $x = 0$. It is a negative sine function for $x < 0$ with the same amplitude and period. As can be seen in Fig. 1, the combination is an even function of x with a slope discontinuity at $x = 0$. This is the location of both leading edges.

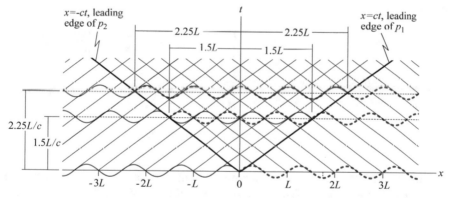

Figure 1. Spatial profiles of pressure showing interference between forward and backward propagaing waves.

For $t > 0$, both waves continue without modification. We draw the characteristics that intersect the origin in order to track the leading edge of each wave. Other features that we could track with characteristic lines are the zeros, maxima and minima. Showing all of these would lead to a rather dense set of lines; showing only the characteristics along which p_1 or p_2 is zero is adequate. Each of these emanates from a location on the x axis at which their profile has a zero.

The next step is to construct the profile of each simple wave at $t = 1.5L/c$. This is done by drawing a line parallel to the x axis at $t = 1.5L/c$. The initial p_1

profile is translated to this line and shifted in the direction of increasing x by $1.5L$, which is the distance the wave propagates in that interval. Thus, the p_1 profile at this instant extends backward from the leading edge at $x = 1.5L$. The p_2 profile is treated similarly, with the only difference being that it is shifted in the direction of decreasing x by $-1.5L$. Figure 1 shows that the zeros of p_1 and p_2 overlap for $-1.5L < x < 1.5L$. In this interval p_1 and p_2 have equal magnitude, but opposite sign. Outside this interval, there is only p_2 for $x > 1.5L$, and only p_1 for $x < -1.5L$. The total pressure is the superposition of the two simple waves, $p = p_1 + p_2$, so $p = 0$ in the region of overlap, and p_1 or p_2 is the pressure outside that interval.

The profile at $t = 2.25L/c$ is treated in the same way. We construct a line parallel to the x axis at $t = 2.25L/c$ and shift the p_1 and p_2 profiles up to that line. Relative to their positions at $t = 1.5L/c$, the p_1 profile is shifted in the positive x direction by $0.75L$, which is the distance p_1 travels in the elapsed time subsequent to $t = 1.5L/c$. Similarly, the p_2 profiles is shifted relative to its earlier position by $0.75L$ in the direction of decreasing x. The region in which p_2 and p_2 overlap is the interval between the leading edges. Now it is $-2.25L < x < 2.25L$. The profiles of p_1 and p_2 are the same in this interval, so the total pressure is twice that in either profile. Inspection of the graph shows that each profile at this instant fits $B \cos(2\pi x/L)$. For x values left of the leading edge of p_2, there is only the p_1 profile, while for x values to the right of the leading edge of p_1 there is only the p_2 profile. These considerations allow us to construct profiles of the total pressure at both instants, which are shown Fig. 2.

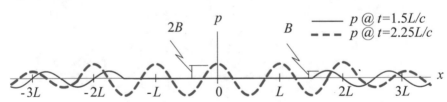

Figure 2. Selected pressure profiles.

The profile at $t = 2.25L/c$ gives a hint regarding when and where the pressure attains a maximum value. The maximum value of either simple wave is B, so the maximum possible value of their superposition is $2B$. It follows that $p = 2B$ whenever and wherever a ξ characteristic tracking $p_1 = B$ intersects an η characteristic tracking $p_2 = B$. At $t = 0$, the peak values of p_2 occur at $x = L/4, 5L/4, \ldots$, and the peak values of p_1 occur at $x = -L/4, -5L/4, \ldots$ Furthermore, at $t = 0$, we know that $\xi = \eta = x$, so we are led to the conclusion that $p = 2B$ if

$$\eta = x + ct = \left(n + \frac{1}{4}\right)L \text{ and } \xi = x - ct = -\left(m + \frac{1}{4}\right)L$$

The simultaneous solution of these conditions is

$$t = \left(\frac{n+m}{2} + \frac{1}{4}\right)\frac{L}{c}, \quad x = \left(\frac{n-m}{2}\right)L$$

Both n and m are arbitrary and non-negative, so $n + m$ may be any integer $j \geq 0$. The value of t is set by assigning j, in which case $n = j - m$. If follows from the preceding conditions that

$$p = 2B \text{ when } t = \left(\frac{2j+1}{4}\right)\frac{L}{c} \text{ at } x = \left(\frac{j-2m}{2}\right)L$$

Thus we are led to the realization that the maximum pressure is observed at odd multiples of $L/(4c)$. The locations at which $p = 2B$ at these instants are separated by L in both directions, alternating between starting at $x = 0$ and at $x = \pm L/2$, depending on j. In any case, the location at which a maxima occurs must fall in the range where p_1 and p_2 overlap, which always is $-ct < x < ct$.

It should be evident from Figure 1 that the minimum value of p is $-2B$, and that it occurs at the same instants that the maxima occur. Furthermore, the minima occur at positions that are halfway between the maxima. It is left to the reader to verify that $p = 0$ in the interval $-ct < x < ct$ when t is an integer multiple of $L/(2c)$. Another interesting observation, which is somewhat more difficult to prove by viewing profiles and characteristics, is that $p = p_1 + p_2 = 0$ at any x that is an odd multiple of $L/4$ in the region of overlap, regardless of t.

It actually is easier to demonstrate each of these features mathematically with the aid of a trigonometric identity. In the region where p_1 and p_2 overlap, the total pressure is

$$p = p_1 + p_2 = B\left[\sin\left(\frac{2\pi(x+ct)}{L}\right) - \sin\left(\frac{2\pi(x-ct)}{L}\right)\right]$$

$$\equiv 2B\cos\left(\frac{2\pi x}{L}\right)\sin\left(\frac{2\pi ct}{L}\right), \quad -ct \leq x \leq ct$$

Thus, $|p| = 2B$ when $2\pi ct/L = \pi/2, 5\pi/2, ...$at $2\pi x/L = 0, \pm 2\pi, ...$ and when $2\pi ct/L = 3\pi/2, 7\pi/2, ...$ at $2\pi x/L = \pm\pi, 3\pi, ...$ We also see that $p = 0$ when $2\pi ct/L = 0, \pi, .$ regardless of x, and at $2\pi x/L = \pm\pi/2, \pm 3\pi/2...$ regardless of t. Each of these conditions is the same as the one identified from the graphical construction.

The dependence of pressure on x and t has many attributes that we will again encounter in our later studies of waveguides that are closed. The signal in such systems will be found to be a standing wave field.

2.4.3 Harmonic Waves

A large part of Chap. 1 was devoted to signals that vary harmonically in time, so it is helpful to consider a plane wave that varies harmonically. Because the d'Alembert solution is a sum of forward and backward propagating waves, it must be that a harmonic signal can only be obtained if each simple wave also is harmonic. Let us consider a forward propagating wave by itself. We know that it must fit $p = F(\xi)$ with $\xi = x - ct$, but the coefficient of t must be the frequency ω. The appropriate form is obtained by multiplying ξ by a constant k, so a sine wave is described as

$$p = A \sin(k(x - ct)) \tag{2.4.18}$$

The coefficient of t is the frequency if $kc = \omega$, in other words,

$$\boxed{k = \frac{\omega}{c}} \tag{2.4.19}$$

This parameter is called the *wavenumber* for reasons that will be discussed. The ratio ω/c arises in the analysis of every type of harmonically varying wave. Its designation as the symbol k is nearly universal.

The profile of a sine wave at an arbitrarily selected t appears in Fig. 2.5a, while the waveform in Fig. 2.5b views the same signal as it would be measured at an arbitrary x. The amplitude A is the maximum absolute pressure in the signal. The spatial interval over which the pattern repeats is the *wavelength* λ, in contrast to the period T, which gives the amount of time that must elapse for the pattern to repeat.

Fig. 2.5 A typical harmonic wave; **a** Spatial profile, **b** Temporal waveform

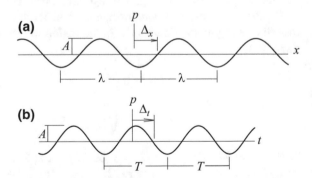

Neither the wavelength nor the period nor the frequency appear explicitly in the basic representation of Eq. (2.4.18), but substitution of the definition of k gives an alternate form

$$\boxed{p = A \sin(kx - \omega t)} \tag{2.4.20}$$

The period T is the time required for the phase of the signal to change by 2π, so

$$\boxed{T = \frac{2\pi}{\omega}} \tag{2.4.21}$$

The wavelength is the distance by which the spatial location x must change (at a fixed t) for a 2π phase change. Thus, we set $k\lambda = 2\pi$.

$$\lambda = \frac{2\pi}{k} \qquad (2.4.22)$$

It can be seen that k plays the same role for the spatial dependence of a harmonic wave as that played by ω for frequency dependence. Just as $\omega/(2\pi) \equiv 1/T$ gives the number of cycles that occur in a unit time interval, $k/(2\pi) \equiv 1/\lambda$ is the number of cycles per unit length. Indeed, a useful alternative form relates the wavelength to the cyclical frequency as

$$\lambda = \frac{c}{f} \qquad (2.4.23)$$

Although the number of wavelengths per unit length is the one that is meaningful from a physical viewpoint, the quantity k is the one that is used for mathematics. For that reason we refer to k, rather than $k/(2\pi)$, as the wavenumber.

The other parameters that appear in Fig. 2.5 but not in Eq. (2.4.18) are the spatial shift Δ_x and time shift Δ_t. To interpret these parameters, we observe that the profile at $t = 0$ is $p(x, 0) = A \sin(kx)$, whereas $p(x, t) = A \sin(k(x - \Delta_x))$ in Fig. 2.5a. Similarly, the waveform at $x = 0$ is $p(0, t) = -A \sin(\omega t)$ whereas $p(x, t) = -A \sin(\omega(t - \Delta_t))$. A comparison of each form to Eq. (2.4.20) might seem to imply that $k\Delta_x = \omega t$ and that $\omega\Delta_t = kx$. Both relations are erroneous because they fail to account for the periodic nature of the signal. In the case of a profile, any point on the curve where the sine function passes through zero and is increasing might correspond to a profile that was a positive sine function at $t = 0$. Similarly, any point on the waveform at which the sine is zero and decreasing could correspond to a waveform that was a negative sine function at $x = 0$. Thus, we only can deduce from Fig. 2.5 that

$$\Delta_x = ct \pm m\lambda$$
$$\Delta_t = x/c \pm nT \qquad (2.4.24)$$

where m and n are integers.

There would be no ambiguity about the amount by which the profile at $x = 0$ is advanced at $t > 0$, and by how much the waveform at $t = 0$ is retarded at $x > 0$ if we see the leading edge of the wave when we begin to observe it. If the leading edge is at the origin when $t = 0$, then we could say that the shifts are $\Delta_x = ct$ and $\Delta_t = x/c$.

2.5 The Method of Wave Images

Although we have obtained the most general form of a plane wave, it is of little use until we can determine the F and G functions. Here we will adapt the d'Alembert

solution to fit several situations in which initial and/or boundary conditions are
imposed. We will begin by establishing how to determine F and G in the case
of an initial value problem for an infinite domain. The next step will develop a relatively simple way to identify spatial profiles at an arbitrary instant, and waveforms
at an arbitrary location. This capability will be the fundamental tool for extension
to systems with one boundary (a semi-infinite domain), and then two boundaries (a
finite domain). In each case the development will lead to an equivalent representation
of the basic solution as the result of an initial value problem in an infinite domain.
Each solution will exemplify basic acoustical phenomena, including reflections, dissipation, reverberation time, normal modes, standing waves, and resonance. The
phenomena occur in a variety of systems, some of which are multidimensional. The
only limitation is that the analyses assume that the boundary has special mechanical
properties. Despite this shortcoming, the ability to follow the evolution of a signal,
and the relatively low level of mathematics required to implement an analysis, enable
us to understand at a fundamental level why various phenomena occur.

2.5.1   Initial Value Problem in an Infinite Domain

Determination of the Wave Functions

Our analyses using characteristics began with the assumption that the F and G functions are known. The key observation here is that there are two functions to identify,
so knowledge of the initial pressure would be inadequate. However, if we know F
and G we can determine the initial velocity. The corollary is that it should be possible
to determine $F(x)$ and $G(x)$ if we know the initial pressure and particle velocity
profiles. Let $P(x)$ and $V(x)$ respectively denote these profiles. It is convenient to
let $t = 0$ be the initial instant.

There are several situations in which a plane wave is encountered. The most
common is a straight tube whose cross section does not change along its length.
An initial temperature distribution can be generated by differential heating along
the length of the tube. If such a distribution is maintained sufficiently long, thermal
equilibrium over the cross section at any x could be achieved, thereby setting up
an initial pressure field $P(x)$. Generation of an initial velocity field $V(x)$ is much
more difficult. One possibility is that an electromagnetic field induces movement of
a magnetic fluid. However, a much greater range of options is obtained by changing
our perspective. Suppose that some excitation had been applied well before $t = 0$,
at which time the excitation ended. If we begin our observation of such a system
at $t = 0$, we would only know the current physical state, which is the pressure and
particle velocity.

Equations (2.4.17) describe the response at any instant. Matching p and v at $t = 0$
to the initial conditions requires that

$$F(x) + G(x) = P(x), \quad F(x) - G(x) = \rho_0 c V(x) \qquad (2.5.1)$$

which leads to

$$
\boxed{
\begin{aligned}
F(x) &= \frac{1}{2}[P(x) + \rho_0 c V(x)], \quad G(x) = \frac{1}{2}[P(x) - \rho_0 c V(x)] \\
p_1 &= F(x - ct), \quad p_2 = G(x + ct), \quad p = p_1 + p_2
\end{aligned}
}
\tag{2.5.2}
$$

An interesting aspect of this result is that it is possible to set up initial conditions that will induce only one of the two simple waves. For example, if $\rho_0 c V(x) = -P(x)$, then only p_2 will be generated. The reason for this is that the condition $p = -\rho_c c v$ is exactly how the backward propagating wave behaves, so the initial conditions in this case are essentially a snapshot of a backward propagating wave that was generated prior to $t = 0$.

Evaluation of Spatial Profiles and Temporal Waveforms

It is assumed henceforth that the functions $F(x)$ and $G(x)$ have been determined. Now the task is to convert the mathematical description of plane wave in Eqs. (2.4.17) and (2.5.2) into spatial profiles of p as a function of x at fixed time, and temporal waveforms depicting p versus t at a specified x. One approach is to evaluate Eq. (2.5.2) directly with mathematical software. To evaluate profiles a sequence of x_n would be defined. The corresponding set of characteristics at any instant are $\zeta_n = x_n - ct$ and $\eta_n = x_n + ct$. Then the pressure at each location is $p_n = F(\zeta_n) + G(\eta_n)$. Similarly a waveform at any x may be found by defining a set of time values t_n then evaluating the corresponding characteristic values. Alternatively, computer graphic capabilities can create a three-dimensional view that shows spatial profiles and waveforms. If x_n and t_j are a set of position and time values, then matrices of characteristic values have elements $\xi_{j,n} = x_n - ct_j$ and $\eta_{j,n} = x_n + ct_j$. Evaluation of the pressure at these values of the characteristics gives a matrix $p_{j,n} = F(\xi_{j,n}) + G(\eta_{j,n})$ in which column n is the waveform at x_n and row j is the profile at t_j. Figure 2.6 is typical of the pictures that can be constructed.

The initial pressure distribution described by Fig. 2.6 is parabolic over an interval of length L. If the initial velocity was such that $\rho_0 c V(x)$ equals $P(x)$, then only a forward propagating wave would be generated, whereas waves in both direction at equal amplitude would be generated if $V(x)$ is zero. The function described here is half way between these alternatives. The result is that the forward propagating wave is three times stronger than the backward wave.

A picture like Fig. 2.6 is useful when the forward and backward waves are distinct. However, it is less apparent that there are distinct waves in the region of x, t where the simple waves overlap, which is $0 < x < L, 0 < ct < L/2$ in the case of the Figure. In each of the topics that we will consider there will be significant contribution from oppositely traveling simple waves. We need a better way of viewing the details of mutual contributions. The general concept to be developed here entails propagating each simple wave separately to identify how each contributes to a profile or waveform. The method will describe procedures in terms of graphical constructions, because such views give a "hands on" experience, but it also provides a foundation for computations.

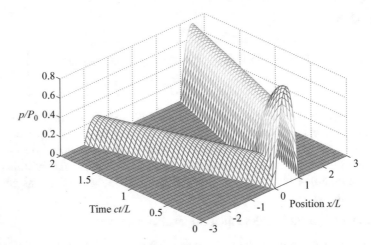

Fig. 2.6 Evolution of a plane wave generated by initial conditions in which $P(x) = 4P_0(x/L)(1 - x/L)$ for $0 < x < L$ and $\rho_0 c V(x) = P(x)/2$

We begin with spatial profiles. A graphical construction begins by drawing $p_1 = F(x)$ and $p_2 = G(x)$. As we saw when the characteristics were plotted in an x, t diagram, a plot of p_1 versus x at some $t > 0$ shows the function $F(x)$ translated, without changing shape, in the positive x direction by the propagation distance ct. Similarly p_2 versus x at that t would be the plot of $G(x)$ translated in the negative x direction by the same propagation distance ct. Thus, the range of x over which the initial profiles are sketched must be sufficiently large that shifting them for a later time does not leave a gap in which no value was computed. After the p_1 and p_2 profiles have been determined, the total pressure is the sum of p_1 and p_2 at a specific x. This sum may be done graphically by adding the height of each profile at a specific x if both waves have the same sign, or taking their difference if their values have opposite signs. The graphical construction is illustrated in Fig. 2.7.

In Fig. 2.7b for time t', the plot of p_1 is $F(x)$ shifted to the right (increasing x) by ct', and the plot of p_2 is $G(x)$ shifted to the left (decreasing x) by ct'. Both $F(x)$ and $G(x)$ were selected as connected straight lines because doing so facilitates the graphical addition of p_1 and p_2. The sum of two functions that are linear in x also is linear in x. Consequently, p also is a set of connected straight lines. The locations where these lines change slope are those where p_1 and/or p_2 change slope. Drawing the profile of p requires that we sum the heights of p_1 and p_2 at these discontinuities. The profile (or waveform) then is obtained by connecting the plotted points with straight lines.

Graphical construction of waveforms at a specific x begins by evaluating the waveform at $x = 0$. At this location, the d'Alembert solution states that

$$p_1(0, t) = F(-ct), \quad p_2(0, t) = G(ct) \tag{2.5.3}$$

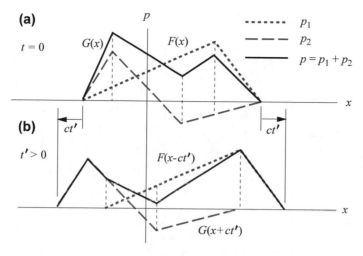

Fig. 2.7 Propagation of pressure profiles as time elapses according to the method of wave images

Let us scale the time axis by c, so that a plotted point is p_1 or p_2 along the ordinate and ct along the abscissa. Then the waveform of p_2 at $x = 0$ has the same appearance as the profile of p_2 at $t = 0$. In contrast, the argument of $F(-ct)$ has the opposite sign from that of $F(x)$. This means that a waveform plot of p_1 at $x = 0$ would assign at each point ct on the abscissa the value of the F function when x is replaced by $-ct$. Consequently, the waveform of p_1 at $x = 0$ appears to be the result of swapping the left and right sides of $F(x)$ relative to the origin. An alternative view is that the p_1 waveform at $x = 0$ appears to be the mirror image of the p_1 profile at $t = 0$ when the mirror is placed at $t = 0$. For this reason the construction methodology shall be referred to as the *method of wave images*. Figure 2.8a shows the construction for the same functions as those in the Fig. 2.7.

To construct a waveform at some location $x_1 > 0$, we recall that constancy of $F(x - ct)$ is obtained if the value of ct is increased by the value of x_1. In other words the waveform of $p_1(t)$ at $x_1 > 0$ is the waveform of p_1 at $x = 0$, shifted forward along the ct axis by distance x_1. An equivalent description is that p_1 is delayed by x_1/c relative to the waveform at $x = 0$. For p_2, we track a constant value of $G(x + ct)$, which requires that the value of ct be decreased by x. Thus, the waveform of p_2 at $x_2 > 0$ is the p_2 waveform at $x = 0$, shifted backward along the ct axis by distance $-x_2$. This corresponds to an advancement in time by x_2/c. The construction is depicted in Fig. 2.8b. It will be noticed that although the waveforms of p_1 and p_2 are depicted for $t < 0$, they have been added only for $t \geq 0$, because $t = 0$ is the initial instant when we began to observe the signal.

The evaluation of a waveform for $x_2 < 0$ follows the some line of reasoning. Constancy of $x - ct$ for p_1 requires that ct be decreased by the amount that x_2 is negative. Thus, the waveform of p_1 at $x_2 < 0$ is obtained by translating the p_1 waveform at $x = 0$ backward along the ct axis by distance $-x_2$. Similarly, the p_2

128   2 Plane Waves: Time Domain Solutions

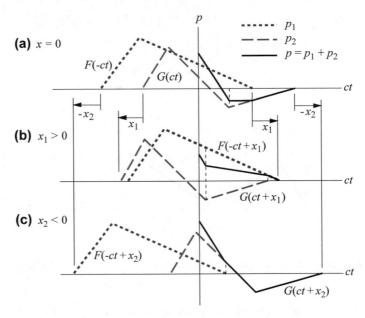

Fig. 2.8 Construction of pressure waveforms at various locations according to the method of wave images

profile at $x_2 > 0$ is shifted forward along the ct axis by $-x_2$. This construction is depicted in Fig. 2.8c, where the total pressure is evaluated only for $t \geq 0$.

EXAMPLE 2.4 Consider the initial forward and backward waves specified in Example 2.3. (a) Construct the spatial profile of the pressure and particle velocity at $t = 0.9L/c$. (b) Construct the waveform of pressure versus time at x locations where the maximum pressure is observed.

Significance

Construction of spatial profiles and temporal waveforms from the initial conditions of a wave will always be a cornerstone of the wave image method, even after it has been extended to treat systems with boundaries. One benefit of using the method of images to solve a problem that was previously analyzed by the method of characteristics is enhanced understanding of both approaches.

Solution

The initial profiles of p_1 and p_2 shown in the first graph of Fig. 1 are the same as in Example 2.3. At $t = 0.9L/c$, the p_1 wave will have propagated forward by $ct = 0.9L$, so its profile is obtained by shifting the initial p_1 profile in the positive

x direction by $0.9L$. Similarly, the p_2 wave is shifted in the negative x direction by $0.9L$. The individual profiles at this instant are shown in the second graph of Fig. 1.

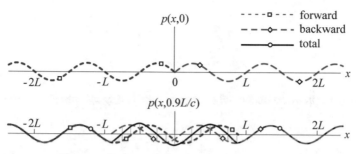

Figure 1. Spatial profiles of pressure at two instants.

A graphical construction was used to describe $p = p_1 + p_2$. Vertical lines were drawn from the x axis to p_2 at a sequence of x values where p_1 and p_2 overlap. Then these lines were translated to the p_1 curve, above it if p_2 is positive or below it if p_2 is negative. The ends of these vertical lines mark p at each location. Before these x, p points were connected, a little mathematics was used. First, it can be seen that because p_1 and p_2 are mirror images, and they translate by equal distances in opposite directions as time elapses, it must be that their sum is an even function of x, that is, $p(-x, t) = p(x, t)$. It follows that after the leading edges move away from $x = 0$, it must be that $\partial p / \partial x = 0$ at $x = 0$. At this juncture, the second basic property comes into play. Both p_1 and p_2 are spatially periodic in an interval L. Furthermore, both p_1 and p_2 are harmonic functions, so their sum must also be harmonic. A harmonic function having zero slope at the origin is a cosine. (This property was identified in Example 2.3 by the application of a trigonometric identity.) Correspondingly, the x, p points that were constructed were connected to resemble a cosine curve. This curve extends over the region where p_1 and p_2 overlap, which is the interval between their leading edges at this instant, $-0.9L < x < 0.9L$. For $x < -0.9L$, the total pressure consists of only p_1 because p_2 has not yet propagated back to that region. Similarly, for $x > 0.9L$ the pressure is p_2 because p_1 has yet to reach that region. Notice that because p_1 and p_2 are both zero at their leading edge, the pressure profile is continuous everywhere.

Now let us turn our attention to waveforms at locations where the maximum pressure is observed. Even without having considered these waves in a previous example, it should be apparent that if lobes of p_1 and p_2 versus x overlap, the maximum pressure $2B$ will be attained. We could draw a few diagrams to identify all possible distances s by which the p_1 and p_2 profiles should be shifted in opposite directions to bring them into alignment. Instead we shall pursue a more formal method. Let us denote as x_j, $j = 1, 2, ...$, the locations at which $p_1 = B$ when $t = 0$. Similarly, x_n, $n = 1, 2, ...$ denotes locations where $p_2 = B$ at $t = 0$. Both sets of locations are readily identified from the first plot in the preceding diagram.

At an arbitrary time t, the x_j values will have increased by ct, and the x_n values will have decreased by that amount. Thus, these locations are given by

$$x_j = -\frac{L}{4} - (j-1)L + ct$$

$$x_n = \frac{L}{4} + (n-1)L - ct$$

The maximum pressure occurs when *any* x_j equals *any* x_n, which occurs when

$$ct = \frac{L}{4}(2j + 2n - 3)$$

Because j and n are any positive integer, the maxima occur when $ct = L/4, 3L/4, 5L/4, \ldots$ Because ct is an odd multiple of $L/4$, the x_j and x_n values are integer multiples of $L/2$. Thus, we conclude that the maxima occur at $x = 0, \pm L/2, \pm L, \ldots$.

Construction of the waveform at $x = 0$, from which the other waveforms are derived, is described in the first graph of Fig. 2. The waveform of p_1 is obtained by reflecting the initial profile of p_1 about the vertical axis through $x = 0$, then relabeling the axis from x to ct. The waveform of p_2 at $x = 0$ is the same as the initial p_2 profile, with the axis relabeled from x to ct. The total pressure is the sum of p_1 and p_2, which can be seen to be twice the contribution of either simple wave.

The second graph in Fig. 2 describes the construction for $x = L/2$ which is the next location where a maximum pressure is observed. The p_1 profile is retarded by shifting it to the right by $\Delta(ct) = L/2$, and the p_2 profile is advanced by shifting it to the left by the same amount. The leading edge for each simple wave is marked by the symbol $\otimes$. These are added to obtain p as a function of t at $x = L/2$.

The construction for $x = -L/2$, which appears in the third graph of Fig. 2, is reversed. Because p_1 is traveling in the positive x direction, it arrives earlier at negative x locations, so we shift the initial profile of p_1 backward (left) along the ct axis by distance $\Delta(ct) = |x|$. Similarly, p_2 is shifted forward (right) along the ct axis by the same distance. It will be noted that the total pressure is the same at $x = -L/2$ and $x = L/2$. This is a consequence of the fact that the initial profiles are symmetric relative to $x = 0$, and the propagation of each wave is a symmetric image of the other. Thus, the same response should be observed at $+x$ and $-x$, so only waveforms at positive x need be considered. The last two graphs of Fig. 2 show the waveforms at larger x where a pressure maximum occurs. Each is obtained by the same procedure that constructed the waveform at $x = L/2$, except that the time by which the p_1 and p_2 profiles are advanced is greater. It is worth noting that the locations at which the maximum pressure is observed also are the locations of minimum (that is, largest negatively) pressure. This is a direct consequence of the overlapping of p_1 and p_2.

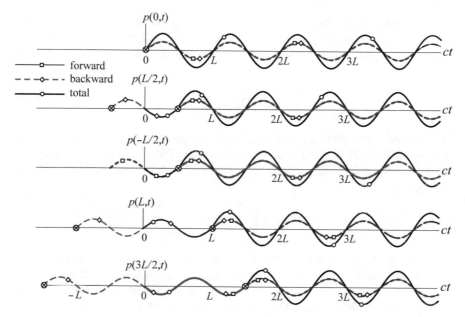

Figure 2. Waveforms at several locations.

2.5.2 Plane Waves in a Semi-infinite Domain

Velocity Excitation

We now turn our attention to a semi-infinite domain in which there is an excitation at the near end, which is defined to be $x = 0$. It is possible that the excitation is a specification of the pressure as a function of time. This case was analyzed by the method of characteristics, which led to Fig. 2.4. However, the usual situation is an imposed velocity at the boundary. A prototype for this condition appears in Fig. 2.9, where there is a piston at one end of a tube.

Fig. 2.9 Continuity of particle velocity at a moving boundary

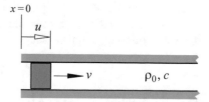

No gaps can open up between the face of the piston and the fluid because the gap would be a vacuum that violates the continuity condition. In the Figure the face of the piston that is exposed to the fluid is at $x = 0$ when the system is in its reference position. The displacement of the piston relative to this position is $u(t)$, which is a

specified function of time. Thus, the instantaneous position of the exposed face is
$x = u$, and the velocity of the piston is du/dt. Continuity at the face of the piston
requires that du/dt equal the fluid's particle velocity v at the interface between the
media, so

$$v\,(x,t)|_{x=u} = \frac{du}{dt} \tag{2.5.4}$$

This statement constitutes a *moving boundary condition*. This type of boundary
condition is difficult to satisfy because the location of the boundary is not constant.
Let us consider simplifying it based on the notion that u probably is small. A Taylor
series gives

$$v\,(x,t)|_{x=0} + u\,(t) \left[\frac{\partial}{\partial x} v\,(x,t) \right]\Bigg|_{x=0} + \cdots = \frac{du}{dt} \tag{2.5.5}$$

We have seen that a linearized analysis requires that $|p| / (\rho_0 c^2)$ be small, and we
know that for each of the simple plane waves $|p_n| = \rho c\,|v_n|$, so it follows that $|v| / c$
and $|p| / (\rho_0 c^2)$ have comparable small orders of magnitude. The above boundary
condition then indicates that v and du/dt have comparable orders of magnitude. If
follows that u is indeed very small, so the second term in the Taylor series expansion
of the boundary condition may be dropped. **An overdot is standard notation for a
time derivative**, so the linearized boundary condition is

$$\boxed{v\,(x,t)|_{x=0} = \dot{u}} \tag{2.5.6}$$

In other words, because the displacement of the boundary is small, it is adequate to
consider fluid particles to contact the wall at the wall's reference position. This is an
approximation that will be seen to be fundamental to any linearized description of
the way a signal in a fluid is affected by a solid boundary.

Implicit to the definition of $\dot{u}$ is the specification that it is zero for $t < 0$, which may
be enforced with a factor $h\,(t)$. It is not difficult to determine the one-dimensional
plane wave induced by the movement of the boundary. The discussion of Fig. 2.4
explained that only forward propagating waves are permissible. A backward propagating wave would have to originate a long time earlier at a great distance from the
boundary, in violation of the principle of causality. Thus, we discard the p_2 wave in
Eq. (2.4.17), so that $v = p_1/\rho_0 c$, and $p_1 = F\,(x - ct)$. Continuity of velocity at the
fluid-wall interface requires that

$$F\,(-ct) = \rho_0 c \dot{u}\,(t)\,h\,(t) \tag{2.5.7}$$

The step function, which is $h\,(t) = 0$ if $t < 0$ and $h\,(t) = 1$ if $t > 0$, as been
appended to $\dot{u}$ to emphasize that the piston motion is zero prior to $t = 0$, which is
when we began to observe the system.

Although the description of the F function has been derived by considering the
boundary condition at $x = 0$, it must be true everywhere because ξ is the only

variable on which F can depend. At $x = 0$, we have $\xi = -ct$, so we replace t on the right side with $-\xi/c \equiv t - x/c$. The result is that the pressure wave is given by

$$p = \rho_0 c \dot{u}\left(t - \frac{x}{c}\right) h\left(t - \frac{x}{c}\right) \tag{2.5.8}$$

This expression states that the pressure is like the velocity of the wall multiplied by the characteristic impedance of the fluid, $\rho_0 c$, and delayed by the propagation time x/c.

Let us consider constructing spatial profiles of this wave. We temporarily ignore the fact that the region $x < 0$ is not physically meaningful because it is behind the wall. At $t = 0$, the signal is $p = \rho_0 c \dot{u}\left(-x/c\right) h\left(-x/c\right)$. The quantity $\dot{u}\left(-x/c\right)$ is the function $\dot{u}\left(t\right)$ evaluated at time $-x/c$. This suggests that the initial profile may be constructed by a procedure that is reminiscent of the one used for the initial value problem in an infinite domain. First we plot the waveform at $x = 0$ as p versus ct, that is, we evaluate p at a specified t and use ct as the abscissa. This is Fig. 2.10a. The initial profile is then obtained by swapping this waveform to the other side of $t = 0$ and changing the abscissa's label from ct to x. Doing so produces Fig. 2.10b. In Fig. 2.10c the waveform at $t > 0$ is constructed. This entails translating to the right the profile in Fig. 2.10b by ct. The portion of the profile that lies in $x < 0$ is blocked out because the fluid is situated in $x > 0$ only.

Fig. 2.10 Conversion of a velocity versus time boundary condition to a spatial pressure profile

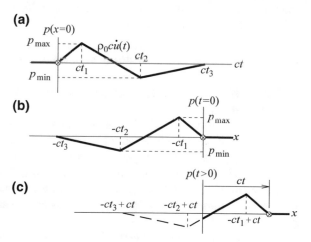

From one perspective it is unnecessary to consider a graphical construction because Eq. (2.5.8) is not difficult to evaluate. The reason we have done so is that the overall objective is to convert all plane wave solutions to an equivalent form defined by an initial profile for an infinite domain. Such a description allows us to employ the techniques in Figs. 2.7 and 2.8 to construct spatial profiles and temporal waveforms in any situation.

Reflection of a Simple Wave

A boundary that has an imposed motion is said to be active, because some outside energy source, like a motor, is required to force that motion to occur. A passive boundary also can move, but it does so in response to the pressure that is applied to it. Thus, the piston in Fig. 2.9 was active, because it was forced to move. If we replace that force with a mechanical spring or dashpot (a device that generates a resistance force that is proportional to velocity), or consider the inertia of the piston, we would have a situation where the piston's velocity $\dot{u}$ depends on the pressure. Our consideration here is limited to the dashpot model of resistance, which states that

$$p|_{x=0} = -\rho_0 c \zeta \dot{u} \tag{2.5.9}$$

The quantity ζ is the called the *specific impedance* because it describes a ratio of pressure to velocity, with "specific" connoting that it is a factor multiplying some reference value, in this case the characteristic impedance of the fluid. The minus sign is inserted into the boundary condition because a positive pressure acting on the boundary pushes the surface in the direction of decreasing x, so it should produce a negative value of $\dot{u}$. The last element of the boundary condition is the continuity requirement, which is the same as the condition in the previous section, where $\dot{u}$ was an imposed motion. In both cases no gap can open up at the interface between the fluid and the boundary. Previously $\dot{u}$ was specified, but it must equal the particle velocity. Thus, the condition to be satisfied is

$$\boxed{p|_{x=0} = -\rho_0 c \zeta \, v|_{x=0}} \tag{2.5.10}$$

This relation is extremely idealized, because all solid surfaces have some resistance features that are elastic (proportional to displacement u) and inertial (proportional to acceleration $\ddot{u}$). Those situations are excluded because the model of a velocity resistance we have adopted will lead to algebraic boundary conditions, whereas any generalization beyond that would lead to boundary conditions that are ordinary differential equations. It is quite difficult, (but not impossible) to obtain time domain solutions for such general conditions. In contrast, restriction to a velocity resistance makes accessible many phenomena that are also encountered in more general cases.

Before we address the response of specific systems, it is useful to examine generally how the presence of an impedance boundary affects a simple wave. We consider a domain $x > 0$ in which a p_2 wave propagates backward until it encounters the boundary at $x = 0$. We say that this wave is *incident* on the boundary. As is true for light waves, a wave is reflected from the surface. This is a p_1 wave because it propagates away from the wall in the direction of increasing x. It follows that p_1 and p_2 must behave according to Eq. (2.4.17), but in the present scenario only $G(x)$ is known, and we must determine $F(x)$.

Substitution of the descriptions of pressure and particle velocity in Eq. (2.4.17) into (2.5.10) gives

$$F(-ct) + G(ct) = -\zeta[F(-ct) - G(ct)], \quad t > 0 \tag{2.5.11}$$

We began with the assumption that we know $G(x)$, but this applies only for $x > 0$ because $x < 0$ is not part of the domain. Hence we know $G(ct)$ for positive t only, which is the reason this range has been specified in the above. We solve this relation for $F(-ct)$, which yields

$$F(-ct) = RG(ct), \quad t > 0 \tag{2.5.12}$$

where R is the reflection coefficient,

$$\boxed{R = \frac{\zeta - 1}{\zeta + 1}} \tag{2.5.13}$$

To understand Eq. (2.5.12) recall that the time dependencies of p_1 and p_2 at $x = 0$ are described by $F(-ct)$ and $G(ct)$, respectively. Thus, the reflection coefficient is a factor of proportionality that gives the instantaneous pressure in the reflected wave at the boundary as a fraction of the incident wave's pressure at that instant.

There are three values of the specific impedance for which the reflection coefficient takes on a special value. Suppose the resistance to motion is enormous, so that the wall barely moves for any pressure. The limiting condition is $\zeta \to \infty$, in which case $R \to 1$. This is the case of a *rigid* boundary, which is a commonly used model, even though it is unattainable in reality. According to Eq. (2.5.12), if $R = 1$, the incident and reflected waves at the boundary have equal pressure at any instant. The consequence is that the total pressure at the wall is double that of the incident wave. The opposite condition is a wall that is so soft that application of a very small pressure induces a very large $\dot{u}$. The limit is $\zeta = 0$, which leads to $R = -1$. The total pressure at the boundary is zero, so this condition is said to represent a *pressure-release surface*. The third special case is a wall whose impedance matches the characteristic impedance of the fluid, which corresponds to $\zeta = 1$. Because $R = 0$ in this case, we say that is a *nonreflecting boundary*. Any value of ζ that is neither zero nor infinite leads to a reflection coefficient whose magnitude is less than one, and therefore attenuates the reflected wave. When R is positive, the pressure in the reflected wave leaves the boundary with the sign of the pressure that was incident, whereas a negative R reverses the sign of the reflected wave relative to the incident one. It is worth noting that the limitation of the reflection coefficient to $-1 \le R < 1$ applies to a passive boundary. A feedback system is active in the sense that it exerts a controlled force. A feedback system could have a gain factor that leads to $|R| > 1$.

A description of the pressure at any location follows from Eq. (2.5.12). In general, F can depend solely on the forward characteristic ξ, and $\xi = -ct$ at $x = 0$. The result of replacing $-ct$ with ξ is $F(\xi) = RG(-\xi)$, that is,

$$F(x - ct) = RG(ct - x) \tag{2.5.14}$$

Equation (2.5.14) is quite general. Let us use it to describe explicitly the field associated with reflection of a backward propagating wave that is incident at the boundary when $t = 0$. The first step is to specialize the G function to account for the transient nature of the incident wave. The interval prior to arrival of the wave at the boundary corresponds to negative t, so the location of the wavefront of p_2 in that interval is $x = -ct$. Furthermore, $p_2 = 0$ ahead of the wavefront, which is the region $x < -ct$, once again, with $t < 0$. The general concept of the method of wave images is to extend the spatial profiles to $-\infty < x < \infty$, which will require that we distinguish between the extended wave and the actual wave. To that end we denote the function describing the incident wave behind its wavefront as $g(\eta)$, whereas $G(\eta)$ denotes the extended wave. To make p_2 be zero ahead of its wavefront we multiply $g(\eta)$ by a step function $h(\eta)$. Thus, we describe the incident wave by setting $G(x + ct) = g(x + ct)h(x + ct)$. To apply Eq. (2.5.14) we change the argument of the G function from $x + ct$ to $ct - x$, and multiply by the reflection coefficient. The result is that the pressure field at any x, t is given by

$$
\boxed{
\begin{aligned}
p_2 &= g(x + ct)\, h(x + ct) \\
p_1 &= Rg(ct - x)\, h(ct - x) \\
p &= p_1 + p_2
\end{aligned}
}
\tag{2.5.15}
$$

It would not be difficult to program these relations, but they also can be interpreted via a graphical construction. As we did in Fig. 2.10, let us temporarily ignore the fact that there is no fluid in $x < 0$. Then the initial pressure profiles of p_1 and p_2 are the distributions obtained from evaluation of Eq. (2.5.15) for $-\infty < x < \infty$ when $t = 0$. The profile of p_2 at this instant is $g(x)$ for $x > 0$ and zero for $x < 0$. The profile of p_1 is zero for $x > 0$, and it is the value of $g(-x)$ multiplied by R for $x < 0$. A different view is that we should take a value of g at some distance from the origin in the positive x direction, multiply that value by R, and assign it to p_1 at the same distance from the origin in the negative x direction. This is the construction on Fig. 2.11a.

In this graph the g function has a maximum $p_a > 0$ at x_a and a minimum $p_b < 0$ at x_b. The situation that is depicted corresponds to $R = -0.6$, so the minimum value of p_1 in the initial profile is Rp_a at $x = -x_a$ and the maximum is Rp_b at $x = -x_b$. If R were plus one, then p_1 would be the mirror image of p_2 with respect to $x = 0$. If R were minus one, p_1 would be the inverted mirror image of p_2. Even though R is neither special value in Fig. 2.11a, we nevertheless refer to the initial p_1 as the image of p_2.

Now that the initial profiles of both simple wave have been defined, they may be propagated in the same manner as we have thus far for initial and boundary value problems. Thus, in Fig. 2.11b the incident wave's profile shifts in the negative x by direction by distance ct, and the reflected wave's profile shifts in the direction of positive x by the same amount. The profile of the total pressure at this instant is the

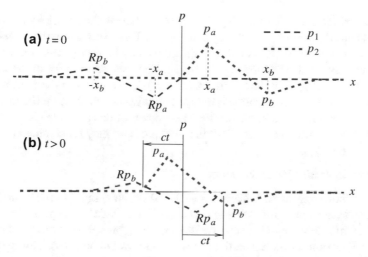

Fig. 2.11 Description by the method of wave images of reflection at a boundary. The incident backward propagating plane wave arrives at $t = 0$ and R is the reflection coefficient

sum of p_1 and p_2 for $x > 0$. In other words, we construct the image of the incident wave as an aid to the study and description of reflection, but ignore any portion of that image that lies behind the boundary when we evaluate the actual pressure.

Just as the construction of spatial profiles follows the procedure developed previously, so too does the construction of waveforms. First the waveforms of both simple waves is created as described by Fig. 2.12a. In this Figure the initial profile of p_2

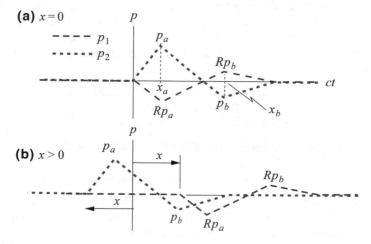

Fig. 2.12 Construction of waveforms of the incident and reflected waves. The reflection coefficient is $R = -0.6$

is replicated from Fig. 2.11a along the ct axis, whereas the initial profile of p_1 is flipped about the vertical axis through the origin. The result is that the waveform of p_1 at $x = 0$ is the waveform of p_2 multiplied by R. The waveforms at any location $x > 0$ are obtained by following Fig. 2.12b. The waveform of p_2 at the origin is advanced by shifting it to the left (decreasing ct). The amount of this shift is x. The p_1 waveform at the same location is delayed by shifting it to the right (increasing ct) relative to its appearance at the origin. The amount of this shift also is x. Once the individual waveforms are identified, the total pressure may be obtained by adding p_1 and p_2 in $ct \geq 0$.

The General Initial Value Problem

The preceding analysis of reflection was based on knowing the $g(x)$ function for the incident wave. Such information is available in an initial value problem. Like the case of an infinite waveguide, suppose that the initial pressure distribution is a given function $P(x)$ and the initial particle velocity is known to be $V(x)$. This means that Eq. (2.5.2) may be used, but only for $x > 0$, that is,

$$F(x) = f(x), \quad G(x) = g(x), \quad x > 0$$
$$f(x) \equiv \frac{1}{2}[P(x) + \rho_0 c V(x)], \quad g(x) \equiv \frac{1}{2}[P(x) - \rho_0 c V(x)] \tag{2.5.16}$$

The $G(x)$ function belongs to the backward propagating wave. Therefore, it is incident at the boundary $x = 0$ starting at time $t = 0$. The previous section described how this wave is reflected. In contrast, the $F(x)$ function belongs to the forward propagating wave. It is unimpeded by the boundary, so it propagates in the sense of increasing x as though the domain had no boundaries.

According to the principle of superposition, the total field within the waveguide is the sum of the field resulting from the incident p_2 wave and the contribution associated with $f(x)$. The former is described by Eq. (2.5.15) and the latter is $p_1 = f(\xi)$, but only for $\xi > 0$. This restriction on p_1 can be enforced with a step function. Thus, adding the two contributions yields the general solution of the initial value problem,

$$\boxed{\begin{array}{c} p_2 = g(x + ct)\, h(x + ct) \\ p_1 = Rg(ct - x)\, h(ct - x) + f(x - ct)\, h(x - ct) \\ f(x) \equiv \frac{1}{2}[P(x) + \rho_0 c V(x)], \quad g(x) \equiv \frac{1}{2}[P(x) - \rho_0 c V(x)] \\ p = p_1 + p_2 \end{array}} \tag{2.5.17}$$

This form is readily described by wave images. Compared to the case where there is only an incident wave, the initial profile of the backward wave is unchanged. The only modification of the forward wave is addition of the f function for $x > 0$, where p_1 previously was zero. After we have constructed the initial profile, evaluation of profiles at later t and waveforms at any x follows the procedure described in Figs. 2.11 and 2.12.

The operations are depicted on Fig. 2.13. The initial pressure $P(x)$ and particle velocity $V(x)$ are described in Fig. 2.13a, where the velocity has been multiplied by the characteristic impedance in order to plot both functions on the same scale. In Fig. 2.13b the initial value functions have been added and subtracted according to Eq. (2.5.17) in order the generate the $f(x)$ and $g(x)$ functions. Figure 2.13c constructs the spatial profiles of p_1 and p_2 at $t = 0$. The portion of p_1 in $x < 0$ is the mirror image of $g(x)$ multiplied by the reflection coefficient, which is set at $R = 0.5$. This image describes the portion of p_1 that is generated by reflection of p_2. In contrast, the portion of p_1 in $x > 0$ is the direct result of the initial conditions being nonzero. It would be observed even if there were no reflection. Figure 2.13d shows the construction of the p_1 and p_2 waveforms at $x = 0$.

From Fig. 2.13c, the spatial profiles at any t may be obtained by shifting the p_1 profile along the x axis by ct to the right, and the p_2 profile to the left by ct. Waveforms at any x may be obtained from Fig. 2.13d by shifting the p_1 wavefront to the right along the ct axis by x, and the p_2 axis to the left by x. The total pressure is obtained by adding the contributions of both simple waves, but only in the intervals $x \geq 0$ and $t \geq 0$.

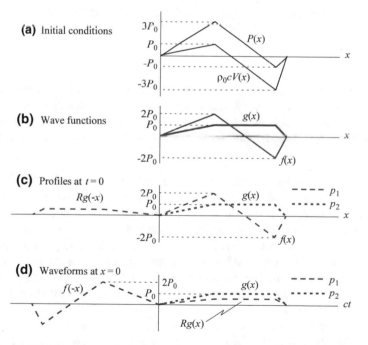

Fig. 2.13 Application of the method of wave images to generate the initial spatial profile and waveform in a semi-infinite waveguide. The initial pressure distribution is $P(x)$ and the initial particle velocity distribution is $V(x)$, and $R = 0.5$

EXAMPLE 2.5 The diagram shows a very long cylindrical tube within which a cylindrical container of length L may move. The container is filled with air at the same density and temperature as the ambient conditions within the tube. Prior to $t = 0$ the cylinder was translating to the left at constant speed v_c, such that the distance to the container's left end from some reference is $-v_c t$. At $t = 0$, a magnet is activated, thereby stopping the container almost instantaneously. Concurrently with halting the container, a high current circuit rapidly melts the plastic wall at the trailing (right) end, thereby merging the air inside and outside the container. The left wall remains intact. Construct spatial profiles of p and v in the region $x > 0$ at $t = 0.75L/c$ and $t = 1.5L/c$. Also construct waveforms of p and v for the interval $t > 0$ at $x = 0.75L$ and $x = 1.5L$.

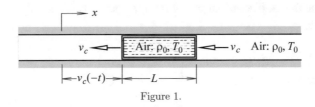

Figure 1.

Significance

The initial state will be seen to be somewhat unusual. This will give rise to waves that are readily described, yet emphasize the fundamental aspects of the combined effects of general initial conditions and reflections. The results will be interesting, and to some extent counterintuitive.

Solution

At the outset, we must assume that $v_c \ll c$. This allows us to avoid treating a moving boundary condition.[4] Subject to this assumption, it is a straightforward matter to determine the field for $t < 0$, which is the interval prior to disintegration of the container's wall. The right side of the container is at $v_c(-t) + L$. It is the near boundary for the semi-infinite wave guide to the right. The container was translating to the left for $t < 0$. Hence the boundary condition is $v = -v_c$ at $x = -v_c t + L$. This motion generates a wave that propagates in the direction of increasing x. There is no wave coming from infinity. Because $p = \rho_0 c v$ for a forward wave, it must be that $p = -\rho_0 c v_c$ and $v = -v_c$ for $x > L - ct$ when $t < 0$.

At $t = 0$, the container's right wall disintegrates, so the fluid within the container merges with the fluid to the right. The pressure within the container for $t < 0$ was the ambient value, and all particles were moving to the left at v_c. Neither of these variables can change instantaneously, so the field within the container at $t = 0$ is

---

[4]The case where movement of the boundary is significant requires consideration of nonlinear effects. This issue is discussed in Chap. 13.

$p = 0$ and $v = -v_c$. Combining this field and the field to the right at $t = 0$ shows that the initial state is

$$p(x, 0) \equiv P(x) = -\rho_0 c_0 v_c h(x - L) \text{ and } v(x, 0) \equiv V(x) = -v_c h(x), \; x > 0 \tag{1}$$

These initial conditions are the profiles in Fig. 2.

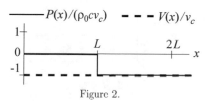

Figure 2.

In addition, it is given that the container comes to rest simultaneously with disintegration of the right end. The left end remains intact, so it becomes a stationary boundary. Therefore, we must satisfy $v = 0$ at $x = 0$ for $t \geq 0$.

This is our first encounter with discontinuous pressure and velocity fields. According to Euler's equation, a discontinuous pressure profile would require infinite acceleration. Also, a discontinuous velocity profile would correspond to an infinite shear rate, which would mean that viscosity cannot be ignored. A proper description of such behavior requires the study of shocks, which is a topic in Chap. 13 on nonlinear effects. For our studies of linear acoustics a discontinuity serves as a simplified model of a large change of state that occurs over an extremely short distance and a very brief time interval.

The profile functions $f(x)$ and $g(x)$ for the forward and backward waves are found by substituting Eq. (1) into (2.5.17). Doing so yields

$$f(x) = -\frac{1}{2}\rho_0 c v_c [h(x) + h(x - L)], \;\; g(x) = +\frac{1}{2}\rho_0 c v_c [h(x) - h(x - L)], \;\; x > 0 \tag{2}$$

The reflection coefficient at $x = 0$ is $R = 1$ because the particle velocity at that location is zero. These considerations complete the preliminaries required to implement the method of wave images.

The first step is to create initial profiles that replace the boundary with wave images. In the physical region $x > 0$ there is a forward propagating wave p_1 whose profile at $t = 0$ is $f(x)$. There also is a backward propagating wave p_2 whose profile at $t = 0$ is $g(x)$. Reflection at $x = 0$ of the backward propagating wave is accounted for with an image in the region $x < 0$. This image belongs to p_1. It is formed by left/right flipping of $g(x)$ relative to the origin, then multiplying the result by R. This forms $Rg(-x)$. Thus, the initial profile of p_1 consists of $f(x)$ in $x > 0$ and $Rg(-x)$ in $x < 0$. The initial profile of p_2 consists of $g(x)$ for $x > 0$ and nothing in $x < 0$. Both are depicted in the upper graph in Fig. 3. The initial velocity profiles in the lower graph are $v_1 = p_1/(\rho_0 c)$ and $v_2 = -p_2/(\rho_0 c)$, so the former replicates p_1 and the latter inverts p_2.

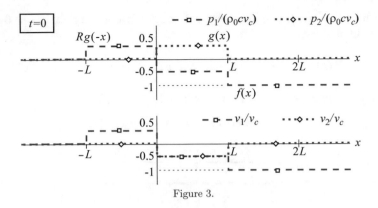

Figure 3.

Spatial profiles at any instant t are obtained by translating the profiles in Fig. 2 by distance ct. The p_1 and v_1 curves are shifted in the direction of increasing x, while the p_2 and v_2 curves are shifted in the direction of decreasing x. The translated curves for $t = 0.75L/c$ are described by the first two graphs in Fig. 4. In the third graph the values of the forward and backward curves are added. Only the portion in the physical domain $x > 0$ is retained. The fact that $p/(\rho_0 c v_c) = v/v_c$ for $x > 0.25L$ comes about because only the forward propagating wave exists in that region.

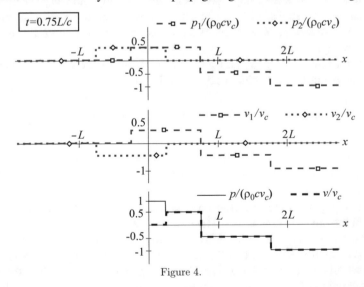

Figure 4.

The same steps as those in Fig. 4 lead to the spatial profiles for $t = 1.5L/c$ in Fig. 5. The shift of the initial profiles in this case is $1.5L$. For any instant after L/c, the p_2 wave has been fully reflected. Consequently, all that remains is the wave that propagates to the right, with $p = \rho_0 c v$.

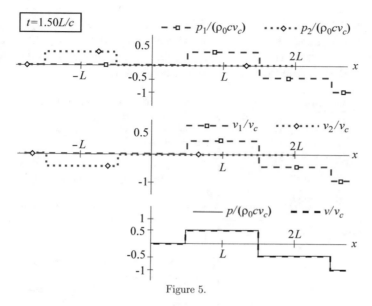

Figure 5.

Construction of waveforms for $x > 0$ follows the standard procedure for an initial value problem in an infinite domain. It begins by converting the initial spatial profiles of p_1 and p_2 in Fig. 3 to waveforms at $x = 0$. The waveforms of p_1 and v_1 are obtained by left/right swapping of the respective initial profiles relative to the origin. The p_2 and v_2 waveforms are obtained by replicating the initial profiles. These operations are shown in Fig. 6.

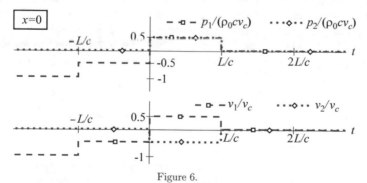

Figure 6.

A waveform at any $x > 0$ is obtained by shifting the p_1 and v_1 waveforms in Fig. 6 in the sense of increasing t by x/c, corresponding to a delayed arrival. The p_2 and v_2 waveforms are shifted by the same amount in the sense of decreasing t, corresponding to an advanced arrival time. The waveforms for the total signal consist of the sum of the individual contributions. The result of these operations for $x = 0.75L/c$ appears in Fig. 7. It is helpful to refer to the initial profiles in Fig. 3 to explain these results. The initial state within the container results in forward and backward waves. The time required for the trailing edge of the backward wave, which initially was at $x = L$, to travel to $x = 0.75L$ is $0.25L/c$. After that time, only the forward wave contributes. At $t = 0.75L/c$, the trailing edge of the p_1 wave, which began at $x = 0$, passes $x = 0.75L$. That also is the arrival time of the leading edge of the reflected wave. Then, at $t = 1.75L$, the trailing edge of the reflected wave passes $x = 0.75L$. After that instant there is no disturbance.

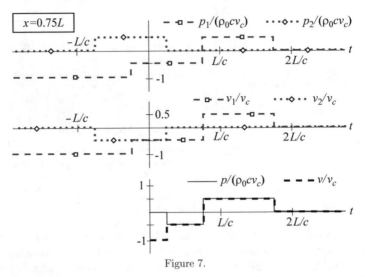

Figure 7.

The same construction steps as those described by Fig. 7, with a time delay of $1.5L/c$ leads to the waveform at $x = 1.5L$. The result is shown in Fig. 8. This location is outside the region that experiences the initial backward wave. Prior to $t = 0.5L/c$ the pressure is unchanged from its initial value. At this instant the phase of the directly generated part of p_1 that initially was at $x = L$ has traveled to $x = 1.5L$. The reflected wave simultaneously arrives. Then at $t = 2.5L$, the last nonzero part of the reflected wave passes $x = 1.5L$. After that instant no disturbance is present.

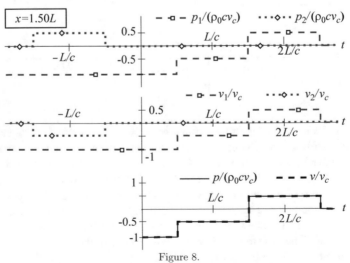

Figure 8.

2.5.3 Plane Waves in a Finite Waveguide

A semi-infinite domain is a highly idealized model. It does not account for the fact that even if the far end of a waveguide is very distant, any forward propagating wave will eventually reach that end, and return. Not only does this observation identify the limitation of the model of a semi-infinite domain, it tells us under what condition the model may be used. Specifically, at a specified location a semi-infinite model is adequate until the wave that is reflected from the far end returns to that location.

Allowance for reflection at the far end leads to a physical process of multiple reflections. A forward propagating wave is reflected at the far end, $x = L$, as a backward propagating wave that is reflected at $x = 0$ as another forward propagating wave. The process continues *ad infinitum*. The description of this multiple reflection process, and its implications for a variety of system configurations, provides closure for our study of time domain solutions for plane waves.

Initial Value Problem

The system consists of a closed waveguide, such as a tube, that is bounded by walls at $x = 0$ and $x = L$. Reflections at $x = 0$ are described by Eq. (2.5.15). We require a similar description of the reflection process at $x = L$. The incident wave in this case is p_1, and the location of the boundary has been shifted to $x = L$. Furthermore, the fluid is situated in $x < L$. Nevertheless, the physical nature of the reflected wave should be unaltered by placing the reflection boundary at a different location. This allows us to introduce a coordinate transformation that makes this end look like $x = 0$, specifically, $x' = L - x$. Thus, we replace x with x' in Eq. (2.5.14) and swap F and G because F represents the incident wave. Doing so gives

$$G\left(x' - ct\right) = R F\left(ct - x'\right) \qquad (2.5.18)$$

Introduction of the coordinate transformation to eliminate x' yields the form we seek

$$G\left(L - x - ct\right) = R_L F\left(x + ct - L\right), \quad 0 \le x \le L \qquad (2.5.19)$$

A form that explicitly gives $G(x)$ will be more useful to the construction of an overall solution. We obtain such a representation by subtracting L from the argument in both sides, and then multiplying each argument by -1. Setting $t = 0$ yields the initial profile of p_2,

$$\boxed{G(x) = R_L F(2L - x), \quad 0 \le x \le L} \qquad (2.5.20)$$

In other words, if we know $F(x)$ to be the incident wave, the initial profile of the reflected wave may be obtained by replacing x with $2L - x$ in the argument and multiplying the function by the reflection factor. Although the mathematical appearance of this rule is different from the one at $x = 0$, the physical and visual aspects when viewed as position in front of, and behind, the boundary, are unchanged.

To see that this is so consider the phase of $F(x)$ that is at a specified distance x from the left boundary. This point is at distance $L - x$ from the right boundary. A point at the same distance behind the right boundary is at $x = L + (L - x)$. Therefore, Eq. (2.5.20) tells us the $G(x)$ is the image of $F(x)$, flipped left to right relative to $x = L$ and multiplied by R.

These observations lead to the recognition that the initial profile of the reflected wave may be visually constructed in two steps, regardless of where the boundary is situated. First a copy of the incident wave's initial profile is mirror-imaged by swapping it left and right relative to the boundary. The second step multiplies the reflected wave's image by the reflection factor.

Now that we can describe how waves are reflected from both ends, we may proceed to satisfy the initial conditions. In the absence of boundaries the d'Alembert solution corresponding to an initial pressure field $P(x)$ and velocity field $V(x)$ is described by Eq. (2.5.2). The p_1 and p_2 waves that are described there undergo the sequence of reflections. Each reflection is described by adding a wave image to the profiles directly generated by the initial conditions. To distinguish waves that are generated from images, let us denote the initial functions as $f(x)$ for p_1 and $p_2(x)$ for p_2. Both are defined only within the confines of the waveguide, $0 < x < L$. The method of wave images extends all profiles infinitely in both directions. We can do so by using a combination of step functions $h(x) - h(x - L)$ as a factor. This factor is one in the physical domain and zero for $x < 0$ and $x > L$. Thus the initial pressure profiles induced by the initial conditions are written as

$$f(x) \equiv \frac{1}{2}[P(x) + \rho_0 c V(x)][h(x) - h(x - L)]$$

$$g(x) \equiv \frac{1}{2}[P(x) - \rho_0 c V(x)][h(x) - h(x - L)]$$

$$(2.5.21)$$

The $f(x)$ portion propagates forward, so it is reflected at $x = L$, whereas the $g(x)$ portion propagates backward and is reflected at $x = 0$. It is easier to track the multiple reflections that ensue if we consider each initial wave individually. Linearity of the governing equations allows us to superpose the individual contributions. The notation we have adopted sets $F(x)$ as the overall initial profile formed from all contributions to the forward propagating wave. Therefore it is logical to designate the initial wave by itself as $F_0(x) = f(x)$.

The concept now is to identify images based on the reflection equations, and trace each image back to $f(x)$. According to Eq. (2.5.20), reflection of $F_0(x)$ at $x = L$ is described by adding image #1 to the profile of p_2. It is described by $G_1(x) = R_L F_0(2L - x) \equiv R_L f(2L - x)$. This reflected wave propagates backward until it is reflected at $x = 0$. This reflection is described by adding image #2 to the initial profile of p_1. According to Eq. (2.5.15), this image is $F_2(x) = R_0 G_1(-x) = R_0[R_L f(2L - (-x))]$. Reflection at $x = L$ of this portion of the forward propagating wave generates image #3 that adds $G_3(x)$ to p_2. The reflection law at $x = L$ gives $G_3(x) = R_L F_2(2L - x) = R_L[R_0 R_L f(2L + (2L - x))]$. This contribution to p_2 is reflected at $x = 0$, which leads to image #4. The reflection law at that end gives $F_4(x) = R_0 G_3(-x) \equiv R_0[R_0 R_L^2 f(4L - (-x))]$.

We could go on generating many more images, but a pattern is now apparent. We construct $F(x)$ by adding to the initial part, $f(x)$, images #2, #4, and their successors. The additional images are obtained by multiplying the previous one by $R_0 R_L$, corresponding to reflection at both ends, and adding $2L$ to the argument of the predecessor, corresponding to the distance the image must travel down the waveguide and back to $x = 0$. Similarly, $G(x)$ consists of images #1, #3, and their successors. The first image is obtained by satisfying the law for reflection of the initial wave $f(x)$ at $x = L$. Successive images also are obtained by multiplying the predecessor by $R_0 R_L$ and adding $2L$ to the argument. A compact representation of the combination of all images is

$$F(x) = \sum_{n=0}^{\infty} (R_0 R_L)^n f(x + 2nL)$$
$$G(x) = \sum_{n=0}^{\infty} (R_0 R_L)^n [R_L f(-x + 2(n+1)L)]$$

(2.5.22)

Figure 2.14 provides a graphic description of this process. It will be observed that because $f(x)$ is defined over an interval whose length is L and each image is shifted by $2L$, in the negative x direction for $F(x)$ and in the positive x direction for $G(x)$, the images are separated by gaps of length L. Also, the front of a reflected wave's image is at the same distance behind the reflecting boundary as the distance from the front of the corresponding incident wave's image to the boundary, in accord with the general discussion of reflection. For example, $F_2(x)$ is reflected at $x = L$, which results in G_3. The front of $G_3(x)$ is at $x = 3L$, and the front of $F_2(x)$ is at $x = -L$, so the distance from each front to $x = L$ is $2L$.

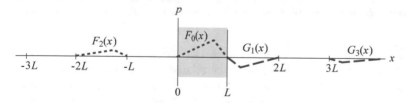

Fig. 2.14 Initial pressure profile in a finite waveguide generated by the portion of the initial wave that propagates in the sense of increasing x

The same reasoning leads to a description of the contribution of $g(x)$ to the initial profiles. This function describes the portion of the p_2 wave generated by the initial conditions, so we denote it as $G_0(x) = g(x)$. This wave is incident at $x = 0$, so its reflection adds image #1 to p_1, with $F_1(x) = R_0 G_0(-x) \equiv R_0 g(-x)$. In turn, $F_1(x)$ is reflected at $x = L$, which leads to image #2 for the p_2 wave that is reflected. This image is $G_2(x) = R_L F_1(2L - x) \equiv R_L [R_0 g(-(2L - x))]$, The p_2 wave described by $G_2(x)$ propagates backward to the origin, where it is reflected, thereby generating image #3 for p_1. This image is $F_3(x) = R_0 G_2(-x) \equiv R_0 [R_L R_0 g(-2L + (-x))]$. The p_1 wave associated with image #3 propagates forward until it reflects at $x = L$, with that reflection being described as $G_4(x) = R_L F_3(2L - x) \equiv R_L \left[R_L R_0^2 g(-2L - (2L - x))\right]$, and so on. A pattern similar to that associated with $f(x)$ has emerged. The first contribution to $G(x)$ is $g(x)$, which is generated by the initial condition. Successive images for $G(x)$ multiply the predecessor by $R_0 R_L$ and subtract $2L$ from the function's argument. The first contribution to $F(x)$ consists of the reflection of the initial wave at $x = 0$, which is $R_0 g(-x)$. Successive images are obtained by multiplying the predecessor by $R_0 R_L$ and subtracting $2L$ from the function's argument. The mathematical representation of the combined set of waves is

$$F(x) = \sum_{n=0}^{\infty} (R_0 R_L)^n \left[R_0 g(-x - 2nL)\right]$$

$$G(x) = \sum_{n=0}^{\infty} (R_0 R_L)^n g(x - 2nL) \tag{2.5.23}$$

This construction is depicted in Fig. 2.15. Once again, the images are separated by intervals of length L. These gaps match the intervals in which $f(x)$ generates images.

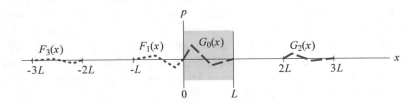

Fig. 2.15 Initial pressure profile in a finite waveguide generated by the portion of the initial wave that propagates in the sense of increasing x

The initial profile that is consistent with the reflection laws and satisfies the initial conditions is the sum of the profiles associated with $f(x)$ and $g(x)$, as depicted in Fig. 2.16. All images behind the boundary at $x = 0$ are part of the forward propagating wave, and all images behind the boundary at $x = L$ are part of the backward propagating wave. The oppositely traveling simple waves initially only overlap in $0 \leq x \leq L$.

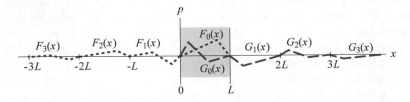

Fig. 2.16 Initial pressure profile in a finite waveguide generated by initial conditions

The general solution is obtained by summing Eqs. (2.5.22) and (2.5.23). Then because $\xi = x$ at $t = 0$, we replace x with $x - ct$ in $F(x)$. Similarly, $\eta = x$ at $t = 0$, so we replace x with $x + ct$ in $G(x)$. For the sake of completeness the definition of the $f(x)$ and $g(x)$ functions is repeated in the description of the full signal.

$$
\begin{array}{c}
p = p_1 + p_2 \\[6pt]
p_1(x, t) = F(x - ct) = \sum_{n=0}^{\infty} (R_0 R_L)^n \left[f(x - ct + 2nL) \right. \\[6pt]
\left. + R_0 g(-x + ct - 2nL) \right] \\[6pt]
p_2(x, t) = G(x + ct) = \sum_{n=0}^{\infty} (R_0 R_L)^n \left[g(x + ct - 2nL) \right. \\[6pt]
\left. + R_L f(-x - ct + 2(n+1)L) \right] \\[6pt]
f(x) \equiv \frac{1}{2} [P(x) + \rho_0 c V(x)] [h(x) - h(x - L)] \\[6pt]
g(x) \equiv \frac{1}{2} [P(x) - \rho_0 c V(x)] [h(x) - h(x - L)]
\end{array}
\qquad (2.5.24)
$$

Image constructions like Fig. 2.16 are a great aid when we wish to interpret why various phenomena occur, but the sketches become difficult if $f(x)$ or $g(x)$ are complicated functions. When we need to obtain quantitative results, the preferable route is to use mathematical software to evaluate Eq. (2.5.24). Both approaches require selection of the number N at which the summation of images contributions may be terminated. To determine this number we observe that halting the summations at N places images in $-2NL \le x \le (2N + 1)L$ at $t = 0$. This is an adequate representation if images outside this range, which have not been constructed, arrive at the location of interest after the maximum time of interest, T_{max}. The distance from the tail of the last image for F or G to the nearest wall is $2NL$, so the time required to propagate the farthest tail to the closest boundary is $2NL/c$, which must be greater than T_{max}. Thus, the index at which the summations in Eq. (2.5.24) are halted must be such that

$$
N \ge \text{ceil}\left(\frac{c T_{max}}{2L} \right)
\qquad (2.5.25)
$$

where "ceil" is a function that gives the smallest integer that is greater than or equal to the argument.

Some interesting phenomena occur in situations where $|R| = 1$ at both ends. The reflection process in such cases does not attenuate the amplitude of a reflected wave relative to the incident, so reflected waves are either the positive or negative mirror image of the incident. We could use results from the study of acoustic energy in Chap. 4 to prove that no energy is dissipated in the reflection, but that property is apparent from the sameness of the waves.

There are three combinations of boundary conditions for which $|R| = 1$ at both ends, because the system is physically the same if $R_0 = 1$ and $R_L = -1$ or vice versa. Typical initial profiles for the case where both walls of the waveguide are rigid, $R_0 = R_L = 1$, are depicted in Fig. 2.17.

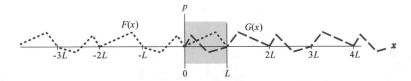

Fig. 2.17 Initial pressure profiles in a finite waveguide whose ends are rigid, $R_0 = R_L = 1$

Because the amplitudes are unchanged, the profiles of $F(x)$ and $G(x)$ can be seen to be periodic, with the pattern repeating in an interval of $2L$. This periodicity property only applies to $x < L$ and $x > 0$, respectively. Furthermore, $F(x)$ for $x < 0$ is the mirror image of $G(x)$ for $x > 0$, and $G(x)$ for $x > L$ is the mirror image of $F(x)$ for $x < L$. These properties are stated mathematically as

$$
\begin{aligned}
F(x - 2L) &= F(x), \quad x < L \\
G(x + 2L) &= G(x), \quad x > 0 \\
F(x) &= G(-x), \quad x < 0 \\
G(x) &= F(2L - x); \quad x > L
\end{aligned}
\tag{2.5.26}
$$

Note that the first and third properties taken together state that $G(-x) = F(x - 2L)$. Changing the sign of the arguments in this relation yields the fourth property, so it is not independent.

A function that is periodic in x may be expanded in a Fourier series in x. Aside from the change of the variables name, the procedures previously used for periodic functions of time are equally applicable here. Furthermore, it follows from the third of Eq. (2.5.26) that changing x to $-x$ in the series for $F(x)$ will give $G(x)$, so that

$$
F(x) = \frac{1}{2} \sum_{n=-\infty}^{\infty} F_n \exp\left(\frac{in\pi x}{L}\right), \quad F_{(-n)} = F_n^*; \quad x < L
$$

$$
G(x) = \frac{1}{2} \sum_{n=-\infty}^{\infty} F_n \exp\left(-\frac{in\pi x}{L}\right); \quad x > 0
$$

The characteristic variables for F and G are ξ and η, respectively, and both equal x at $t = 0$. Thus replacing x with the appropriate characteristic yields descriptions of the p_1 and p_2 waves,

$$
p_1(x, t) = F(x - ct) = \frac{1}{2} \sum_{n=-\infty}^{\infty} F_n \exp\left(\frac{in\pi}{L}(x - ct)\right)
$$

$$
p_2(x, t) = G(x + ct) = \frac{1}{2} \sum_{n=-\infty}^{\infty} F_n \exp\left(-\frac{in\pi}{L}(x + ct)\right)
\tag{2.5.27}
$$

No specification of the range of x and t is indicated in this representation because the series for F and G are always applicable in the physical domain, $0 < x < L$.

The total pressure is the sum of the pressures in the forward and backward waves, so we have

$$p(x,t) = \frac{1}{2} \sum_{n=-\infty}^{\infty} F_n \left[\exp\left(\frac{in\pi x}{L}\right) + \exp\left(-\frac{in\pi x}{L}\right) \right] \exp\left(-\frac{in\pi ct}{L}\right) \quad (2.5.28)$$

Our convention for description of a harmonic oscillation uses a complex exponential with a positive frequency. We can convert the preceding to such a form by setting $n = -n'$. The prime symbol may be deleted after the substitution, and we know that $F_{(-n)} = F_n^*$. The representation of the pressure field becomes

$$p(x,t) = \frac{1}{2} \sum_{n=-\infty}^{\infty} F_n^* \left[2\cos\left(\frac{n\pi x}{L}\right) \right] \exp\left(\frac{in\pi ct}{L}\right) \quad (2.5.29)$$

Each term for negative n is the complex conjugate of the term for positive n, so the pressure reduces to

$$p(x,t) \equiv F_0 + \sum_{n=1}^{\infty} \text{Re}\left[2F_n^* \exp\left(\frac{in\pi ct}{L}\right) \right] \cos\left(\frac{n\pi x}{L}\right) \quad (2.5.30)$$

This is an extremely important representation of the signal. Consider a specific n in the summation. Because $\cos(n\pi x/L)$ is real, we conclude that the pressure contribution of the nth term consists of an oscillation at frequency $n\pi c/L$, and that the pressure at all locations is in-phase, with an amplitude that is proportional to $\cos(n\pi x/L)$. Each term represents a *standing wave,* which is an oscillation in which the profile maintains a fixed spatial pattern while its magnitude fluctuates. The initial conditions only influence the value of the complex amplitudes F_n, from which it follows that the harmonic time dependence and the cosinusoidal spatial variation of each term are fundamental properties of the waveguide. In fact, we say that the $n\pi c/L$ values are the system's *natural frequencies* ω_n, and the $\cos(n\pi x/L)$ functions constitute its *natural modes* $\phi_n(x)$. Thus, the method of images has led us to the recognition that the signal generated by initial conditions within a closed, rigid-rigid, waveguide is composed of an infinite series of modes whose spatial distribution is an integer multiple of half-cosine segments, with natural frequencies that are integer multiples of the fundamental frequency $\pi c/L$, that is,

$$\boxed{\begin{array}{c} \text{If } R_0 = R_L = 1, \text{ then} \\ \omega_n = n\frac{\pi c}{L} \text{ and } \phi_n = \cos\left(\frac{n\pi x}{L}\right), \quad n = 0, 1, 2, ... \end{array}} \quad (2.5.31)$$

Each natural mode function has the property that $d\phi_n/dx = 0$ at $x = 0$ and $x = L$. According to Euler's equation, this means that each mode's contribution to the particle velocity is such that $dv/dt = 0$ at both ends. Integration of a zero acceleration gives $v = 0$ at $x = 0$ and $x = L$, which is the condition imposed by rigidity. Clearly, the $n = 0$ mode, whose distribution is constant and whose natural frequency is zero, fits this description.

Typical initial profiles for the case where both ends are pressure-release are depicted in Fig. 2.18.

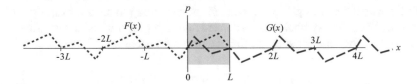

Fig. 2.18 Initial pressure profiles in a finite waveguide whose ends are pressure-release, $R_0 = R_L = -1$

The periodicity and mirror image properties exhibited by this figure are

$$
\begin{aligned}
F(x - 2L) &= F(x); \quad x < L \\
G(x + 2L) &= G(x); \quad x > 0 \\
F(x) &= -G(-x); \quad x < 0 \\
G(x) &= -F(2L - x), \quad x > L
\end{aligned}
\tag{2.5.32}
$$

Here too, the fourth property is redundant, because it can be obtained from the first and third properties.

These periodicity conditions lead to a Fourier series that is different from Eq. (2.5.31) for a waveguide with rigid ends. The task in Homework Exercise 2.17 is to demonstrate that the initial condition response in a waveguide whose end are pressure-release is

$$
p(x, t) \equiv \sum_{n=1}^{\infty} \text{Re} \left[\frac{2}{i} F_n^* \exp\left(\frac{in\pi ct}{L}\right) \right] \sin\left(\frac{n\pi x}{L}\right)
\tag{2.5.33}
$$

It follows that the natural frequencies are the same as those in the rigid case, except that there is no zero frequency mode. The natural modes now consist of an integer number of lobes of a sine function, specifically

$$
\boxed{
\begin{array}{c}
\text{If } R_0 = R_L = -1, \text{ then} \\
\omega_n = n\frac{\pi c}{L} \text{ and } \phi_n = \sin\left(\frac{n\pi x}{L}\right), \quad n = 1, 2, \ldots
\end{array}
}
\tag{2.5.34}
$$

The fact that the natural modes are such that $\phi_n = 0$ at $x = 0$ and $x = L$ shows that the solution is consistent with the pressure-release condition at each end.

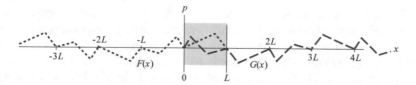

Fig. 2.19 Initial pressure profiles in a finite waveguide in which one end is rigid and the other is pressure-release, $R_0 = 1$, $R_L = -1$

A typical initial profile of $F(x)$ and $G(x)$ for the case where $x = 0$ is rigid and $x = L$ is pressure release appears in Fig. 2.19. Both functions are periodic, but the period now is $4L$. This change relative to the previous configurations stems from the fact that reflection of the wave at the pressure-release end inverts it. Two trips back and forth are required to recover the original sign. The periodicity and mirror image properties now are

$$
\begin{aligned}
F(x - 4L) &= -F(x - 2L) = F(x), \quad x < L \\
G(x + 4L) &= -G(x + 2L) = G(x), \quad x > 0 \\
F(x) &= G(-x), \quad x < 0 \\
G(x) &= -F(2L - x), \quad x > L
\end{aligned}
\tag{2.5.35}
$$

Once again, the fourth property is redundant.

The Fourier series that results from these properties are like Eq. (2.5.30), except that the period is doubled and only odd harmonics are nonzero,

$$
p(x, t) = \sum_{\substack{n=-\infty \\ n \text{ odd}}}^{\infty} \operatorname{Re}\left[2F_n^* \exp\left(\frac{in\pi ct}{2L}\right)\right] \cos\left(\frac{n\pi x}{2L}\right)
\tag{2.5.36}
$$

Here, the image construction leads to the recognition that the modes consist of an odd multiple of quarter-cosine segments, with the natural frequencies being that multiple of the fundamental $\pi c/2L$, that is,

$$
\boxed{
\begin{array}{c}
\text{If } R_0 = 1 \text{ and } R_L = -1, \text{ then} \\
\omega_n = (2n-1)\dfrac{\pi c}{2L} \text{ and } \phi_n = \cos\left(\dfrac{(2n-1)\pi x}{2L}\right), \quad n = 1, 2, \ldots
\end{array}
}
\tag{2.5.37}
$$

An overview of the three cases where the boundary conditions do not dissipate energy shows that the repetitive nature of the reflection process, which causes the reflected wave to be either a positive or negative mirror image of the incident wave, leads to spatial periodicity of the initial profiles of the forward and backward waves. Such behavior is manifested by representation of the response to any set of initial conditions as a set of natural modes, which represent standing waves patterns, each of which features a pressure oscillation at a natural frequency. Modal decomposition is an essential feature of the acoustical analysis of many systems, especially enclosures. Such decompositions are usually identified by solving partial differential equations.

When either end is dissipative, there is no spatial periodicity in the initial profile functions $F(x)$ and $G(x)$. This situation was depicted in Fig. 2.16. When a wave leaves one end, then undergoes reflections at the far end and then returns to the end from which it departed, it is attenuated by the product of the reflection coefficients at both ends. This factor applies to the profile in each interval of $2L$. This observation leads to the conclusion that

$$F(x) = R_0 R_L F(x + 2L); \quad x < L$$
$$G(x) = R_0 R_L G(x - 2L); \quad x > 0 \tag{2.5.38}$$

Because of their similarity to the periodicity conditions for nondissipative ends, it would be appropriate to refer the preceding as periodic attenuation conditions. They describe a decay given by

$$F(x) = (R_0 R_L)^{n_F} F_0(x + 2n_F L); \quad x < L$$
$$G(x) = (R_0 R_L)^{n_G} G_0(x - 2n_G L); \quad x > 0 \tag{2.5.39}$$

where $F_0(x)$ and $G_0(x)$ are the functions on the first $2L$ intervals.

General descriptions of the forward and backward simple waves are obtained by replacing x in $F(x)$ and $G(x)$ with the characteristic associated with each function. Furthermore, the physical domain is restricted to $0 < x < L$, so the argument of each function passes to the next interval as t increases. We indicate this behavior by associating n_F and n_G with the time dependence, with the result that Eq. (2.5.39) becomes

$$p_1(x, t) = (R_0 R_L)^{n_F} F_0\left(x - c\left(t - n_F \frac{2L}{c}\right)\right)$$
$$p_2(x, t) = (R_0 R_L)^{n_G} G_0\left(x + c\left(t - n_G \frac{2L}{c}\right)\right) \tag{2.5.40}$$

Note that it is no longer necessary to indicate the range of x and t for which these expressions apply, because our interest now is restricted to $0 < x < L$ and $t > 0$.

Evaluation of p_1 and p_2 at a specified x, t requires knowing what values of n_F and n_G should be used. In Eq. (2.5.39), the integer n_F is selected to maintain the argument of the F_0 function in the range $-L < x < L$, while n_G is selected to maintain the argument of the G_0 function in the range $0 < x < 2L$. Because the range of x in Eq. (2.5.40) is restricted, the arguments of F_0 and G_0 in this equation go out of range if t increases too much. For example in the case of F_0, we would find that if $ct > L + x$, then $x - ct < -L$. The number of time intervals to subtract from $ct - x - L$ increase as t increases. The argument of G_0 behaves in a similar manner. Both numbers can be determined from the integer function floor(u), which is the largest integer that is less than u. The specific expressions are

$$n_F = \text{floor}\left(\frac{ct - x - L}{2L}\right), \quad n_G = \text{floor}\left(\frac{x + ct}{2L}\right) \tag{2.5.41}$$

The combination of Eqs. (2.5.40) and (2.5.41) state that the waveforms at a specific x are obtained by multiplying the pressure in an initial time interval by an attenuation factor. Let us use t' to denote the elapsed time in this interval, so that

$$t' = t - n_F \frac{2L}{c} \tag{2.5.42}$$

Also, if we are willing to ignore the step-wise manner in which the pressure changes, the floor functions for both integers may be approximated as the continuous function $ct/(2L)$. This allows Eq. (2.5.40) to be approximated as

$$p_1(x, t) = \text{sign}(R_0 R_L)\left(|R_0 R_L|\right)^{ct/(2L)} F_0(x - ct')$$
$$p_2(x, t) = \text{sign}(R_0 R_L)\left(|R_0 R_L|\right)^{ct/(2L)} G_0(x + ct') \tag{2.5.43}$$

where the sign function gives plus one if its argument is positive, and minus one if its argument is negative. The behavior exhibited above is an exponential decay, as can be seen by setting

$$\left(|R_0 R_L|\right)^{ct/(2L)} = e^{-\beta t} \tag{2.5.44}$$

This replacement is correct if the decay constant is

$$\beta = -\frac{c}{2L} \ln\left(|R_0 R_L|\right) \tag{2.5.45}$$

The corresponding representation of the waveforms of each simple wave at any location is

$$p_1(x, t) = \text{sign}(R_0 R_L) e^{-\beta t} F_0(x - ct')$$
$$p_2(x, t) = \text{sign}(R_0 R_L) e^{-\beta t} G_0(x + ct') \tag{2.5.46}$$

Thus, the waveforms in intervals of duration $2L/c$ may be evaluated by multiplying the waveform in the first interval by the exponential attenuation factor.

In addition to providing a general picture for the decay associated with dissipation, the preceding leads to the concept of *reverberation time* T_{rev}, which is defined as the time interval required for the mean-squared pressure to be attenuated by a factor of 10^{-6}. This corresponds to a decay of 0.001 for the instantaneous pressure amplitude. Equating the exponential factor in Eq. (2.5.44) to 0.001 leads to

$$T_{\text{rev}} = -\frac{6 \ln(10)}{\ln\left(|R_0 R_L|\right)} \frac{L}{c} \tag{2.5.47}$$

This form is reminiscent of the Norris–Eyring reverberation time.[5] Reverberation time is one of the primary factors affecting the acoustical quality of rooms and auditoria. A room with a very large reverberation time hinders speech comprehension, because the sound of a previously spoken word would still be audible when the next word is spoken. In contrast, a very small reverberation time for a symphony hall would make the music seem stifled and dull.

EXAMPLE 2.6 A tube of length L is closed at one end by a resistive liner whose impedance is $\zeta = 2.5$, and the other end is covered by a mylar membrane that is only slightly tensioned, so it effectively is a pressure-release termination. Differential heating of the walls of the tube has created an initial pressure distribution that is parabolic, $p(x,0) = 4\varepsilon\left(x/L - x^2/L^2\right)$, with the fluid at rest. Use the method of wave images as the basis for a graphical construction of spatial profiles of the pressure in the forward and backward propagating waves at $t = 5.25L/c$. Then write a computer program to evaluate spatial profiles of the directly generated waves and their reflected images, as well as temporal waveforms at arbitrarily selected locations within the waveguide. Use this program to verify the spatial profile at $t = 5.25L/c$, and also evaluate the waveform at $x = L/2$.

Significance

This example provides additional practice in the graphical construction of pressure profiles and waveforms and their interpretation. However, quantitative evaluations for complicated initial conditions are best done with computer software. Minor modifications of the program that is developed would provide a general tool to explore other configurations.

Solution

The graphical construction follows the procedure employed in Figs. 2.14, 2.15 and 2.16. We are free to take the pressure-release condition to be either end; we will let it be $x = L$. We set $R_0 = 0.427$ corresponding to $\zeta = 2.5$, and $R_L = -1$. The initial velocity field is zero, so $f(x) = g(x) = 2\varepsilon\left(x/L - x^2/L^2\right)[h(x) - h(x - L)]$. Before we construct the initial profiles, we must determine how many images are required. To answer this, we observe that the profile of p_1 at an arbitrary instant t will be shifted in the direction of increasing x by distance ct. For a complete picture, the point in the initial profile at the most negative x must still be in the region $x < 0$ when $t = 5.25L/c$. (Otherwise, it would be as though the p_1 wave was cutoff in $0 < x < L$, even though its images are endless.) If N is the total number of images for p_1 (from both the $f(x)$ and $g(x)$ functions) then the smallest x for our picture of p_1 will be $-NL$. A similar consideration applies to p_2 except distances now are measured beyond $x = L$. Thus, we require that $NL > ct$, so we set $N = 6$.

[5]C.F. Eyring, "Reverberation time in 'dead' rooms," J. Acoust. Soc. Am. **1**, 217–241 (1930). See also A.D. Pierce, *Acoustics* (McGraw-Hill, New York, reprinted by the Acoustical Society of America, Melville, NY), 20–22, 263–265 (1981).

Figure 1 was generated with the aid of drawing software that scaled and joined the various images. Like Fig. 2.14, the first graph describes the images associated with $f(x)$. Images for reflections at $x = L$ are obtained by inverting and flipping the incident image about the vertical axis. (If one does this construction using graphics software, the shape of the new image can be obtained by applying scale factors of -1 for the x direction and $R_L = -1$ for the y direction.) The new image is placed such that its leading edge is at the same distance behind $x = L$ as the distance from the leading edge of the incident wave to that boundary. Each image beyond $x = L$ begets another image behind $x = 0$. These images are obtained by multiplying the incident wave's height by R_0 and flipping the incident image about the vertical axis. (For graphics software the scale factors would be -1 in the x direction and R_0 in the y direction.) The distances for placing these images is based on distances to leading edges measured from $x = 0$. The second graph describes the images associated with the reflections of the waves generated by $g(x)$. In all respects the construction procedure is the same as in the first graph. The third graph assembles both sets of images, thereby forming $p_1(x, 0)$ and $p_2(x, 0)$. The fourth graph obtains the profiles at the desired instant by shifting $p_1(x, 0)$ in the positive x direction and $p_2(x, 0)$ in the negative x direction. The distance for both shifts is ct.

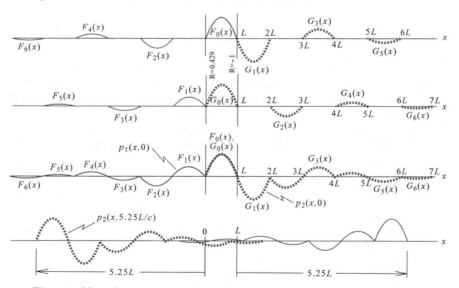

Figure 1. Manual construction of pressure profiles at $ct/L = 0$ and $ct/L = 5.25$.

The computer program we shall create uses nondimensional variables, that is, x represents x/L, t represents ct/L, and p represents p/ε. Thus the only parameters for a computation are R_0, R_L and the number of images N. Equation (2.5.24) contains four summations. Profiles call for computing these sums for a range of x values at fixed t, whereas waveforms entail their evaluation for a range of t at fixed x. In view of the repetitive nature of this computation it is useful to perform the main computations within a subprogram. In Matlab this can be done with a function M-file, which we will call p_f_g_ic.m. It returns the four sum values corresponding to input vectors of ξ and

η, the reflection coefficients, and the number of images to be computed. The following listing begins by defining the f and g functions corresponding to the specified initial conditions. These definitions are in the form of anonymous functions in order to evaluate them for various arguments. This is the only part of this subprogram that requires change to handle a different set of initial conditions. After that, the sums are initialized and then accumulated by successive passes through a loop. The specific program is

```
function [p_1_F  p_1_G  p_2_F  p_2_G] = p_f_g_ic(xi,eta,R_0,R_L,N_image)
f_image = @(xi) 2 * (xi - xi.^2).*((xi > 0) - (xi > 1));
g_image = @(eta) 2 * (eta - eta.^2).*((eta > 0) - (eta > 1));
p_1_F=zeros(length(xi));   p_1_G=p_1_F;   p_2_F=p_1_F;  p_2_G=p_1_F;
for n=0:N_image;
        p_1_F=p_1_F + (R_0*R_L)^n*f_image(xi+2*n);
        p_2_G=p_2_G + (R_0*R_L)^n*g_image(eta-2*n);
        p_1_G=p_1_G + (R_0*R_L)^n*(R_0*g_image(-xi-2*n));
        p_2_F=p_2_F + (R_0*R_L)^n*(R_L*f_image(-eta+2*(n+1)));
end
```

Preceding operators with a period in Matlab implements the operation on an element by element basis.

To evaluate profiles the value of t and a set of x locations are defined in the main program. The waveguide is situated between zero and one in terms of the nondimensional x, and that interval is augmented by the extent of images at either end. Thus we set `x=[-N_image:0.01:N_image+1];`. The nondimensional characteristics are `xi=x-t;   eta=x+t;`. This data is the input to the `p_f_g_ic` function. The output `p_1_F`, etc., may be plotted individually, then combined to evaluate the forward and backward waves according to `p_1=p_1_F+p_1_G;` `p_2=p_2_F+p_2_G;`. Addition of these waves yields the total pressure between the boundaries, `p=(p_1+p_2)*((x > 0)-(x > 1));`. There is no need to display the individual p_1 and p_2 because they closely match Fig. 1 after compensation for different scaling factors. Figure 2 is the profile that results from evaluating $p = p_1 + p_2$, then eliminating the part that is not in $0 < x < L$.

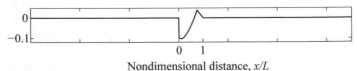

Figure 2. Computer generated profile of total pressure for $0 < x/L < 1$ when $ct/L = 5.25$

Waveforms are obtained by defining a vector of nondimensional time values. The time interval in which the construction is valid begins at $t = 0$ and ends when the last image has transited the waveguide. The computation ceases to be valid when the tail of the most distant image enters the domain of the waveguide. The distance to the tail of the last image on both sides NL, so the longest permissible window for a waveform is $0 \le t < NL/c$. The corresponding vector of nondimensional time instants is `t=[0:0.01:N_image];`. The value of x in the nondimensional range $0 \leq x/L \leq 1$ is set in the main program and used to evaluate vectors of characteristic values `xi=x-t;    eta=x+t;`. These vectors are inputs to the `p_f_g_ic` function.

The program results are displayed in Fig. 3. The first graph shows p_1 and p_2 versus t at $x = 0$, which are the starting point in the graphical construction of waveforms, but are not needed for computations. Aside from differences in scale factors, this and the other plots are consistent with the waveforms that would be obtained if the graphical method were applied to the initial profiles in Fig. 1. The second graph depicts the pressure in the p_1 and p_2 waves at $x/L = 0.5$. It is evident that relative to the first graph, p_1 is retarded by $ct/L = x/L = 0.5$, and p_2 is advance by that amount. The total pressure is plotted in the third graph. This plot indicates that the reverberation time is very small.

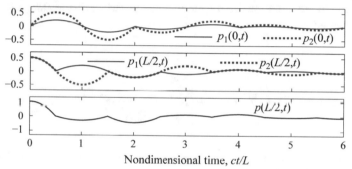

Figure 3. Computer generated waveforms at $x/L = 0$ and $x/L = 0.5$.

Excitation at a Boundary

The response within a waveguide resulting from nonzero initial conditions is not typically a concern. Rather, the important question usually is what pressure waves are induced in a finite length waveguide when the velocity at one end is a specified function of time $\dot{u}(t)$? Before we extend the wave image formulation to address this question, it is useful to contemplate what happens from a physical viewpoint. The boundary at which v is specified is taken to be $x = 0$, which is the same as the setup described by Eq. (2.5.6). Thus, movement of the boundary induces a forward propagating wave. The boundary condition at $x = L$ is an impedance $\rho_0 c \zeta_L$, so Eq. (2.5.19) governs the reflected wave. This is a backward propagating wave, so it returns to $x = 0$ where it is reflected again. The question is: What boundary condition governs this reflection? To answer this we observe that the first part of the p_1 wave, which is the one induced by movement of the boundary, already has a particle velocity at $x = 0$ that equals $\dot{u}$. Hence, there can be no additional contributions to the fluid's particle velocity from the presence of additional waves. In other words, all waves that are incident at $x = 0$ as a result of their prior reflection at $x = L$ see $x = 0$ as being rigid. This means that their reflection is governed by the reflection law in Eq. (2.5.15) with $R_0 = 1$.

These considerations allow us to extend the wave image formulation. The forward propagating wave generated by the boundary motion is described by Eq. (2.5.8), because it is a direct response to the excitation, and it is unaffected by the presence of a far boundary. Hence, the first contribution to the initial spatial profile is

$$F_0\left(x\right) = \rho_0 c \dot{u}\left(-x/c\right) h(-x) \tag{2.5.48}$$

Successive waves are obtained by constructing images whose profiles alternately satisfy the reflection laws at $x = L$ and $x = 0$. This construction is like that in Fig. 2.16, with two differences: Here, the F_0 function is nonzero only for $x < 0$. The other difference is that all images previously occupied a distance L in the solution to the initial value problem, whereas here F_0, and therefore all of the images, have lengths of ct_F, where t_F is the largest time at which $\dot{u}$ is nonzero. If $ct_F > 2L$, images will overlap. Such a situation is depicted in Fig. 2.20, where the leading edge of each image is marked by a dot.

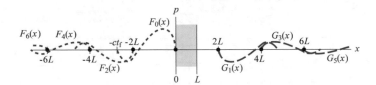

Fig. 2.20 Construction of the initial pressure profiles resulting from movement of the boundary at $x = 0$ in a finite waveguide, $R_L = -0.6$

A mathematical representation of the general response is obtained by following steps like those for the analysis of the incident wave case. We begin by constructing the initial spatial profile. The directly generated wave $F_0\left(x\right)$ is reflected at $x = L$, thereby generating a backward propagating wave described by $G_1\left(x\right) = R_L F_0\left(2L - x\right)$. This wave is reflected at $x = 0$, which generates a reflected forward propagating wave. The reflection factor for waves that are incident at $x = 0$ was shown previously to be $R_0 = 1$ because the directly generated wave already matches the wall motion of the boundary. Thus the first image for the forward propagating wave is $F_2\left(x\right) = G_1\left(-x\right) = R_L F_0\left(2L - (-x)\right)$. Reflection of this wave at $x = L$ generates another contribution to the backward propagating wave, which is described by image $G_3\left(x\right) = R_L F_2\left(2L - x\right) = \left(R_L\right)^2 F_0\left(2L + (2L - x)\right)$, *ad infinitum*. The initial profiles of p_1 and p_2 are the sum of the respective images, so that

$$\begin{aligned} p_1\left(x, 0\right) &= \sum_{n=0}^{\infty} \left(R_L\right)^n F_0\left(x + 2nL\right) \\ p_2\left(x, 0\right) &= \sum_{n=1}^{\infty} \left(R_L\right)^n F_0\left(-x + 2nL\right) \end{aligned} \tag{2.5.49}$$

To obtain general descriptions for arbitrary x, t we recall that p_1 depends solely on the forward characteristic ξ, and that p_2 depends solely on the backward characteristic η. The preceding expressions apply for $t = 0$, at which time $\xi = x$

and $\eta = x$. Hence, expressions for arbitrary t result from replacing x by ξ in p_1, while x in p_2 is replaced by η. The next step is to use Eq. (2.5.48) to describe the F_0 function. The occurrence of this function in the expression for p_1 is $F_0\,(\xi + 2nL) = \rho_0 c \dot{u}\,(-(\xi + 2nL)/c)$, whereas the term in the summation for p_2 is $F_0\,(-\eta + 2nL) = \rho_0 c \dot{u}\,(-(-\eta + 2nL)/c)$. Replacing the characteristics by their definitions in terms of x and t leads to

$$
\left.
\begin{aligned}
p &= p_1 + p_2 \\
p_1 &= \rho_0 c \sum_{n=0}^{\infty} (R_L)^n\, \dot{u}\,(t - x/c - 2nL/c) \\
p_2 &= \rho_0 c \sum_{n=1}^{\infty} (R_L)^n\, \dot{u}\,(t + x/c - 2nL/c)
\end{aligned}
\right\}, \quad 0 < x < L,\ \ t > 0
\qquad (2.5.50)
$$

This representation tells us that successive images that contribute to p_1 and p_2 depend on a retarded time that is the true time less the time required to propagate the leading edge of each image to position x.

This representation is useful for computer evaluations, but we can learn much from the pictorial representation of the initial spatial profile. One general feature already is apparent. Figure 2.20 depicts the situation for a pulse excitation. We know that each reflection at $x = L$ attenuates the amplitude of the reflected wave by the factor R_L. Even though individual images overlap, the result is that the both p_1 and p_2 decrease with increasing t. The manner in which this decrease occurs is the same as for the case where initial conditions are specified. Consequently, unless $x = L$ is rigid or pressure-release, the overall p_1 and p_2 waves essentially decay exponentially at large t, with the decay constant being β in Eq. (2.5.45). A corollary is that the notion of a reverberation time and the associated Norris–Eyring formula, Eq. (2.5.47), are equally applicable here. This is physically what should happen, because it would not make sense that sound produced by movement of the boundary behaves differently from sound that was generated prior to $t = 0$.

These assertions are only true if the excitation $\dot{u}$ at the boundary ceases at $t > t_F$. Such an excitation is said to be *transient*. A continuous excitation goes on forever, or at least, later than the time at which we cease our observations. A special case is a periodic excitation, which can produce some interesting phenomena. In Chap. 1 we found that Fourier series and transfer functions may be used to determine the response to such excitations. Here we will use the method of wave images. We consider a wall motion $\dot{u}$ that is periodic in time for $t > 0$, such that $\dot{u}\,(t + T) = \dot{u}\,(t)$. Because the image $F_0\,(x)$ directly generated by this type of motion extends to $x \to -\infty$, all subsequent images associated with reflection of the forward and backward propagating waves also extend to $|x| \to \infty$, so the number of images that overlap will increase as we look at smaller $x < 0$ for the forward propagating waves, or at larger $x > L$ for the backward waves. This will make it difficult to pick out distinctive features in a single image. An aid to doing so is to use a very simple form for $\dot{u}$, specifically, a scalene triangle lasting for an interval τ less than a period, followed by an interval of zero. The one to be used here is

$$\dot{u} = \dot{u}_{\max}\left[3\frac{t}{\tau}h\left(t\right) - \frac{9}{2}\left(\frac{t}{\tau} - \frac{1}{3}\right)h\left(t - \frac{\tau}{3}\right) + \frac{3}{2}\left(\frac{t}{\tau} - 1\right)h\left(t - \tau\right)\right], \quad 0 \leq t < T$$

$$(2.5.51)$$

where $\dot{u}_{\max}$ is the amplitude of this motion. The value of τ for all evaluations will be $0.75T$.

The directly generated initial profile is

$$F_0\left(x\right) = \rho_0 c\dot{u}\left(-x/c\right)h\left(-x\right) \tag{2.5.52}$$

This profile is spatially periodic in $x < 0$, with its pattern repeating over distance cT. The spacing between images for successive reflections matches the arrival times for the fronts of the incident and reflected waves, so it is $2L$ behind each boundary. This suggests that a crucial issue is whether cT equals $2L$. It does not for Fig. 2.21a. Successive images grow smaller because $|R_L| < 1$. The interference between images is not constructive because their spatial spacing is $2L$ but their individual wavelength is $cT = 1.42L$. These observations suggest that the amplitude of the total pressure in the steady state condition will be comparable to that of the directly generated wave, which is $\rho_0 c\dot{u}_{\max}$.

The situation in Fig. 2.21b, for which $cT = 2L$, is quite different. Behind their front, all forward propagating images are in-phase with $F_0\left(x\right)$, and all backward propagating images behind their front are in-phase with $G_0\left(x\right)$. Even though successive images are attenuated, Fig. 2.21b suggests that the combined amplitude will be substantially larger than the amplitude $\rho_0 c\dot{u}_{\max}$ of the directly generated wave. We can predict the amplitudes of p_1 and p_2 without constructing profiles. A $2L$ increase in $|x|$ adds another image to each profile. For the forward propagating wave, in the interval $-2\left(n + 1\right)L < x < -2nL$ there is $F_0\left(x\right)$ plus $n - 1$ images associated with reflections at $x = 0$. Because $x = 0$ is effectively rigid for reflections, each of these images is attenuated by a factor R_L relative the previous image, so that the jth image is $\left(R_L\right)^j F_0\left(x\right) = \left(R_L\right)^j \rho_0 c\dot{u}\left(-x/c\right)$ in this interval. It follows that the initial profile of p_1 is such that

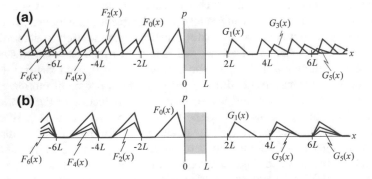

Fig. 2.21 Initial spatial profiles in a finite waveguide resulting from a periodic vibration of the boundary at $x = 0$, $R_L = 0.6$. **a** $T = 1.42L/c$, **b** $T = 2L/c$

$$p_1(x, 0) = \left[\sum_{j=0}^{n} (R_L)^j \right] \rho_0 c \dot{u} (-x/c), \quad -2(n+1) L < x < -2nL \quad (2.5.53)$$

A general expression for p_1 results when x is replaced in the above with the characteristic associated with forward propagation, $\xi = x - ct$, which gives

$$p_1(x, t) = \left[\sum_{j=0}^{n} (R_L)^j \right] \rho_0 c \dot{u} \left(t - \frac{x}{c} \right), \quad -2(n+1) L < x - ct < -2nL$$

$$(2.5.54)$$

The steady-state response may be obtained by letting $n \to \infty$ in the above expression. In this limit the summation is the binomial series expansion of $1/(1 - R_L)$, so that

$$\lim_{t \to \infty} p_1(x, t) = \frac{1}{1 - R_L} \rho_0 c \dot{u} \left(t - \frac{x}{c} \right), \quad 0 < x < L, \quad t > 0 \quad (2.5.55)$$

The images forming the initial profile of p_2 behave in a similar manner. As exhibited by Fig. 2.21b, there is no direct wave, and the first image of $F_0(x)$ is $G_0(x) - R_L F_0(2L - x) - R_L \rho_0 c \dot{u} (-(2L - x)/c)$. Each subsequent image multiplies the previous image by R_L and shifts it by distance $2L$ farther behind $x = L$. Thus, we find that the backward propagating waves initial profile is given by

$$p_2(x, 0) = \left[\sum_{j=1}^{n} (R_L)^j \right] \rho_0 c \dot{u} \left(\frac{x - 2L}{c} \right), \quad 2nL < x < 2(n+1) L, \quad n > 0$$

$$(2.5.56)$$

Replacement of x in the initial profile of p_2 with $\eta = x + ct$ leads to yields the general description,

$$p_2(x, 0) = \left[\sum_{j=1}^{n} (R_L)^j \right] \rho_0 c \dot{u} \left(t + \frac{x - 2L}{c} \right), \quad 2nL < x + ct < 2(n+1) L, \quad n > 0$$

$$(2.5.57)$$

Aside from the absence of the $n = 0$ term, this summation is the same as that for the p_1 profile. Thus, as $n \to \infty$, the summation approaches $1/(1 - R_L) - 1 \equiv R_L/(1 - R_l)$.

$$p_2(x, t) \to \frac{R_L}{1 - R_L} \rho_0 c \dot{u} \left(t + \frac{x - 2L}{c} \right), \quad 0 < x < L, \quad t > 0 \quad (2.5.58)$$

This analysis is confirmed by computations of waveforms. Figure 2.22 depicts the waveforms of p_1 and p_2 at $x = 0$, and the waveform of the total pressure at $x = L/2$. The period for this case is $cT = 1.42L$, which is not expected to lead to significant enhancement of the pressure. The steady-state amplitudes of p_1 and p_2 are substantially less than $\rho_0 c \dot{u}_{max}$. The pressure amplitude obtained from their sum at $x = L/2$ is approximately $\rho_0 c \dot{u}_{max}/2$. The oscillation is relative to a mean value that results from $\dot{u}$ having a nonzero mean.

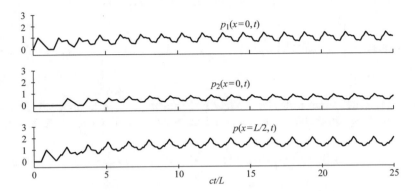

Fig. 2.22 Waveforms in a finite waveguide generated by vibration of the boundary at $x = 0$ when $T = 1.42L/c$, $R_L = 0.6$

The situation for $cT = 2L$ is described by Fig. 2.23. The waveforms of p_1 and p_2 attain their steady-state value in the vicinity of $ct = 12L$, and the steady-state amplitudes of each agree with Eqs. (2.5.55) and (2.5.58). The result is a much larger amplitude for the pressure at $x = L/2$.

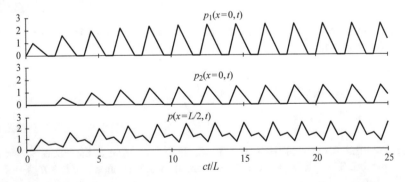

Fig. 2.23 Waveforms in a finite waveguide generated by vibration of the boundary at $x = 0$ when $T = 2L/c$, $R_L = 0.6$

An interesting question now arises. Suppose $R_L = 1$ and $cT = 2L$. According to Eqs. (2.5.55) and (2.5.58), this means that the late time versions of p_1 and p_2 will be infinite. Can this be correct? Before we view computed waveforms, let us consider how letting $R_L = 1$ would change Fig. 2.21b. The image for every reflected wave would be situated as it is in this figure, except but all images would be the same as that of $F_0(x)$ for the directly generated wave. Thus, the spatial profiles of p_1 and p_2 would grow with increasing $|x|$ as more and more images are added. This means that the waveforms of these waves would grow in magnitude essentially linearly with increasing t. This line of reasoning leads us to the conclusion that the waves will approach infinite amplitude, but only as $t \to \infty$. Figure 2.24 confirms this expectation.

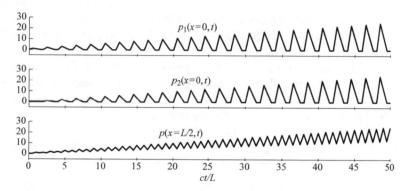

Fig. 2.24 Waveforms in a finite waveguide generated by vibration of the boundary at $x = 0$ when $T = 2L/c$, $R_L = 1$

Of course, this is a highly idealized situation. If the pressure amplitudes were to grow to large levels, nonlinear effects would become significant, thereby invalidating the entire analysis. Furthermore, no energy is dissipated in the model of a rigid wall, which obviously is unrealistic. If R_L is extremely close to one, the denominator in Eqs. (2.5.55) and (2.5.58) will be finite, but very small. In that case the late time versions of p_1 and p_2 will have a very large, but finite, steady-state amplitude. The situation where R_L is close to or equal to one and $cT = 2L$ is said to be a *resonance*. We previously encountered this phenomenon in Example 1.1 of Chap. 1. There the system had a single natural frequency, and the excitation was periodic. Resonance occurred if the frequency of any of the Fourier harmonics matched the system's natural frequency. In the case of a finite length waveguide with forced motion at $x = 0$, waves that are incident at that end perceive the reflection coefficient to be $R_0 = 1$. The specific system we have considered has $R_L = 1$, so its analog for an initial value problem is a waveguide with rigid ends. Hence, the natural frequencies of this system are given by Eq. (2.5.31). The fundamental (that is, lowest) natural frequency, is $\pi c/L$. The fundamental frequency of $\dot{u}$ is $2\pi/T$. Thus, the condition $cT = 2L$ corresponds to an unabated system resonance. If we reduce R_L slightly, some energy is removed in each reflection at $x = L$, thereby inhibiting unlimited growth.

A corollary of this discussion is that resonance can be expected to occur in systems for which R_L is close to one whenever the frequency of any harmonic in the Fourier series for $\dot{u}$ matches any natural frequency. That is, a resonance will occur if there are two integers j and n for which $j(2\pi/T) = n(\pi c/L)$. These observations lead to the question of whether resonance is unique to the case where $x = L$ is rigid? The answer is no! Its occurrence requires that there be little dissipation, which is satisfied if $|R_L|$

is close to one. Thus, the pressure-release condition also can lead to resonance. The effective reflection coefficient at $x = 0$ is $R_0 = 1$ because the wall motion is specified, so setting $R_L = -1$ yields the rigid/pressure-release configuration described by Eq. (2.5.37). Like the behavior of the system in the previous case, resonance of this system occurs if any harmonic of periodic $\dot{u}$ has a period that matches one of the system's natural frequencies, that is, if $j\,(2\pi/T) = (2n - 1)\,\pi c/2L$. The lowest resonance corresponds to $j = n = 1$, which gives $cT = 4L$. Another configuration that can exhibit resonance is $R_0 = R_L = -1$. Motion at $x = L$ cannot be imposed in this case, but a resonance could be attained if the excitation at $x = 0$ were a specified pressure as a function of time.

In most situations resonance is a phenomenon to be avoided. For example, in a symphony hall instruments that play notes in the vicinity of resonant frequencies will be too loud relative to other instruments. Furthermore, these instruments will not sound the way they should, because one or more of the harmonics they generate will be too large relative to the others. However, the resonance phenomenon can be useful. It is essential to the response characteristics of every musical instrument, and it is exploited to make some highly efficient transducers.

EXAMPLE 2.7 A waveguide, whose length is L, is filled with water. The specific impedance at one end is ζ, while the other end executes a periodic square wave vibration in which the velocity is $v\,(t) = \varepsilon c$ if $0 < t < T/2$, $v\,(t) = -\varepsilon c$ if $T/2 < t < T$, $v\,(t) = v\,(t - T)$ if $t > T$. Parameters for the system are $L = 2$ m, $\rho_0 = 1000$ kg/m^3, $c = 1480$ m/s, and $\varepsilon = 10^{-5}$. (a) Determine the values of T for which the pressure within the waveguide will exhibit a resonance. (b) Write a computer program to evaluate waveforms at any x. (c) Use the computer program to evaluate the waveform at $x = 0$. Cases of interest are $T = 4L/\,(3c)$ and $\zeta = 0$; $T = 4L/\,(3c)$ and $\zeta = 0.02$; $T = (0.99)\,4L/\,(3c)$ and $\zeta = 0$. For each T and ζ combination examine the waveform to determine the time at which the pressure excursion, positive or negative, will first exceed 400 Pa.

Significance

The computer program that we will develop may be used as an independent tool for the exploration of a variety of phenomena. Its application here will provide a more detailed picture of how the excitation frequency and dissipation affect the severity of resonances.

Solution

Part (a) In each case the impedance at $x = L$ is very small, so the end is effectively pressure-release. Because the velocity at $x = 0$ is specified, that end is rigid for considerations pertaining to natural frequencies and modes. Hence, we can anticipate that resonance will occur when the any multiple of the excitation frequency equals any natural frequency of a rigid/pressure-release waveguide whose length is L. Equation

(2.5.37) indicates that the natural frequencies of a waveguide with $R_0 = 1$ and $R_L = -1$ are $\omega_n = (2n-1)\pi c/(2L)$. The square wave is periodic in intervals of T, so its fundamental frequency is $2\pi/T$. The frequencies of the harmonics in its Fourier series are $j(2\pi/T)$. Equating one of each type of frequency gives

$$\frac{2\pi j}{T} = \frac{(2n-1)\pi c}{2L} \Longrightarrow \frac{cT}{L} = \frac{4j}{(2n-1)}$$

The first harmonic of the excitation ($j = 1$) matches a natural frequency of the waveguide if $T = 4L/c, 4L/(3c), \ldots = 5.405, 1.802, \ldots$ ms. Furthermore, the jth harmonic matches a natural frequency if T is j times any of these values. The period that is specified is either equal or close to $4L/(3c)$. Thus, the specified period corresponds to the first harmonic of the excitation resonantly exciting the second mode of the waveguide.

The preceding resonance criterion applies for an arbitrary periodic $\dot{u}$, but another condition must be met. The preceding resonance condition assumes that the jth Fourier series coefficient of $\dot{u}$ is nonzero. In the case of the specified square wave, the Fourier coefficients for even j are identically zero. Thus, resonances occur only if the first, third, fifth, etc. harmonic of $\dot{u}$ matches the frequency of any mode.

Part (b) The issue here is computerized evaluation of waveforms, so there is no need to determine the initial profile from which the graphical procedure begins. It will be necessary to compute $\dot{u}$ for many different sets of arguments, and the computation is best done in a two step process. A separate function is most suitable to this purpose. The square wave is periodic, with the first half of each period being $\dot{u} = \varepsilon c$ and the second half being $\dot{u} = -\varepsilon c$. The argument at which $\dot{u}$ is to evaluated might fall into any period. It is placed in an interval $0 \le t' < T$ by employing the modulus function, whose definition is $\text{mod}(a, b) \equiv a - b\,\text{floor}(a/b)$. Also, we must set $\dot{u} = 0$ if $t < 0$. In Matlab, these operations may be implemented as a function M-file, u_dot.m, whose code is

```
function v_boundary=u_dot(t,T_period)
t_prime=mod(t, T_period);
v_boundary=(t > =0).*((t_prime > 0)-2*(t_prime > T_period/2));
```

Let us temporarily forego consideration of the number of images that must be incorporated to the analysis. The basic algorithm for evaluating p_1 and p_2 is like the one for an initial value problem in the previous Example. Here the task is to evaluate a single waveform corresponding to each set of parameters, so all operations can be done in the main program. The two sums in Eq. (2.5.50) are computed in a loop. In the following program fragment the variable tau_1 is $t - x/c$ and tau_2 is $t + x/c$. The summation for p_1 is initialized with the $n = 0$ term, and the loop increments the sum for each $n = 1, \ldots, N$. Similar steps evaluate p_2. There is no $n = 0$ term in the summation for p_2, so it is initialized by a set of zeros. Matlab code that performs these operations is:

```
tau_1=t-x/c; tau_2=t+x/c;
p_1=eps*rho*c*u_dot(tau_1,T);   p_2=0*p_1;
for n=1:N
        p_1=p_1+eps*rho*c*R_L^ n*u_dot(tau_1-2*n*L,T_period);
        p_2=p_2+eps*rho*c*R_L^ n*u_dot(tau_2-2*n*L,T_period);
end
```

Prior to arriving at this program fragment, a vector of values of t for the waveform, as well as the values of x, R_L and T, must be defined.

The maximum time for computation of waveforms is specified indirectly by the requirement that we should determine the time at which the pressure rises to 400 Pa. Let this time be t_{max}, and let the time increment be Δt. The latter should be sufficiently small to represent the discontinuities of the square wave. The results that follow were obtained with $\Delta t = T/1000$. The instants at which the waveform will be evaluated are $t_j = j\Delta t$. The number of images N must be sufficiently large that t_{max} for evaluation of a waveform is less than the time required for the leading edge of image $N + 1$, which is the first to be omitted, to arrive at the waveguide. The leading edge of image n for p_1 is at $2nL$ behind $x = 0$, whereas image n for p_2 is $(2n - 1) L$ beyond $x = L$. Thus, we require that $2(N + 1) L/c > t_{max}$, which means that $N > ct_{max}/(2L) - 1$. Trial and error was used to identify the value of t_{max} for which the pressure reaches the 400 Pa. Eventually, it was extended in order to exhibit some interesting effects.

Part (c) The reflection coefficient is $R_L = (\zeta - 1) / (\zeta + 1)$, so the three parameter combinations each correspond to R_L being close to minus one. The first combination, $T = 4L/(3c)$ and $\zeta = 0$, corresponds to an uncontrolled resonance because R_L is exactly minus one. Consequently, the images are not attenuated. The result is Fig. 1. It can be seen that the pressure at $x = 0$ grows in amplitude linearly with elapsed time. The instant when $|p|$ at $x = 0$ first exceeds 400 Pa may be found by scanning the output, which is more accurate than identifying that value in the graph. The result is 37.85 ms.

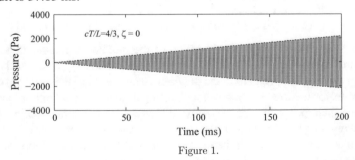

Figure 1.

Another case where the fundamental frequency of $\dot{u}$ exactly equals the second natural frequency is described by Fig. 2, for which $\zeta = 0.02$. The reflection coefficient is $R_L = -0.961$, so successive images are attenuated. Initially, the pressure amplitude increases, but at a slower rate than it does in Fig. 1. The pressure first surpasses 400 Pa at $t = 52.26$ ms. Eventually, the amplitude approaches a steady value.

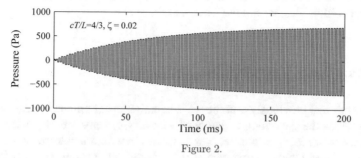

Figure 2.

The behavior when $R_L = -1$ and a harmonic of $\dot{u}$ is very close to, but not equal to, a natural frequency is unlike either of the previous cases. Indeed, even experienced individuals find it surprising. Figure 3 shows a waveform that is reminiscent of a beating signal. Indeed, this is the process, but proving it requires a different line of analysis. The beating occurs between many harmonics of $\dot{u}$ and many modes. Note in this regard that because $cT/L = 4/3$, the resonance criterion identified above states that resonances occur if $3j = 2n - 1$. Recall that j must be odd because the even harmonics of $\dot{u}$ are zero. Thus, the resonances are mode 2 by harmonic 1, mode 5 by harmonic 3, mode 8 by harmonic 5, and so on. The fact that the envelope of the beat is diamond shaped is a consequence of the simultaneous occurrence of these multiple resonances. If $\dot{u}$ were a sine function at the same pressure, the envelope would also be a sine. A 400 Pa pressure is first reached 38.21 ms, which is only slightly later than the value for Fig. 1.

An overview of these three cases suggests methods to limit acoustic resonances. One possibility is to cut off the excitation before an unacceptable state is attained. However, the pressure builds up quickly, so one must be prepared to act before the pressure level is unacceptable. If doing so is not feasible, the only recourse is to alter the parameters. The easiest alteration is to shift the excitation's period sufficiently to reduce the steady-state pressure amplitude to an acceptable value. As evidenced by Fig. 3, the detuning of the excitation frequency relative to a natural frequency must be reasonably substantial. However, in some situations the excitation's frequency is

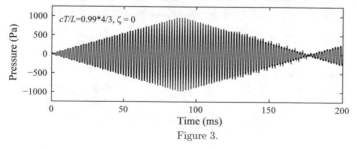

Figure 3.

mandated by other considerations. In that case raising the amount of dissipation is the only recourse. This is done in general by using dissipative liner materials, like acoustical ceiling tiles, within the waveguide.

2.6 Analogous Vibratory Systems

It was mentioned previously that the wave equation is not unique to acoustics. Two occurrences as the governing equation of vibratory systems are of particular interest because in some situations acoustic response in a fluid is intertwined with their vibration. A cable that is prestretched by a large tensile force, and an elastic bar provide accessible prototypes for investigating the interaction of acoustic waves in fluids and the vibration of solid bodies. It therefore is appropriate that we understand the commonality of these systems, which superficially appear to be quite dissimilar. Identification of the analogies between these systems and acoustic signals in a waveguide will provide a global perspective for the wave equation.

2.6.1 Stretched Cable

Field Equations

Linear vibration theory for solid media is like linear acoustics, in that both rely on some core assumptions and approximations. One difference in the approach to the derivation is it is possible to track the particles of a solid object, for example, by marking points with ink. Figure 2.25 shows a segment of a cable bounded by two interior locations x_1 and x_2 that are arbitrarily selected. The cable is assumed to displace solely in the transverse y direction, which is the first primary assumption on which the derivation lies. We seek an equation of motion governing the displacement field $w(x, t)$ measured from the x axis where the cable is situated when it is in static equilibrium. The situation addressed here is one in which the cable carries a very large tensile force, even if it is at static equilibrium. The angle θ locates the tangent to the curve formed by plotting w as a function of x at an arbitrary t.

Fig. 2.25 Isolated segment of a stretched cable showing the internal tensile force

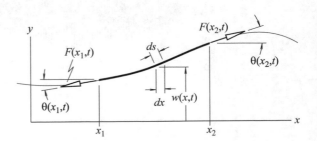

We begin with the time derivative of the momentum-impulse equation for the system of particles of which the isolated cable segment is composed. Isolating the cable segment in Fig. 2.25 exposes the tensile force F applied by the other portions of the cable. Allowance is made for the tensile force to be variable, but we will soon see that the actual situation is much simpler. Gravity is excluded from the formulation, which can be justified either on the basis that the motion occurs in the horizontal plane, or more generally by recognizing that gravity has a static effect that would merely add a time-invariant displacement that is known as the sag. In regard to inertial effects, the mass per unit length of the cable is the constant value μ. The infinitesimal arclength ds is measured along the cable, so dx is the projection of ds onto the x axis. Thus, $ds = dx / \cos (\theta)$ and the differential element of mass is $\mu dx / \cos (\theta)$.

The acceleration of a differential element is $\partial^2 w / \partial t^2$ in the y direction. The acceleration in the x direction is zero, so there must be a force balance in that direction,

$$\sum F_x = [F \cos (\theta)]|_{x_2} - [F \cos (\theta)]|_{x_1} = 0 \tag{2.6.1}$$

The second assumption is introduced here. Specifically, the maximum of the displacement is taken to be much smaller than the distance between the cable's ends. In that condition a profile of w as a function of x would show that the cable everywhere is close to the x axis, so θ is much less than one in magnitude. Consequently, we may use

$$\cos (\theta) \approx 1, \quad \sin (\theta) \approx \theta \tag{2.6.2}$$

The approximation for $\sin \theta$ will soon be used for the y momentum equation. The approximation for $\cos (\theta)$ converts the force balance in the x direction to a statement that $F (x_2, t) = F (x_1, t)$. Because x_1 and x_2 were selected arbitrarily, it must be that the same tensile force occurs at all locations at any instant. The third assumption simplifies the situation further. It was postulated at the outset that the cable carries a large tensile force F_0 when it is situated in its undisplaced configuration along the x axis. Such a large force would stretch the cable considerably before any displacement occurs. A displacement field $w (x, t)$ would stretch the cable further, but we have restricted our considerations to situations where $w (x, t)$ is very small. In that case the additional stretch is also small, so we can take the tensile force at any instant to be unchanged from the force within the cable at static equilibrium,

$$F (x, t) \approx F_0 \tag{2.6.3}$$

We use this approximation to formulate the y momentum equation, which is

$$[F_0 \sin (\theta)]|_{x_2} - [F_0 \sin (\theta)]|_{x_1} = \int_{x_1}^{x_2} \left(\mu \frac{dx}{\cos (\theta)} \right) \frac{\partial^2 w}{\partial t^2} \tag{2.6.4}$$

Equation (2.6.2) yields a linearized version of this equation of motion. We also invoke the one-dimensional divergence theorem to convert the terms at x_1 and x_2 to an integral, specifically,

$$[F_0 \sin (\theta)]|_{x_2} - [F_0 \sin (\theta)]|_{x_1} \approx [F_0 \theta]|_{x_2} - [F_0 \theta]|_{x_1} = \int_{x_1}^{x_2} F_0 \frac{\partial \theta}{\partial x} dx \quad (2.6.5)$$

The result of these operations is

$$\int_{x_1}^{x_2} \left[F_0 \frac{\partial \theta}{\partial x} - \mu \frac{\partial^2 w}{\partial t^2} \right] dx = 0 \quad (2.6.6)$$

The integral will vanish for any combination of x_1 and x_2 only if the integrand vanishes. The result is

$$\boxed{F_0 \frac{\partial \theta}{\partial x} - \mu \frac{\partial^2 w}{\partial t^2} = 0} \quad (2.6.7)$$

Another equation is needed to eliminate θ. For an acoustic signal, Euler's equation is supplemented by the continuity condition. The latter is a kinematical requirement that prevents the occurrence of gaps. Similarly, θ is eliminated by a kinematical condition. If θ were discontinuous, there would be a kink, that is a sharp bend, in the shape of the $w (x, t)$ curve. This possibility is eliminated by recognizing that θ is the angle of the tangent to the $w (x, t)$ curve from the x axis, but that angle can be found from the slope $\partial w / \partial x$. Thus,

$$\tan (\theta) \approx \theta = \frac{\partial w}{\partial x} \quad (2.6.8)$$

Substitution of this relation into the momentum-impulse equation leads to the field equation we seek,

$$\boxed{F_0 \frac{\partial^2 w}{\partial x^2} - \mu \frac{\partial^2 w}{\partial t^2} = 0} \quad (2.6.9)$$

This is the one-dimensional wave equation for the vibration of a cable.

Analogies

The process of identifying how the equations governing motion of a stretched cable are analogous to acoustic relations entails comparison of similar types of equations. Our first step compares the momentum equation, Eq. (2.6.7), and Euler's equation, which may be written as

$$\frac{\partial p}{\partial x} + \rho_0 \frac{\partial v}{\partial t} = 0 \quad (2.6.10)$$

Because $\partial w / \partial t$ is a velocity, the two have analogous forms, with $F_0 \theta$ being like $-p$ if we consider μ to be analogous to ρ_0. (The difference in sign may be traced to the definition of F_0 as a tension, whereas positive p is compressional.)

The analogies for the continuity equation are recognizable by differentiating Eq. (2.6.8) with respect to t, then multiplying the equation by F_0. The relation for acoustics is given by Eq. (2.2.11). The comparable forms are

$$\frac{\partial}{\partial t}(F_0 \theta) - F_0 \frac{\partial}{\partial x}\left(\frac{\partial w}{\partial t}\right) = 0 \iff \frac{\partial p}{\partial t} + \rho_0 c^2 \frac{\partial v}{\partial x} = 0 \qquad (2.6.11)$$

The analogy of $-F_0 \theta$ to p holds if we identify F_0 as the bulk modulus $K = \rho_0 c^2$. This also reinforces identification of the transverse velocity of the cable, $\partial w / \partial t$, as the analog to the fluid particle velocity v. The analogies we have established are summarized in the follow table.

Acoustic fluid	p	v	K	ρ_0
Flexible cable	$-F_0 (\partial w / \partial x)$	$\partial w / \partial t$	F_0	μ

A basic difference between the one-dimensional wave equation for a cable and for a fluid is that the former governs displacement, whereas the latter governs a force-like quantity. Either equation is readily transformed to the type of variable for the other equation. For example, if we differentiate Eq. (2.6.9) with respect to x, then multiply it by F_0, we obtain a wave equation governing the transverse force component, $F_0 (\partial w / \partial x)$. Comparing that equation to the equation for a fluid leads to

$$\frac{\partial^2}{\partial x^2}\left(F_0 \frac{\partial w}{\partial x}\right) - \frac{\mu}{F_0}\frac{\partial^2}{\partial t^2}\left(F_0 \frac{\partial w}{\partial x}\right) = 0 \iff \frac{\partial^2 p}{\partial x^2} - \frac{1}{c^2}\frac{\partial^2 p}{\partial t^2} = 0 \qquad (2.6.12)$$

These forms are similar. The phase speed of transverse waves is found by matching the equations, which leads to

$$c = \left(\frac{F_0}{\mu}\right)^{1/2} \qquad (2.6.13)$$

Boundary conditions describe the manner in which the cable is supported at its end. The physical arrangement must be such that the axial tension is sustained. This requirement is inherently met for a fixed end, at which the cable is anchored to an immovable object, like a wall. An anchor prevents displacement. Because $w = 0$ for all t at that end, it must be that $\partial w / \partial t = 0$, which is analogous to a rigid termination in an acoustic waveguide. It also is possible that the end is shaken in the transverse direction while the tension is provided by pulling the end in the axial direction. This would correspond an acoustic waveguide in which the particle velocity is specified at an end.

Other types of boundary conditions are more difficult to obtain. This is so because they must allow for transverse movement of the end concurrently with resisting the axial force. The analogy of a pressure-release end in a waveguide is $F_0\theta = 0$. A realization of this condition can be obtained with a roller that moves within a groove that is aligned transversely. This is the configuration depicted in Fig. 2.26. The string is tied to the axle of the roller. The displacement of the roller from the reference position is $w\,(\xi = 0, t)$ and the slope is $\theta\,(0, t)$. Based on the assumption that the roller's inertia is negligible, the vertical component of the force exerted by the cable on the roller must be zero. In other words, $F_0\theta = 0$ at this end.

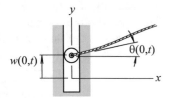

Fig. 2.26 A roller support at the end of a cable that produces the analog to a pressure-release boundary condition

A small modification of the arrangement in Fig. 2.26 leads to an impedance boundary condition. In an acoustical waveguide this condition is attained if the pressure applied to a wall produces a proportional velocity. The analog for a cable is a transverse velocity that is proportional to the vertical component of the tensile force. The nonzero force component may be obtained by attaching a viscous dashpot to the roller. Such a configuration is depicted in Fig. 2.27 for the end at $x = 0$, wherein the end of the cable is tied to the center of a roller.

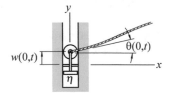

Fig. 2.27 A dashpot whose viscous resistance is η leads to an impedance boundary condition for a stretched cable

The movement of the roller is opposed by the dashpot, whose resistance (force/velocity) is denoted as χ. Upward movement of the roller is $w\,(x = 0, t)$. The force exerted by the dashpot on the roller is $\chi \partial w\,(0, t)\,/\partial t$, downward if the velocity is upward. If the inertia of the roller is negligible, the cable must exert an upward force that balances the force applied by the dashpot. The resulting boundary condition is $\chi(\partial w/\partial t) = F_0\theta$ at $x = 0$. In an acoustical waveguide, a specific impedance ζ at $x = 0$ leads to $p = -\rho_0 c \zeta v$ at that end. That this is analogous to the condition in the cable is evident by the analogy of p to $-F_0\theta$. The dashpot resistance χ is analogous to $\rho_0 c \zeta$. If χ is infinite, it must be that $\partial w/\partial t = 0$ at that end for all t, which leads to $w = 0$ at that end. Thus, any boundary condition governed by a velocity impedance may be written as

$$\chi_0 \frac{\partial w}{\partial t} = F_0 \frac{\partial w}{\partial x} \text{ at } x = 0, \ \chi_L \frac{\partial w}{\partial t} = -F_0 \frac{\partial w}{\partial x} \text{ at } x = L, \ \chi_0, \chi_L \geq 0 \quad (2.6.14)$$

Application of the Method of Wave Images

It is awkward to implement the analogies when we wish to use the method of wave images. This is so because the typical initial conditions are specification of the cable's displacement and transverse velocity. To use the analogies we would need to work with the cable's slope, rather than its displacement. For this reason we shall rederive the basic formulation. We decompose the displacement of the cable into forward and backward waves,

$$w = f(x - ct) + g(x + ct) \quad (2.6.15)$$

The corresponding description of the transverse velocity is

$$\frac{\partial w}{\partial t} = -cf'(x - ct) + cg'(x + ct) \quad (2.6.16)$$

Let $W(x)$ and $V(x)$ respectively denote the initial displacement and velocity profiles. Matching these functions to the preceding general solution evaluated at $t = 0$ gives

$$W(x) = f(x) + g(x), \quad V(x) = -cf'(x) + cg'(x) \quad (2.6.17)$$

If we are to implement the d'Alembert solution in the same manner as we did for a pressure field, we must determine the f and g functions, rather than their derivatives. To that end we integrate the velocity initial condition, which we may do because its sole independent variable is x. Doing so yields

$$U(x) = \int V(x)dx + C = -cf(x) + cg(x) + C \quad (2.6.18)$$

where C is a constant of integration. The constant of integration C is not arbitrary. Rather it must be selected such that the displacement is continuous at the leading edges, which are the junction of the disturbed and undisturbed regions. This consideration is addressed in the next Example. The last step prior to constructing images is to solve the equations relating W and U to f and g. The result is

$$f(x) = \frac{1}{2}\left[W(x) - \frac{1}{c}U(x)\right], \quad g(x) = \frac{1}{2}\left[W(x) + \frac{1}{c}U(x)\right] \quad (2.6.19)$$

If the cable's extent were infinite, these relations would be sufficient to evaluate the displacement at arbitrary x and t. Other configurations require that we consider reflections. To do so we require that Eq. (2.6.15) satisfy the general boundary condition at $x = 0$ in Eq. (2.6.14). Thus,

$$\chi_0 \frac{\partial w}{\partial t}\bigg|_{x=0} = F_0 \frac{\partial w}{\partial x}\bigg|_{x=0} \implies \chi_0 c[-f'(-ct) + g'(ct)] = F_0[f'(-ct) + g'(ct)]$$
(2.6.20)

The incident wave is $g(x + ct)$, so we solve for the reflected wave,

$$f'(-ct) = R_0 g'(ct)$$
(2.6.21)

where the reflection coefficient is

$$\boxed{R_0 = \frac{c\chi_0 - F_0}{c\chi_0 + F_0}}$$
(2.6.22)

To generalize the reflection rule to describe the forward wave at arbitrary x, we observe that in general f is solely a function of $\xi = x - ct$. At $x = 0$, $\xi = -ct$, so we replace $-ct$ with $x - ct$ in the reflection equation, which leads to

$$f'(x - ct) = R_0 g'(ct - x)$$
(2.6.23)

This is the same relation as that which governs the forward and backward pressures in a waveguide. However, the analysis is not complete because we wish to describe the displacement. This requires that we integrate the preceding. Because $(\partial/\partial x) f(x - ct) = f'(x - ct)$, whereas $(\partial/\partial x) g(ct - x) = -g'(ct - x)$, the integration leads to

$$\boxed{f(x - ct) = -R_0 g(ct - x)}$$
(2.6.24)

It might seem as though the law is reversed, in that a fixed end corresponds to $R_0 = 1$, which means that the reflected wave is inverted. In contrast, a free end, for which $R_0 = -1$, results in upright reflection. This apparent paradox is a consequence of the analogy of p to the slope $\partial w/\partial x$, whereas we are interested in w.

EXAMPLE 2.8 A cable is fixed at one end. Its other end is extremely distant, so that reflections from that end are irrelevant in the time interval of interest. Thus, the cable may be considered to be semi-infinite. At $t = 0$, an initial velocity field is imparted to it, such that $\partial w/\partial t = v_0 \sin(\pi x/L)$ if $0 \le x \le L$, and $\partial w/\partial t = 0$ if $x > L$. The initial displacement is zero. (a) Determine the profiles of the forward and backward displacement waves at $t = 0$. (b) Determine the profile of the displacement at $ct = L/3$ and $ct = 4L/3$. (c) Determine the displacement waveform at $x = L/3$ and $x = 4L/3$.

Significance

The similarities and differences of the method of wave images for stretched cables and acoustic waveguides will be evident as the analysis proceeds.

Solution

The given initial velocity field is $V(x)$. Integrating it leads to

$$U(x) = -\frac{Lv_0}{\pi} \cos\left(\frac{\pi x}{L}\right) [h(x) - h(x - L)] + C$$

The initial displacement is zero, so $f(x) = -U(x)/c$ and $g(x) = U(x)/c$. If C were zero, both functions would be discontinuous at $x = 0$ and $x = L$, so we must select it to yield a displacement that is continuous. The forward wave, which corresponds to to $f(x)$, is the sole contributor to the displacement field in $x > L$. The displacement immediately ahead of this wave is zero according to the stated initial conditions. Continuity requires $f(x)$ be zero at its leading edge. This leads to the condition that $U(L) = 0$, which is met be selecting C. Doing so leads to

$$U(x) = -\frac{Lv_0}{\pi}\left[1 + \cos\left(\frac{\pi x}{L}\right)\right] [h(x) - h(x - L)] = -cf(x) = cg(x)$$

It is evident that $f(x)$ and $g(x)$ are continuous at $x = L$. Their discontinuity at $x = 0$ is addressed in the reflection process.

In the method of wave images the function representing the initial profiles for all values of x, negative and positive, are denoted as $F(x)$ and $G(x)$. The preceding descriptions of $f(x)$ and $g(x)$ respectively represent these function for $x > 0$, that is,

$$F(x) = f(x) \text{ and } G(x) = g(x) \text{ if } x > 0$$

Because $f(x)$ represents a wave that propagates in the direction of increasing x, it is not reflected at $x = 0$. Therefore, it does not lead to an image behind the boundary. This is contrasted by $g(x)$, which represents a wave that is reflected at $x = 0$. This reflection is described by adding an image to $F(x)$. Equation (2.6.24) describes this image. The reflection coefficient for a rigid termination is $R_0 = 1$. Hence, the image in $x < 0$ is

$$F(x) = -R_0 g(-x) = -\frac{1}{c}U(-x), \quad x < 0$$

Thus, we have established that wave functions for the initial profiles of the forward and backward waves are

$$F(x) = \frac{1}{c}[-U(x)h(x) - U(-x)h(-x)]$$
$$G(x) = \frac{1}{c}U(x)h(x)$$

Figure 1 graphs both functions.

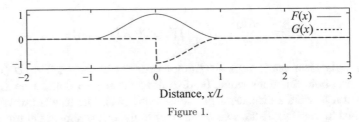

Figure 1.

The remainder of the analysis proceeds as it would for an acoustic wave. A profile at a specified t is obtained by translating $F(x)$ by ct in the positive x direction, which produces $F(x - ct)$. Similarly, $G(x)$ is translated by ct in the direction of decreasing x, thereby producing $G(x + ct)$. The displacement is obtained by adding these functions, then deleting the region $x < 0$, which is not part of the actual cable. The result of this process for the two requested instants is depicted in Fig. 2.

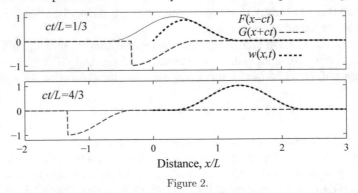

Figure 2.

Construction of waveforms begins by graphing $G(ct)$ and $F(-ct)$ along the ct axis. These are the contributors to the displacement waveform at $x = 0$. According to the given initial conditions, this waveform should be zero. This attribute is readily verified. Waveforms at a specified $x > 0$ are obtained by translating $F(-ct)$ by $+x$, and $G(ct)$ by $-x$, then adding the shifted functions. The result for $x = L/3$ and $x = 4L/3$ appear in Fig. 3.

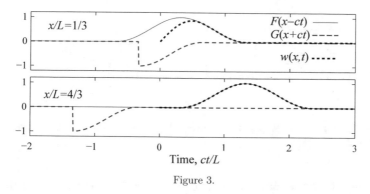

Figure 3.

The preceding analysis illustrates the main alterations required to employ the method of wave images to describe wave propagation along a stretched cable. The simplest is reversal of sign for the rule governing reflections for the comparable acoustic waveguide. The sign change in the formulas for the initial wave functions $f(x)$ and $g(x)$ relative to the acoustic relations also is easy to introduce. The task requiring careful consideration is integration of the initial velocity field consistent with continuity conditions. After the initial profiles have been constructed by combining the wave images, evaluation of displacement of a cable proceeds in the same way as it does for pressure in a waveguide.

2.6.2 Extensional Waves in an Elastic Bar

Field Equations

Our concern here is with the axial displacement $u(x, t)$ that occurs within a straight elastic bar when the internal axial stress σ is unsteady. It might seem that this system is directly analogous to a one-dimensional acoustic waveguide, because axial stress is like pressure. However, the dependent variable of primary interest for a bar is the displacement of cross-sections, so the equations will be closer to those for a stretched cable.

We begin with the momentum-impulse law for a segment of the bar between two arbitrary locations. In Fig. 2.28 a tensile force $F = \sigma A$ acts at each end. At an arbitrary instant a cross section that was at axial position x has displaced by $u(x, t)$ to its current location. The velocity of this cross section is $\partial u / \partial t$.

The mass per unit length of the bar is the product of its density ρ_0 in the unstretched state and the cross-sectional area A. The element of mass is contained between two cross-sections that were separated by distance dx prior to displacement. Equating the rate of change of the momentum of all mass elements to the force acting on this system yields

$$\sigma A|_{x_2} - \sigma A|_{x_1} = \int_{x_1}^{x_2} (\rho_0 A\, dx) \frac{\partial^2 u}{\partial t^2} \qquad (2.6.25)$$

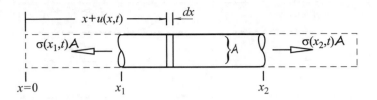

Fig. 2.28 Free body diagram describing the longitudinal forces acting on an isolated segment of an elastic bar

Application of the one-dimensional divergence theorem converts this relation to

$$\int_{x_1}^{x_2} \left(A \frac{\partial \sigma}{\partial x} - \rho_0 A \frac{\partial^2 u}{\partial t^2} \right) dx = 0 \qquad (2.6.26)$$

As was asserted previously, the integral relation will hold for any x_1 and x_2 only if the integrand vanishes, so the momentum equation becomes

$$\boxed{\frac{\partial \sigma}{\partial x} - \rho_0 \frac{\partial^2 u}{\partial t^2} = 0} \qquad (2.6.27)$$

We seek a field equation for u, which means that σ must be eliminated. If the strain is not excessive, many materials behave elastically according to a linear stress-strain law. In the one-dimensional case of a bar the relation is Hooke's law, $\sigma = E\varepsilon$, where E is Young's modulus and ε is the axial strain. A small stress implies that the strain also is small, so we may use the linearized definition of strain $\varepsilon = \partial u/\partial x$. Thus the role of the continuity equation is fulfilled here by the combination of the stress-strain and strain-displacement equations, which give

$$\sigma = E \frac{\partial u}{\partial x} \qquad (2.6.28)$$

Substitution of this relation into the momentum equation produces the wave equation for extensional waves in a bar,

$$\boxed{E \frac{\partial^2 u}{\partial x^2} - \rho_0 \frac{\partial^2 u}{\partial t^2} = 0} \qquad (2.6.29)$$

The phase speed of extensional waves is

$$\boxed{c = \left(\frac{E}{\rho_0} \right)^{1/2}} \qquad (2.6.30)$$

A consequence of the similarity of the equation for elastic bars and stretched cables is the different way the reflection coefficient is applied. As was true for a cable, an incident displacement wave is reflected with the negative of the coefficient for a pressure wave. Thus, reflection at a fixed end leads to an inverted displacement wave, and reflection at a free end leads to an upright reflected wave. An interesting aspect is that it is easier to implement physically a free or impedance boundary condition for a bar. This is so because the bar can maintain its shape without external action. In contrast, the axial force must be applied at the end of a cable, regardless of other aspects of the end condition. An impedance boundary condition is quite realizable for an elastic bar. Suppose a dashpot aligned axially resists movement of the end $x = 0$ of an elastic bar. A positive strain corresponds to elongation. Thus if the end is moving in the positive x direction, so that $\partial u/\partial t$ is positive, the dashpot will exert a tensile force $\chi\,(\partial u/\partial t)$ on the bar. The factor χ is the resistance; its units are force/velocity. The stress resultant σA at this end must match the externally applied force, so the boundary condition

$$\sigma A = \chi \frac{\partial u}{\partial t} \tag{2.6.31}$$

Very small χ is analogous to a pressure-release termination of a waveguide, and extremely large χ is like a rigid end.

Analogies

The equations governing extensional waves and those for a stretched cable have the same mathematical structure. Consider the momentum equation. The relation for a bar is Eq. (2.6.27). It can be obtained by changing the labels for the dependent variables and the parameters in Eq. (2.6.7). Specifically, $F_0\theta$ becomes σ, w becomes u, and μ becomes ρ_0. Note that this is not merely a mathematical similarity. Rather, the quantities are physically alike. Both u and w are displacements. The stress σ is the horizontal force per unit area acting on a cross section, whereas $F_0\theta$ is the vertical force at a cross section. Correspondingly, ρ_0 is a mass per unit volume, while μ is the mass per unit length.

Multiplication of the continuity equation for a stretched cable, Eq. (2.6.8), by F_0 gives $F_0\theta = F_0\partial w/\partial x$. A comparison of this relation and Eq. (2.6.28) for a bar reinforces the identification that σ is like $F_0\theta$, and u is like w. Then it must be that E is like F_0. This is true because F_0 was identified as being analogous to the bulk modulus $K = \rho_0 c^2$ of a fluid. Note that F_0 is a force, whereas E has units of force per unit area, which is consistent with the different units of $F_0\theta$ and σ.

The final element in identification of the analogy of stress waves in a bar and transverse vibration of a stretched cable comes from consideration of the respective equations for the phase speed. Recognition that they are analogous follows from the identification that E is like F_0 and ρ_0 is like μ. It is interesting to note that the phase speeds of extensional waves in a bar and transverse waves in a stretched cable usually have vastly different magnitudes, and that both speeds differ significantly from the speed of sound in fluids. For example, E for steel is approximately 200

GPa and $\rho_0 = 7800$ kg/m^3, which gives a phase speed in excess of 5 km/s. This is much greater than what is observed in liquids. To compare these properties to those for a cable, consider a 20 mm diameter steel cable whose internal stress is half the yield strength for some types of steel, $\sigma = 140$ MPa. The corresponding values are $F_0 = \sigma A = 44000$ N and $\mu = \rho_0 A = 25$ kg/m, which gives a phase speed of approximately 40 m/s. This is far slower than the speed of stress waves in a bar or the speed of sound in a fluid.

In view of these observations, extensional waves may be incorporated into the prior tabulation of analogies. We now have

Acoustic fluid	p	v	K	ρ_0
Flexible cable	$-F_0\,(\partial w/\partial x)$	$\partial w/\partial t$	F_0	μ
Elastic bar	$-\sigma$	$\partial u/\partial t$	E	ρ_0

This tabulation has the paradoxical effect of clarifying how the systems are related, while also confusing implementation of the method of wave images. For a cable the displacement usually is the state variable of interest. The displacement is the time integral of velocity, and it is the spatial integral of the slope. These differences lead to a reversal of sign of the reflection coefficients relative those for an acoustics problem. For stress waves in an elastic bar, there are two alternatives. If we seek to determine the displacement of cross sections, then the formulation for a cable is suitable. In that case we begin with initial conditions for the transverse displacement w and velocity $\partial w/\partial t$.

Young's modulus for solids is quite large, so bar displacements are quite small. Instrumentation to measure stress facilitates investigating this variable. For this reason, stress typically is the primary variable of interest. (Of course, the other variable can be found if either the displacement or stress field is known.) It is possible to determine the stress field directly from the method of images for acoustic waves. In such a formulation, the appropriate initial conditions are σ and $\partial u/\partial t$ as functions of x. The development must account for the sign difference in the analogy of pressure and stress. Let us decompose the stress into forward and backward waves. The stress-strain law states that $\sigma = E\,(\partial u/\partial x)$ and the particle velocity is $\partial u/\partial t$. The consequence is that the particle velocities have reversed signs when they are related to the stress in the respective waves, that is,

$$\begin{aligned} &\sigma_1 = F\,(x - ct)\,, \quad \sigma_2 = G\,(x + ct)\,, \quad \sigma = \sigma_1 + \sigma_2 \\ &v_1 = -\frac{c}{E} F\,(x - ct)\,, \quad v_2 = \frac{c}{E} G\,(x + ct)\,, \quad v = v_1 + v_2 \end{aligned} \tag{2.6.32}$$

Note that $c^2 = E/\rho_0$, so that $E/c = \rho_0 c$, which is analogous to the fluid impedance. The change of sign is manifested by an alteration in the relations for the wave functions in terms of the initial conditions. In particular, Eq. (2.5.2) become

$$F\,(x) = \frac{1}{2}\left[\sigma - \rho_0 c \frac{\partial u}{\partial t}\right]_{t=0}, \quad G\,(x) = \frac{1}{2}\left[\sigma + \rho_0 c \frac{\partial u}{\partial t}\right]_{t=0} \tag{2.6.33}$$

2.7 Closure

We began this chapter by deriving basic equations governing one-dimensional acoustic response in the time domain. The field equation describing the pressure as a function of position and time is the wave equation, and the Euler equation provides the means by which particle velocity and pressure may be related. The only solution of the wave equation that we derived was the general one attributed to d'Alembert. It tells us that the signal within the fluid consists of two waves that propagate in opposite directions without changing the shape of their spatial profile and temporal waveforms. Such waves are said to be simple, but they also are known as nondispersive waves because they do not spread out or contract as they propagate. In the case where the domain is unbounded, these waves are set by the initial conditions. We extended the d'Alembert solution to describe the response of semi-infinite and bounded waveguides to initial condition and boundary excitation. These extensions were achieved by implementing the method of wave images. We encountered a wealth of phenomena, including reflections, attenuation, reverberation time, natural frequencies and modes, and multiple resonances. These phenomena also are encountered in other situations, as we saw in analyses of the transverse displacement of a stretched cable and extensional deformation of an elastic bar. Similar phenomena also occur in far more complicated acoustical systems. Unfortunately the relative simplicity of the method of wave images stems from restricting our attention to a highly idealized model of a reflecting boundary, in which the velocity at a termination is proportional to the pressure that is applied to it. Generalization of the boundary condition requires that we pursue frequency domain solutions. Nevertheless, the present developments provide a unique perspective that will enhance our understanding of the response features of more complicated systems.

2.8 Homework Exercises

Exercise 2.1 Consider a fluid whose ambient state consists of a homogeneous uniform flow at constant velocity U in the x direction, with corresponding pressure p_0, density ρ_0, and temperature T_0. Let the acoustic perturbation of this flow be represented by pressure p', density ρ', and particle velocity v', with $p' \ll \rho_0 c^2$, $\rho' \ll \rho_0$, and $v' \ll U$. Derive the modified linearized wave equation governing the dependence of pressure on position x and time t.

Exercise 2.2 Consider an isentropic process in an ideal gas. Derive an expression for the temperature change T' from the ambient value T_0 when the pressure increases from p_0 to $p_0 + p'$. Then linearize that relation.

Exercise 2.3 Some gases when compressed greatly show a relationship between pressure, density, and pressure that deviates significantly from the ideal gas law. Laws that account for these deviations are known as virial equations. Derive an expression for the speed of sound in the case where the virial equation describes T explicitly as a function of p and ρ, that is, $T = F(p, \rho)$.

Exercise 2.4 An important dissipation mechanism for many fluids is relaxation. The linearized equation of state for such a fluid is

$$(p - p_0) - c_0^2 (\rho - \rho_0) + \tau \left(\frac{\partial p}{\partial t} - c_\infty^2 \frac{\partial \rho}{\partial t} \right) = 0$$

where τ is a constant called the relaxation time, and c_0 and c_∞ are two different sound speeds, usually close together. The continuity and momentum equations may be taken to be the same as those for an ideal fluid. Show that the corresponding wave equation is

$$\frac{\partial^2 p'}{\partial t^2} - c_0^2 \frac{\partial^2 p'}{\partial x^2} + \tau \left[\frac{\partial^3 p'}{\partial t^3} - c_\infty^2 \frac{\partial}{\partial t} \left(\frac{\partial^2 p'}{\partial x^2} \right) \right] = 0, \quad p' = p - p_0$$

Exercise 2.5 Consider a situation in which the walls of a circular tube are porous. The radius of the tube is a. The ambient conditions in the exterior are the same as those inside. The consequence is that there is a mass flow per unit length through the tube's wall that is proportional to the acoustic pressure. The result is that the mass flowing out of the wall between x to $x + dx$ in an interval dt is $\dot{m}_{wall} dx dt$, where $\dot{m}_{wall} = \sigma p(x, t)$. If this flow is sufficiently small, it is permissible to assume that the acoustical pressure and particle velocity within the tube depend only on x and t, as in the case where the tube is impermeable. (a) Determine how the porosity affects the continuity and momentum equations. (b) Derive the linearized field equation governing pressure.

Exercise 2.6 The signal within a duct consists of a pair of oppositely propagating plane waves, as described by the d'Alembert solution. Suppose the acoustic pressure and particle velocity waveforms are measured at $x = 0$ as a function of time, so that $p(0, t) = p_A(t)$ and $v(0, t) = v_A(t)$ are known. Derive an expression for $p(x, t)$ in terms of the functions p_A and v_A.

Exercise 2.7 When $t = 0$ the pressure within a tube is observed to depend on the distance along the tube in the manner shown in the graph. The fluid within the tube is not moving at $t = 0$. Sketch spatial profiles of the pressure and particle velocity at $t = 0.5L/c$ and $1.5L/c$.

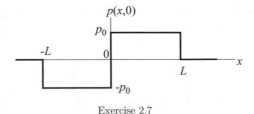

Exercise 2.7

Exercise 2.8 The initial pressure and velocity distributions for a plane wave are depicted below. (a) For the instant when $t = 0$, draw a graph showing the x dependence of the initial acoustic pressure in the individual left and right propagating plane waves generated by this set of initial conditions. (b) Determine the spatial profile of the pressure when $t = 0.5L/c$ and $t = L/c$. (c) Determine the pressure waveforms at $x = L/2$ and $x = L$.

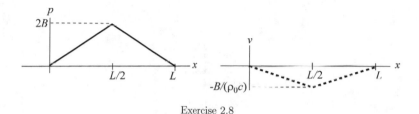

Exercise 2.8

Exercise 2.9 The initial velocity particle velocity in an infinite waveguide is $v(x, 0) = V \sin(\pi x/L)$ for $-L < x < L$. (a) Determine the pressure and particle velocity profiles at $t = L/c$. (b) Determine the waveforms at $x = 0, \pm L$.

Exercise 2.10 When $t = 0$ the initial particle velocity field within an infinitely long tube is the N-wave shown below. The initial pressure is zero. Sketch p and v as functions of x when $t = L/2c$. Then sketch the waveforms of p and v as functions of t at $x = L/2$.

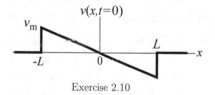

Exercise 2.10

Exercise 2.11 A piston at the left end of a tube filled with water is given the acceleration $\dot{v} = a(t)$ graphed below. (a) Determine the pressure, particle velocity, and particle displacement temporal waveforms at the location $x = c\tau$. (b) Determine the spatial profiles of pressure, particle velocity, and particle displacement at $t = 2\tau$. (c) The cross-sectional area of the tube is A and the mass of the piston is m. Determine the force $F(t)$ that must be applied to the piston to achieve this motion.

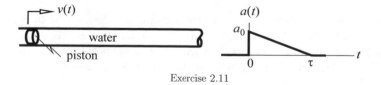

Exercise 2.11

Exercise 2.12 At $t = 0$ the initial pressure perturbation in a semi-infinite waveguide, $0 \leq x < \infty$, depends on the distance from the closed end in the manner described by the graph. The end $x = 0$ is terminated by a material whose impedance is $Z = 0.20\rho_0 c$, independent of frequency. The initial particle velocity is zero. (a) Sketch graphs of the pressure perturbation and particle velocity as functions of x when $t = s/2c$. (b) Sketch a graph of the pressure perturbation as a function of time t at $x = s/2$. Be sure to mark critical values in the graphs.

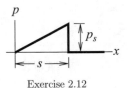

Exercise 2.12

Exercise 2.13 If the termination of a waveguide is actively controlled, it is possible to attain a negative impedance. Consider a waveguide for which the specific impedance at $x = 0$ is $\zeta = -0.5$. The initial pressure distribution is $p(x,0) = A \exp(-\beta x)$ for $x \geq 0$. Determine the pressure distribution when $t = 1/(\beta c)$ and the pressure waveform observed at $x = 1/\beta$.

Exercise 2.14 The pressure and particle within a semi-infinite waveguide at $t = 0$ are $p = -\rho_0 c v = A[h(x - L) - h(x - L/2)]$. It also is known that the pressure distribution at an unspecified instant after the wave has reflected is $p = 0.2[h(x - L/2) - h(x - L)]$. (a) Determine the time at which the reflected signal was measured. (b) Determine the reflection coefficient. (b) Determine the waveforms at $x = L/2$ and $x = L$.

Exercise 2.15 A wave $p = g(t + x/c)$ propagates to the left inside the tube. The left end, $x = 0$, is closed by a diaphragm that exerts a restoring force ku to the left when it displaces u to the right. This force must balance the pressure force exerted by the incident and reflected waves. Derive a differential equation whose solution would yield the time dependence of the pressure wave that is reflected from the membrane.

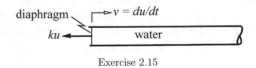

Exercise 2.15

Exercise 2.16 The termination of a semi-infinite tube has a specific impedance $\zeta = 3$. The reflected waveform observed at $x = 1.5L$ is the function in the graph. The incident wave is not known. (a) Determine the spatial profile of the incident wave at $t = 0$. (b) Determine the waveform of the total pressure (incident plus reflected) at $x = 1.5L$.

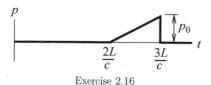

Exercise 2.16

Exercise 2.17 Both ends of a waveguide whose length is L are pressure-release. Prove that Eq. (2.5.33) is the Fourier series describing the response of this system to arbitrary initial conditions.

Exercise 2.18 The end $x = 0$ of a waveguide is pressure-release and $x = L$ is rigid. Prove that Eq. (2.5.36) is the Fourier series describing the response of this system to arbitrary initial conditions. Then use that equation to identify the natural frequencies and modes for the case where the ends are swapped, so that $p = 0$ at $x = 0$ and $v = 0$ at $x = L$.

Exercise 2.19 The initial pressure in a finite waveguide is $p = B[h(x - L/2)]$. The particles are initially at rest. The end at $x = 0$ is pressure-release and the end at $x = L$ is rigid. Graphically construct the pressure profile at $t = L/c$ and $t = 2L/c$.

Exercise 2.20 Perform the analysis requested in Exercise 2.19 for the case where both ends are pressure-release.

Exercise 2.21 The end at $x = 0$ of a tube is rigid, while the reflection coefficient at $x = L$ is $R_L = 0.5$. Membranes stretched across the tube at $x = L/4$ and $x = 3L/4$ hold the pressure perturbation in the region between them at value B, while the pressure in $0 < x < L/4$ and $3L/4 < x < L$ is zero. The fluid is at rest in this condition. At $t = 0$ the membranes are destroyed. (a) Determine the spatial pressure profile over $0 < x < L$ at $t = L/c, 2L/c, 3L/c$. (b) Determine the waveform at $x = L/2$ for $0 < t < 4L/c$.

Exercise 2.22 It hypothetically is possible to induce an arbitrary initial velocity field in a magnetic fluid. Such a velocity field in a waveguide at $t = 0$ appears in the graph, where $v > 0$ indicates that a particle is moving in the positive x direction. The initial pressure perturbation relative to ambient conditions is zero. The end $x = 0$ is rigid and the impedance at $x = L$ is $Z_L = 0.5\rho_0 c$. (a) Determine the pressure profile at $t = 1.5L/c$. (b) Determine the pressure waveform at $x = L$ for $0 \leq t \leq 3L/c$. Use graphical constructions to address both tasks.

Exercise 2.22

Exercise 2.23 Use mathematical software to solve Exercise 2.22.

Exercise 2.24 The particle velocity at $x = 0$ in a waveguide is a pulse that is a ramp function whose duration is L/c, that is, $v(0, t) = at[h(t) - h(t - L/c)]$. The end at $x = L$ is rigid. Use graphical construction methods to determine the pressure profile at $t = 2L/c$ and the pressure waveforms at $x = 0$ and $x = L$.

Exercise 2.25 Use mathematical software to solve Exercise 2.24.

Exercise 2.26 The end at $x = L$ of a waveguide is rigid. Beginning at $t = 0$, the pressure at $x = 0$ decays exponentially according to $p = A \exp(-\beta t)$. (a) Derive an expression for the pressure within the waveguide. (b) Create a program that will evaluate pressure profiles and waveforms. (c) Evaluate pressure profiles when $t = 2L/c$ and $t = 2.5L/c$. (d) Evaluate waveforms at $x = 0$, $L/2$, and L. The decay constant is $\beta = 2c/L$.

Exercise 2.27 Perform the analysis requested in Exercise 2.26 for the case where the impedance at $x = L$ is $\zeta_L = 0.4$.

Exercise 2.28 The particle velocity at $x = 0$ in a waveguide is zero until $t = 0$. Then it varies harmonically at $V_0 \sin(\omega t)$. The impedance at $x = L$ is ζ. Develop a computer program to evaluate the pressure waveform at an arbitrary location. Use it to determine the waveform at $x = L/2$ for the following combination of parameters: (a) $\omega = \pi c/(3L)$ and $\zeta = 0.95$, (b) $\omega = \pi c/(2.95L)$ and $\zeta = 1$, (c) $\omega = \pi c/(3L)$ and $\zeta = 1$.

Exercise 2.29 Consider the situation where a waveguide is excited at $x = L$ and the reflection coefficient at $x = 0$ is R_0. Modify Eq. (2.5.50) to describe the pressure field at arbitrary x, t (a) if the velocity at $x = L$ is specified to be an arbitrary function $v_L(t)$, (b) if the pressure at $x = L$ is specified to be an arbitrary function $p_L(t)$.

Exercise 2.30 The end $x = 0$ of a waveguide is rigid. The pressure at $x = L$ is zero until $t = 0$, at which time it is made to vary as a sequence of positive lobes of a sine function, so that $p(x = L, t) = 0$ if $t < 0$, $p(x = L, t) = B \sin(2\pi t/T)$ if $0 < t < T/2$, $p(x = L, t) = 0$ if $T/2 < t < T$, $p(x = L, t) = p(x = L, t - T)$ if $t \geq T$. (a) Develop a software program to evaluate waveforms at an arbitrary x. (b) Cases of interest are $T = 2L/c$ and $T = 4L/c$. Validate the program by verifying that, for both values of T, the result for the pressure waveform at $x = L$ is the specified function. (c) Evaluate the waveform at $x = L/2$ for both values of T.

Exercise 2.31 The A string of a cello is tensioned such that its fundamental natural frequency is 220 Hz. The length of the string is 695 mm and its mass per unit length is 2.2 gm/m. What is the tensile force?

Exercise 2.32 A cable is tensioned, then tied between two fixed posts. The sting is pulled upward at the quarter-point and prevented from moving at the midpoint, so the initial displacement is the triangular distribution graphed below. The string is held stationary in this state, then released. (a) Determine the displacement of the cable at $t = 3.5L/c$. (b) Determine the displacement waveform at the midpoint of the cable for $t \leq 4L/c$.

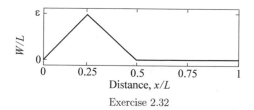

Exercise 2.32

Exercise 2.33 Solve Exercise 2.32 in the case where the cable at $x = L$ is free to move vertically while the tension force is applied. The end $x = 0$ is fixed.

Exercise 2.34 A 600 mm long steel rod has a cross-sectional diameter of 20 mm. The end at $x = 0$ is welded to a rigid wall, and the right end is attached to a dashpot that is aligned with the longitudinal axis of the rod. Determine the viscous resistance coefficient of the dashpot for which the displacement reflection coefficient at $x = L$ is -0.5.

Exercise 2.35 A pulse excitation applied to one end of an aluminum rod consists of a 1 kN axial stress whose duration is 2 ms. The reflection coefficient for displacement at the far end is 0.9. The rod has a square cross section 40 mm × 40 mm, and the length of the rod is 4 m. Determine the stress waveform observed at the midpoint and far end.

Exercise 2.36 The elastic bar in the sketch is sliding at speed v_0 along the floor with negligible frictional resistance. At $t = 0$, it strikes the wall, which may be considered to be rigid. Contact with the wall is maintained as long as the stress at the end is compressive. Thus, the interval of contact extends from $t = 0$ until $t = t_f$ when the compressive stress at the end is zero. After t_f, the end condition becomes zero stress, corresponding to the bar rebounding from the impact. (a) Determine t_f. (b) Determine the stress and particle velocity profile at an instant that slightly precedes t_f. From that, describe how the bar rebounds from the collision.

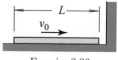

Exercise 2.36

Chapter 3
Plane Waves: Frequency-Domain Solutions

The frequency domain now becomes the lens for our study of plane waves. We will not rely on prior developments in the time domain, such as the d'Alembert solution and the ensuing wave image construction technique. The primary reason for pursuing the analysis in a self-contained manner is that it will prepare us for the study of multidimensional phenomena, for which a time-domain solution might not be available as a guide.

There are other reasons to analyze plane waves in the frequency domain. Some system properties often are known solely as transfer functions. Such is the case for polymeric materials used to enhance acoustical performance. These materials are viscoelastic, which means that rather than having a simple stress–strain relation like Hooke's law for elastic materials, their stress and strain are related through a differential equation. Construction of an accurate time-domain model of a system that contains such materials would be quite challenging. In contrast, there are a variety of tests by which one can measure a transfer function between the complex amplitudes of stress and strain as a function of frequency. An important aspect is that viscoelastic materials dissipate energy, as do some nonideal effects in the fluid, such as viscosity. Physical models that account for dissipation in the fluid are much more amenable to analysis in the frequency domain, as will be evident later in this chapter.

Another benefit derived from viewing plane waves in the frequency domain is the greater ease in which analyses may be performed. Time derivatives became frequency factors, so the one-dimensional wave equation becomes an ordinary differential equation governing the spatial dependence of pressure. In some cases, this equation can be solved to obtain point-to-point transfer functions that may be fitted together to describe waveguides that are joined at junctions, such as piping systems and ducts. These concepts will carry over into our later study of wave propagation through layered media.

Electronic supplementary material The online version of this chapter (DOI 10.1007/978-3-319-56844-7_3) contains supplementary material, which is available to authorized users.

3.1 General Solution

We begin by representing the acoustic pressure, particle velocity, and density perturbation in a harmonic plane wave as complex exponentials,

$$
\begin{aligned}
p\,(x,t) &= \mathrm{Re}\left(P\,(x)\,e^{i\omega t}\right)\\
v\,(x,t) &= \mathrm{Re}\left(V\,(x)\,e^{i\omega t}\right)\\
\rho'\,(x,t) &= \mathrm{Re}\left(\hat{\rho}\,(x)\,e^{i\omega t}\right)
\end{aligned}
\tag{3.1.1}
$$

(Although the symbol V was used in the previous chapter to describe the initial
distribution of particle velocity, there is no context in which that quantity could be
confused with its usage here as a complex amplitude.)

The basic equations of linear acoustics are continuity, momentum, and the equation of state. Substitution of the frequency-domain representation of the state variables yields an assortment of terms, each of which is the real part of a complex
exponential. Each equation must be satisfied at every instant, which can only occur
if the complex amplitudes on either side of each equality match. For example, the
continuity equation requires that

$$
\mathrm{Re}\left[\left(i\omega\hat{\rho} + \rho_0\frac{dV}{dx}\right)e^{i\omega t}\right] = 0
\tag{3.1.2}
$$

where an ordinary derivative of V appears because x now is the only independent
variable. The complex amplitude that is the factor of $\exp\,(i\omega t)$ must be zero because
the exponential cannot yield a zero value at every instant. Thus, the frequency-domain
continuity equation is

$$
i\omega\hat{\rho} + \rho_0\frac{dV}{dx} = 0
\tag{3.1.3}
$$

The linear equation of state is a proportionality that must be satisfied by the respective
complex amplitudes, so that

$$
P = c^2\hat{\rho}
\tag{3.1.4}
$$

With this relation, we can eliminate $\hat{\rho}$ from the continuity equation, which yields

$$
i\omega P = -\rho_0 c^2\frac{dV}{dx}
\tag{3.1.5}
$$

The frequency-domain version of the linearized momentum equation is

$$
i\omega\rho_0 V = -\frac{dP}{dx}
\tag{3.1.6}
$$

This is *Euler's equation*. It has been highlighted because it will serve as a boundary condition in the typical circumstance where we know the velocity of some surface. For example, if there is a piston at $x = 0$, it tells us what the pressure gradient at $x = 0$ must be. Equation (3.1.5) is not useful for that purpose because it requires knowing the gradient of V, which is not what the piston imposes.

In principle, we could endeavor to seek a solution by solving simultaneously the frequency-domain continuity and momentum equations. However, as was the case for the time-domain wave equation, it is easier to combine them into a single equation for the pressure, whose general solution is quite accessible. Rather than obtaining the desired field equation by eliminating the particle velocity from these basic equations, a more expedient derivation merely substitutes the harmonic representation of p' into the wave equation, which then requires that

$$\boxed{\frac{d^2 P}{dx^2} + k^2 P = 0}$$ (3.1.7)

This is the one-dimensional *Helmholtz equation*.[1] The coefficient k appearing here is the wavenumber, which was defined in Eq. (2.4.20),

$$\boxed{k = \frac{\omega}{c}}$$ (3.1.8)

A general plane solution of the one-dimensional Helmholtz equation is quite accessible because it is an ordinary differential equation. In fact, it is the fundamental equation describing the response of a one-degree-of-freedom vibratory system, except that the independent variable here is x rather than time. Let us derive the general solution from basic principles. The complementary solution to any ordinary differential equation with constant coefficients depends exponentially on its independent variable. In anticipation that the signal is oscillatory, we adopt a trial solution whose form is

$$P = B e^{i\mu x}$$ (3.1.9)

where the coefficient B may be complex because $P(x)$ is the complex amplitude of pressure, as defined by Eq. (3.1.1). Substitution of this ansatz (or synonymously, this trial solution) into Eq. (3.1.7), combined with the requirement that the equation be satisfied for all x, leads to

$$\mu = \pm k$$ (3.1.10)

Each possible sign for μ is associated with a different solution, so the most general solution is

---

[1] Hermann Ludwig Ferdinand von Helmholtz (1821–1894), who was German, made many fundamental contributions to the science of acoustics.

$$\boxed{P = B_1 e^{-ikx} + B_2 e^{ikx}} \tag{3.1.11}$$

The associated particle velocity may be found from Euler's equation, which gives

$$\boxed{V = \frac{1}{\rho_0 c}\left(B_1 e^{-ikx} - B_2 e^{ikx}\right)} \tag{3.1.12}$$

The time-domain representation that results from substituting P and V into Eq. (3.1.1) and replacing k with ω/c is

$$p(x,t) = p_1 + p_2, \quad v(x,t) = v_1 + v_2$$
$$p_1 = \rho_0 c v_1 = \text{Re}\left[B_1 e^{i\omega(-x/c+t)}\right] \tag{3.1.13}$$
$$p_2 = -\rho_0 c v_2 = \text{Re}\left[B_2 e^{i\omega(x/c+t)}\right]$$

The coefficients B_1 and B_2 are arbitrary in regard to satisfaction of the Helmholtz equation. Their values will be obtained from other requirements, which usually will be boundary conditions. The expressions for p and v are decomposed into two parts to emphasize that this solution is consistent with the d'Alembert solution. Specifically, p_1 solely depends on the combination of variables that define the forward characteristic $x-ct$, while the variables for the second term are those of the backward characteristic, $x + ct$.

Harmonic waves were discussed in Sect. 2.4.3. It was shown there that the wavenumber k describes spatial dependence analogously to the way that ω describes time dependence. Both harmonic waves in Eq. (3.1.13) are spatially periodic, with the interval of repetition being the wavelength, which is

$$\lambda = \frac{2\pi}{k} \tag{3.1.14}$$

The relation between λ and k is analogous to the temporal period being $T = 2\pi/\omega$. Similarly, the cyclical frequency is $1/T = \omega/(2\pi)$, and the number of wavelengths per unit length is $1/\lambda = k/(2\pi)$.

Some individuals prefer to see the signal represented in real form. This may be obtained by applying Euler's identity to each complex exponential in Eq. (2.4.3). Let β_1 and β_2 be the phase angles of B_1 and B_2, so that $B_n = |B_n|\exp(i\beta_n)$. Then, Eq. (2.4.3) give

$$p(x,t) = |B_1|\cos(\omega t - kx + \beta_1) + |B_2|\cos(\omega t + kx + \beta_2) \tag{3.1.15}$$

If it is preferred to use sine functions, such a representation may be obtained by defining the phase angles such that $B_n = -i|B_n|\exp(i\beta_n)$.

A different picture emerges if we first apply Euler's identity to Eq. (3.1.11), which gives

$$P = (B_1 + B_2)\cos(kx) - i(B_1 - B_2)\sin(kx) \tag{3.1.16}$$

The result of using this expression to form the pressure signal in Eq. (3.1.1) is

$$p(x, t) = p_a + p_b$$
$$p_a = B^+ \cos(kx) \cos\left(\omega t + \beta^+\right), \quad p_b = B^- \sin(kx) \sin\left(\omega t + \beta^-\right) \quad (3.1.17)$$

where the amplitudes and phase angles are obtained from the sum and difference of B_1 and B_2, according to

$$B_1 + B_2 = B^+ e^{i\beta^+}, \quad B_1 - B_2 = B^- e^{i\beta^-} \quad (3.1.18)$$

Each term in Eq. (3.1.17) consists of a harmonic time function whose real amplitude factor is $B_+ \cos(kx)$ or $B_- \sin(kx)$. Thus, the spatial profiles of p_a and p_b vary in magnitude from one instant to another, but the shape of each profile is invariant. The individual signals are *standing waves*.

The consequence of this discussion is that we may view a general harmonic wave in one of three ways, either as a signal whose complex amplitude is position dependent, or a pair of waves that travel in opposite directions, or a pair of standing waves whose nodal points are separated by one-quarter of a wavelength. A pictorial comparison of the three perspectives is provided by Fig. 3.1, which displays spatial profiles at four instants separated by one-quarter of the period T. Figure 3.1a, which does not decompose the signal, indicates that at any instant, the locations at which the pressure is zero are separated by $\lambda/2$, while adjacent maxima or minima at any instant are separated by λ. These features neither move in one direction, as they would if the signal was a simple propagating wave, nor do they occur at stationary locations, which is a basic property of standing waves. Rather, they shift pack and forth with elapsed time. Figure 3.1b decomposes the signal into forward and backward simple waves described in Eq. (3.1.15). Tracking the progression indicated by the label of each wave from t_1 to t_3 makes evident the propagation in opposite directions. Although the signal labeled by t_4 seems to not follow the correct progression, that perception fails to recognize that the periodic nature of each wave means that the label could have been shifted by a wavelength to make it consistent with the others. Figure 3.1c shows the incarnation of Eq. (3.1.17). Each standing wave has maxima and minima at any instant that are separated by λ, and its zeroes are separated by $\lambda/2$. Each of these features occur at fixed locations.

Although the representations of pressure in terms of real functions are useful for understanding the nature of a signal, the complex form in Eq. (3.1.11) is more suitable for analytical work because it allows us to follow the general concept of a frequency-domain formulation, which suppresses the time dependence. One use is to evaluate the mean-squared pressure. The result is

$$\left(p^2\right)_{av} = \frac{1}{2} P P^* = \frac{1}{2} |B_1|^2 + \frac{1}{2} |B_2|^2 + \text{Re}\left(B_1 B_2^* e^{-2ikx}\right)$$
$$= \frac{1}{2} |B_1|^2 + \frac{1}{2} |B_2|^2 + |B_1| |B_2| \cos\left(2kx - \text{angle}(B_1) + \text{angle}(B_2)\right)$$
$$(3.1.19)$$

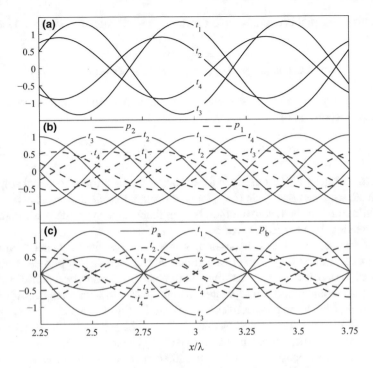

Fig. 3.1 Different representations of the general solution for harmonic plane waves: **a** the total signal; **b** propagating waves traveling in opposite directions; and **c** two standing wave patterns. $B_1 = 0.25 - 0.5i$, $B_2 = 1$, $t_1 = 0$, $t_2 = 0.25T$, $t_3 = 0.5T$, $t_4 = 0.75T$

Thus, the superposition of the oppositely traveling waves results in a mean-squared pressure that fluctuates spatially about the mean value $\left(|B_1|^2 + |B_2|^2\right)/2$. The fluctuating part has an amplitude of $|B_1||B_2|$ and is periodic in a distance that is $\lambda/2$. It follows that the mean-squared pressure is invariant in x only if one of the waves is not present.

EXAMPLE 3.1 A harmonic signal is measured at $x = 0$ and $x = 1.2$ m in a waveguide. The speed of sound is 340 m/s, and the ambient density within the waveguide is 1.2 kg/m³. (a) Determine the complex amplitudes of the pair of oppositely traveling plane waves that compose this signal. Specifically, determine the values of k, ω, B_1, and B_2. (b) Determine the amplitude of the particle velocity at $x = 0$ and $x = 1.2$ m. (c) Determine and graph the RMS pressure as a function of x.

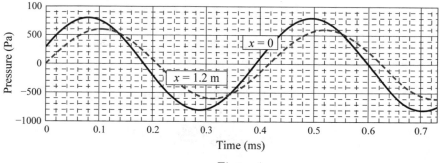

Figure 1.

Significance

The inverse process required to extract the basic parameters of a signal from its measured behavior serves to emphasize the meaning of the fundamental equations. We will see in Sect. 3.2.4 that the process followed here forms the foundation for the two-microphone technique by which certain material properties are identified.

Solution

Our approach is to match the mathematical form of a harmonic function to the properties of each of the given waveforms. The frequency is found by measuring the time between zeros. For the signal at $x = 0$, the first zero is (approximately) at $t = 0.181$ ms, and the third zero is at $t = 0.6$ ms. Adjacent zeros are separated by $T/2$, so we estimate that $2(T/2) = (0.60 - 0.181)$ ms. For the signal at $x = 1.2$ m, the first zero is at $t = 0$ and the fourth zero is at $t = 0.622$ ms, so the estimate is $3(T/2) = 0.622$ ms. The average of these estimates is $T = 0.4168$ ms, which corresponds to a cyclical frequency of $f = 1/T = 2399$ Hz. Thus, the circular frequency is

$$\omega = 2\pi f = 14.84 \left(10^3\right) \text{ rad/s}$$

(For reference, the actual cyclical frequency used to generate the signals is 2.4 kHz.) From this, we find that the wavenumber is

$$k = \frac{\omega}{c} = 44.33 \text{ m}^{-1}$$

The amplitudes of the signals obtained from the graph are 800 Pa at $x = 0$ and 600 Pa at $x = 1.2$ m. To determine the phase angle for each signal, we use an instant when the pressure is zero to identify the corresponding time delay. The signal at $x = 0$ is decreasing as it crosses the first zero at $t = 0.181$ ms, so we use $t - 0.000181$ as the elapsed time for a negative sine function,

$$p(0, t) = -800 \sin(\omega(t - 0.000181)) \text{ Pa}$$

The signal at $x = 1.2$ m appears to be a pure sine function, so we have

$$p\,(1.2, t) = 600 \sin{(\omega t)}$$

These observed signals are converted to complex exponentials in order match them to the general solution, as shown in Eq. (3.1.11)

$$p\,(0, t) = \mathrm{Re}\left[-\frac{800}{i}e^{-i\omega(0.000181)}e^{i\omega t}\right]$$

$$p\,(1.2, t) = \mathrm{Re}\left[\frac{600}{i}e^{i\omega t}\right]$$

The complex amplitude of each signal must match the value of $P\,(x)$ in Eq. (3.1.11) evaluated at the respective locations. Thus, it must be that

$$P\,(0) = B_1 + B_2 = -\frac{800}{i}e^{-i\omega(0.000181)}$$

$$P\,(1.2) = B_1 e^{-ik(1.2)} + B_2 e^{ik(1.2)} = \frac{600}{i}$$

The simultaneous solution of these equations is

$$B_1 = 3379.1 + 402.1, \quad B_2 = -3057.8 - 1134.4i \ \ \mathrm{Pa}$$

To check the analysis, the signals obtained from substituting these coefficients into Eq. (3.1.13) are compared in Fig. 2 to the given data. The agreement is excellent.

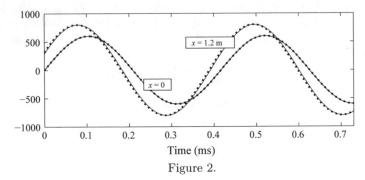

Figure 2.

The complex amplitude of the particle velocity is given in Eq. (3.1.12). The complex amplitudes of the forward and backward propagating waves at the two locations are

$$P_1\,(0) = B_1 = 3379.1 + 402.1i, \quad P_2\,(0) = B_2 = -3057.8 - 1134.8i$$

$$P_1(1.2) = B_1 e^{-ik(1.2)} = -3225.3 - 1085.1i,$$

$$P_2(1.2) = B_2 e^{ik(1.2)} = 3225.3 + 485.1i$$

The corresponding complex velocity at each location is

$$V(0) = \frac{P_1(0) - P_2(0)}{\rho_0 c} = 15.7767 + 3.7669i \text{ m/s}$$

$$V(1.2) = \frac{P_1(1.2) - P_2(1.2)}{\rho_0 c} = -15.8104 - 3.8483i \text{ m/s}$$

The RMS pressure is the square root of $\left(p^2\right)_{av}$, which is described in Eq. (3.1.19). As a check of that equation, the result is compared in Fig. 3 to the values obtained from a direct computation, according to which gives $\left(p^2\right)_{av} = |P_1(x) + P_2(x)|^2 /2$. There is no perceptible difference.

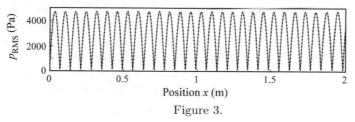

Figure 3.

3.2 Waveguides with Boundaries

At this juncture, we have established the manner in which plane harmonic waves propagate. Our next objective is to determine how they are generated and modified by boundaries. For discussion purposes, the system is considered to be a tube that has an arbitrary cross-sectional shape. Nevertheless, all developments are equally valid for any situation in which a plane wave encounters a wall-like boundary whose plane is perpendicular to the direction of propagation. It is possible to generate a signal at $x = 0$ by creating a pressure fluctuation, which would be Re $\left(P_0 e^{i\omega t}\right)$. The more common situation is an excitation generated by a vibrating transducer, which is modeled as a piston. The piston's axial velocity is Re $\left(V_0 e^{i\omega t}\right)$, and continuity requires that the fluid's particles at $x = 0$ move in unison with the boundary at that location.

We begin with the case of a semi-infinite waveguide occupying $x \geq 0$. The most general signal consists of two oppositely traveling waves, but the boundary condition at $x = 0$ is a single equation. The second condition is obtained from the frequency-domain version of the principle of causality. It is argued that there can be no backward propagating wave, because it could only be generated at the end that is infinitely distant. The sound speed is finite, so even if such a wave was created, it would never arrive at a finite x. Thus, the signal within the waveguide must fit

$$P = \rho_0 c V = B e^{-ikx} \tag{3.2.1}$$

Matching the particle velocity to what is imposed at $x = 0$ yields

$$B = \rho_0 c V_0 \qquad (3.2.2)$$

This result is not surprising, in that it is a manifestation of the fact that the characteristic impedance is the ratio of pressure to velocity in a simple wave.

3.2.1 Impedance and Reflection Coefficients

When the length of the waveguide is finite, any wave that is generated at one end will be reflected at the other end. The frequency-domain formulation enables us to generalize the *surface impedance*, rather than limiting it to be a proportionality of pressure and particle velocity. The quantity Z now is defined as the ratio of the complex amplitude of pressure at the boundary to the complex amplitude of particle velocity into the boundary, which is the sense in which a positive pressure pushes on the boundary. If $\bar{n}$ is the normal direction at the boundary *pointing into the fluid*, then

$$Z = \left. \frac{P}{-\bar{n} \cdot \bar{V}} \right|_{\text{boundary}} \qquad (3.2.3)$$

(Other quantities bearing the name "impedance" occur later in this chapter. Unless there is a possibility of confusing Z with other quantities, the adjective "surface" will be omitted.)

In some cases, the *specific impedance* is given. It is defined as the impedance divided by the characteristic impedance of the fluid,

$$\zeta = Z / (\rho_0 c) \qquad (3.2.4)$$

The virtue of using ζ is that it is a dimensionless parameter that slightly simplifies formulas. However, its usage masks an essential feature, which is that Z is a property of the boundary. This is evident from the fact that we could imagine using a stiff metal plate to uniformly distribute a known harmonic force over the surface of the boundary. In this scenario, the pressure applied to the boundary is the force divided by the area of the plate. It is exerted at the interface between the boundary and the plate. It follows that the velocity of the boundary would not depend on the fluid's properties.

The impedance may be complex and frequency dependent. A simple mechanical model demonstrates how such a behavior might arise. Consider a termination that consists of a spring K and dashpot D that act in opposition to movement of a mass M, like the system in Fig. 3.2.

The displacement is $u = \mathrm{Re}\left(U e^{i\omega t}\right)$, so the complex velocity amplitude, positive in the sense of increasing x, is $V = i\omega U$. We take $x = 0$ to be the location of the

Fig. 3.2 A
one-degree-of-freedom
oscillator that is the
boundary of a waveguide

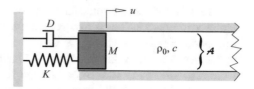

piston face, so the equation of motion for the piston is

$$M\ddot{u} + D\dot{u} + Ku = - p|_{x=0}\,\mathcal{A} \tag{3.2.5}$$

The equivalent frequency domain equation is

$$\left(K + i\omega D - \omega^2 M\right) U = - P|_{x=0}\,\mathcal{A} \tag{3.2.6}$$

The unit vector pointing into the fluid is $\bar{n} = \bar{e}_x$, and the particle velocity vector is $\bar{V} = \bar{e}_x \operatorname{Re}(V(x)\exp(i\omega t))$. Hence, continuity at the piston–fluid interface requires that $\bar{n} \cdot \bar{V} = i\omega U$. When this relation is used to eliminate U in the preceding, the boundary impedance is found to be

$$Z = \frac{P|_{x=0}}{-\bar{n} \cdot \bar{V}} = \frac{D}{\mathcal{A}} + i \left(\frac{M\omega}{\mathcal{A}} - \frac{K}{\mathcal{A}\omega}\right) \tag{3.2.7}$$

The terminology used to describe impedance originates in the study of AC circuits. The real part of Z, which is proportional to the dashpot constant, is called the resistive part, or more briefly, the *resistance*. The *reactance* is the imaginary part of the impedance; it has two contributors. If the mass of the piston is dominant, then $\operatorname{Im}(Z) > 0$. This represents a pressure that is proportional to the acceleration because it leads the velocity by 90°. The value of $\operatorname{Im}(Z)$ in this case is the *inertance*. If the spring effect is dominant, then $\operatorname{Im}(Z) < 0$. Rather than referring to $-\operatorname{Im}(Z)$ in this case as the stiffness, the convention is to say that $-1/\operatorname{Im}(Z)$ is the *compliance*.

A spring that is very compliant displaces greatly when a force is applied to it, which is consistent with the definition of compliance as reciprocal to the the stiffness. However the terminology can be confusing.

An interesting situation occurs if $\omega = (K/M)^{1/2}$, which is the natural frequency ω_{nat} of the mechanical system. At ω_{nat}, the restorative force of the spring Ku exactly equals $M\ddot{u}$ of the piston. The consequence is that if the piston was in a vacuum and the dashpot was not present, the oscillator would vibrate at the natural frequency without an external force. At ω_{nat}, Eq. (3.2.7) gives $\operatorname{Im}(Z) = 0$, that is, there is no reactance to oppose the pressure. It follows that if the dashpot D also is very small, then Z will be small at ω_{nat}. Thus, one can approximate a pressure-release termination at a specific ω by selecting the oscillator's natural frequency to equal ω, concurrently with minimizing the dashpot D. The opposite case of a rigid termination can be approximated by making K, M, or D as large as possible, because an infinite value for either property would require an infinite pressure to induce a finite V.

Although the surface impedance Z usually is frequency dependent, the trends displayed in Eq. (3.2.7) seldom are encountered. Specifically, the reactive part seldom is proportionally to ω when ω is large, nor does it vary reciprocally to ω when ω approaches zero. Its value is defined by the relationship between stress and strain in the case where the surface has a rigid backing. For a viscoelastic material, the complex modulus $\hat{E}$ is the extension to viscoelastic materials of Young's modulus for an ideal elastic material. The value of $\hat{E}$ is the ratio of complex amplitudes of stress and strain in a process where both vary harmonically. If a material sample's length L is much less than the wavelength, then the stress and strain may be considered to be the same everywhere in the sample. Application of a pressure Re $(P \exp (i\omega t))$ at one end with the opposite end fixed induces a displacement Re $(U \exp (i\omega t))$ at that end. The compressive stress within the material equals the applied pressure and the strain is Re $((U/L) \exp (i\omega t))$. The complex modulus is $\hat{E} = P/(U/L)$. The velocity amplitude at the location where the pressure is applied is $V = i\omega U$, from which it follows that

$$\hat{E} = \frac{PL}{U} = \frac{i\omega PL}{V} = i\omega LZ \tag{3.2.8}$$

The value of $\hat{E}$ can be measured experimentally in a dynamic testing machine, from which Z can be evaluated. An alternative is to measure Z in an inverse acoustic test using an impedance tube, which we soon will study. Then, the value of the complex modulus can be found from the preceding. An aspect that is not apparent is that Z cannot be selected arbitrarily, because doing so is likely lead to reflection processes that are inconsistent with the principle of causality, as we saw in Example 1.12. This attribute is a consequence of $\hat{E}$ being a transfer function.

To ascertain how the presence of an impedance closure affects a wave, let us suppose we know the complex amplitude of a backward propagating plane wave that is incident at $x = 0$ whose impedance is Z. We wish to determine the reflected wave. The unit vector into the fluid is $\bar{n} = \bar{e}_x$, so the impedance boundary condition requires that

$$P|_{x=0} = -Z \, V|_{x=0} \tag{3.2.9}$$

The general harmonic signal within a waveguide is described in Eqs. (3.1.11) and (3.1.12). In the current, scenario we know B_2 and wish to determine B_1. At $x = 0$, these relations state that $P = B_1 + B_2$ and $V = (B_1 - B_2)/(\rho_0 c)$. Hence, the boundary condition requires that

$$B_1 + B_2 = -\frac{Z}{\rho_0 c} (B_1 - B_2) \tag{3.2.10}$$

It follows that the reflected wave's complex amplitude is

$$\boxed{B_1 = R B_2, \quad R = \frac{Z - \rho_0 c}{Z + \rho_0 c}} \tag{3.2.11}$$

The quantity R is the *reflection coefficient*. Its definition here is the same as Eq. (2.5.13) for reflection in the time domain, except that here R may be complex. In terms of the specific impedance, it is

$$R = \frac{\zeta - 1}{\zeta + 1} \tag{3.2.12}$$

In some situations, we would like to know the surface impedance that produces a known reflection coefficient. Solution of the definition of R yields

$$Z = \rho_0 c \frac{1 + R}{1 - R} \tag{3.2.13}$$

Another general relation that sometimes is useful results from explicitly representing ζ as a complex number, $\zeta = r + is$. Rationalizing R makes the denominator real, with the result that

$$R = \frac{r^2 + s^2 - 1 + 2is}{(r + 1)^2 + s^2} \tag{3.2.14}$$

This description shows that the imaginary part of the reflection coefficient will have the same sign as the reactive part of the impedance. The value of R is real only if ζ is real, that is, purely resistive.

EXAMPLE 3.2 The fact that the impedance is a complex number obscures recognition of how the reflection coefficient is affected by the magnitudes of the resistance and reactance, and whether the reactance is an inertance or compliance. To answer these questions, consider a boundary for which the specific impedance is $\zeta = \Gamma \exp(i\chi)$. The phase angle χ must be in the range $-\pi/2 \le \chi \le \pi/2$ because the resistance cannot be negative. Determine and graph the magnitude and phase angle of the ratio of the reflected to incident complex pressure amplitudes at the boundary as a function of χ. Perform this evaluation for cases where $\Gamma = 0.1, 0.9, 1.1, 2$, and 5.

Significance

A quantitative understanding of the manner in which the impedance properties affect the reflected wave can be quite useful when one must select acoustical materials. This is especially so because some behaviors are counterintuitive.

Solution

The reflection factor is the ratio of the complex amplitude of the reflected wave relative to the incident. In terms of the given polar representation of ζ, the expression is

$$R = \frac{\zeta - 1}{\zeta + 1} = \frac{\Gamma e^{i\chi} - 1}{\Gamma e^{i\chi} + 1}$$

The value of R for all parameter combinations may be placed in a single rectangular matrix $[R]$ whose number of columns is the number of values of Γ. The number of rows is set by the fineness of the increment in χ. The graphs appearing here correspond to a one-degree increment. The computation may be vectorized by placing the χ values in a column vector and assigning the column of R values for a specific Γ to a column of $[R]$.

The upper graph in Fig. 1 describes how $|R|$ depends on χ for each Γ value. The first noteworthy feature is that $|R| = 1$ if $\chi = \pm 90°$, regardless of $|\zeta|$. A material having this property is purely reactive. In Chap. 5, we will see that there is no energy loss in the reflection process for purely reactive materials. This is evidenced by the mechanical model in Fig. 3.2, where $\chi = \pm \pi/2$ corresponds to absence of the dashpot. The spring and piston store mechanical energy, rather than dissipating it. An interesting feature is that a small value of $|R|$ can only be obtained if $|\zeta| \approx 1$ and Im $(\zeta) \approx 0$. Such materials have an impedance that is close to the characteristic impedance of the fluid, that is, $Z \approx \rho_0 c$.

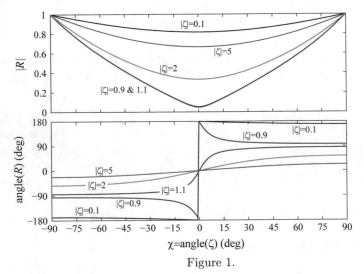

Figure 1.

This jump in the phase angle of R might appear to be an anomaly because the other curves are smooth. In fact, it is an artifice resulting from placing the phase angle in the range $-\pi \leq \text{angle}(R) < \pi$. Any range covering 2π would have been equally correct. Regularizing the phase angle dependence is a process of unwrapping it. Some software provides a routine to perform the operations. Manually, unwrapping can be performed by assembling the phase angles sequentially in a column vector. Then, each element is compared to the preceding one. If the angle$(R_n)-$angle$(R_{n-1}) > \pi$, then 2π is subtracted from all values of angle(R_j) for $j \geq n$. Sometimes, the jump might be an abrupt decrease. This is corrected by testing if angle$(R_n)-$angle$(R_{n-1}) < -\pi$, in which case 2π is added to all values of angle(R_j) for $j \geq n$. The unwrapped phase angle dependence is described in Fig. 2. All curves now are continuous.

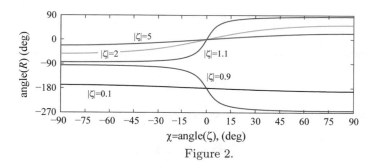

Figure 2.

The usual terminology is to say that the boundary is *soft* if $|\zeta| < 1$, and *hard* if $|\zeta| > 1$. Furthermore, the reactive part of the impedance is a compliance if $\chi < 0$. Inspection of Fig. 2 reveals some general trends. A soft surface that is compliant leads to a value of R in the third quadrant of the complex plane, whereas a soft inert surface leads to R in the second quadrant. For a hard surface, compliance leads to R in the fourth quadrant, while inertance leads to R in the first quadrant. Furthermore, the figure indicates that angle$(R) = 0$ if ζ is real and greater than one (hard surface), while angle$(R) = 180°$ if ζ is real and less than one (soft surface).

If it is desired to control noise, it is highly desirable that $|R|$ be close to zero. The present results indicate that such a condition cannot be attained if the surface is very soft or very hard. Rather, $|\zeta|$ must be close to one. In addition, angle(ζ) must be close to zero. In other words, if one wishes to minimize the amplitude of the reflected wave, the desired material for the boundary must have an impedance that is essentially resistive with a magnitude that is close to the characteristic impedance of the fluid.

EXAMPLE 3.3 The spring-mass system in Fig. 3.2 serves as a prototype for an impedance boundary condition, but it also is the simplest model of a structural dynamic system. A general question for such systems is the effect of fluid loading when a force is applied to a structure. Suppose a known harmonic excitation Re $\left(F e^{i\omega t} \right)$ is applied to the piston and the waveguide is infinite. Derive an expression for the displacement of the piston, and compare it to the response that the piston would have if the waveguide was replaced with a vacuum.

Significance

Acoustic radiation from vibrating structures is a vital issue for a broad range of applications. This example will highlight some fundamental phenomena without requiring complicated structural models.

Solution

We begin with the equation of motion, Eq. (3.2.5), to which we add the excitation force. Setting the displacement to Re $\left(U e^{i\omega t} \right)$ as before leads to

$$\left(K + i\omega D - \omega^2 M\right) U = F - P|_{x=0}\,\mathcal{A}$$

The waveguide is infinitely long so there is no backward propagating wave. Thus, we set

$$P\left(x\right) = Be^{-ikx}$$

The particle velocity is $V = P/\rho c$. Matching this to the piston's velocity, $i\omega U$ gives

$$B = \rho_0 c\left(\iota\omega U\right)$$

We use this relation to eliminate the pressure term in the equation of motion, which leads to

$$U = \frac{F}{\left(K - \omega^2 M\right) + i\omega\left(D + \rho_0 c\mathcal{A}\right)}$$

The case where the waveguide is replaced by a vacuum is obtained by setting $\rho_0 = 0$. Thus, it is apparent from the standpoint of the structure that the fluid adds to the dissipative effect represented by the dashpot. This effect is called *radiation damping*. This term actually is a misnomer because damping implies a loss of energy. Radiation damping is merely a transfer of energy from the structure into the fluid. Because the acoustic wave propagates away from the structure, it is never recovered.

The fluid loading of the structure in this system is a pressure that is in-phase with the structure's velocity. In more complicated situations, the acoustic signal will not be a simple wave. In such cases, the surface pressure and velocity will not be in-phase, which means that the fluid loading will also add to the stiffness or mass of the structure. One way of capturing all effects that act counter to forces applied to a structure is to form the *structural radiation impedance* Z_{rad}, which is the ratio of the excitation force to the structure's velocity at its interface with the fluid. Note that the units of Z_{rad} are N-s/m, not Rayl. For the present system, this quantity is

$$Z_{rad} = \frac{F}{i\omega U} = \left(D + \rho_0 c\mathcal{A}\right) + i\left(\omega M - \frac{K}{\omega}\right) = \rho_0 c\mathcal{A} + \mathcal{A}Z_{str}$$

The structure's innate ability to oppose a force is represented by $\mathcal{A}Z_{str}$. Radiation damping adds to that ability so individuals who think in terms of analogies say that the fluid acts in parallel to the structure's in vacuo impedance.

A quantitative picture emerges if we use the standard definitions of a one-degree-of-freedom's natural frequency ω_{nat} and ratio of critical damping ξ to describe Z_{rad}. These properties are

$$\omega_{nat} = \left(\frac{K}{M}\right)^{1/2}, \quad \xi = \frac{D}{2M\omega_{nat}}$$

When these definitions are used to eliminate D and K, the result is

$$Z_{rad} = M\omega_{nat}\left(2\xi + \frac{\rho_0\mathcal{A}}{M}\frac{c}{\omega_{nat}}\right) + i\omega M\left(1 - \frac{\omega_{nat}^2}{\omega^2}\right) \qquad (3.2.15)$$

The first term is the resistance, and the second term is the reactance. For most structural systems, ξ is much less than one. To estimate the importance of the other contribution to resistance, we observe that the structural mass per unit surface area is $M/\mathcal{A}$. The acoustic wavelength λ at the structure's natural frequency is $2\pi c/\omega_{nat}$, so $\rho_0 c/\omega_{nat} = \rho_0 \lambda/(2\pi)$ may be regarded as an effective fluid mass per unit surface area. Thus, the magnitude of the fluid loading's contribution to the resistance depends on the ratio of the effective fluid mass per unit surface area to the structure's mass per unit surface area. For gases, the ratio typically is small, so that radiation damping and internal dissipation, as described by ξ, are comparable in their influence. In liquids, notably water, the effective fluid mass per unit surface area usually is comparable to that of the structure's, so that radiation damping is the dominant effect. Indeed, it is not uncommon to find that it leads to overdamping, in which the combined resistance is equivalent to what would be obtained in a vacuum with $\xi > 1$.

3.2.2 Evaluation of the Signal

A common configuration features a waveguide that has an impedance closure at one end and some type of excitation at the other. It usually is preferable that $x = 0$ be the active end, because then the phase of the pressure field relative to the excitation will be evident. Thus, we set $x = L$ as the impedance boundary condition. In this situation, the normal pointing into the fluid is $\bar{n} = -\bar{e}_x$, so it must be that

$$P|_{x=L} = Z \, V|_{x=L} \tag{3.2.16}$$

In comparison with the situation where the boundary is at $x = 0$, the roles of the waves are reversed because the forward propagating wave is incident and the backward wave is reflected. The relation of their complex amplitudes could be obtained by substituting the general solution, Eqs. (3.1.11) and (3.1.12), into the preceding boundary condition. An alternative is to use the earlier analysis, which is done by defining a coordinate system whose x' axis is reversed, such that x' measures distance into the fluid, that is, $x' = L - x$. The general solution, Eq. (3.1.11), may be written as

$$P = B_1 e^{-ikx'} + B_2 e^{ikx'} \tag{3.2.17}$$

The situation relative to x' is the same as it previously was. The first term is a wave that propagates in the direction of increasing x, so it is the reflected wave. It follows that $B_1 = R_L B_2$, where the subscript "L" serves to remind us that it is the reflection coefficient at $x = L$. Eliminating x' in this expression for P gives

$$P = B_1 \left[R_L \, e^{-ik(L-x)} + e^{ik(L-x)} \right] \tag{3.2.18}$$

The fact that the first exponential's argument increases linearly with x confirms that this term represents a backward propagating wave. Based on this observation, the particle velocity obtained by combining the contributions of each simple wave is

$$V = \frac{B_1}{\rho_0 c}\left(-R_L\, e^{-ik(L-x)} + e^{ik(L-x)}\right) \tag{3.2.19}$$

The usual situation is that the excitation is a velocity input V_0 at $x = 0$. Equating the above expression for V at $x = 0$ to V_0 yields

$$B_1 = \frac{\rho_0 c V_0}{e^{ikL} - R_L e^{-ikL}} \tag{3.2.20}$$

Correspondingly, the preceding expressions become

$$\boxed{\begin{aligned} P &= \rho_0 c V_0 \left(\frac{R_L\, e^{-ik(L-x)} + e^{ik(L-x)}}{e^{ikL} - R_L e^{-ikL}}\right) \\[2ex] V &= V_0 \left(\frac{-R_L e^{-ik(L-x)} + e^{ik(L-x)}}{e^{ikL} - R_L e^{-ikL}}\right) \end{aligned}} \tag{3.2.21}$$

For comparison, if the excitation was a specified pressure $\hat{P}_0$ at $x = 0$, then the boundary condition would be $P = \hat{P}_0$ at $x = 0$, which is satisfied by

$$\boxed{\begin{aligned} P(x) &= \hat{P}_0 \left(\frac{e^{ik(L-x)} + R_L e^{-ik(L-x)}}{e^{ikL} + R_L e^{-ikL}}\right) \\[2ex] V(x) &= \frac{\hat{P}_0}{\rho_0 c}\left(\frac{e^{ik(L-x)} - R_L e^{-ik(L-x)}}{e^{ikL} + R_L e^{-ikL}}\right) \end{aligned}} \tag{3.2.22}$$

Some experiments require that there be a single simple wave. A forward propagating wave that is generated at $x = 0$ when $t = 0$ will not reach the other end until $t = L/c$, but this interval is quite short. To have only the forward propagating wave for a long time interval requires that $R_L = 0$. As we saw in Example 3.2, attaining this condition with a viscoelastic material requires that Z for the far end be very close to $\rho_0 c$. No material will fit this specification across a broad range of frequencies. Another possible way to avoid generation of a reflected wave is to gradually taper the end. The end is loosely packed with a porous material such as sand. The idea here is to absorb the forward propagating wave, until there is nothing left to reflect. Another alternative is to make $x = L$ active. A feedback control system senses the downstream pressure, which is used to activate a servomotor that vibrates a piston at $x = L$. If the piston's velocity is coerced to match the particle velocity of the incident wave, then continuity condition at that end is satisfied without an accompanying reflected wave.

A useful tool for interpreting the behavior of a waveguide is the *acoustic impedance*, which is the ratio of complex amplitudes of pressure and particle velocity,

$$Z_{\text{ac}}(x) \equiv \frac{P(x)}{V(x)} \qquad (3.2.23)$$

This definition of acoustic impedance differs from the standard definition that has generally been used, especially in other textbooks, in which the particle velocity is replaced by the volume velocity $\mathcal{A}V(x)$. Inclusion of the area factor eliminates that parameter from some relations that arise later in the study of interconnected waveguides. However, the standard definition creates a quantity for fluid response that is called impedance, but does not have the unit of Rayl. The acoustic impedance defined above is called the specific acoustic impedance in those treatments. The terminology adopted here is intended to avoid confusion about what impedance means.

Acoustic impedance is most useful for understanding how waveguides behave when they are joined together. In such situations, there might be individual segments in which there is only a wave that propagates in the positive direction because reflection at the far end is suppressed. The acoustic impedance of such a waveguide is the characteristic impedance, $Z_{\text{ac}} = \rho_0 c$. Another case in which we may ascertain the acoustic impedance independently of an analysis of the interaction of waves is certain situations in which the waveguide is very short, $kL \ll 1$. In that case, the pressure within the waveguide varies very little from point to point, so the waveguide's behavior may be captured by a single parameter. Such a representation is said to be a lumped parameter model. If the waveguide is not short, Z_{ac} at any location is a computed quantity. For example, the acoustic impedance in a waveguide that has an impedance termination at $x = L$ is found from the ratio of P to V in Eq. (3.2.22),

$$Z_{\text{ac}}(x) = \rho_0 c \left(\frac{e^{-ik(x-L)} + R_L e^{ik(x-L)}}{e^{-ik(x-L)} - R_L e^{ik(x-L)}} \right) \qquad (3.2.24)$$

Recall that Eqs. (3.2.18) and (3.2.19) resulted solely from satisfying the impedance boundary condition at $x = L$. It follows that the acoustic impedance in a single waveguide does not depend on whether the excitation at $x = 0$ is an imposed pressure or velocity oscillation.

EXAMPLE 3.4 A waveguide is driven at $x = 0$. Two cases are of interest: a pressure excitation in which $p(0, t) = \hat{P} \sin(\omega t)$ Pa and a velocity excitation for which $\rho_0 c v(0, t) = \hat{P} \sin(\omega t)$ m/s. The specific impedance at $x = L$ is $\zeta_L = 10 - 5i$. Determine and graph as a function of frequency the value of $|P|$ at the midpoint, $x = L/2$, in both cases. Then, use the acoustic impedance at $x = 0$ and $x = L/2$ to explain the similarities and differences of the two cases. The fluid within the waveguide is air, $L = 800$ mm, and $\hat{P} = 50$ Pa.

Significance

From an analytical viewpoint, there is not much difference between a pressure excitation and the more common velocity excitation. However, our computations will
reveal that there is a fundamental physical difference between the two cases. The
acoustic impedance will help us understand these differences.

Solution

The signal in the case of a pressure excitation is described in Eq. (3.2.22) with
$\hat{P}_0 = \hat{P}/i$, whereas Eq. (3.2.21) with $\rho_0 c V_0 = \hat{P}/i$ describes the case of a velocity excitation. The acoustic impedance in both cases is given in Eq. (3.2.24). The
reflection factor for the given material is $R = 0.8493 - 0.0685i$. Thus, although the
reactance at $x = L$ is substantial, R is essentially real.

Figure 1 displays the pressure amplitude at the midpoint as a function of frequency.
The alternative excitations produce drastically different responses, with resonance
peaks at vastly different frequencies. If the resistive part of ζ was smaller, these peaks
would be much higher.

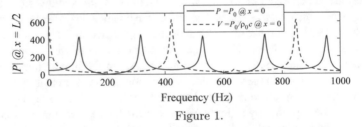

Figure 1.

We saw in our time-domain studies using the method of wave images that resonance occurs when a system is excited at one of its natural frequencies. Natural
frequencies are properties of a system, not of the excitation, so how can it be that the
resonant peaks for a velocity excitation do not occur at the same frequencies as those
for a pressure excitation? The explanation lies in the nature of each system when the
excitation is removed. If we remove an imposed pressure, we are left with an open
end. If we cease a velocity excitation, we will hold that end fixed. It is proven in the
next section that the present system does not have natural frequencies and modes
because $|R| < 1$. However, $|R|$ is close to one, which suggests that $x = L$ is close to
rigid. In turn, this suggests that the peak midpoint pressures in the case of a pressure
excitation occur at frequencies that are close to the natural frequencies of a pressure-
release/rigid waveguide, whereas the pressure peaks for a velocity excitation should
occur close to the natural frequencies of a rigid/rigid waveguide. The natural frequencies of both configurations were determined as a by-product of our exploration
of the method of wave images, and they are derived differently in the next section.
Equations (2.5.31) state that the natural frequencies of a waveguide whose ends are
rigid are $\omega_n = n\pi c/L$, which are 0, 212.5, 425.0, 637.5, 850.0 Hz, ... for the present
system. The natural frequencies of a rigid/pressure-release waveguide are stated
by Eq. (2.5.37) to be $\omega_n = (2n - 1)\pi c/(2L)$, which gives 106.3, 318.8, 531.3,

743.8 Hz,... Each frequency corresponds to a peak in the pressure at $x = L/2$, although these features are barely noticeable for the velocity excitation case at 212.5 Hz and 637.5 Hz.

This explanation leads to new questions: Why, in the velocity excitation case, are some peak amplitude values much smaller than others? If each peak indicates a resonance at a natural frequency, why do the resonances for each type of excitation not behave similarly? The acoustic impedance provides a perspective that helps to answer these questions. The upper graph in Fig. 2 is the result of evaluating Eq. (3.2.24) at $x = 0$. This data is said to be the *drive point impedance* because it describes the behavior at the location of the excitation. The drive point acoustic impedance exhibits nearly equal magnitude peaks at each of the natural frequencies of the waveguide with rigid terminations at both ends. Another confirmation that these peaks of the drive point acoustic impedance are resonances is the transition of the reactance as the frequency is increased. Specifically, the reactance is zero at a natural frequency because inertial and stiffness effects cancel at those frequencies.

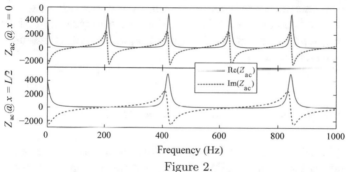

Figure 2.

The frequencies at which the drive point impedance has a peak are closely aligned with those of the peak midpoint pressure in the velocity excitation case, but that does not explain why the peaks in Fig. 1 for velocity excitation are so much lower at 212.5 and 637.5 Hz than the others. The answer may be found in the lower graphs in Fig. 2, which depicts the acoustic impedance at $x = L/2$ computed from Eq. (3.2.24). This quantity is very small in the vicinity of 212.5 Hz and 637.5 Hz. Smallness of Z_{ac} at $x = L/2$, but not at $x = 0$, means that $x = L/2$ is close to being a pressure node at those frequencies. Hence, although the drive point impedance indicates that a velocity excitation induces a high overall pressure at 212.5 Hz and 637.5 Hz, the corresponding pressure at $x = L/2$ is suppressed because the oppositely traveling waves are nearly $180°$ out-of-phase at those frequencies.[2] Resonances would have

---

[2]The modes of a rigid/rigid waveguide occur alternately as even and odd functions relative to the midpoint. The former, which correspond to $n = 0, 2, 4, ...$, are said to be symmetric modes. They have a maximum value at $x = L/2$. The modes for $n = 1, 3, 5, ...$ are antisymmetric modes. A continuous function that is odd with respect to $x = L/2$ must be zero at that location. Thus, if R_L for this system was exactly one, the midpoint pressure would not have shown any peak at 212.5 and 637.5 Hz.

been observed at these frequencies if we had evaluated $|P|$ at some other location, like $x = 300$ mm.

The drive point impedance also provides an explanation of the response for a pressure excitation. Let us switch the way in which we regard the excitation in this case. Rather than considering it to be a constant amplitude pressure excitation $\hat{P}/i$, we may consider it to be a frequency-dependent velocity amplitude $\hat{V}_0 = (\hat{P}/i)/Z_{ac}\,(x = 0)$. At frequencies for which the drive point impedance is very large (resonances of the fixed/fixed waveguide), $\hat{V}_0$ will be very small, so the overall pressure field will be small. Conversely, $\hat{V}_0$ is large at those frequencies where the drive point impedance is small. Equation (3.2.20) tells us that if $\hat{V}_0$ is large, then the amplitude of the forward and backward propagating waves also will be large, which, in turn, means that the overall pressure field will be large. Nevertheless, the pressure at $x = 0$ remains at the relatively small value P_0 at these frequencies because the phase difference between the forward and backward propagating waves at that location is almost, but not exactly, 180°. In other words, the largest pressure amplitudes generated be a pressure excitation occur at frequencies where the drive point impedance is minimized. This is the situation at frequencies that are midway between the rigid/rigid resonances. These mid-frequencies are the natural frequencies in the pressure-release/rigid case.

3.2.3 Modal Properties and Resonances

Equations (3.2.18) and (3.2.19) describe the relationship that must exist between the forward and backward propagating waves when there is an impedance boundary condition at $x = L$, regardless of the condition at $x = 0$. Rather than considering a signal that is generated by an excitation at $x = 0$, let us seek the conditions under which a harmonic signal can persist without any excitation. Such a situation would arise if the excitation that induced a steady-state field was suddenly switched off. To obtain a general perspective, we shall allow for the possibility that both ends are described by impedance boundary conditions.

The reflection coefficients are R_0 and R_L at the respective ends. As noted, Eq. (3.2.18) describes the combination of forward and backward waves that satisfies the condition at $x = L$. The complex amplitudes at $x = 0$ are $B_1 \exp(ikL)$ for the forward wave and $B_1 R_L \exp(-ikL)$ for the backward wave. At $x = 0$, the backward wave is incident and the forward wave is reflected. By definition, the ratio of the reflected to incident complex amplitudes is the reflection coefficient R_0, so it must be that $B_1 \exp(ikL) = R_0[B_1 R_L \exp(-ikL)]$. This reduces to

$$\boxed{e^{2ikL} = R_L R_0} \tag{3.2.25}$$

Because no conditions remain to be satisfied, we may assign any value to the complex factor B_1.

In the complex plane, $\exp(2ikL)$ is a vector whose magnitude is one. Thus, this equality is possible only if the magnitude of $R_0 R_L$ also is one. The common cases where this occurs are rigid or pressure-release ends. This condition also is met if the ends are purely reactive. If $|R_0 R_L| = 1$, then Eq. (3.2.25) is the *characteristic equation* for the waveguide. Because $\exp(2ikL)$ is a periodic function of kL, the characteristic equation has an infinite number of roots, k_n. Each root gives a frequency $\omega_n \equiv c k_n$ at which a signal may exist in a waveguide without any external excitation. By definition, these are the natural frequencies.

We examined natural frequencies and mode functions in Chap. 2 from the perspective of wave images. The frequency-domain formulation follows a different path to the same destination. The mode function is the pressure distribution in the waveguide at the natural frequency. According to Eq. (3.2.18), it is described in

$$\phi_n = B_1 \left[R_L\, e^{-k_n(L-x)} + e^{ik_n(L-x)} \right] \tag{3.2.26}$$

Because B_1 is arbitrary, let $B_1 = 1$, so that

$$\boxed{\begin{aligned} \phi_n &= R_L\, e^{-k_n(L-x)} + e^{ik_n(L-x)} \\ &= (R_L + 1)\cos\left(k_n\,(L-x)\right) + i\,(1 - R_L)\sin\left(k_n\,(L-x)\right) \end{aligned}} \tag{3.2.27}$$

It might seem from the preceding that the mode function does not depend on R_0, but that perception ignores the fact that R_0 appears in the characteristic equation, from which the value of k_n is extracted. Indeed, the situation would be changed if we had defined B_1 differently. Let $\exp(ik_n L)\, B_1 = 1$. The characteristic equation can be manipulated to state that $R_L \exp(-2ik_n L) = 1/R_0$, so factoring $\exp(-ik_n L)$ from the above complex exponential form of ϕ_n converts it to

$$\begin{aligned} \phi_n &= e^{k_n x} + \frac{1}{R_0} e^{-ik_n x} \\ &= \frac{(R_0 + 1)}{R_0}\cos\left(k_n x\right) + i\,\frac{(R_0 - 1)}{R_0}\sin\left(k_n x\right) \end{aligned} \tag{3.2.28}$$

Both representations of ϕ_n are equally valid.

Idealized Models of End Conditions

Any combination in which the ends are rigid or pressure-release leads to $R_L R_0 = \pm 1$. Equating the roots of the characteristic equation to this product requires that all $\exp(2ik_n L)$ values lie on the real axis. Hence, it must be that $2kL$ is some multiple of π. The specific cases are

$$R_0 = R_L = 1 : \omega_n = \frac{n\pi c}{L}, \quad P(x) = \phi_n(x) = \cos\left(\frac{n\pi x}{L}\right), \quad n = 0, 1, \dots$$

$$R_0 = R_L = -1 : \omega_n = \frac{n\pi c}{L}, \quad P(x) = \phi_n(x) = \sin\left(\frac{n\pi x}{L}\right), \quad n = 1, 2, \dots$$

$$R_0 = 1, \ R_L = -1 : \ \omega_n = \frac{(2n-1)\pi c}{2L},$$

$$P(x) = \phi_n(x) = \cos\left(\frac{(2n-1)\pi x}{2L}\right), \quad n = 1, 2, \dots$$

$$R_0 = -1, \ R_L = 1 : \ \omega_n = \frac{(2n-1)\pi c}{2L},$$

$$P(x) = \phi_n(x) = \sin\left(\frac{(2n-1)\pi x}{2L}\right), \quad n = 1, 2, \dots$$

$$(3.2.29)$$

Note that each mode function is real. (The expressions derived from Eq. (3.2.28) have been divided by 2 or $-2i$, which is allowable because it merely is a redefinition of the arbitrary B_1.) Real functions correspond to a pressure field that is a standing wave. In fact, the classical definition of a mode is a standing wave that exists without external excitation.

The lowest modes for each case are plotted in Fig. 3.3. The first two sets are fundamentally different, even though their nonzero natural frequencies coincide. In the first case, where both ends are rigid, the modal pressures are one or more half-periods of a cosine function. According to the Euler equation, the zero slope of the

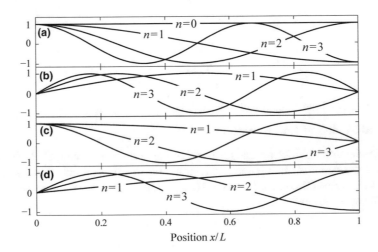

Fig. 3.3 Mode functions for simple end conditions. **a** Rigid at both ends; **b** pressure-release at both ends; **c** rigid at $x = 0$ and pressure-release at $x = L$; **d** pressure-release at $x = 0$ and rigid at $x = L$

function at both ends is consistent with the dictum that $V = 0$ at each end. The $n = 0$ mode is a uniform pressure field in which the particle velocity is zero. In the second case, the pressure at both ends is zero, which is fit by integer multiples of a half-period of a sine function. There is no $n = 0$ mode in this case. The natural frequencies in the third and fourth cases coincide because they merely swap end conditions—they are identical from a physical viewpoint. This is evident when the graphs are compared, and it can be proven by introducing the substitution $x' = L - x$ into one of the mode functions.

Reactive Ends

The rigid end of a termination considers the reactance and/or resistance of the boundary to be infinite. In real materials, the boundary might be very massive and stiff, but nevertheless movable. Furthermore, it is likely the resistance will be much smaller in magnitude than the reactance. Thus, an improved model of a rigid end is one for which the reactance is large but finite, with no dissipation. In regard to the pressure-release model, it can be used if an end of a waveguide is open to a large body of the same fluid, like the open end of a tube. However, we will see that such a model can be corrected with an impedance that accounts for inconsistencies resulting from taking the pressure to be zero. A first-order correction of either idealized condition is a purely reactive end.

The values of $k_n L$ are not difficult to find if the impedances are constant, independent of frequency. Then, the value of $R_0 R_L$ is described by a constant polar angle, which is denoted as θ. The lowest frequency is found directly, and the periodicity of the complex exponential gives the higher values. The positive values are

$$e^{2ik_n L} = R_0 R_L \equiv e^{i\theta}$$

$$k_n L = \begin{cases} \dfrac{\theta}{2} + (n-1)\,\pi \text{ if } \theta > 0 \\[2mm] \dfrac{\theta}{2} + n\pi \text{ if } \theta < 0 \end{cases}, \quad n = 1, 2, \ldots \tag{3.2.30a}$$

As an example, suppose that $\zeta_0 \equiv Z_0/(\rho_0 c) = si$ with s real, and $R_L = 1$. In that case, Eq. (3.2.14) gives $R_0 = (s^2 + 1 + 2is)/(s^2 + 1)$, so that $\theta = \tan^{-1}\left(2s/(s^2 + 1)\right)$. An inertial impedance, $s > 0$, corresponds to $\theta > 0$, whereas a compliant impedance, $s < 0$, corresponds to $\theta < 0$. Because $s = 0$ corresponds to a pressure-release end, these trends tell us that adding inertance raises the natural frequencies from the values for an open end, whereas adding compliance lowers them.

A different perspective results if the natural frequencies are compared to those for a system with two rigid ends. A boundary can be converted from one that is rigid to one that is movable by making it less massive. This scenario corresponds to s being a large positive number, rather than infinite. The angle θ in that case will be a small positive value. Consequently, an end that is very massive will lead to higher natural frequencies than those for a rigid end. Alternatively, a rigid end can be converted to one that is movable by increasing its compliance from zero. (Recall that $-\text{Im}(Z)$ is the reciprocal of the compliance if Im (Z) is negative.) This corresponds to s being negatively large, in which case θ is a small negative value. The natural frequencies are lower than those for the case of rigid ends. In other words, a system that has an end whose compliance is small will have lower natural frequencies than it would if the end was rigid. However, because frequency must be positive, the $n = 0$ mode of a rigid waveguide, for which $\omega_0 = 0$, disappears when an end has a compliance. A corollary of this behavior is that a very low fundamental natural frequency can only be obtained with a massive termination.

The preceding discussion is based on the reflection coefficients being constants. This seldom is the case. As was noted previously, real materials have frequency-dependent properties. In that case, $R_0 R_L$ will be a function of frequency. By definition $\omega = kc$, so the characteristic equation, Eq. (3.2.25), in this situation, is a transcendental function of kL. The roots $k_n L$ can only be found by numerical methods.

The mode functions corresponding to reactive boundary conditions constitute standing waves, as they do for rigid and pressure-release boundaries. To demonstrate this feature, we set $\zeta_0 = si$. Substitution of the corresponding R_0 into Eq. (3.2.28) gives

$$\phi_n = e^{-ik_n x} + \frac{si + 1}{si - 1} e^{ik_n x} \tag{3.2.31}$$

The polar representations of the numerator and denominator are

$$1 + si = \left(s^2 + 1\right) e^{i\psi}, \quad 1 - si = \left(s^2 + 1\right) e^{-i\psi}, \quad \psi = \tan^{-1}(s) \tag{3.2.32}$$

The corresponding description of a mode function is

$$\phi_n = e^{i\psi} \left[e^{-(ik_n x + \psi)} - e^{i(k_n x + \psi)} \right] \equiv -ie^{i\psi} \sin(k_n x + \psi) \tag{3.2.33}$$

The complex factor may be folded into the arbitrary factor B_1 that was eliminated previously. Doing so converts ϕ_n to a real function. The result describes a field in which the pressure at all locations oscillates in-phase; this is the definition of a standing wave.

Graphical Explanation of Modal Properties

The emphasis thus far has been the evaluation of natural frequencies and mode functions. Here, we shall develop a pictorial approach to these tasks. The notion is that the construction will provide a physical picture to accompany the mathematical operations.

Any pressure field in a finite length waveguide must consist of forward and backward propagating harmonic waves. The combination of a pair of waves oppositely traveling waves will be a standing wave only if their amplitudes are equal. Hence, we have established that any mode must fit $P = B \sin(k(x - x_0))$.

Let us draw a graph of P as a function of kx. We do not know x_0 at the outset, so let us replace $x - x_0$ with a relative coordinate x'. With respect to this coordinate, the mode is a pure sine. The wavelength along the kx' axis is 2π, as shown in the upper graph in Fig. 3.4.

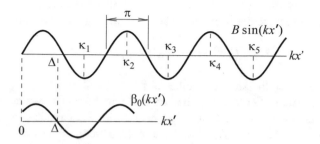

Fig. 3.4 Graphical construction leading to the modal properties of a waveguide whose specific impedance at the left end is $\zeta_0 = -i/2$, and whose right end is rigid. The *upper graph* is a plot of B $\sin(kx')$, and the *lower graph* is a plot of $\beta_0(kx')$, which is zero when the left boundary condition is satisfied

We locate the left boundary in this graph by seeking the location closest to $x' = 0$ at which the boundary condition is satisfied. If the left boundary is rigid, then Euler's equation gives $dP/dx' = -i\omega\rho_0 V = 0$. In this case, we search for the first location in the plotted sine curve at which the slope is zero. The value of kx' at this location is marked as Δ. Thus, $\Delta = \pi/2$ for a rigid end. Another possibility is that the left boundary is pressure-release. That case is marked by Δ being the smallest value of kx' at which $P(kx)$ is zero. This would give $\Delta = 0$. The case of an impedance boundary condition, $P = -Z_0 V$ at $x = 0$, is slightly more complicated. When Euler's equation is used to eliminate V, the boundary condition may be written as

$$P - \frac{\zeta_0}{ik}\frac{dP}{dx} = 0 \qquad (3.2.34)$$

Substitution of the graphed representation, $P = B \sin(kx')$, converts the preceding to

$$\beta_0(kx') = \sin(kx') + i\zeta_0 \cos(kx') = 0 \qquad (3.2.35)$$

In this case, the value of Δ is the smallest positive kx' that is a root of this equation. [As an aside, it is worth noting that this root will be real only if ζ_0 is purely imaginary. This is a different proof that a standing wave mode exists only if the resistive part of the impedance is zero.] The parameter Δ is the smallest value of kx' at which the plot of $\beta(kx')$ is zero. This identification is depicted by the lower graph in Fig. 3.4 for the case where the impedance is a compliance, $\zeta_0 = -i/C$. (The value used to construct the graph is $C = 2$.)

By definition, the left boundary is located at $x = 0$, and $kx' = kx - kx_0$. We have found that $kx' = \Delta$ at the location where the boundary condition is satisfied, so $kx_0 = -\Delta$. Therefore, a mode must fit $P = B \sin(kx + \Delta)$.

The next task is to identify where the right boundary is located. Like the situation for the left boundary, we search for a location where P has zero value (pressure-release), or a zero slope (rigid), or where a boundary function is zero (reactive impedance). This function is derived by noting that the normal to the boundary at the right end is opposite to the normal at the left end, so the boundary function would be

$$\beta_L\left(kx'\right) = \sin\left(kx'\right) - i\zeta_L \cos\left(kx'\right) \tag{3.2.36}$$

We denote as κ_1 the smallest value of $k'x \geq \Delta$ at which the right boundary condition is satisfied. We then continue on to the next location where the boundary condition is satisfied. The value of kx' at that location is marked as κ_2. The process can continue as far as we wish, so we have found a hierarchical sequence of values $\kappa_1 < \kappa_2 < \dots$. As was true for the left boundary condition, satisfaction of the right boundary condition depends only on the shape of P, so the amplitude factor B truly is arbitrary in regard to the existence of modes. The construction in Fig. 3.4 describes this process for the case where $x = L$ is rigid.

The natural frequency and modal wavenumbers follow from the observation that each κ_n value marks a possible location for the right boundary. Because $x' = x - x_0$ and the right boundary is situated at $x = L$, it must be that a mode exists whenever $\kappa_n = kL - kx_0$. Previously, we found that $kx_0 = -\Delta$. This leads to recognition that the wavenumber at the nth natural frequency is

$$k_n L = \kappa_n - \Delta \tag{3.2.37}$$

The corresponding natural frequencies are

$$\omega_n = \frac{c}{L}\left(\kappa_n - \Delta\right) \tag{3.2.38}$$

The pressure distribution for each κ_n is a mode, so we have found that

$$\phi_n = \sin\left(k_n x'\right) = \sin\left((\kappa_n - \Delta)\frac{x}{L} + \Delta\right) = \sin() \tag{3.2.39}$$

A comparison of this description of the standing wave to Eq. (3.2.33) gives $\psi = \Delta$. Thus, this graphical construction has demonstrated why a mode has the properties that were identified analytically. The construction could be used as an alternative to the analysis if precise values of κ_n and Δ are not required. However, such an evaluation may be pursued only if the impedance at both boundaries is independent of the frequency.

Resonances

The modes of a waveguide are possible responses when there is no excitation, but they also play an important role when a waveguide is excited by an external source. The descriptions of the field generated by velocity excitation, Eq. (3.2.21), and by pressure excitation, Eq. (3.2.22), reference the waves to $x = L$. A comparable description of the mode functions is the complex exponential form in Eq. (3.2.27).

The characteristic equation also will enter the analysis, so Eq. (3.2.25) is restated here,

$$\Delta \left(kL, R_0, R_L \right) \equiv e^{2ikL} - R_0 R_L = 0 \tag{3.2.40}$$

We begin with the case of a velocity excitation. The complex pressure amplitude in this case may be written as

$$P = \rho_0 c \hat{V}_0 \mathcal{N}_v \left(kL \right) F \left(kx, kL, R_L \right) \tag{3.2.41}$$

where

$$\mathcal{N}_v \left(kL \right) = \frac{1}{e^{ikL} - R_L e^{-ikL}} = \frac{1}{e^{-ikL} \Delta \left(kL, R_0 = 1, R_L \right)} \tag{3.2.42}$$
$$F(kx, kL, R_L) = R_L \, e^{-ik(L-x)} + e^{ik(L-x)}$$

Suppose kL is one of the nondimensional natural frequencies $k_n L$ of a waveguide for which the reflection coefficients are $R_0 = 1$ and R_L. This means that $\Delta \left(k_n L, R_0 = 1, R_L \right) = 0$. Consequently, $\mathcal{N}_v \left(k_n L \right)$ is infinite. Concurrently, when $kL = k_n L$, the spatial function $F \left(kx, kL, R_L \right)$ is the mode function ϕ_n in Eq. (3.2.27). Thus, we have found that imposing a particle velocity at $x = 0$ will lead to a resonance if the excitation is at the natural frequencies of that waveguide with a rigid termination at $x = 0$. The spatial pressure field at that resonance will have the pattern of the mode corresponding to that natural frequency, and the computed pressure amplitude is infinite.

Now let us consider the case of a pressure excitation at $x = 0$. The description in Eq. (3.2.22) may be written as

$$P = \hat{P}_0 \mathcal{N}_p \left(kL \right) F \left(kx, kL, R_L \right) \tag{3.2.43}$$

where

$$\mathcal{N}_p \left(kL \right) = \frac{1}{e^{ikL} + R_L e^{-ikL}} = \frac{1}{e^{-ikL} \Delta \left(kL, R_0 = -1, R_L \right)} \tag{3.2.44}$$

Infinite values of $\mathcal{N}_p \left(kL \right)$ correspond to kL being a root of $\Delta \left(kL, R_0 = -1, R_L \right) = 0$, but that is the characteristic equation for a waveguide whose boundary at $x = 0$ is pressure-release. Furthermore, $F \left(kx, kL, R_L \right)$ at any of these natural frequencies is the corresponding mode function. Thus, we arrive at a similar conclusion regarding resonances. Specifically, if a waveguide is excited by a specified pressure oscillation at $x = 0$, then resonances occur at the natural frequencies of that waveguide with a pressure-release termination at $x = 0$. The spatial distribution of pressure will be proportional to the resonant mode, and the pressure amplitude will be infinite.

Both excitation cases lead to a true resonance when the excitation frequency equals a natural frequency of the waveguide. The natural frequencies and modes in each case correspond to the free responses that may exist when the respective excitation is set to zero. In a resonance, the spatial pressure distribution is proportional to the mode function associated with that frequency. A corollary is that a true resonance, in which the amplitude is predicted to be infinite, can only occur if $|R_L| = 1$.

These conclusions are analogous to resonances of a mechanical oscillator or electrical circuit. They also are representative of the properties of any linear dynamic system that does not dissipate energy. In addition, they parallel the observations in Example 3.4, where the amplitude was finite at all frequencies because the wall impedance had a resistive part. Furthermore, one should bear in mind the behavior that was demonstrated in Example 2.7 in the situation where a system truly had no dissipative mechanisms. Excitation exactly at a natural frequency of an ideal system is manifested as a pressure oscillation whose amplitude grows with time, rather than an infinite steady-state amplitude.

To explore the significance of a small amount of dissipation, let us consider a frequency sweep for a system where R_L differs slightly from the value it would have if $\mathrm{Re}\,(Z_L)$ were zero. The function $\mathcal{N}_v\,(kl)$ defined above represents the scale by which the overall pressure field is magnified relative to the reference value $\rho_0 c\,|V_0|$. For this reason, $|\mathcal{N}_v\,(kL)|$ is sometimes referred to as a *magnification factor*. Similarly, $|\mathcal{N}_p\,(kL)|$ is the magnification factor for a pressure excitation. Suppose we have determined the modal properties for the ideal case where the impedance at $x = L$ is purely reactive. Let this idealized impedance be $(Z_L)_{\mathrm{ideal}}$. The corresponding reflection coefficient shall be denoted as $(R_L)_{\mathrm{ideal}}$. We wish to determine how the addition of a small amount of dissipation affects the response. To do so, we set $Z_L = \rho_0 c\,(\zeta_L)_{\mathrm{ideal}}\,(1 + \varepsilon_Z)$, where ε_Z is real and $|\varepsilon_Z| \ll 1$. A Taylor series expansion of the reflection coefficient shows that

$$R_L = (R_L)_{\mathrm{ideal}} + \varepsilon_R, \quad (R_L)_{\mathrm{ideal}} = \frac{(\zeta_L)_{\mathrm{ideal}} - 1}{(\zeta_L)_{\mathrm{ideal}} + 1}, \quad \varepsilon_R = \frac{2\,(\zeta_L)_{\mathrm{ideal}}}{\left[(\zeta_L)_{\mathrm{ideal}} + 1\right]^2}\varepsilon_Z \quad (3.2.45)$$

Most materials that are used for acoustical treatment have frequency-dependent impedances, but the variation is seldom drastic over the interval where the pressure field is maximized. Thus, we may consider ε_R to be constant in the vicinity of a natural frequency. We will examine the behavior of the magnification factor for velocity excitation, but the case of a pressure excitation shows similar behavior. To indicate proximity to the natural frequency of the purely reactive system, we let $kL = k_n L + \sigma$. Then, the magnification factor is

$$\mathcal{N}_v\,(k_n L + \sigma) = \frac{1}{e^{ik_n L}e^{i\sigma} - [(R_L)_{\mathrm{ideal}} + \varepsilon_R]\,e^{-ik_n L}e^{-i\sigma}} \quad (3.2.46)$$

Smallness of ε and σ allows us to simplify this expression by applying a two-term Taylor series to $\exp\,(i\sigma)$, followed by dropping terms that are a product of σ and ε_R. These operations yield

$$\mathcal{N}_v\,(k_n L + \sigma) \approx \frac{1}{\left[e^{ik_n L} - (R_L)_{\mathrm{ideal}}\,e^{-ik_n L}\right] + i\sigma\left[e^{ik_n L} + (R_L)_{\mathrm{ideal}}e^{-ik_n L}\right] - \varepsilon_R e^{-ik_n L}}$$
$$(3.2.47)$$

Because $k_n L$ satisfies the characteristic equation, we have $\exp(ik_n L) = (R_L)_{\text{ideal}}$ $\exp(-ik_n L)$. This property reduces the magnification factor to

$$\mathcal{N}_v(k_n L + \sigma) \approx \frac{1}{\left[2i\sigma(R_L)_{\text{ideal}} - \varepsilon_R\right]e^{-ik_{nL}}} \tag{3.2.48}$$

Our interest is the magnitude of this quantity. Both ε_R and $(R_L)_{\text{ideal}}$ are complex, so that

$$|\mathcal{N}_v(k_n L + \sigma)| \approx \frac{1}{\left\{4\sigma^2|(R_L)_{\text{ideal}}|^2 + |\varepsilon_R|^2 + 4\sigma\,\text{Im}\left(\varepsilon_R(R_L)_{\text{ideal}}\right)\right\}^{1/2}} \tag{3.2.49}$$

Thus, the occurrence of a true resonance requires that $\varepsilon_R = 0$ and $\sigma = 0$, in other words, no dissipation at the boundary and exact tuning of the excitation frequency. Otherwise, the maximum magnification factor will be $|\mathcal{N}_v(k_n L)| = 1/|\varepsilon_R|$ at $\sigma = 0$. Because all materials dissipate some energy when they vibrate, it follows that a true resonance cannot be observed in actual systems. Also, although most materials may be considered to have constant properties in a narrow band around a natural frequency, the variation between natural frequencies is sufficiently large that ε_R probably will depend on which natural frequency is under consideration. Thus, a scan of magnification factor as function of frequency typically shows peak values that vary from one natural frequency to the next.

EXAMPLE 3.5 The magnitude and polar angle of the complex impedance Z_L affect the magnification $\mathcal{N}_v(kL)$ in different ways. Evaluate the magnitude and polar angle of $\mathcal{N}_v(kL)$ when kL is close to the the fundamental nonzero frequency of a rigid/rigid waveguide, first for $\zeta_L \equiv Z_L/(\rho_0 c) = 10, 40$, and 160 and then for $\zeta_L = 40e^{i\chi}$ with $\chi = -\pi/4, 0$, and $\pi/4$. Explain any noteworthy features of the result.

Significance

In the course of the analysis, we will see that a large pure resistance, for which R is close to one, is fundamentally different from a rigid boundary, for which $R = 1$. Some of the results will be especially enlightening for anyone who is only familiar with resonance phenomena in mechanical and electrical systems.

Solution

This analysis entails a straightforward implementation of Eq. (3.2.21) using standard mathematical software to evaluate and graph the data. The fundamental frequency of a rigid-rigid waveguide is $k_1 L = \pi$, which is the center frequency for evaluation of the frequency response. The first set of results, depicted by the graphs in the left side of Fig. 1, compare pure resistance cases, for which $\zeta_L = 10$ $(R_L = 0.818)$, $\zeta_L = 40$ $(R_L = 0.951)$, and $\zeta_L = 160$ $(R_L = 0.988)$. The results for a rigid boundary are also depicted there. The smallest ζ_L does not quite fit the prescription that $|R| \approx 1$. This is evidenced by the fact that a peak value of $|\mathcal{N}_v (kL)|$ is barely evident, which means that dissipation is quite substantial. At the same time, even the largest ζ_L leads to significant differences from those for infinite ζ_L.

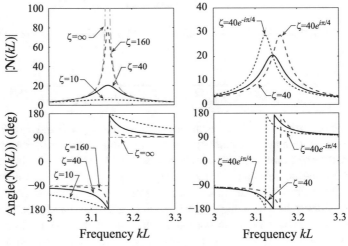

Figure 1.

We see that increasing values of a purely resistive impedance raises the peak of $|\mathcal{N}_v (kL)|$ without broadening its shoulder, thereby giving the peak a more slender appearance. Some individuals are surprised by this trend because it means that raising $\text{Re}(Z_L)$, which increases the resistance, results in a decrease in the amount of dissipation. Continuing to increase $\text{Re}(Z_L)$ in the limit leads to a rigid end, which does not dissipate energy and therefore shows a true resonance. The decrease in the height of the peak at $kL = \pi$ as real ζ is decreased stems from the fact that this no longer is a natural frequency. Indeed, in the limit $\zeta \to 0$, the end is pressure-release and the natural frequencies are $k_n L = 0.5\pi, 1.5\pi, \dots$.

The phase angle of $N_v (kL)$ shows an abrupt jump when ω equals a natural frequency. In the case of each finite impedance, the plot of the angle as a function of frequency becomes smooth if the data is unwrapped. Doings so yields a phase angle that transitions from $-90°$ well below the natural frequency, through $-180°$ at resonance, to $-270°$ degrees well above the natural frequency. When the boundary is rigid, this transition is abrupt.

The graphs on the right side of Fig. 1 correspond to adjusting the polar angle of the impedance with the magnitude held fixed. Because it must be that $\text{Re}(Z) \geq 0$ for a passive boundary, this angle must lie between $-\pi/2$ and $\pi/2$. The cases considered here are $R_L = 0.9647 - 0.0341i$ for $\arg(\zeta) = -\pi/4$, $R_L = 0.9512$ for $\arg(\zeta) = 0$, and $R_L = 0.9647 + 0.0341i$ for $\arg(\zeta) = \pi/4$. Unwrapping the phase angle of $\mathcal{N} (kL)$ shows that the behavior for complex Z_L is similar to the behavior for real Z_L. Specifically, the angle transitions gradually from $-90°$ for low frequencies to $-270°$ at high frequencies. In each case, the angle is $-180°$ at the frequency for which $|\mathcal{N}_v (kL)|$ is a maximum.

Among the three polar angles considered here, the greatest resistance with $|\zeta|$ fixed is obtained at $\arg(\zeta) = 0$, so the smallest peak magnification factor occurs in this case. The case where $\arg(\zeta) = -\pi/4$ gives $\text{Im}(\zeta) < 0$. Compliance is inversely proportional to $-\text{Im}(Z_L)$, so $\arg(\zeta) = -\pi/4$ means that the boundary is less compliant, that is, stiffer, and has lower resistance. Lower resistance means less dissipation, so the peak value is higher than it was for $\arg(\zeta) = 0$. The reduction in compliance also has the effect of decreasing the frequency at which the peak occurs. If $\arg(\zeta) = \pi/4$, the inertance is larger and the resistance is smaller than it is for $\arg(\zeta) = 0$. In this case, the peak value also is increased, but the frequency at which the peak occurs is increased.

An overview of the results shows that regardless of how closely the reflection coefficient approaches one, the fact that its magnitude is less than one inherently marks a boundary as being dissipative. This is important for numerical simulations that use an impedance model. Often, it is desirable to check the formulation by comparing an analysis with a previously known solution for a rigid boundary. This limiting analysis should be done by either raising the inertance or lowering the compliance, while simultaneously reducing dissipation to a very small value.

3.2.4 Impedance Tubes

Thus far, we have devoted our attention to evaluation of the pressure field in a waveguide whose termination has a known impedance. How do we know this impedance? This is an important question because flexible materials that are typically used for acoustic systems have a wide range of other applications, of which their use as the sole of athletic shoes is probably the most common. These materials are foamy polymers; that is, they have tiny bubbles finely interspersed within a rubbery plastic. These bubbles sometimes have an irregular shape in order to give the material some unusual properties, such as a negative Poisson's ratio. Current analytical capabilities are not adequate to predict the mechanical properties of these materials. Consequently, the impedance of a material must be derived from measurements.

If might seem that a testing machine that applies a harmonic force to a test specimen could be used for this purpose. However, this is a problematic approach, in part because it often is necessary to determine the impedance over a broad frequency range, which introduces issues regarding the dynamic behavior of the testing machine. Fortunately, there is an alternative procedure that extracts Z in an inverse manner from measurement of the acoustic field within a closed waveguide. The basic concept is to trace properties of the pressure field back to the reflection coefficient, from which the impedance is readily calculated.

The inverse identification process uses the mean-squared pressure as a function of x. When the complex pressure amplitude is described in Eqs. (3.2.41) and (3.2.42), this quantity is

$$\left(p^2\right)_{\text{av}} = \frac{1}{2}\,\text{Re}\,(PP^*)$$

$$= \frac{1}{2}\,(\rho_0 c\,|V_0|)^2\,|\mathcal{N}_v\,(kL)|^2\left[1+|R_L|^2+2\,\text{Re}\left(R_L e^{-2ik(L-x)}\right)\right] \tag{3.2.50}$$

A more useful form represents the reflection factor in polar form as

$$R_L = |R_L|\,e^{i\delta} \tag{3.2.51}$$

which gives

$$\left(p^2\right)_{\text{av}} = \frac{1}{2}\,(\rho_0 c\,|V_0|)^2\,|\mathcal{N}_v\,(kL)|^2\left[1+|R_L|^2+2\,|R_L|\cos\left(2k\,(L-x)-\delta\right)\right] \tag{3.2.52}$$

The dependence of $\mathcal{N}_v\,(kL)$ on kL is not known because it depends on R_L, but the value of $\mathcal{N}_v\,(kL)$ does not affect ratios of the values of $\left(p^2\right)_{\text{av}}$ at various x. The locations that are used are those at which $\left(p^2\right)_{\text{av}}$ has a maximum or minimum. The cosine function oscillates between plus and minus one, and the ratio of these values is denoted as S^2. Thus, we have

$$S^2 \equiv \frac{\max\left(p^2\right)_{\text{av}}}{\min\left(p^2\right)_{\text{av}}} = \frac{(1+|R_L|)^2}{(1-|R_L|)^2} \tag{3.2.53}$$

The value of $|R_L|$ that results is

$$|R_L| = \frac{S-1}{S+1} \tag{3.2.54}$$

The evaluation of δ also uses the properties of maxima and minima, but here, it is the phase angle. Let x_{max} denote any location where a maximum value of $\left(p^2\right)_{\text{av}}$ is observed. At this location, the cosine term in Eq. (3.2.52) is one, so that its phase must be a multiple of 2π. Therefore, the position x_{max} must be such that

$$2k\,(L-x_{\text{max}})-\delta = \pm 2n\pi,\ n=0,1,2,... \tag{3.2.55}$$

Similarly, at any position x_{min} where $\left(p^2\right)_{\text{av}}$ is a minimum, the phase should be an odd multiple of π, so that

$$2k\,(L-x_{\text{min}})-\delta = \pm\,(2n+1)\,\pi,\ n=0,1,2,... \tag{3.2.56}$$

The value of n in each case could be selected to place δ in a 2π range. However, our only use for δ is to evaluate R_L according to Eq. (3.2.42). Adding or subtracting multiples of 2π to δ does not alter that result. Thus, we can set $n=0$ for both locations, so that

$$\delta = 2k\,(L-x_{\text{max}})$$
$$\delta = 2k\,(L-x_{\text{min}})-\pi \tag{3.2.57}$$

The overall concept is to scan the pressure field to locate where $(p^2)_{av}$ has extreme values. The magnitude of the reflection coefficient is found from Eq. (3.2.54). Identification of a value of x_{min} or x_{max} permits determination of the reflection coefficient's polar angle δ. The corresponding reflection coefficient R_L is computed from Eq. (3.2.51). The impedance corresponding to the identified value of this coefficient is found from Eq. (3.2.13),

$$Z_L = \rho_0 c \frac{1 + R_L}{1 - R_L} = \frac{1 + |R_L| \cos \delta + i |R_L| \sin \delta}{1 - |R_L| \cos \delta - i |R_L| \sin \delta} \qquad (3.2.58)$$

The impedance tube technique has a fundamental limitation. It requires a fairly long waveguide because at least one maximum and one minimum must be observed. The period of the cosine term in Eq. (3.2.52) is a half-wavelength, which means that the length of the tube should be at least this long. Because $\lambda/2 = 0.5c/f$, this criterion sets the low-frequency limit of a given apparatus. For example, if we wish to use an impedance tube filled with air at standard conditions to determine Z_L upward from 20 Hz, we would need a waveguide that is 8.5 m long. Viewed differently, a 10-m-long impedance tube filled with water has a lower frequency limit of 74 Hz. Furthermore, this is a minimal condition because having multiple maxima and minima in a longer waveguide would enable us to average the estimates from each, which would tend to negate various measurement errors.

Another design issue pertains to the fact that the equations assume a planar standing wave field. This requires that the walls of the waveguide not vibrate. If they do, then the walls will have a velocity perpendicular to the axis of the waveguide, which is incompatible with the purely axial velocity in a plane wave. This requirement is especially difficult to meet because the pressure in the vicinity of resonances might be quite high. Sustaining a plane wave in such conditions requires that the walls be very massive and stiff, in order that structural resonances occur at much higher frequencies than the resonances of the acoustic field. (A colleague of the author mentioned that he was considering using the barrel of a 16 in. cannon taken off a battleship.)

An alternative technique for measuring impedance is a two-microphone technique. Its basic concept uses the time-domain representation of the pressure field as forward and backward propagating waves. The two unknown complex amplitudes B_1 and B_2 are extracted from measurement of the waveform at two locations. The value of R_L then is obtained as B_2/B_1. The fundamental theory for the technique was developed in Example 3.1. Some of the difficulties and errors associated with this technique have been discussed by Boden and Abom (1986).[3]

EXAMPLE 3.6 The solid line curve in the graph describes the analytical dependence of $(p^2)_{av}$ in a 2-m-long impedance tube, whereas "$\times$" marks data

---

[3]H. Bodén and M. Åbom, "Influence of errors on the two-microphone method for measuring acoustic properties in ducts," J. Acoust. Soc. Am. **79**, pp. 541–549 (1986).

points that would be obtained if the field were scanned in increments of 50 mm. This sampled data will be taken to be the result of measurements. The test material is at $x = 2$ m. The frequency is 600 Hz, and the fluid is water, with $c = 1480$ m/s and $\rho = 1000$ kg/m^3. Determine the specific impedance of the material at this frequency.

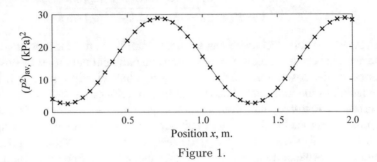

Figure 1.

Significance

The impedance calculation is straightforward if the pressure is a known function of position, but performing the same calculations with pressure data requires decisions about how that data should be used. This example introduces some of the issues.

Solution

Identification of locations at which the mean-squared pressure has an extreme value is the essential starting point. One could build an automated apparatus that will scan back and forth to accurately perform this task. Here, we assume a simpler scheme in which data is taken at a fixed spatial increment. The graph indicates that this scan is reasonably fine, but none of the sampled locations is exactly a maximum or minimum. How should we proceed without using the solid curve? One approach is to perform a spline fit to create a smooth curve that interpolates between the measurements. Another approach is to use a least-squares procedure to find the parameters in Eq. (3.2.52) that best fit the data. However, both approaches are problematic if there is significant noise, due to either ambient conditions or the measurement apparatus. We will take a simplistic approach in which we search around each peak or valley of the data for the location at which the extreme value occurs. How to use these multiple values is an issue to be resolved.

A scan of the sampled data set, which consists of values of $\left(p^2\right)_{\text{av}}$ at 41 locations, indicates that the maxima and minima are as tabulated in the first two rows. The third and fourth rows describe the evaluation of the polar angle of the reflection coefficient according to Eq. (3.2.57).

The two maximum values of the mean-squared pressure are not equal, nor are the two minima. We could form S^2 by evaluating the ratio of each maximum of $\left(p^2\right)_{\text{av}}$ to each minimum, which would give four S values that could be averaged. Another idea is to use the largest maximum and smallest minimum, based on the reasoning

Type	min	max	min	max
x (meter)	0.10	0.70	1.35	1.95
$\left(p^2\right)_{\mathrm{av}}$ (kPa)^2	2.628	28.843	2.708	28.857
$2k\,(L-x)$	9.6795	6.6228	3.3114	0.2547
δ (rad)	6.5379	6.6228	0.1698	0.2547

that a finer scan would get closer to the extreme value. That would be good idea for this analytical data set. However, the presence of noise detracts from this approach because the "maximum maximum" and "minimum minimum" could merely be the consequence of alignment of the signal with the noise. The approach we shall adopt is to use average values of the maxima and minima. Thus,

$$S = \left[\frac{(28.843 + 28.857)\,/2}{(2.6280 + 2.7080)\,/2}\right]^{1/2} = 3.288$$

(For comparison, if we had averaged the values obtained from the ratios of $\left(p^2\right)_{\mathrm{av}}$ at $x = 1.96$ and $x = 0.70$ to the values at $x = 1.30$ m and $x = 0.10$ m, we would have obtained $S = 3.289$.) Evaluation of Eq. (3.2.54) gives $|R_L| = 0.5336$.

The value of δ for each extreme value of $\left(p^2\right)_{\mathrm{av}}$ is the last line of the tabulation. It is evident that even without noise, the polar angle of R_L is more sensitive to error than is S. Each δ gives a reflection coefficient as $R_L = |R_L|\exp(i\delta)$, so we obtain four estimates of R_L. The average of these values is

$$R_L = 0.5155 + 0.1342i$$

The corresponding specific impedance is

$$\zeta = \frac{Z_L}{\rho_0 c} = \frac{1 + R_L}{1 - R_L} = 2.8336 + 1.0619i$$

The value of ζ used to generate the pressure data is $3e^{i0.4} = 2.763 + 1.168i$. The procedure worked reasonably well for the magnitude of ζ, which is estimated as $|\zeta| = 3.02$. The difference between the estimate of ζ and the actual value is primarily due to an imprecise identification of where $\left(p^2\right)_{\mathrm{av}}$ has an extreme value, which has a direct impact on the estimate of δ.

3.3 Effects of Dissipation

Energy is dissipated at the boundary whenever the magnitude of the reflection coefficient is less than one. Dissipation also occurs in the fluid field. There are multiple sources of this phenomenon. The development here will introduce each dissipative effect and explain how each leads to a modified wave equation. In each case, this equation will be analyzed for the propagation properties of a simple plane wave. The texts by Pierce[4] and Blackstock[5] detail the derivation of the field equation, and how the concepts may be extended to more general types of signals.

A general analysis of dissipation would be a formidable task. For example, inclusion of viscous effects would entail solving the Navier-Stokes equations for a compressible fluid. Several simplifying factors in the present context enable us to proceed. We will only examine one-dimensional waves in the frequency domain. This will reduce the problem to an ordinary differential equation. Another simplification will come from the fact that dissipation effects are relatively weak for most fluids of everyday interest. This property will simplify account for the combined effects of the various dissipation mechanisms.

The various sources of dissipation are manifested by similar mathematical descriptions of the pressure. Harmonic variation is represented by a factor $\exp(i\omega t)$. Each of the dissipation models for a plane wave will lead to a linear ordinary differential equation for the spatial dependence of the complex pressure. The general form of the pressure in a plane harmonic wave that propagates in the sense of increasing x will be seen to be

$$p = \text{Re}\left[Be^{i(\omega t - k'x)} \right] \tag{3.3.1}$$

We shall refer to k' as the *complex wavenumber*. For propagation in the sense of increasing x, its real part κ must be positive. Furthermore, because dissipation must attenuate the signal, the imaginary part of k' must be negative, in which case the plane wave may be represented as

$$p = \text{Re}\left[Be^{-i\alpha x} e^{i(\omega t - \kappa x)} \right]$$
$$\kappa = \text{Re}\left(k'\right), \quad \alpha = -\,\text{Im}\left(k'\right) \tag{3.3.2}$$

The parameter κ is the *propagation wavenumber*, which corresponds to the wavelength being $2\pi/\kappa$. (A strict interpretation indicates that the wave is not spatially periodic because it is attenuated. However, it is correct in any case to say that π/k is the distance between zeros.) The corresponding phase speed is

$$c_{\text{phase}} = \frac{\omega}{\kappa} \tag{3.3.3}$$

If α is nonzero, the pressure amplitude decreases exponentially. At a distance of n wavelengths, the amplitude will be reduced relative to its value at $x = 0$ by a factor of $\exp(-2\pi n\alpha/\kappa)$. If $n > (2/\pi)(\kappa/\alpha)$, this factor is less than 0.02. Attenuation generally is weak, in which case $\alpha \ll \kappa$. It follows that attenuation will be noticeable only after the signal has propagated many, many wavelengths. We refer to α as the

---

[4]A.D. Pierce, *Acoustics,* McGraw-Hill, Chap. 10 (1981). Reprinted by the Acoustical Society of America (1989).

[5]D.T. Blackstock, *Fundamentals of Physical Acoustics,* John Wiley $ Sons, Chap. 2 and Appendix C (2000).

absorption coefficient. The units of α are "neper/meter textquotedblright although *neper* is a dimensionless unit, analogous to a radian. (A neper may be regarded as a logarithmic measure defining relative magnitude. It differs from the decibel scale by using natural logarithms. Thus, $\ln\left(|P\left(x\right)|\,/\,|P\left(x+s\right)|\right) = \alpha s$ nepers.)

The relation between pressure and particle velocity in a plane wave is altered by the complex nature of the wavenumber. In the case of a plane wave in an ideal fluid, we identified this relation by satisfying Euler's equation. Whether we may implement the same procedure now depends on the dissipation mechanism. Two that we shall consider, thermal transport and molecular relaxation, feature alterations of the equation of state. Euler's equation remains valid for those models. Two other mechanisms we shall examine are viscosity and friction at the side wall of a waveguide. Both effects are manifested by a stress field that has components other than a pressure. Euler's equation is not valid in such cases. Fortunately, the equation of state is unaltered in those models. This allows us to use the frequency-domain continuity equation, Eq. (3.1.3). Note that this equation is not valid for a nonideal equation of state unless the speed of sound c that appears there is replaced by the derivative of the actual equation of state evaluated at ambient conditions. It will be easier to follow the development if we perform the analysis individually for each dissipation effect. Nevertheless, each will lead to proportionality of the complex pressure and particle velocity, but the factor of proportionality will be complex.

Although the various dissipative processes are manifested similarly in the mathematical form of the pressure, each process leads to a different equation relating k' and ω, which means that both κ and α are frequency-dependent. The consequence is that a signal composed of several harmonics will change the shape of its spatial profile and waveform as it propagates. Frequency dependence of α results in the amplitude of each harmonic being attenuated differently. Frequency dependence of κ results in each harmonic having a different phase speed. This is manifested by the higher speed harmonics running ahead of the slower harmonics, which is the process of *dispersion.* An interesting aspect is that in most fluids the frequency dependence of κ is very minor. This means that the phase speed is essentially constant, so that dispersion is negligible. The major effect of dissipation in that case is attenuation. Nevertheless, the equation relating k' to ω (or to k) is said to be a *dispersion equation.* (This terminology is used because an equation relating a wavenumber to frequency occurs in other systems, where the primary effect is frequency dependence of $\mathrm{Re}\left(k'\right)$.)

3.3.1 Viscosity

If the propagation wave truly is planar, then the shear stress is zero. However, viscous effects also arise as a frictional resistance to expansion and contraction. This effect is proportional to $\partial^2 v/\partial x^2$, which can be traced back to the rate of change of the volume per unit axial length in a plane wave. This leads to a stress that is manifested in the linear momentum equation as an additional term, specifically

$$\rho_0 \frac{\partial v}{\partial t} = -\frac{\partial p}{\partial x} + (\lambda + 2\mu)\frac{\partial^2 v}{\partial x^2} \tag{3.3.4}$$

In a shear flow, the shear stress is proportional to the *shear viscosity coefficient* μ. The friction effect associated with expansion and contraction depends on λ, as well as μ, so λ is said to be the *dilatation viscosity coefficient*.

Like the derivation of the wave equation for an ideal fluid, a field equation for the pressure is obtained by combining the equation of state with the momentum and continuity equations. For the first, we retain the adiabatic equation of state whose linearized form states that $\rho - \rho_0 = p/c^2$. We do so even though the existence of a viscous effect obviates the possibility that entropy is constant. The justification for this approximation lies in a formal analysis that demonstrates that entropy changes due to viscosity are nonlinear terms in the energy equation. The linearized equation of state is used to eliminate the density perturbation from the linearized continuity equation. The result is the same as it is for an ideal fluid,

$$\frac{1}{c^2}\frac{\partial p}{\partial t} + \rho_0 \frac{\partial v}{\partial x} = 0 \tag{3.3.5}$$

The desired field equation is obtained by differentiating the momentum equation with respect to x, then substituting $\partial v/\partial x$ from the continuity equation. The result is

$$\boxed{\frac{\partial^2 p}{\partial x^2} + \frac{(\lambda + 2\mu)}{\rho_0 c^2}\frac{\partial^3 p}{\partial x^2 \partial t} - \frac{1}{c^2}\frac{\partial^2 p}{\partial t^2} = 0} \tag{3.3.6}$$

Although this equation does not appear to be much more complicated than the linear wave equation, the presence of a mixed time-space derivative makes any time-domain analysis quite difficult. In contrast, removal of the time dependence in a frequency-domain analysis makes a solution quite accessible. We begin with the ansatz in Eq. (3.3.1). The dispersion equation, which is obtained by canceling the complex exponential, is

$$\left(k'\right)^2 = \frac{\omega^2}{c^2 + i\omega\,(\lambda + 2\mu)} \tag{3.3.7}$$

This form tells us little about the dependence of k' on the fluid's properties and frequency. However, we have not yet exploited a basic property of the fluids that are commonly encountered, which is that dissipation due to viscosity is a relatively small effect. If that is the case, then a three-term Taylor series should yield a good approximation to the square root of the denominator. The result is

$$k' \approx \frac{\omega}{c}\left[1 - i\frac{\omega}{2c^2}\,(\lambda + 2\mu) - \frac{\omega^2\,(\lambda + 2\mu)^2}{8c^4} + O\!\left(\frac{\omega^3\,(\lambda + 2\mu)^3}{c^6}\right)\right] \tag{3.3.8}$$

Comparison of this expression and the second of Eq. (3.3.2) shows that the propagation wavenumber and absorption coefficient are

$$\kappa \approx \frac{\omega}{c} - \frac{\omega^3\,(\lambda + 2\mu)^2}{8c^5}, \quad \alpha = \frac{\omega^2}{2c^3}\,(\lambda + 2\mu) \tag{3.3.9}$$

For most fluids, the second term in the expression for κ is negligible, which means that $c_{\text{phase}} = c$. The fact that the absorption coefficient increases as the square of the frequency means that the high- frequency components of a multiharmonic signal will be substantially diminished well before the lower frequency contributions.

For the reasons discussed in the opening discussion, we use the continuity equation to determine the relation of pressure and particle velocity. The complex pressure and particle velocities for a forward propagating wave are described by $P(x) = B \exp(-ik'x)$ and $V(x) = C \exp(-ik'x)$. Substitution of this representation into Eq. (3.3.9) leads to

$$B = \rho_0 c^2 \frac{k'}{\omega} C \implies P = \rho_0 c \frac{k'}{k} V \qquad (3.3.10)$$

The occurrence of a minus sign is the only change that would result from considering a plane wave propagating backward. Thus, we have established that

$$\boxed{V^+ = \left(\frac{k}{k'}\right) \frac{P^+}{\rho_0 c}, \quad V^- = -\left(\frac{k}{k'}\right) \frac{P^-}{\rho_0 c}} \qquad (3.3.11)$$

The net effect is that the characteristic impedance now is $\rho_0 c \left(k'/k\right)$. This quantity is complex and frequency dependent, which means that the particle velocity in either simple plane wave is not in-phase with the pressure.

3.3.2 Thermal Transport

If conditions are such that heat is conducted from particle to particle, the isentropic equation of state does not apply. A revised equation of state results from combining an energy conservation equation that accounts for heat transport with Fourier's law of heat conduction. The last element in the derivation is the ideal gas law, but the result is readily adapted to describe a liquid.

The derivation of the energy equation employs a control volume whose cross-sectional area is $\mathcal{A} = 1$ and whose length is dx. This control volume moves with the fluid. The rate at which the total energy of the particles contained within this region increases must equal the net power input by the pressure and the rate at which heat is transported into the domain. The one-dimensional nature of the system means that the power input is $(p\mathcal{A})v$, positive at the trailing cross section and negative at the leading cross section. Similarly, the heat flowing into the domain is $q\mathcal{A}$, positive at the trailing cross section where it flows into the domain. The total energy of the particles is the sum of the kinetic, potential, and thermal energies per unit volume, and the volume is $\mathcal{A}dx$. We will derive expressions for the potential and kinetic energies in the next chapter. The internal energy in a linearized approximation is $C_v (T - T_0)$, where C_v is the specific heat at constant volume and T_0 is the (absolute) ambient temperature.

Balancing the energies leads to a differential equation relating the particle velocity v, pressure p, heat flow q, and temperature T. The heat flow is eliminated with the aid of Fourier's law of heat conduction, which states that $q = -\beta_T \partial T/\partial x$, where β_T is the coefficient of heat conduction. The temperature is eliminated with the aid of the ideal gas law, which states that $p + p_0 = R\rho T$. The eventual result is an equation of state whose form is much more complicated than that for ideal conditions,

$$\frac{\partial}{\partial t}\left[p - c^2\left(\rho - \rho_0\right)\right] = \frac{\beta_T}{\rho_0 C_v}\frac{\partial^2}{\partial x^2}\left(p - \frac{c^2}{\gamma}\left(\rho - \rho_0\right)\right) \qquad (3.3.12)$$

The governing equations at this stage are continuity and momentum, which are the same as those for an ideal fluid and the above equation of state. The last step in the derivation is to eliminate the density and particle velocity. This may be done by employing the mathematical operator technique explained in Example 2.1. The result is a field equation for the pressure,

$$\boxed{\frac{\beta_T}{\rho_0 C_p}\frac{\partial^2}{\partial x^2}\left(\frac{\partial^2 p}{\partial x^2} - \frac{\gamma}{c^2}\frac{\partial^2 p}{\partial t^2}\right) - \frac{\partial}{\partial t}\left(\frac{\partial^2 p}{\partial x^2} - \frac{1}{c^2}\frac{\partial^2 p}{\partial t^2}\right) = 0} \qquad (3.3.13)$$

where $C_p \equiv \gamma C_v$ is the specific heat at constant pressure. The term inside the second pair of parentheses is the wave equation for an ideal fluid. The term inside the first pair is the wave equation with sound speed $c/\gamma^{1/2} \equiv (p_0/\rho_0)^{1/2}$. This is the speed of sound that results if the propagation is taken to be an isothermal process.

We can identify the relative importance of these trends by considering how a harmonic wave propagates. The dispersion equation that results from substitution of Eq. (3.3.1) into (3.3.13) is

$$\frac{\beta_T}{\rho_0 C_p}\left(k'\right)^2\left[\left(k'\right)^2 - \frac{\gamma\omega^2}{c^2}\right] + i\omega\left[\left(k'\right)^2 - \frac{\omega^2}{c^2}\right] = 0 \qquad (3.3.14)$$

A minor rearrangement of terms leads to

$$\boxed{\delta_{\text{thermal}}\left[\left(\frac{k'}{k}\right)^4 - \gamma\left(\frac{k'}{k}\right)^2\right] + i\left[\left(\frac{k'}{k}\right)^2 - 1\right] = 0} \qquad (3.3.15)$$

where the *coefficient of thermal conduction* is

$$\delta_{\text{thermal}} = \frac{\beta_T\omega}{\rho_0 c^2 C_p} \qquad (3.3.16)$$

Because δ_{thermal} is proportional to frequency, it is not a basic property of the fluid, but it does set the importance of thermal conduction. At very low frequencies, the first bracketed term in Eq. (3.3.15) will be negligible, in which case $k' \approx k$. In other words, thermal conduction is quite unimportant for low-frequency waves. The opposite situation is an extremely high frequency, so that $\delta_{\text{thermal}} \gg 1$. In that case, the first term in Eq. (3.3.15) is dominant, and there are two roots. The lower one is $k' \approx 0$, whose meaning for arbitrary δ_{thermal} we will examine. The other value is $k' \approx \gamma k$, which is equivalent to $c_{\text{phase}} = c/\gamma^{1/2}$. This is the sound speed in the isothermal approximation. Thus, we conclude that at very high frequencies wave propagation is isothermal, rather than adiabatic. In other words, the early prediction for the speed of sound based on the isothermal approximation would not have been incorrect if measurements of sound speed at very high frequencies had been available.

In both, asymptotic limits k' is real, which means that the absorption coefficient is zero. However, such is not the case in the frequency range between these limits. The behavior at arbitrary ω may be identified by solving the dispersion equation, which is quadratic on (k'/k),

$$
\begin{aligned}
\left(\frac{k_1'}{k}\right)^2 &= \frac{(1 + i\gamma\delta_{\text{thermal}})}{2i\,\delta_{\text{thermal}}}\left\{1 - \left[1 - \frac{4i\,\delta_{\text{thermal}}}{(1 + i\gamma\delta_{\text{thermal}})^2}\right]^{1/2}\right\} \\
\left(\frac{k_2'}{k}\right)^2 &= \frac{(1 + i\gamma\delta_{\text{thermal}})}{2i\,\delta_{\text{thermal}}}\left\{1 + \left[1 - \frac{4i\,\delta_{\text{thermal}}}{(1 + i\gamma\delta_{\text{thermal}})^2}\right]^{1/2}\right\}
\end{aligned}
\tag{3.3.17}
$$

Given the fluid properties and ω, it would not be difficult to evaluate these roots. If this route is taken, care must be exercised to place k' in the fourth quadrant because $\text{Re}\left(k_n'\right) = \kappa_n$ and $\text{Im}\left(k_n'\right) = -\alpha_n$.

The typical realm of acoustics is such that $\delta_{\text{thermal}} \ll 1$. This allows us to apply the binomial series formula to expand the fraction within the discriminant and then the discriminant itself, in binomial series, with the result that

$$
\begin{aligned}
\left(\frac{k_1'}{k}\right)^2 &= 1 - i\,(\gamma - 1)\,\delta_{\text{thermal}} - (\gamma - 1)\,(\gamma + 2)\,\delta_{\text{thermal}}^2 + O\left(\delta_{\text{thermal}}^3\right) \\
\left(\frac{k_2'}{k}\right)^2 &= \frac{1}{i\,\delta_{\text{thermal}}} + (\gamma - 1)\,(1 + i\delta_{\text{thermal}}) + O\left(\delta_{\text{thermal}}^2\right)
\end{aligned}
\tag{3.3.18}
$$

The smaller root k_1' is close to k. Indeed, another binomial series expansion leads to

$$
\begin{aligned}
\kappa_1 &= k\left[1 - \frac{3}{8}\,(\gamma - 1)\,(\gamma + 3)\,\delta_{\text{thermal}}^2 + O\left(\delta_{\text{thermal}}^4\right)\right] \\
\alpha_1 &= \frac{1}{2}\,(\gamma - 1)\,k\delta_{\text{thermal}} + O\left(\delta_{\text{thermal}}^4\right)
\end{aligned}
\tag{3.3.19}
$$

Thus, the propagation wavenumber differs little from k. Because $k = \omega/c$ and δ_{thermal} is proportional to ω, the absorption coefficient α_1 is essentially proportional to ω^2. This is the same trend as the frequency dependence of α due to viscosity. Indeed, it often is the case that absorption due to viscosity and heat conduction are quantitatively close.

The binomial series for the second root gives

$$
\frac{k_2'}{k} = (1 - i)\left(\frac{1}{2\delta_{\text{thermal}}}\right)^{1/2}\left[1 + \frac{i}{4}\,(\gamma - 1)\,\delta_{\text{thermal}}\right] + O\left(\delta_{\text{thermal}}^{3/2}\right) \tag{3.3.20}
$$

The values of κ_2 and α_2 extracted from this expression are proportional to $k/(\delta_{\text{thermal}})^{1/2}$ and therefore much greater than κ_1 and α_1 for the k_1' wave. The k_2' wave therefore oscillates spatially over a very small scale and decays rapidly. At a distance of one wavelength of the k_1' wave, it will be attenuated to a negligible value.

The solution associated with k_2' is a thermal wave. The most general solution is a linear combination of the k_1' and k_2' waves. However, the contribution of the thermal wave to the acoustic field is negligible, so it usually may be ignored. One circumstance where the thermal wave must be considered occurs when sound is generated by thermal oscillations at a boundary. In that case, solutions for p and T are required in order to satisfy the boundary conditions for particle velocity and temperature on the boundary. The solution for the temperature may be obtained by substituting the solution for pressure into the equation of state.

The relation between the complex amplitudes P and V for a plane wave may be derived from Euler's equation because it is the same as the relation of p and ρ for an ideal fluid. As we did for the analysis of viscosity, we set $P(x) = B \exp(-ik'x)$ and $V(x) = C \exp(-ik'x)$. Substitution of this representation into Eq. (3.1.6) leads to

$$C = \frac{k'}{\omega \rho_0} B$$

Here too, a backward propagating wave gives rise to the same relation, other than the occurrence of a minus sign. Thus, we have established that simple plane waves are characterized by

$$\boxed{V^+ = \left(\frac{k'}{k}\right)\frac{P^+}{\rho_0 c}, \quad V^- = -\left(\frac{k'}{k}\right)\frac{P^-}{\rho_0 c}} \tag{3.3.21}$$

The effective characteristic impedance now is $\rho_0 c\,(k/k')$. The resemblance of this quantity to the expression in Eq. (3.3.11) for viscosity can cause confusion, but any other approach would lead to a complicated form for one effect or the other.

3.3.3 Molecular Relaxation

Anyone who is familiar with the mechanical properties of a viscoelastic solid will recognize the analogous behavior of a relaxation process in a fluid. It is associated with molecular vibration states, of which there are several corresponding to different types of motion. For a solid, the elastic stress–strain relations are generalized by allowing the stress and its time derivatives to depend linearly on the strain and its derivatives. For a fluid, the generalization entails allowing the equation of state to be a differential equation in which the independent variables are pressure and density, and the derivatives are with respect to t. In a solid, this relationship might feature several orders of derivatives of stress and strain, but the situation for a fluid is much less complicated. The equation of state is taken to be

$$\left(\tau \frac{\partial}{\partial t} + 1\right) p = \left[c_\infty^2 \tau \frac{\partial}{\partial t} + c_0^2\right] (\rho - \rho_0) \tag{3.3.22}$$

Representation of the equation of state in operator form emphasizes its similarity to the stress–strain laws for viscoelastic solids, for which either operator order might be higher than first. Indeed, it is conceivable that a highly engineered liquid developed someday would also require higher order operators.

Experiments that measure c_0, c_∞, and τ indicate that $c_\infty > c_0$. A hint at the motivation for the subscripts may be obtained by considering alternative extreme conditions. If the pressure and density vary very slowly, which corresponds to extremely low frequencies, then the time derivatives are negligible. The equation of state reduces to $p = c_0^2 (\rho - \rho_0)$, which is the linearized form we have used thus far for an ideal fluid whose speed of sound is c_0. At the opposite extreme, the time derivatives are the dominant terms if the pressure and density vary rapidly. Integrating the equation of state in that case leads to $p = c_\infty^2 (\rho - \rho_0)$. This is the linearized equation of state for an ideal fluid whose sound speed is c_∞. The time constant τ weights these two trends.

What is not evident is how the equation of state affects a signal whose frequency is intermediate to the zero and infinite frequency limits. This can be ascertained by solving the wave equation in the frequency domain. Later, we will address the question of what happens when more than one dissipation mechanism is present. Here, we shall assume that viscosity and thermal transport are negligible. In that case, the momentum and continuity equations for an ideal fluid are applicable. A field equation for the pressure is obtained by using the differential operator approach to eliminate the density and particle velocity from these equations and Eq. (3.3.22). The result has the appearance of a combination of the wave equation for fluids whose sound speed is c_0 and c_∞,

$$\tau \frac{\partial}{\partial t} \left(\frac{\partial^2 p}{\partial t^2} - c_\infty^2 \frac{\partial^2 p}{\partial x^2}\right) + \left(\frac{\partial^2 p}{\partial t^2} - c_0^2 \frac{\partial^2 p}{\partial x^2}\right) = 0 \tag{3.3.23}$$

The dispersion equation that results from substitution of the complex exponential description of a plane wave, Eq. (3.3.1), is

$$\omega^2 (1 + i\omega\tau) - (k')^2 (c_0^2 + i\omega\tau c_\infty^2) = 0 \tag{3.3.24}$$

Thus, the complex wavenumber is given in

$$k' = \frac{\omega}{c_0} \left[\frac{1 + i\omega\tau}{1 + i\omega\tau \left(\frac{c_\infty}{c_0}\right)^2}\right]^{1/2} \tag{3.3.25}$$

Note that $c_\infty > c_0$, so the imaginary part in the denominator is greater than the imaginary part in the numerator. Consequently, the complex quantity inside the square root lies in the fourth quadrant, as required to have positive values of k and α.

We could obtain analytical expressions for κ and α by expressing the numerator and denominator of Eq. (3.3.25) in polar form. We shall not do so because such a description would not help us discern the frequency dependence of k'. Instead, let us consider the trends for low and high frequency. If $\omega\tau \ll 1$, the series expansion is

$$
k' = \frac{\omega}{c_0} \left\{ 1 - \frac{i}{2} \left(\left(\frac{c_\infty}{c_0}\right)^2 - 1 \right) \omega\tau - \frac{1}{8} \left[3 \left(\frac{c_\infty}{c_0}\right)^4 \right. \right.
$$
$$
\left. \left. - 2 \left(\frac{c_\infty}{c_0}\right)^2 + 1 \right] \omega^2 \tau^2 + O\left(\omega^3 \tau^3\right) \right\}
$$

(3.3.26)

It follows from this expression that

$$
c_{\text{phase}} = \frac{\omega}{\text{Re}\,(k')}
$$
$$
= c_0 \left\{ 1 + \frac{1}{8} \left[3 \left(\frac{c_\infty}{c_0}\right)^4 - 2 \left(\frac{c_\infty}{c_0}\right)^2 + 1 \right] \omega^2 \tau^2 + O\left(\omega^3 \tau^3\right) \right\}
$$

(3.3.27)

$$
\alpha = -\,\text{Im}\,(k') = \left(\left(\frac{c_\infty}{c_0}\right)^2 - 1 \right) \frac{\omega^2 \tau}{c_0}
$$

The last relation indicates that the absorption coefficient increases from zero as the square of the frequency and the phase speed increases from c_0 proportionally to the square of the phase speed.

At very low frequencies, the state of the fluid changes sufficiently slowly that molecules can remain at their equilibrium state. Correspondingly, c_0 is said to be the *equilibrium phase speed*. At high frequencies, we have $1/(\omega\tau) \ll 1$. To examine the behavior, we rewrite Eq. (3.3.25) as

$$
k' = \frac{\omega}{c_\infty} \left[\frac{1 - \dfrac{i}{\omega\tau}}{1 - \left(\dfrac{c_0}{c_\infty}\right)^2 \dfrac{i}{\omega\tau}} \right]^{1/2}
$$

(3.3.28)

The series expansion in powers of $1/(\omega\tau)$ is

$$
k' = \frac{\omega}{c_\infty} \left\{ 1 - \frac{i}{2\omega\tau} \left[1 - \left(\frac{c_0}{c_\infty}\right)^2 \right] + \frac{1}{8\omega^2\tau^2} \left[1 + 2 \left(\frac{c_0}{c_\infty}\right)^2 \right. \right.
$$
$$
\left. \left. - 3 \left(\frac{c_0}{c_\infty}\right)^4 \right] + O\left(\frac{1}{\omega^3\tau^3}\right) \right\}
$$

(3.3.29)

The corresponding phase speed and absorption coefficient are

$$c_{\text{phase}} = c_\infty \left\{ 1 - \frac{1}{8\omega^2\tau^2} \left[1 + 2\left(\frac{c_0}{c_\infty}\right)^2 - 3\left(\frac{c_0}{c_\infty}\right)^4 \right] + O\left(\frac{1}{\omega^4\tau^4}\right) \right\}$$

$$\alpha = \frac{1}{2c_\infty\tau} \left[1 - \left(\frac{c_0}{c_\infty}\right)^2 \right] + O\left(\frac{1}{\omega^2 c_\equiv \tau^3}\right)$$

(3.3.30)

Thus, as the frequency increases, α approaches a constant value and the phase speed approaches c_∞ from below. (Recall that $c_0 < c_\infty$.) These trends are evident in Fig. 3.5.

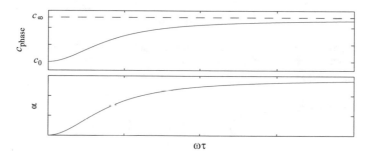

Fig. 3.5 Typical dispersion curves for a relaxing fluid

It is useful to examine quantitatively the parametric dependence of c_{phase} and α. It is evident from Eq. (3.3.25) that $k'/(\omega/c_0)$ depends only on two nondimensional parameters: $\omega\tau$ and c_∞/c_0. In turn, this means that c_{phase}/c_0 and $\alpha/(\omega/c_0)$ depend solely on these parameters. Figure 3.6 describes the (nondimensional) frequency dependence for a range of values of c_∞/c_0. Each ratio is only slightly larger than unity because that is the nature of most fluids.

An important feature is that for any ratio c_∞/c_0 there is a specific value of $\omega\tau$ at which $\alpha/(\omega/c_0)$ is a maximum. This value of ω is the *relaxation frequency* ω_r. Dissipation is greatest in the frequency range surrounding ω_r and the overall amount of dissipation increases monotonically with increasing c_∞/c_0. Less obvious is the decrease of ω_r as this ratio increases.

Blackstock[6] examined the dispersion equation based on c_∞/c_0 being very close to one, which is generally true. The outcome of that analysis is a simple expression for the absorption coefficient,

$$\tau = \frac{1}{\omega_r}, \quad \alpha = A \frac{\omega_r\omega^2}{\omega_r^2 + \omega^2}$$

(3.3.31)

[6]D.T. Blackstock, *Physical Acoustics*, Wiley (2000) p. 321.

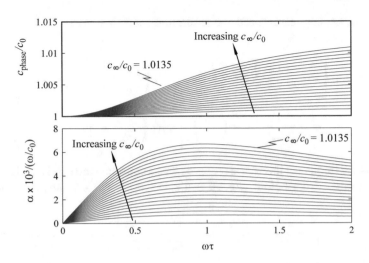

Fig. 3.6 Frequency dependence of the phase speed and attenuation coefficient of a relaxing fluid. Each curve corresponds to a fixed value of c_∞/c_0

where the coefficient A is

$$A = \frac{1}{2c_0} \left(\frac{c_\infty^2}{c_0^2} - 1 \right)$$ (3.3.32)

The value of A typically is identified experimentally. For example, the measured value of $\alpha c_0/\omega$ is plotted as a function of frequency. The maximum of this quantity occurs at $\omega = \omega_r$. Let $\alpha_{max} c_0/\omega_r$ be this maximum. According to Eq. (3.3.31), $\alpha_{max} c_0/\omega_r = A c_0/2$. The value of c_0 presumably has been identified previously, so A is set. From this and the value of ω_r, the ratio c_∞/c_0 may be found from Eq. (3.3.32) and τ has been identified as $1/\omega_r$.

The determination of the effect of relaxation on the relation between pressure and particle velocity follows the same path as it did for the case of thermal transport. For both, the equation of state is altered, but Euler's equation is the same as it is for an ideal fluid. Consequently, Eq. (3.3.21) with the appropriate value of k' describes both dissipation mechanisms.

3.3.4 Absorption in the Atmosphere and Ocean

The preceding examined the phase speed and absorption coefficient associated with distinct dissipation mechanisms. We found that viscosity and heat conduction lead to similar frequency dependencies. It is difficult to measure their individual contributions, so they usually are referred to jointly as the *thermoviscous effect*. Relaxation leads to a very different frequency dependence, and two relaxation processes coexist in the media of greatest interest: the atmosphere and the oceans. It would be quite challenging to solve field equations that simultaneously account for all dissipation processes.

Fortunately, the parameters δ, $\delta_{thermal}$, and $c_\infty/c_0 - 1$ characterizing the magnitude of each dissipation effect typically are very small. If that is the case, the total absorption coefficient is essentially the sum of the contributions of each process. The preceding analyses identify the general form of these contributions. The coefficients in such expressions are extracted from experimental data, in part, because their dependence on environmental conditions is quite intricate. Air is composed primarily of nitrogen and oxygen, each of which has an associated relaxation process. The ambient conditions affecting α are temperature, pressure, and humidity. Empirical formulas for the coefficients describing each relaxation process and the thermoviscous effect are available in an ANSI standard,[7] as well as in Blackstock's text.[8] In those formulas, the frequency appears in units of Hz as f. A typical case is air at the surface of the earth at $20°C$ and 20% relative humidity. The sound speed in this state is 343 m/s. The corresponding relaxation frequencies are 289 Hz for nitrogen and 1387 Hz for oxygen. The absorption coefficient for these conditions is found to be

$$\alpha_{air} = \frac{0.0033191 f^2}{f^2 + 8.3521\left(10^4\right)} + \frac{0.0084840 f^2}{f^2 + 1.92274(10^6)} + 1.84\left(10^{-11}\right) f^2 \text{ ncpcr/m}$$

$$(3.3.33)$$

The first two terms, respectively, are due to relaxation of nitrogen and oxygen molecules, and the last term is the thermoviscous effect. The contribution of each of these effects and their combination is depicted in Fig. 3.7. At the lower end of the audible range, relaxation of nitrogen molecules is the dominant process, slightly above 1 kHz, oxygen relaxation becomes the largest contributor, and thermoviscosity is dominant at extremely high frequencies.

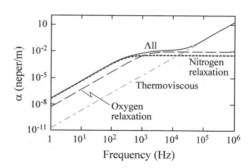

Fig. 3.7 Absorption coefficient for air at zero altitude, 20 °C, and 20% relative humidity

Two relaxation processes also exist in the ocean. Dissolved boric acid has a low relaxation frequency, and magnesium sulfate has a high value. Their contributions

[7] ANSI S1.26-1995, "American National Standard method for calculation of the absorption of sound by the atmosphere".

[8] D.T. Blackstock, *Physical Acoustics*, pp. 513–516.

appear, respectively, as the first two terms below. The last term is the thermoviscous effect for purified water. The environmental factors affecting α are the temperature, salinity, and depth. Blackstock provides formulas for arbitrary conditions. At 15 °C and a depth of 10 m with a salinity of 35 grams of salt per kg of water, the relaxation frequencies are found to be 1.331 and 100.511 kHz. The corresponding absorption coefficient is

$$\alpha_{\text{ocean}} = \frac{1.78277 \left(10^{-5}\right) f^2}{f^2 + 1.77208 \left(10^6\right)} + \frac{9.94914 \left(10^{-3}\right) f^2}{f^2 + 1.01158 \left(10^{10}\right)} + 2.98906 \left(10^{-14}\right) f^2$$

(3.3.34)

Fig. 3.8 describes the individual contributions and their combined effect. Relaxation of the boron molecules is the dominant contributor at low frequencies, but magnesium sulfate relaxation takes over beyond the boron relaxation frequency. The relaxation frequency for magnesium sulfate is very high, with the result that thermoviscous attenuation only becomes dominant in the megahertz range. A comparison of the curves for α in air and the ocean shows that the latter is much less dissipative at all frequencies.

Fig. 3.8 Absorption coefficient for the ocean at a depth of 10 m when the temperature is 15 °C and the salinity is 35 parts per thousand

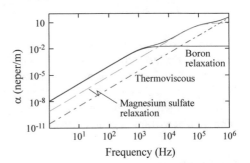

An overview shows that each dissipation process , thermoviscosity in the mainstream, relaxation, and wall friction, has little effect on the phase speed in air and water, including the ocean. The same is not true for the attenuation coefficient. In contrast, viscosity and thermal transport result in α increasing as the square of frequency. Attenuation due to relaxation increases monotonically with frequency, but its dependence is more complicated than that of thermoviscosity. The relaxation effect is important somewhat below the relaxation frequency and beyond.

EXAMPLE 3.7 A 20 kHz square wave is generated in air. The amplitude at its source is 20 Pa. Determine the waveforms at a distances of 10 and 20 m from the source, and compare each to what would be obtained if air had ideal properties. The ambient conditions are the same as those for α in Eq. (3.3.33).

Significance

The effect of dissipation on dispersion usually is not important for most fluids. Long-range propagation, especially if the signal is multiharmonic, is an exception. This example will provide a quantitative picture of the effect.

Solution

The square wave has a period $T = 0.05$ ms. It may be represented as a Fourier series as described in Eq. (1.4.7) with the coefficients described in Eq. (1.4.9). The steady-state signal must sustain this periodicity as it propagates, but each series coefficient is the complex amplitude of a simple wave whose complex wavenumber is $k' = \kappa - i\alpha$. Our plan is to evaluate the coefficients at the source and then propagate each harmonic based on the value of k' for that frequency. This operation will yield the Fourier coefficients of the waveform at $x = 10$ and 20 m. Synthesis of the corresponding Fourier series will yield the waveform at each location.

The instant at which the source square wave begins is arbitrary. We shall situate it such that the upward jump in the value of p occurs at $t = 0$. Thus, we set p at $x = 0$ equal to $\hat{P}[-h(t + T/2) + 2h(t)]$ for $-T/2 < t < T/2$, where $\hat{P} = 20$ Pa. The corresponding Fourier series coefficients are

$$
P_n = \frac{2}{T} \hat{P} \left[\int_{-T/2}^{0} (-1) e^{-i2n\pi t/T} dt + \int_{0}^{T/2} (+1) e^{-i2n\pi t/T} dt \right]
$$
$$
= \frac{2}{in\pi} \left[1 - (-1)^n \right], \quad n = 1, 2, \dots \tag{1}
$$
$$
P_0 = 0, \quad P_{(-n)} = P_n^*
$$

The frequency of harmonic n is $\omega_n = n(2\pi/T)$, so the corresponding complex wavenumber is $k' = \kappa(\omega_n) - i\alpha(\omega_n)$. The $\alpha_n(\omega_n)$ values are given in Eq. (3.3.33) with $f = \omega_n/(2\pi) = n/T$. The values for the first five harmonics appear in the tabulation. The attenuation coefficients increase to large values monotonically with increasing n. As a result, the higher harmonics will decrease much more rapidly as x increases.

n	1	2	3	4	5
f_n (kHz)	20	40	60	80	100
α_n (neper/m)	0.0192	0.0412	0.0780	0.1296	0.1958
k_n (m)	366.4	732.7	1099.1	1465.5	1831.8

The behavior of κ_n for air has not been discussed. Examination of Fig. 3.7 shows that at all frequencies the value of α is much less than ω/c, where c is the speed of sound in the absence of dissipation. This trend is verified by the tabulation. The nominal wavenumber for each harmonic is $k_n = 2\pi f_n/c$, with $c = 343$ m for the earth's surface for 20°C. The corresponding α_n is several orders of magnitude smaller than k_n. The analysis of the dispersion equation for each dissipation effect indicated that κ_n is essentially equal to k_n if $\alpha_n/k_n \ll 1$. In other words, the dispersion effect is not important.

As was mentioned, each harmonic propagates as a simple wave at wavenumber $\kappa_n - i\alpha_n$. The linear superposition of these waves is

$$p(x,t) = \text{Re}\left[\sum_{n=1}^{\infty} P_n e^{-\alpha_n x} e^{i(\omega_n t - \kappa_n x)}\right]$$

Substitution of ω_n and $\kappa_n = \omega_n/c$ shows that this is a Fourier series,

$$p(x,t) = \text{Re}\left[\sum_{n=1}^{\infty} B_n e^{i2\pi nt/T}\right] \tag{2}$$

The series coefficients depend on the propagation distance according to

$$B_n = P_n e^{-\alpha_n x} e^{-i2n\pi x/(cT)} \tag{3}$$

To synthesize a waveform, we must evaluate Eq. (2) at a sequence of t covering one period. The highest harmonic N must be very large to capture the discontinuity in the waveform at the source. Sampling of the waveform at the Nyquist rate requires that the number of samples per period be $2N$. The data appearing below was obtained with $N = 256$, which corresponds to sample instants at $t_j = (0, 1, ..., 2N-1)T/(2N)$. The matrix algorithm was used for the calculation. It defines a rectangular array $[E]$ whose elements are $E_{j,n} = \exp\left(-i2\pi f_n t_j\right)$. Then, a column vector of pressures $p(t_j)$ is found by evaluating $\{p\} = \text{Re}\left([E]\{B\}\right)$. The graphs in Fig. 1 periodically replicate the data in order to more clearly visualize the trends.

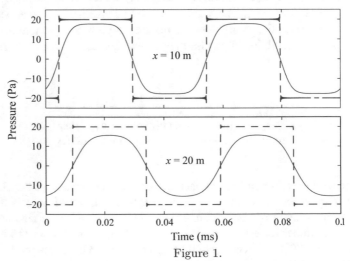

Figure 1.

Reduction in the overall amplitude is primarily due to attenuation of the fundamental harmonic. The reduction in the slope at the discontinuities results from increasing attenuation as the harmonic number increases. At farther locations, the latter effect will make the higher harmonics negligible in comparison with the first. The result will be a greatly reduced sinusoidal waveform.

3.3.5 Wall Friction

Friction at the wall of a waveguide is another source of dissipation. We shall consider situations in which the cross-sectional dimension is sufficiently large that the velocity profile consists of a thin boundary layer in the vicinity of a wall, with the rest being a nearly uniform axial velocity v. Viscous stresses are taken to be significant only within the boundary layer, whose thickness is Δ_{bl}. This effect was approximated in Example 2.1 by adding a viscous force to the momentum equation. The resulting field equation adds to the wave equation a term that depends on the rate at which the pressure changes, specifically,

$$\frac{\partial^2 p}{\partial t^2} + \chi \frac{\mu}{\rho_0} \frac{C}{A} \frac{\partial p}{\partial t} - c^2 \frac{\partial^2 p}{\partial x^2} = 0 \tag{3.3.35}$$

where μ is the dynamic viscosity (Pa-s), μ/ρ_0 is the kinematic viscosity, and C is the circumference of the cross section whose area is A.

The description of χ is obtained from basic flow physics. The shear stress at the wall must be proportional to the particle velocity and μ. The particle velocity vector is in the propagation direction, which is labeled x. The particle velocity away from the wall is taken to be a plane wave, but the effect of viscosity is to reduce it to zero at a wall. Thus, at fixed x, the velocity near the wall depends on the transverse distance y. The shear strain rate at the wall is $\dot{\gamma}_{xy} = (\partial v_x/\partial y)_w$, so the shear stress at the wall is

$$\tau = \mu \left(\frac{\partial v_x}{\partial y} \right)_w \tag{3.3.36}$$

Further progress requires the results of boundary layer studies in fluid mechanics. Many analyses have considered a harmonically varying particle velocity.[9] Those works scaled the distance y from the wall by the boundary layer thickness, so that y/Δ_{bl} is order one within the boundary layer. Correspondingly, the complex amplitude of the particle velocity is found to be the product of the mainstream velocity V and a complex function $F(y/\Delta_{bl})$, that is,

---

[9]D.T. Blackstock, *Fundamentals of Physical Acoustics*, John Wiley & Sons, pp. 322–325 (2000), provides a thorough discussion of boundary layer effects in an otherwise planar oscillating flow. Analysis of this problem may be traced back to G. Kirchhoff's efforts in 1868.

$$v(x, y, t) = \text{Re}\left[F\left(\frac{y}{\Delta_{\text{bl}}}\right) V e^{i\omega t}\right] \tag{3.3.37}$$

(The complex nature of F is a consequence of the fact that the pressure oscillation outside the boundary layer induces a wave that propagates in the y direction toward the wall.) The shear strain rate at the wall is found by differentiating this expression with respect to y and then evaluating it at the wall. Thus, the shear stress at the wall is

$$\tau = \text{Re}\left(\frac{F'_w}{\Delta_{\text{bl}}} \mu V e^{i\omega t}\right) \tag{3.3.38}$$

where F'_w denotes the derivative of the F function with respect to its argument evaluated at the wall. A comparison of this relation to the one used in Example 2.1 shows that

$$\chi = \frac{F'_w}{\Delta_{\text{bl}}} \tag{3.3.39}$$

Note that F is a dimensionless function, so the dimensionality of $F'_w / \Delta_{\text{bl}}$ is reciprocal length, as required for the definition of χ.

The boundary layer analyses related Δ_{bl} and F'_w to the basic parameters. The boundary layer thickness depends on the frequency according to

$$\Delta_{\text{bl}} = \left(\frac{2\mu}{\rho_0 \omega}\right)^{1/2} \tag{3.3.40}$$

(Strictly speaking, this is the distance measured from the wall, where $v_x = 0$, to a point at which the particle velocity amplitude is $V_x = |v|\,(1 - 1/e) = 0.632\,|v|$. This distinction is not important here because only the value of $F'_w / \Delta_{\text{bl}}$ appears in the analysis.) The value of F'_w associated with the above value of Δ_{bl} and the corresponding value of χ are

$$F'_w = 1 + i \quad \Longrightarrow \quad \chi = e^{i\pi/4}\left(\frac{\rho_0 \omega}{\mu}\right)^{1/2} \tag{3.3.41}$$

If extreme accuracy is required, F'_w may be considered to be an empirical factor derived from measurements of the particular system of interest.

Now that χ is known, we may proceed to determine the acoustical signal. The basic complex exponential representation of a plane wave in Eq. (3.3.1) is applicable. That ansatz satisfies Eq. (3.3.35) if

$$(k')^2 = k^2 - \left(\chi \frac{\mu}{\rho_0 c^2} \frac{C}{A}\right) \omega i \tag{3.3.42}$$

According to Eq. (3.3.41), χ depends on $\omega = kc$. Consequently, the dispersion equation that results is

$$\boxed{\left(k'\right)^2 = k^2 - e^{i3\pi/4}\sigma k^{3/2}} \tag{3.3.43}$$

where

$$\sigma = \left(\frac{\mu}{\rho_0 c}\right)^{1/2}\frac{\mathcal{C}}{\mathcal{A}} \tag{3.3.44}$$

For arbitrarily large σ, the complex wavenumber corresponding to propagation and decay in the sense of increasing x could be found by converting $k^2 + i\sigma k^{3/2}$ to polar form. However, a more revealing picture is obtained by exploiting the fact that wall friction, like the other dissipation mechanisms, is relatively weak for the common fluids. The kinematic viscosity μ/ρ_0 typically is small. For example, $\mu/\rho_0 = 1.5\left(10^{-5}\right)$ m²/s for air at 20°C and $\mu/\rho_0 = 0.98\left(10^{-6}\right)$ m²/s for water. (Honey, which is the fluid for Example 3.8, is an exception, and there are others, such as motor oil.) A small kinematic viscosity leads to $\sigma/k^{1/2}$ being much less than one. To prove this property, consider the definition of σ in Eq. (3.3.44). The magnitude of F'_w is $O\left(1\right)$ and $\mathcal{C}/\mathcal{A}$ is inversely proportional to the size a of the cross section ($\mathcal{C}/\mathcal{A} = 1/\left(\pi a\right)$ for a circle). From Eq. (3.3.40), $\sigma = \left(k/2\right)^{1/2}\Delta_{\text{bl}}\mathcal{C}/\mathcal{A}$, so $\sigma/k^{1/2}$ is $O\left(\Delta_{\text{bl}}/a\right)$, and the analysis is only valid if a is large compared to Δ_{bl}.

Smallness of $\sigma/k^{1/2}$ leads to approximation of the square root in Eq. (3.3.43) by the first few terms in a binomial series, so that

$$k' = k\left(1 - e^{i3\pi/4}\frac{\sigma}{k^{1/2}}\right)^{1/2} \approx k - \frac{1}{2}e^{i3\pi/4}\sigma k^{1/2} - \frac{1}{8}i\sigma^2 \tag{3.3.45}$$

The real part of k' is the propagation wavenumber κ, and the absorption coefficient α is the negative of the imaginary part. The second term represents the first-order correction to the inviscid approximation. It contains both real and imaginary parts, so it affects both κ and α, such that

$$\boxed{\begin{aligned} \kappa &= \text{Re}\left(k'\right) \approx \frac{\omega}{c} + \frac{1}{8^{1/2}}\sigma k^{1/2} = \frac{\omega}{c} + \left(\frac{\mu\omega}{8\rho_0 c^2}\right)^{1/2}\frac{\mathcal{C}}{\mathcal{A}} \\ \alpha &= -\,\text{Im}\left(k'\right) \approx \frac{1}{8^{1/2}}\sigma k^{1/2} = \left(\frac{\mu\omega}{8\rho_0 c^2}\right)^{1/2}\frac{\mathcal{C}}{\mathcal{A}} \end{aligned}} \tag{3.3.46}$$

The correction to both parameters is proportional to the square root of the frequency and decreases as the cross section becomes larger. In practice, the correction to the wavenumber is negligible, but the fact that the absorption coefficient is nonzero becomes evident with increasing propagation distance.

Viscous forces appear here as they do in the model of viscosity in the overall fluid. In both models, viscosity is manifested by addition of a term to the Euler equation for an ideal fluid, but the equation of state is unmodified. Therefore, we use the continuity equation to relate the pressure and particle velocity. The result is that Eq. (3.3.11) is applicable here, so the effective characteristic impedance is $\rho_0 c\left(k'/k\right)$.

EXAMPLE 3.8 A waveguide is filled with honey, whose properties are $\rho_0 = 1420$ kg/m^3, $c = 2030$ m/s, and $\mu = 25$ Pa-s. The length of the waveguide is 1.5 m and its diameter is 40 mm. The diameter is believed to be sufficiently small that friction at the wall is an important effect. The velocity imposed at $x = 0$ is $v_0 \cos(\omega t)$, and $x = L$ is rigid. Determine the frequency greater than 200 Hz at which the overall amplitude of the pressure in the waveguide is maximized. Then, evaluate the spatial dependence of the pressure amplitude at that frequency, and compare the distribution to the field that would be obtained if honey were inviscid.

Significance

A primary issue is how an analytical description of state variables for an inviscid fluid may be modified to account for dissipation. The result of the analysis will be demonstration of significant dissipation effects.

Solution

The kinematical viscosity of honey is almost 2000 times greater than it is for water. Consequently, it probably would not be advisable to employ the approximate solution of the dispersion equation, which was based on σ/k being small. We shall determine the complex wavenumber k' by solving the fundamental dispersion equation.

Equation (3.2.21) with $R_L = 1$ would give the pressure field if wall friction was not an issue. Relations describing pressure and particle velocity in a waveguide filled with ideal fluid are readily adapted by implementing Eq. (3.3.11). Accordingly, we modify of Eq. (3.2.21) by changing k to k' and setting the characteristic impedance to $\rho_0 c \left(k'/k\right)$. Making these adjustments and setting $R_L = 1$ lead to

$$P(x) = \rho_0 c \frac{k'}{k} V_0 \left(\frac{e^{ik'(L-x)} + e^{-ik'(L-x)}}{e^{ik'L} - e^{-ik'L}} \right)$$

By definition, the pressure is the product of the characteristic impedance, which is constant, the magnification factor, which is frequency dependent, and a function of position. The magnification factor in the preceding is

$$\mathcal{N}_v(kL) = \frac{k'L}{kL} \left(\frac{1}{e^{ik'L} - e^{-ik'L}} \right)$$

The complex nature of this function and the complicated form of the dispersion equation make it difficult to identify analytically conditions for a maximum. Mathematical software greatly facilitates a numerical evaluation. Our approach will be to select a frequency, to solve the dispersion equation, Eq. (3.3.43), for k' at that frequency, and then to use that value to evaluate $|\mathcal{N}_v(kL)|$. This computation is repeated over a range of frequencies, and the data set is scanned to identify the frequency corresponding to max $(|\mathcal{N}_v(kL)|)$. To select the frequency range for this computation, we observe that if μ were zero, the fundamental (nonzero) frequency of the waveguide with rigid ends would be $\pi c/L = 4252$ rad/s. In order to be sure that the coverage is adequate, the scan is expanded to twice that frequency. Figure 1 shows the result of the computation. Each peak is analogous to the resonances that would be observed if the fluid were inviscid.

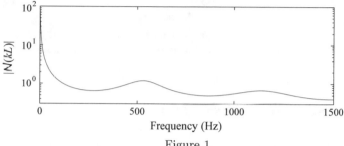

Figure 1.

Inspection of the numerical data reveals that the largest peak value of $|\mathcal{N}(kL)|$ above 200 Hz occurs at 529 Hz. The complex wavenumber obtained from Eq. (3.3.43) is $k' = 2.0432 - 0.3376i$ m^{-1}. The corresponding phase speed and attenuation factor are

$$c_{\text{phase}} = \frac{\omega}{\text{Re}(k')} = 1626.8 \text{ m/s}, \quad \alpha = -\text{Im}(k') = 0.3376 \text{ neper/m}$$

Note that at 529 Hz, $k = 1.6373$ m^{-1}, so $\text{Re}(k')/k = 1.248$. The fact that this ratio is significantly greater than unity confirms that it would not have been correct to use the approximate dispersion equation, Eq. (3.3.45). The wavelength is $\lambda = 2\pi/\text{Re}(k') = 3.072$ m, which corresponds to $\alpha\lambda = 1.0383$. Consequently, in the distance of a single wavelength, the pressure in a propagating plane wave would decay by a factor greater than $1/e$; this is a case of strong dissipation.

Figure 2 describes $P(x)$ at 529 Hz. The value of $\text{Im}(P)$ is small and nearly constant, which means that the response is very close to a standing wave. In contrast, if μ were zero, the field would be standing wave in which $P(x)$ is purely imaginary. The plots of $|P|$ indicate that the pressure amplitude has a minimum in the vicinity of $x = 0.75$ m, whereas the field in the inviscid case has a true node near $x = 0.5$ m. From an overall perspective, the most significant aspect of the analysis is that it reveals that the maximum pressure amplitude that would be observed at any frequency above 200 Hz is approximately $2\rho_0 c |V_0|$. In the inviscid case, there would be a true resonance at the natural frequency.

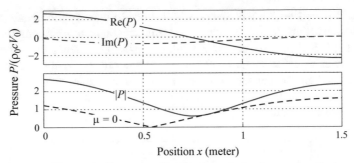

Figure 2. Pressure distribution at the frequency that
maximizes the pressure magnfication.

3.4   Acoustical Transmission Lines

Many applications feature interconnected waveguides. The system might be as simple as two waveguides joined end to end, or it might be the intricate network that is encountered in heating and ventilation systems, exhaust systems for internal combustion engines, and some musical instruments. Such an arrangement is often referred to as an *acoustical transmission line*. The general problem is to determine the pressure signal within the network when there is an excitation at one end. In order to avoid ambiguity, we will say that the downstream direction corresponds to propagation of the signal away from the end where the excitation is applied, as though it was flowing through the network.

Our consideration here is limited to configurations in which each segment has a constant cross section, with transitions from one cross section to another occurring over a very short distance relative to the wavelength. The waves in each segment are assumed to be planar. (This behavior cannot be sustained exactly at junctions, as we will see.) The development will identify algorithms for assembling the algebraic equations describing pressure within the transmission line. These algorithms can be extended to describe systems in which individual segments have cross sections that change gradually, but doing so would require introduction of the model of a horn, which is taken up in Chap. 9.

3.4.1   Junction Conditions

Figure 3.9 shows a section of length s that is used to transition the cross section from area $\mathcal{A}_1$ to $\mathcal{A}_2$. The situation depicted there is an expansion, $\mathcal{A}_1 < \mathcal{A}_2$, but a contraction, $\mathcal{A}_1 > \mathcal{A}_2$, also is allowed. If there is no internal barrier, the fluids in each section will be identical. However, it is conceivable that the ambient temperature will vary between segments. To allow for such a possibility, each segment is

Fig. 3.9 A transition section joining two waveguides having different cross sections

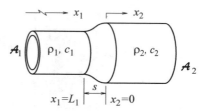

allowed to have arbitrary density and sound speed. Figure 3.9 shows two segments, but ultimately we will consider many. Location within each segment is described by x measured in the downstream direction. A subscript denotes the segment number, so that $0 \leq x_n \leq L_n$ and a junction has the property that $x_n = L_n$ adjoins $x_{n+1} = 0$.

Before we proceed with any analysis, let us consider conceptually what will happen. Suppose a wave is generated at $x_1 = 0$. It propagates downstream until it is incident at $x_1 = L_1$. It is reasonable to anticipate that the change in the cross section will cause some part of the incident wave to be reflected and that some part will be transmitted into the second waveguide. When the latter passes through the junction, the fluid particles must expand into the wider area. They can only do so by having a particle velocity component that is transverse to the x-axis. It follows that the flow cannot be a plane wave everywhere. However, if the length s of the transition is very small compared to the wavelength, we may consider the nonplanar aspects to occur over a very small interval. Doing so will enable us to replace an analysis of the actual behavior within the transition with continuity conditions relating plane waves.

Our later studies of multidimensional modes in a waveguide, Chap. 9, serve to justify the assumption of plane wave behavior. Those studies will reveal that below a certain frequency, called the cutoff frequency, only planar signals can propagate if the walls are stiff, whereas nonplanar signals decay rapidly. In the alternate case, where the frequency exceeds this cutoff value, it is indeed possible that nonplanar aspects will be present. Nevertheless, even at frequencies above the cutoff value, planar waves are important contributors to the overall signal. Hence, an analysis of high-frequency propagation based on purely planar signals should be regarded as an approximation of the manner in which sound propagates through the system.

The notation we will use to describe the pressure and particle velocity in segment n is $p^{(n)}(x_n, t)$ and $v^{(n)}(x_n, t)$. The corresponding complex amplitudes for a frequency-domain analysis will be $P^{(n)}(x_n)$ and $V^{(n)}(x_n)$. Differential equations describing conservation of mass and momentum for planar waves were derived in Chap. 2. It was inherent to both derivations that the signal changes gradually because the cross section is constant. Such is not the case here. Rather, a finite change of the cross section over a limited distance makes it possible that the state of the fluid can change significantly over the short length s of the junction. Thus, the signal at the end of waveguide 1 consists of pressure $p^{(1)}(L_1, t)$ and particle velocity $v^{(1)}(L_1, t)$, while the signal at the beginning of waveguide 2 is pressure $p^{(2)}(0, t)$ and particle velocity $v^{(2)}(0, t)$. For now, we ignore the decomposition into forward and backward propagating waves, so pressure and particle velocity are taken to be independent values.

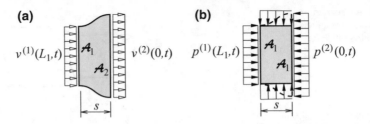

Fig. 3.10 Control volumes used to analyze state changes across a transition between segments of a waveguide having different cross sections. **a** Control volume for examination of conservation of mass, **b** Control volume for the impulse-momentum principle

Figure 3.10 shows the conceptual model we shall use. The control volume in Fig. 3.10a is used to derive the equation describing conservation of mass. Its choice is dictated by the fact that regardless of the details of the flow within the transition region, no fluid crosses the side wall. The rate at which mass enters the control volume at $x_1 = L_1$ is $\rho_1 v^{(1)}(L_1, t) \mathcal{A}_1$, and the rate at which it exits at $x_2 = 0$ is $\rho_2 v^{(2)}(0, t) \mathcal{A}_2$. Our interest is the limiting behavior as $s \to 0$, so the mass contained in the control volume is zero. Thus, the rate at which mass leaves the control volume must equal the rate at which it enters, so that

$$\rho_1 v^{(1)}(L_1, t)\, \mathcal{A}_1 = \rho_2 v^{(2)}(0, t)\, \mathcal{A}_2 \tag{3.4.1}$$

This is a statement of *mass flow continuity*. The usual situation is that all segments are filled with the same fluid, in which case $\rho_2 = \rho_1$. The quantity $\mathcal{A}v$ is sometimes referred to as the *volume velocity* because the volume that is displaced in a unit time is $\mathcal{A}(v\, dt)$. Thus, there must be continuity of the volume velocity at cross-sectional transitions in a waveguide filled with a homogeneous fluid.

The control volume in Fig. 3.10a is not suitable for a formulation of the impulse-momentum principle because the pressure distribution on the sidewall, which is unknown, has a resultant force in the x direction. The control volume in Fig. 3.10b is the cylindrical region whose cross section is $\mathcal{A}_1$. (If the transition was a contraction, the control volume would be formed from $\mathcal{A}_2$.) The pressure on the side of the control volume in Fig. 3.10b acts transversely to the x-axis, so it does not contribute to the forces that act in the x direction. By virtue of taking $s \to 0$, the mass of this control volume is negligible. Consequently, the resultant of all forces acting in the x direction must be zero, but the force acting at each cross section is the pressure at that location multiplied by the area $\mathcal{A}_1$ (or $\mathcal{A}_2$ in the contraction case). The sum of these forces must be zero, so we find that

$$p^{(1)}(L_1, t) = p^{(2)}(0, t) \tag{3.4.2}$$

A different line of reasoning leading to the same result argues that when $s \to 0$, fluid particles on either face will contact, and the pressure exerted between contacting particles must be the same.[10]

The development may be generalized to describe junctions at which more than two branches are joined. In Fig. 3.11, the junction corresponds to $x_1 = L_1$, $x_2 = x_3 = 0$.

Fig. 3.11 Junction of multiple waveguides

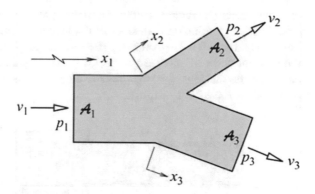

Increasing x for each waveguide defines the sense of positive particle velocity on each face. For the velocity vectors defined in the figure, mass enters the control volume at $\mathcal{A}_1$ and departs at $\mathcal{A}_2$ and $\mathcal{A}_3$. We take the control volume to be sufficiently small to justify taking its mass to be zero, which leads to

$$\rho_1 v^{(1)}(L_1, t)\,\mathcal{A}_1 = \rho_2 v^{(2)}(0, t)\,\mathcal{A}_2 + \rho_3 v^{(3)}(0, t)\,\mathcal{A}_3 \qquad (3.4.3)$$

The geometry of the branch junction makes it problematic to relate the pressures by following the procedure associated with Fig. 3.10b. However, it remains valid to reason that if the length of the junction is sufficiently small, points on each exposed cross section will be in extremely close proximity. This leads to the conclusion that the pressure in all segments at a junction must be equal,

$$p^{(1)}(L_1, t) = p^{(2)}(0, t) = p^{(3)}(0, t) \qquad (3.4.4)$$

These relations are readily modified to describe junctions of more than three waveguides. The frequency-domain version of these condition is obtained by replacing all time-dependent variables with the respective complex amplitudes.

---

[10]The junction conditions developed here are approximations based on assumptions of planar wave behavior and shortness of the transition distance s relative to the wavelengths $2\pi c^{(n)}/\omega$. An improved approximation for a frequency-domain analysis consistent with the planar approximation replaces pressure continuity at a junction with an impedance relation that asserts that the pressure difference is proportional to the average velocity, $P_1(L_1) - P_2(0, t) = Z_{\text{junction}}[V_1(L_1) + V_2(0)]/2$. A good starting point for exploring this refinement is the reference by Karal (1953).

3.4.2   Time Domain

There is much similarity between the effect of a junction and termination of a waveguide with a resistive impedance. This similarity is most evident in a time-domain analysis. The forward propagating plane wave $F_1(t - x/c_1)$ in Fig. 3.12 is incident at the transition between two segments having different fluids and cross-sectional areas. (It is convenient for this development to use a single position variable x to describe position on both sides of the junction. Thus, $x < 0$ is upstream of the junction and $x > 0$ is downstream.) Because the properties change discontinuously, the incident wave is reflected as a backward propagating wave $G_1(t + x/c_1)$. Concurrently, the incident wave disturbs the fluid in segment 2, giving rise to a forward propagating wave $F_2(t - x/c_2)$.

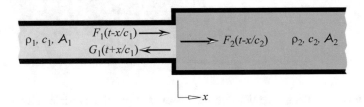

Fig. 3.12   Incident, reflected, and transmitted plane waves at a junction

The junction conditions state that the pressure and volume flow must be continuous from one side to the other. The particle velocity is obtained by dividing the pressure function by the characteristic impedance, with the sign selected according to the direction of propagation. Thus, the junction conditions require that

$$F_1(t) + G_1(t) = F_2(t)$$
$$\rho_1 A_1 \left(\frac{F_1(t) - G_1(t)}{\rho_1 c_1} \right) = \rho_2 A_2 \left(\frac{F_2(t)}{\rho_2 c_2} \right) \tag{3.4.5}$$

Presumably, the incident wave is known, in which case these equations tell us that $G_1(t)$ and $F_2(t)$ are proportional to $F_1(t)$. These factors of proportionality are the reflection coefficient R and the *transmission coefficient* T. They are

$$\boxed{\begin{aligned} R &= \frac{c_2 A_1 - c_1 A_2}{c_2 A_1 + c_1 A_2} \\[2mm] T &= \frac{2c_2 A_1}{c_2 A_1 + c_1 A_2} \end{aligned}} \tag{3.4.6}$$

A simple identity satisfied by these expressions is $T = 1 + R$. The result is that the pressure within each segment is given by

$$p_1 = F_1 (t - x/c_1) + R F_1 (t + x/c_1), \quad p_2 = T F_1 (t - x/c_2) \qquad (3.4.7)$$

Several trends are worth noting. The case where $c_2 A_1 \gg c_1 A_2$, for example, air into water, is like a wave in a light fluid arriving at a heavy fluid. Then, $R \approx 1$ and $T \approx 2$. In contrast, $c_2 A_1 \ll c_1 A_2$, like water into air, leads to $R \approx -1$ and $T \approx 0$. An interesting possibility is that $c_2/c_1 = A_2/A_1$. This condition gives rise to $R = 0$ and $T = 1$, as though the junction was not present.

The reflection coefficient affects the reflected wave in the same manner as it does for waves in a single closed waveguide with a resistive termination. This similarity allows the method of wave images to be modified to treat waveguides that contain multiple segments.

EXAMPLE 3.9 A waveguide consists of two segments. The first, whose cross-sectional area is A_1, has a piston at $x = 0$ that induces a velocity $v_0 (t)$. The junction at $x = L$ connects the left segment to a very long branch whose cross-sectional area is A_2. This branch is terminated at its distant end in a manner that eliminates reflections. Consider the case where v_0 is a ramp function whose duration is τ, $v_0 = V_0 t/\tau$ for $0 < t < \tau$ with $\tau < L/c_1$. Construct the initial spatial profile of images whose processing will lead to the spatial profiles at later t and the waveforms at arbitrary x. The reflection and transmission coefficients at the junction of this segment are $R = 0.4$ and $T = 1.4$, and the sound speeds are $c_2 = 1.25c_1$.

Significance

The wave image method will disclose how a junction influences the signal without masking the effect with intricate mathematics.

Solution

There are two boundaries at which waves are incident. Because many reflections will occur in an extended interval, it is useful to track these events by viewing the characteristics of the leading and trailing edges of the pulse. The slope of these lines is $dt/dx = \pm 1/c_1$ for the left segment and $dt/dx = +1/c_2$ for the right. (There are no characteristic lines with negative slope in segment 2 because only forward propagating waves exist there.) The construction begins with lines that track the actual pulse. They begin at the line $x = 0$ at $t = 0$ and $t = \tau$. These lines end at their intersection with the line $x = L$. Characteristic lines that describe transmission of the direct pulse into segment 2 emanate from these intersections. In addition, characteristic lines with negative slope, which describe the reflection of the direct pulse back into segment 1, emanate from these intersections. The reflected characteristic lines end at their intersection with the line $x = 0$. Positive slope characteristics depart from this set of intersections. These lines represent a reflected forward propagating wave. From there, the process repeats. This construction is depicted in Fig. 1.

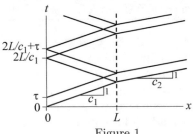

Figure 1.

The characteristic curves enable us to anticipate what wave images will be required. For segment 2, all pairs of characteristics are essentially the same. Thus, each reflection will be represented in the initial profile by an image in $x < L$, just as the actual pulse is represented in the initial profile by an image in $x < 0$. In contrast, for segment 1, each reflection at $x = L$ will require an image in $x > L$, and each reflection at $x = 0$ will require an image in $x < 0$. This is like the process observed in a finite length waveguide that is driven at $x = 0$.

In order to avoid confusing the images for each segment, we will draw them along separate x-axis. A minor complication arises because of the difference of the sound speeds. The basic length unit for segment 1 is L. An image in this fluid requires time L/c_1 to travel that distance. Images in fluid 2 must arrive simultaneously, but their propagation speed is c_2. Hence, the basic length unit in fluid 2 is $c_2 L/c_1$. It is convenient to denote the sound speed ratio as $r_c = c_2/c_1$. Also, the maximum velocity of the piston is V_0, so the maximum pressure for the initial image is $\rho_1 c_1 V_0$, which we will denote as P_0.

Image $F_{1,0}$, directly behind $x = 0$, is the initial profile for the actual pulse. It rises linearly with $-x$, terminating at $x = -c_1 \tau$ where the pressure is P. The reflection of $F_{1,0}$ at $x = L$ is represented by image $G_{1,1}$ that propagates in the negative x direction. This wave must arrive at $x = L$ at the same time as $F_{1,0}$, so it begins at distance L to the right of $x = L$. Its shape is $F_{1,0}$ multiplied by R and flipped left to right. During an interval of duration τ, a wave in segment 1 will travel a distance $c_1 \tau$. Therefore, that distance is the width of each image in segment 1. The first transmission also begins when $F_{1,0}$ arrives at $x = L$, which is $t = L/c_1$. The transmitted wave that results is represented by image $F_{2,1}$. The distance this image must travel from $t = 0$ until $t = L/c_1$ is $r_c L$. For this reason, the leading edge of $F_{2,1}$ is situated at distance $r_c L$ to the left of $x = L$. The maximum $F_{2,1}$ value is $T P_0$. The width of this image is $c_2 \tau$. The images describing the first reflection are as shown in Fig. 2.

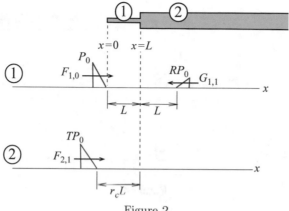

Figure 2.

The next reflection occurs when $G_{1,1}$ arrives at $x = 0$. This boundary is stationary after $t = \tau$, which corresponds to a unit reflection coefficient. Therefore, incidence of $G_{1,1}$ at $x = 0$ results in a forward propagating image $F_{1,1}$ whose maximum value is the maximum of $G_{1,1}$. Because $G_{1,1}$ is $2L$ to the right of $x = 0$, it must be that $F_{1,1}$ is $2L$ to the left of $x = 0$. The addition of this image is described in Fig. 3.

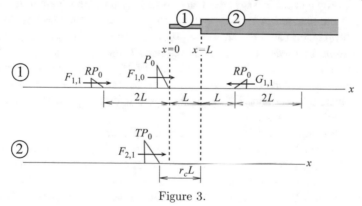

Figure 3.

The next reflection occurs when $F_{1,1}$ arrives at $x = L$. The reflection of this pulse back into segment 1 is represented by image $G_{1,2}$. Its maximum is R times the maximum in $F_{1,1}$. It is located a distance $3L$ to the right of $x = L$ because $F_{1,1}$ is $3L$ to the left of $x = L$. Incidence of $F_{1,1}$ at $x = L$ also gives rise to transmission of a pulse into segment 2, which is represented by image $F_{2,2}$. The maximum of $F_{2,2}$ is T times the maximum of $F_{1,1}$. Image $F_{1,1}$ travels distance $3L$ from its initial location before it arrives at the boundary. The image $F_{2,2}$ must arrive simultaneously with $F_{1,1}$, which requires that $F_{2,2}$ be at $3r_cL$ to the left of $x = L$. This is the situation on Fig. 4.

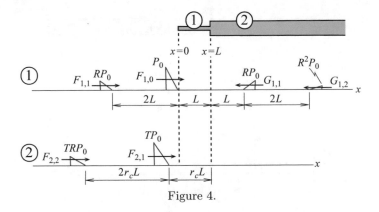

Figure 4.

The construction process could continue to track many more reflections, but the general trends already are evident. Once the initial spatial profile has been finalized, the techniques described in Sect. 2.5 would allow us to construct spatial profiles at later instants and waveforms at any x.

For segment 1, the pulses that travel back and forth are the same as those that would occur if $x = L$ was a resistive termination with reflection coefficient R. The images for propagation in segment 1 are spaced by $2L$. The images for forward propagation are situated behind $x = 0$, starting at $x = 0$, whereas the images for backward propagation are beyond $x = L$, starting at $x = 2L$. Each image is a factor R smaller than the preceding image. In segment 2, the images are solely forward propagating pulses. They are spaced at intervals of $2r_c L$ behind $x = 0$, with the first at $x = -(r_c - 1)L$. Each of these images is a factor RT smaller than the preceding image. Because $|R| < 1$ and $|RT| < 1$, the signal in both segments eventually will decay to a negligible value.

The signal in segment 2 is that of a semi-infinite waveguide that is driven by a sequence of pulses. Unlike a locally reacting end, wherein energy is transported into the wall, the energy lost in segment 1 at each reflection is transferred into segment 2. This is like a fundamental phenomenon in acoustic radiation, which is a topic we shall take up in Chap. 7. Because energy is lost from one system, the process is referred to as *radiation damping*, even though the energy is transferred, not dissipated.

A further generalization of the analysis would extend the wave image formulation to treat situations where both segments of the waveguide have finite length. Closure of segment 2 in the present system would require consideration of reflections within that segment, as well as transmission of signals from segment 2 into segment 1, which would generate additional reflections in segment 1. Although such generalizations are possible, the intricacies we encountered here would be greatly magnified. Fortunately, a system whose segments are all finite is readily analyzed in the frequency domain.

3.4.3 Frequency-Domain Formulation for Long Segments

Time-domain analysis of connected waveguides is useful because it sheds light in how signals evolve. As we did for a single waveguide, we could use it to explore issues such as modes and resonances. However, it has limitations beyond handling terminations that are not locally reacting. The most important of these is that it becomes quite cumbersome when applied to networks that consist of multiple segments. Although we will not fully address the issue, another limitation of time-domain analysis is that it cannot be modified to handle dissipative effects, such as thermoviscosity and wall friction. Frequency-domain analysis does not suffer from these limitations.

Each segment of a network is a waveguide like those we have analyzed, so the signal within each consists of forward and backward waves, both of which are described in terms of the local coordinate system for that segment. Because the frequency ω is common to all segments, the starting ansatz is

$$p^{(n)} = \mathrm{Re}\left[P^{(n)}\left(x_n\right)e^{i\omega t}\right], \quad v^{(n)} = \mathrm{Re}\left[V^{(n)}\left(x_n\right)e^{i\omega t}\right]$$

$$P^{(n)} = B_1^{(n)}e^{-ik_n x_n} + B_2^{(n)}e^{ik_n x_n}, \quad k_n = \frac{\omega}{c_n}$$

$$V^{(n)} = \frac{1}{\rho_n c_n}\left[B_1^{(n)}e^{-ik_n x_n} - B_2^{(n)}e^{ik_n x_n}\right]$$

In a short segment, the length L_n is so small that $\exp\left(\pm ik_n x_n\right)$ is essentially one. Such configurations are treated as lumped elements in Sect. 3.5. Our interest here is segments whose lengths are sufficiently large that the field within each segment must be described in terms of waves. In addition to the basic relations derived from the properties of the forward and backward waves in each segment, it is necessary to satisfy the junction conditions, Eqs. (3.4.3) and (3.4.4), as well as boundary conditions at the beginning and end of the transmission line .

3.4.3.1 Direct Method

The first approach we shall develop is based on treating the complex amplitudes of the forward and backward waves in each branch as the basic variables. The method does not require that one remember many details, and it may be implemented directly on a case-by-case basis. Let us consider the simplest situation, which is two connected waveguides. The respective wavenumbers are $k_1 = \omega/c_1$ and $k_2 = \omega/c_2$. Subscripts 0 and L shall denote that a pressure or velocity is evaluated at one end or another. For segments $n = 1$ or 2, superposition of the internal forward and backward simple waves gives the pressure and particle velocity at the ends. Thus, we have

$$P_0^{(n)} = B_1^{(n)} + B_2^{(n)}, \quad V_0^{(n)} = \frac{1}{\rho_n c_n}\left(B_1^{(n)} - B_2^{(n)}\right)$$

$$P_L^{(n)} = B_1^{(n)}e^{-ik_n L_n} + B_2^{(n)}e^{ik_n L_n}, \quad V_L^{(n)} = \frac{1}{\rho_n c_n}\left(B_1^{(n)}e^{-ik_n L_n} - B_2^{(n)}e^{ik_n L_n}\right)$$

$$\tag{3.4.8}$$

Continuity conditions at the junction, where $x_1 = L_1$ and $x_2 = 0$, must be satisfied. In addition, there is the boundary condition at $x_1 = 0$, which is considered to be a specified velocity excitation $v = \text{Re}\left(\hat{V}_0 \exp(i\omega t)\right)$. It also is necessary that the signal satisfies the impedance condition at the closed end, $x_2 = L_2$. These conditions are stated as

$$V_0^{(1)} = \hat{V}_0$$

$$P_L^{(1)} = P_0^{(2)}, \quad \rho_1 \mathcal{A}_1 V_L^{(1)} = \rho_2 \mathcal{A}_2 V_L^{(2)} \tag{3.4.9}$$

$$P_L^{(2)} = Z_L V_L^{(2)}$$

Substitution of the state variables in Eq. (3.4.8) into the preceding yields a set of four simultaneous equations for the four complex amplitudes. Each equation is linearly independent, so we may proceed to their solution. This usually is done with mathematical software, which is aided by collecting the wave amplitudes in a column vector. The coefficients of the algebraic equations form a square matrix $[E(\omega)]$. The result is

$$[E(\omega)]\left[B_1^{(1)} \; B_2^{(1)} \; B_1^{(2)} \; B_2^{(2)}\right]^{\text{T}} = [1 \; 0 \; 0 \; 0]^{\text{T}} \hat{V}_0 \tag{3.4.10}$$

where

$$[E(\omega)] = \begin{bmatrix} \dfrac{1}{\rho_1 c_1} & -\dfrac{1}{\rho_1 c_1} & 0 & 0 \\ e^{-ik_1 L_1} & e^{ik_1 L_1} & -1 & -1 \\ \dfrac{\mathcal{A}_1 e^{-ik_1 L_1}}{c_1} & -\dfrac{\mathcal{A}_1 e^{ik_1 L_1}}{c_1} & -\dfrac{\mathcal{A}_2}{c_2} & \dfrac{\mathcal{A}_2}{c_2} \\ 0 & 0 & e^{-ik_2 L_2}\left(1 - \dfrac{Z_L}{\rho_2 c_2}\right) & e^{ik_2 L_2}\left(1 + \dfrac{Z_L}{\rho_2 c_2}\right) \end{bmatrix}$$

$$\tag{3.4.11}$$

Thus, at each frequency, we would evaluate k_1 and k_2, form $[E(\omega)]$, solve for the four unknown wave amplitudes, and then use the solution to evaluate any desired properties of the acoustic field.

If there are N segments, then the pressures within the network are represented by a total of $2N$ amplitudes. The junction and termination conditions will extend Eq. (3.4.10) to $2N$ equations. For example, suppose there are three branches with a single junction as in Fig. 3.11. Conservation of mass at the junction gives one equation. Two more equations are obtained from setting $P_L^{(1)} = P_0^{(2)} = P_0^{(3)}$. Three more equations are obtained from the condition at the termination of each branch, and the last imposes the excitation.

3.4.3.2 A General Algorithmic Method

An alternative formulation takes the junction pressures to be the basic variables. It is somewhat more difficult to remember, but it lends itself to automated analysis of intricate networks. The first step is determination of the relationship between velocity and pressure at the ends of any segment. These relations are obtained by solving the first and third of Eq. (3.4.8) for the values of $B_1^{(n)}$ and $B_2^{(n)}$. Doing so leads to

$$B_1^{(n)} = \frac{1}{2i \sin{(k_n L_n)}} \left(P_0^{(n)} e^{ik_n L_n} - P_L^{(n)} \right)$$

$$B_2^{(n)} = -\frac{1}{2i \sin{(k_n L_n)}} \left(P_0^{(n)} e^{-ik_n L_n} - P_L^{(n)} \right)$$

(3.4.12)

These expressions are used to eliminate $B_1^{(n)}$ and $B_2^{(n)}$ from the second and fourth of Eq. (3.4.8), with the result that

$$V_0^{(n)} = \frac{1}{i \rho_n c_n \sin{(k_n L_n)}} \left(P_0^{(n)} \cos{(k_n L_n)} - P_L^{(n)} \right)$$

$$V_L^{(n)} = \frac{1}{i \rho_n c_n \sin{(k_n L_n)}} \left(P_0^{(n)} - P_L^{(n)} \cos{(k_n L_n)} \right)$$

(3.4.13)

It is useful to write these relations in matrix form as

$$\left\{ \begin{matrix} V_0^{(n)} \\ V_L^{(n)} \end{matrix} \right\} = [D_n] \left\{ \begin{matrix} P_0^{(n)} \\ P_L^{(n)} \end{matrix} \right\}$$

$$[D_n] = \frac{1}{i \rho_n c_n \sin{(k_n L_n)}} \begin{bmatrix} \cos{(k_n L_n)} & -1 \\ 1 & \cos{(k_n L_n)} \end{bmatrix}$$

(3.4.14)

A segment whose response is known in terms of the pressures (or particle velocities) at each end is said to be a *two-port element*. Correspondingly, $P_0^{(n)}$ and $P_L^{(n)}$ are the *port pressures* of that element, and $V_0^{(n)}$ and $V_L^{(n)}$ are the *port velocities*. The inverse of $[D_n]$ gives the pressures corresponding to a set of velocities, so it is a two-port acoustic impedance matrix, while $[D_n]$ is called the *two-port acoustic mobility*. (The term follows from the fact that large $[D_n]$ means that a small pressure will induce a high particle velocity, so the particles are highly mobile.)

Continuity of pressure requires that a port pressure equals the pressure in the junction where that port is located. Let us use letter subscripts to denote the junction pressures. For the system that was just analyzed, in which two waveguides are joined, the junction pressures are P_a at the left end where there is a velocity source, P_b at the junction of the segments, and P_c at the closure where there is an impedance condition. The assembly of the pressures at all junctions constitutes the *junction pressure vector*. Matching each port pressure to the corresponding junction pressure gives

$$P_0^{(1)} = P_a, \quad P_L^{(1)} = P_0^{(2)} = P_b, \quad P_L^{(2)} = P_c \qquad (3.4.15)$$

A generic way of describing these relations is

$$\left\{ \begin{array}{c} \left\{ \begin{array}{c} P_0^{(1)} \\ P_L^{(1)} \end{array} \right\} \\ \left\{ \begin{array}{c} P_0^{(2)} \\ P_L^{(2)} \end{array} \right\} \end{array} \right\} = [S] \left\{ \begin{array}{c} P_a \\ P_b \\ P_c \end{array} \right\}, \quad [S] = \begin{bmatrix} 1 & 0 & 0 \\ 0 & 1 & 0 \\ 0 & 1 & 0 \\ 0 & 0 & 1 \end{bmatrix} \qquad (3.4.16)$$

Note that the port pressure vector is assembled by stacking the two-port pressures sequentially. The elements of the port velocity vector are sequenced in the same manner. The manner in which the ports are joined at junctions is described by $[S]$, so it is called the *connectivity matrix*. Its definition is unique to a specific interconnection topology.

The connectivity matrix takes care of enforcing pressure continuity. It remains to enforce continuity of mass flow at the junctions, as well as the boundary condition at any end that is not connected to other segments. There is one equation for mass flow at each junction. For the two joined branches of our example, there is only one such equation. In matrix form, it is

$$\begin{bmatrix} \rho_1 A_1 & -\rho_2 A_2 \end{bmatrix} \left\{ \begin{array}{c} V_L^{(1)} \\ V_0^{(2)} \end{array} \right\} = \left\{ \begin{array}{c} 0 \\ 0 \end{array} \right\} \qquad (3.4.17)$$

The boundary conditions entail matching the particle velocity at $x_1 = 0$ to $\hat{V}_0$ and enforcing the impedance boundary at the far end. The corresponding equations are

$$\begin{array}{c} V_0^{(1)} = \hat{V}_0 \\ P_L^{(2)} = Z_L V_L^{(2)} \end{array} \qquad (3.4.18)$$

The full set of junction and boundary equations are assembled into a matrix equation in which coefficient matrices $[\Lambda]$ and $[\Gamma]$ multiply column vectors of port velocities and port pressures, respectively. For the example system, the first row describes mass continuity at the junction, the second row matches the particle velocity to the excitation, and the third row is the impedance boundary condition at the far end. The form of the assembled set of equations is

$$[\Lambda] \left\{ \begin{array}{c} \left\{ \begin{array}{c} V_0^{(1)} \\ V_L^{(1)} \end{array} \right\} \\ \left\{ \begin{array}{c} V_0^{(2)} \\ V_L^{(2)} \end{array} \right\} \end{array} \right\} + [\Gamma] \left\{ \begin{array}{c} \left\{ \begin{array}{c} P_0^{(1)} \\ P_L^{(1)} \end{array} \right\} \\ \left\{ \begin{array}{c} P_0^{(2)} \\ P_L^{(2)} \end{array} \right\} \end{array} \right\} = \left\{ \begin{array}{c} 0 \\ V_0 \\ 0 \end{array} \right\} \qquad (3.4.19)$$

where $[\Gamma]$ is a 3×4 matrix whose only nonzero element is $\Gamma_{3,4} = 1$, and

$$[\Lambda] = \begin{bmatrix} 0 & \rho_1 A_1 & -\rho_2 A_2 & 0 \\ 1 & 0 & 0 & 0 \\ 0 & 0 & 0 & -Z_L \end{bmatrix} \qquad (3.4.20)$$

The definitions of the port pressure and velocity vectors as partitions expedite substitution of the two-port acoustic mobility equation for each branch. Each such relation is assembled to form a mobility matrix that gives all port velocities in terms of all port pressures. Conformably sized partitions behave like single elements in matrix operations, so the assembled mobility equation is

$$\left\{ \begin{Bmatrix} V_0^{(1)} \\ V_L^{(1)} \end{Bmatrix} \\ \begin{Bmatrix} V_0^{(2)} \\ V_L^{(2)} \end{Bmatrix} \right\} = \begin{bmatrix} [D^{(1)}] & [0] \\ [0] & [D^{(2)}] \end{bmatrix} \left\{ \begin{Bmatrix} P_0^{(1)} \\ P_L^{(1)} \end{Bmatrix} \\ \begin{Bmatrix} P_0^{(2)} \\ P_L^{(2)} \end{Bmatrix} \right\} \qquad (3.4.21)$$

This form is substituted into Eq. (3.4.19), with the result that

$$\left[[\Lambda] \begin{bmatrix} [D^{(1)}] & [0] \\ [0] & [D^{(2)}] \end{bmatrix} + [\Gamma] \right] \left\{ \begin{Bmatrix} P_0^{(1)} \\ P_L^{(1)} \end{Bmatrix} \\ \begin{Bmatrix} P_0^{(2)} \\ P_L^{(2)} \end{Bmatrix} \right\} = \begin{Bmatrix} 0 \\ V_0 \\ 0 \end{Bmatrix} \qquad (3.4.22)$$

A solvable set of equations results when the connectivity matrix is used to eliminate the port pressures in favor of the junction variables. The equation to be solved is

$$\left[[\Lambda] \begin{bmatrix} [D^{(1)}] & [0] \\ [0] & [D^{(2)}] \end{bmatrix} + [\Gamma] \right] [S] \begin{Bmatrix} P_a \\ P_b \\ P_c \end{Bmatrix} = \begin{Bmatrix} 0 \\ V_0 \\ 0 \end{Bmatrix} \qquad (3.4.23)$$

This procedure is readily extended to any system of connected segments. The port pressures and port velocities are collected into column vectors $\{P\}$ and $\{V\}$ that are sequenced by the branch number, that is,

$$\boxed{\begin{aligned} \{P\} &= [P_0^{(1)} \;\; P_L^{(1)} \;\; P_0^{(2)} \;\; P_L^{(2)} \;\; \cdots]^{\mathrm{T}} \\ \{V\} &= [V_0^{(1)} \;\; V_L^{(1)} \;\; V_0^{(2)} \;\; V_L^{(2)} \;\; \cdots]^{\mathrm{T}} \end{aligned}} \qquad (3.4.24)$$

If there are N segments, then each vector has $2N$ elements. The port velocities and pressures in each segment are related through the two-port acoustic mobility. The assembly of such relations for all branches is described in

$$\boxed{\{V\} = [D]\{P\}}$$ (3.4.25)

where $[D]$ is block diagonal. Each block is a two-port mobility $[D_n]$ evaluated according to Eq. (3.4.14) and placed sequentially by segment number, that is,

$$[D] = \begin{bmatrix} [D_1] & [0] & [0] & \cdots \\ [0] & [D_2] & [0] & \cdots \\ [0] & [0] & [D_3] & \cdots \\ \vdots & \vdots & \vdots & \ddots \end{bmatrix}$$ (3.4.26)

The definitions of $\{V\}$, $\{P\}$, and the associated $[D]$ are independent of the manner in which the branches are connected. Description of these features begins by denoting alphabetically the pressures at the junctions and at the ends where the system terminates. (Any sequence is acceptable, but proceeding through the system consistently with the numbering of the branches will expedite the formulation.) These pressures form the junction pressure vector

$$\boxed{\left\{\hat{P}\right\} = [P_a \ P_b \ P_c \ \cdots]^{\mathrm{T}}}$$ (3.4.27)

The specific way in which the branches join is examined to identify which element of $\left\{\hat{P}\right\}$ is the same as a selected element of $\{P\}$. These identities lead to a mapping whose form is

$$\boxed{\{P\} = [S]\left\{\hat{P}\right\}}$$ (3.4.28)

Each row of the connectivity matrix $[S]$ consists of zeros except for a single element that is one. For example, suppose that the end $x_4 = L_4$ of the fourth segment is at junction c. The definition of $\{P\}$ places $P_L^{(4)}$ in the eighth element, while P_c is the third element of $\left\{\hat{P}\right\}$. This would be described by setting $S_{8,3} = 1$, while all the other elements of the eighth row of $[S]$ would be zero.

Satisfaction of Eq. (3.4.28) assures that the pressure will be continuous across the junctions, but it still remains to enforce the conservation of mass condition at each junction, as well as the boundary condition that applies at each termination. These equations are formulated in terms of the port variables, which are the elements of $\{P\}$ and $\{V\}$. The equations are assembled into a matrix form whose general form will be

$$\boxed{[\Lambda]\{V\} + [\Gamma]\{P\} = \{F\}}$$ (3.4.29)

where $\{F\}$ contains any excitations that are imposed. Note that each junction will have a mass conservation equation, and each termination will have a boundary condition, so the number of equations, which is the size of $\{F\}$, will equal the number of elements contained in $\{\hat{P}\}$. A solvable set of equations is obtained by using the mobility matrix in Eq. (3.4.25) to eliminate $\{V\}$, after which the connectivity matrix in Eq. (3.4.28) eliminates $\{P\}$. The result is an algebraic set of equations

$$[[\Lambda][D] + [\Gamma]][S]\{\hat{P}\} = \{F\} \tag{3.4.30}$$

After $\{\hat{P}\}$ has been computed at a specific frequency, the port velocities may be evaluated according to

$$\{V\} = [D][S]\{\hat{P}\} \tag{3.4.31}$$

Determination of the port responses makes it possible to evaluate the field in a branch between junctions. To that end, we recall Eq. (3.4.12), which describe the forward and backward wave amplitudes in an segment n in terms of the port pressure. The result is that the pressure at an arbitrary distance x_n is

$$\begin{aligned}
P^{(n)}(x_n) &= B_1^{(n)} e^{-ik_n x_n} + B_2^{(n)} e^{ik_n x_n} \\
&= \frac{P_0^{(n)} \cos(k_n(L_n - x_n)) - P_L^{(n)} \cos(k_n x_n)}{i \sin(k_n L_n)}
\end{aligned}$$

It often is necessary to perform these computations over a frequency band. Doing so requires recognition that some elements of $[D]$ are sines and cosines, whose arguments are the $k_n L_n$ values for the various branches. The increment in frequency should be sufficiently fine to assure that the values of these elements do not change drastically from one frequency to the next. A rule of thumb is to select $(\Delta\omega/c)L$ to be a small fraction of 2π, for example, $\Delta\omega = 0.5\pi c/L$.

If one is interested in a system that has a small number of segments, the direct formulation, in which the forward and backward wave amplitudes are the fundamental unknowns, might be preferable. This is so because that formulation works directly with the junction and boundary conditions, so there is little to remember in an analysis. However, if the task is identification of a sound propagation model for an automotive exhaust system or the HVAC system of a large factory, the algorithmic formulation features approximately half the number of unknowns. For example, if N segments are connected end to end, there are $2N$ wave amplitudes, but only $N + 1$ junctions.

3.4.3.3 A Specialized Algorithm

A different algorithmic formulation does not entail solving many simultaneous equations. The reason it was not the first to be discussed is that it is best suited to a system in which the segments are connected serially, that is, end to end. If there are N segments, then the serial connection leads to $x_n = L_n$ being the same location as $x_{n+1} = 0$ for $n = 1, ..., N - 1$. Let us consider such a system. Rather than solving for the two-port velocities in terms of the pressures, let us use Eq. (3.4.8) to solve for the pressure and velocity at $x_n = 0$ in terms of the variables at $x_n = L_n$. The result is

$$P_0^{(n)} = \cos(k_n L_n) P_L^{(n)} + i \sin(k_n L_n) \rho_n c_n V_L^{(n)}$$
$$\rho_n c_n V_0^{(n)} = i \sin(k_n L_n) P_0^{(n)} + \cos(k_n L_n) \rho_n c_n V_L^{(n)} \tag{3.4.32}$$

The mass flow rate at the ends is $\rho_n A_n V_0^{(n)}$ and $\rho_n A_n V_L^{(n)}$. According to the preceding relations, the pressure and flow rate at $x_n = 0$ are related to the values at $x_n = L_n$ by

$$\left\{ \begin{array}{c} P_0^{(n)} \\ \rho_n A_n V_0^{(n)} \end{array} \right\} = [T_n] \left\{ \begin{array}{c} P_L^{(n)} \\ \rho_n A_n V_L^{(n)} \end{array} \right\} \tag{3.4.33}$$

where

$$[T_n] = \begin{bmatrix} \cos(k_n L_n) & i\dfrac{c_n}{A_n} \sin(k_n L_n) \\ i\dfrac{A_n}{c_n} \sin(k_n L_n) & \cos(k_n L_n) \end{bmatrix} \tag{3.4.34}$$

The $[T_n]$ matrix may be regarded as a set of backward transmission coefficients.

The continuity equations for the end-to-end configuration require that

$$\left\{ \begin{array}{c} P_L^{(n-1)} \\ \rho_n A_n V_L^{(n-1)} \end{array} \right\} = \left\{ \begin{array}{c} P_0^{(n)} \\ \rho_n A_n V_0^{(n)} \end{array} \right\}, \quad n = 2, 3, ..., N \tag{3.4.35}$$

where N is the number of segments. We begin at the beginning of the network, where $x_1 = 0$, and alternately apply Eq. (3.4.33) and (3.4.35) to proceed to the other end. Thus, we have

$$\left\{ \begin{array}{c} P_0^{(1)} \\ \rho_1 A_1 V_0^{(1)} \end{array} \right\} = [T_1] \left\{ \begin{array}{c} P_L^{(1)} \\ \rho_1 A_1 V_L^{(1)} \end{array} \right\} = [T_1] \left\{ \begin{array}{c} P_0^{(2)} \\ \rho_2 A_2 V_0^{(2)} \end{array} \right\}$$
$$= [T_1][T_2] \left\{ \begin{array}{c} P_L^{(2)} \\ \rho_2 A_2 V_L^{(2)} \end{array} \right\} = [T_1][T_2] \left\{ \begin{array}{c} P_0^{(3)} \\ \rho_3 A_3 V_0^{(3)} \end{array} \right\}$$
$$= [T_1][T_2][T_3] \left\{ \begin{array}{c} P_L^{(3)} \\ \rho_3 A_3 V_L^{(3)} \end{array} \right\} = [T_1][T_2][T_3] \left\{ \begin{array}{c} P_0^{(4)} \\ \rho_3 A_3 V_0^{(4)} \end{array} \right\} \tag{3.4.36}$$

This sequence of operation continues until the far end, $x_N = L_n$, is reached. The ultimate result is that

$$\boxed{\left\{ \begin{array}{c} P_0^{(1)} \\ \rho_1 A_1 V_0^{(1)} \end{array} \right\} = [\tilde{T}] \left\{ \begin{array}{c} P_L^{(N)} \\ \rho_N A_N V_L^{(N)} \end{array} \right\}} \tag{3.4.37}$$

where $\left[\tilde{T}\right]$ is the product of the individual transmission matrices

$$\left[\tilde{T}\right] = [T_1][T_2] \cdots [T_N] \tag{3.4.38}$$

The last step is to apply the end conditions. The impedance boundary condition at $x_N = L_N$ may be written as $\rho_N A_N V_L^{(N)} = \left(\rho_N A_N / Z_L^{(N)}\right) P_L^{(N)}$. The input condition at $x_1 = 0$ also must be satisfied. If the particle velocity there is specified to be $\mathrm{Re}\left(\hat{V}_0 e^{i\omega t}\right)$, we require that $\rho_1 A_1 V_0^{(1)} = \rho_1 A_1 \hat{V}_0$. The second scalar equation in Eq. (3.4.37) gives $\rho_1 A_1 V_0^{(1)}$ in terms of the variables at the far end. Thus, we have

$$\boxed{\rho_1 A_1 \hat{V}_0 = \tilde{T}_{2,1} P_L^{(N)} + \tilde{T}_{2,2}\left(\rho_N A_N V_L^{(N)}\right) = \left(\tilde{T}_{2,1} + \tilde{T}_{2,2}\frac{\rho_N A_N}{Z_L^{(N)}}\right) P_L^{(n)}} \tag{3.4.39}$$

After this equation has been solved for $P_L^{(n)}$, the value of $V_L^{(N)}$ may be found from the impedance condition. The value of $P_0^{(1)}$ may then be found from the first scalar equation in Eq. (3.4.37). With the quantities for junction a known, the pressure and mass flow rates at intermediate junctions may be computed by returning to Eq. (3.4.36).

The reason this formulation is quite simple is the same as the reason its applicability is limited. Statement of the continuity conditions as Eq. (3.4.35) is essential. Major modifications are required to handle network topologies that are not serial, such as branches to several segments and short segments that are described as lumped elements in Sect. 3.5. On the other hand, when we study acoustic transmission through layered media, we will encounter a transmission matrix formulation that is similar to Eq. (3.4.37).

The effects of dissipation may be incorporated into any of the procedures developed here. The only modifications required to do so are replacement of the real wavenumber $k_n = \omega_n/c_n$ and real specific impedance $\rho_n c_n$ with the complex wavenumber and the effective characteristic impedance. Another possible modification addresses situations where the cross section of a segment changes at a relatively gradual rate. Such a configuration is said to be a horn, whose study opens Chap. 9. The analysis of horns would lead to replacement of Eq. (3.4.8) with a modified set equations.

3.4.3.4 Acoustic Impedance Relations

The acoustic impedance Z_{ac} was defined in Sect. 3.2.2 as the ratio of complex amplitudes of pressure and particle velocity at some location within a waveguide. Equation (3.2.24) describes this quantity in terms of the reflection coefficient R_L at the downstream end. That relation remains valid for each segment of a network. The perspective for the following developments is that we know the acoustic impedance, for example, as an experimentally measured quantity.

The reciprocal of the acoustic impedance is used to replace each port velocity with a port pressure in the equation for mass flow continuity through the junction, Eq. (3.4.3). Because there is only one pressure at the junction, it factors out, so that

$$\frac{\rho_1 A_1}{Z_{ac}^{(1)}(L_1)} = \frac{\rho_2 A_2}{Z_{ac}^{(2)}(0)} + \frac{\rho_3 A_3}{Z_{ac}^{(3)}(0)} \tag{3.4.40}$$

Note that setting $1/Z_{ac}^{(3)}(0) = 0$ is the only change required to treat a junction of two branches, while junctions of more than three branches are handled by placing all contributions from segments on the upstream side to the left of the equality and contributions from the downstream side to the right.

One use of the acoustic impedance is to decompose the signal within a branch into its constituent forward and backward waves. Consider branch 1 in the three-branch junction, as shown in Fig. 3.11. Let P_1^+ and P_1^- be the complex amplitudes of the forward and backward propagating waves in that branch at the junction. The standard relations for planar waves tell us that

$$P_1^+ + P_1^- = Z_{ac}^{(1)}(L_1)\left(\frac{P_1^+ - P_1^-}{\rho_1 c_1}\right) \tag{3.4.41}$$

This is essentially the same as the impedance boundary condition for an isolated segment, so we can assert directly that the reflection coefficient for segment 1 at the junction is

$$P_1^- = R_L^{(1)} P_1^+, \quad R_L^{(1)} = \frac{Z_{ac}^{(1)}(L_1) - Z_1}{Z_{ac}^{(1)}(L_1) + Z_1} \tag{3.4.42}$$

where $Z_1 = \rho_1 c_1$. Similar relations apply for the reflection coefficients of other branches at the junction. In some situations, it is desirable to extract $R_L^{(1)}$ from the acoustic impedances of branches 2 and 3. The requisite expression may be derived by multiplying the numerator and denominator of the preceding by $(\rho_1 A_1)/(Z_1 Z_{ac}^{(1)}(L_1))$ and then introducing Eq. (3.4.40). The result is

$$R_L^{(1)} = \frac{\left(\dfrac{\rho_1 A_1}{Z_1} - \dfrac{\rho_2 A_2}{Z_{ac}^{(2)}(0)} - \dfrac{\rho_3 A_3}{Z_{ac}^{(3)}(0)}\right)}{\left(\dfrac{\rho_1 A_1}{Z_1} + \dfrac{\rho_2 A_2}{Z_{ac}^{(2)}(0)} + \dfrac{\rho_3 A_3}{Z_{ac}^{(3)}(0)}\right)} \tag{3.4.43}$$

The fact that the pressure at the junction is $P_{\text{junction}} = P_1^+ + P_1^-$ then leads to

$$P_{\text{junction}} = (1 + R_1) P_1^+ = \frac{2Z_{\text{ac}}^{(1)}(L_1)}{Z_{\text{ac}}^{(1)}(L_1) + Z_1} P_1^+$$

$$= \frac{2\dfrac{\rho_1 A_1}{Z_1}}{\left(\dfrac{\rho_1 A_1}{Z_1} + \dfrac{\rho_2 A_2}{Z_{\text{ac}}^{(2)}(0)} + \dfrac{\rho_3 A_3}{Z_{\text{ac}}^{(3)}(0)}\right)} P_1^+ \qquad (3.4.44)$$

Sometimes, it is possible to determine an acoustic impedance based on the intrinsic properties of a segment, rather than as a quantity that is computed after the response has been determined. One such case is an downstream segment that has a nonreflecting termination, which could be achieved with the absorbing terminations discussed at the end of Sect. 3.2.2. The signal in such a segment is a simple wave that propagates downstream. The acoustic impedance everywhere in such a segment is the corresponding characteristic impedance, and the port pressure at the junction is the complex amplitude of the simple wave.

This property may be exploited to gain insights when a segment does not have a nonreflecting termination. If we replace $Z_{\text{ac}}^{(n)}$ for each downstream segment, $n = 2$ and $n = 3$, with $\rho_n c_n$, then the value of $P_{\text{junction}}/P_1^+$ obtained from Eq. (3.4.44) is the amplitude of the simple wave that propagates downstream in each segment as a consequence of incidence of a unit wave in segment 1. These ratios are *transmission coefficients*. Because $Z_1 \equiv \rho_1 c_1$, it follows that the transmission coefficient for that three-port junction is

$$T_2 = T_3 = \frac{2\dfrac{A_1}{c_1}}{\left(\dfrac{A_1}{c_1} + \dfrac{A_2}{c_2} + \dfrac{A_3}{c_3}\right)} \qquad (3.4.45)$$

An application of this idea is the special case where the same fluid fills all branches. Because $Z_1 \equiv \rho_1 c_1$, replacement of the acoustic impedances with $\rho_1 c_1$ in Eq. (3.4.43) leaves only the cross-sectional areas. This leads to $R_L^{(1)} = 0$ if the area of the exiting segments equals the area of the entering segment, that is, $A_2 + A_3 = A_1$. Furthermore, in this condition, the preceding equation indicates that $T_2 = T_3 = 1$. From this, we conclude that balancing the cross-sectional areas at a junction maximizes the transmission into branches 2 and 3 because none of the signal entering the junction is reflected back. We might have surmised this to be the case based on thinking about flow through pipes, but it is good to have our intuition confirmed.

EXAMPLE 3.10 The air-filled waveguide in Example 3.4 is modified by inserting a branch at the midpoint. All cross sections are circular, and the diameters and lengths are as shown in the sketch. The specific impedance at the right end is $\zeta_L = 10 - 5i$, as previous, while the termination of the added branch is rigid. The excitation is the same pressure source at $x = 0$, $p(0, t) = 50 \sin(\omega t)$ for any ω. Determine $|P|$ at the junction and $Z_{\text{ac}} = P/V$

of the three ports at the junction. Graph each quantity as a function of ω over
an interval from zero to 40% greater than the third highest resonance.

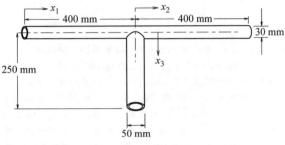

Figure 1.

Significance

This example demonstrates how the algorithmic formulation using junction pressures
is adapted to a specific system. Furthermore, interrogation of the acoustic impedances
will be seen to greatly enhance our ability to explain phenomena observed in a
computation.

Solution

The sense in which x is measured along each branch has been defined in the figure.
We begin by writing the junction conditions, which are

$$
\begin{aligned}
P^{(1)}\left(L_1\right) &= P^{(2)}\left(0\right) = P^{(3)}\left(0\right) \\
A_1 V^{(1)}\left(L_1\right) &= A_2 V^{(2)}\left(0\right) + A_3 V^{(3)}\left(0\right)
\end{aligned}
\tag{1}
$$

Note the second equation is posed in terms of volume velocity, because all branches
have the same density. The boundary conditions at the ends are the given pressure at
$x_1 = 0$, the impedance condition at $x_2 = L_2$, and a rigid condition at $x_3 = L_3$,

$$
\begin{aligned}
P^{(1)}\left(0\right) &= P_0 \\
P^{(2)}\left(L_2\right) &= Z_L V^{(2)}\left(L_2\right) \\
V^{(3)}\left(L_3\right) &= 0
\end{aligned}
\tag{2}
$$

If we were to follow the formulation that uses the complex amplitudes of the
forward and backward waves, we would substitute the appropriate variables from
Eq. (3.4.8) into the six junction and boundary conditions of Eqs. (1) and (2). There
are three branches and two unknown amplitudes per branch, so the simultaneous
equations that result would be solvable. We shall instead follow the formulation that
treats junction pressures as the fundamental variables because doing so will enhance
our proficiency for intricate systems.

The column vectors of port pressures and velocities are defined in accord with Eq. (3.4.24),

$$\{P\} = \left[P_0^{(1)} \quad P_L^{(1)} \quad P_0^{(2)} \quad P_L^{(2)} \quad P_0^{(3)} \quad P_L^{(3)} \right]^{\mathrm{T}}$$
$$\{V\} = \left[V_0^{(1)} \quad V_L^{(1)} \quad V_0^{(2)} \quad V_L^{(2)} \quad V_0^{(3)} \quad V_L^{(3)} \right]^{\mathrm{T}}$$

These are related by Eq. (3.4.25), with $[D]$ composed of three blocks on its diagonal, so that

$$\{V\} = \begin{bmatrix} [D_1] & [0] & [0] \\ [0] & [D_2] & [0] \\ [0] & [0] & [D_3] \end{bmatrix} \{P\} \tag{3}$$

Each $[D_n]$ is computed according to Eq. (3.4.14).

The vector of junction pressures is

$$\left\{ \hat{P} \right\} = [P_a \quad P_b \quad P_c \quad P_d]^{\mathrm{T}}$$

where the locations are point a: $x_1 = 0$; point b: the junction; point c: $x_2 = L_2$,; and point d: $x_3 = L_3$. Matching each port pressure to one of the junction values leads to identification of the connectivity matrix, which is

$$\{P\} = [S] \left\{ \hat{P} \right\} \tag{4}$$

where $[S]$ is a 6×4 matrix whose nonzero elements are $S_{1,1} = S_{2,2} = S_{3,2} = S_{4,3} = S_{5,2} = S_{6,4} = 1$.

Enforcement of Eq. (4) assures that the pressure continuity conditions in Eq. (1) are identically satisfied. The remaining conditions are

$$P_0^{(1)} = \hat{P}_0$$
$$\mathcal{A}_1 V_L^{(2)} - \mathcal{A}_2 V_0^{(2)} - \mathcal{A}_3 V_0^{(3)} = 0$$
$$P_L^{(2)} - Z_L V_L^{(2)} = 0 \tag{5}$$
$$V_L^{(3)} = 0$$

where $\hat{P}_0 = 50/i$ Pa is the pressure excitation. Matching these equations to the standard representation in Eq. (3.4.29) leads to identification of the 4×6 matrices $[\Lambda]$ and $[\Gamma]$, and the 4×1 vector $\{F\}$. The nonzero elements are $\Lambda_{2,2} = \mathcal{A}_1$, $\Lambda_{2,3} = -\mathcal{A}_2$, $\Lambda_{2,5} = -\mathcal{A}_3$, $\Lambda_{3,4} = -Z_L$, $\Lambda_{4,6} = 1$, $\Gamma_{1,1} = 1$, $\Gamma_{3,4} = 1$, $F_1 = P_0$.

This operation completes the definition of the constituent matrices required to formulate Eq. (3.4.30), so the equation is solved for $\{\hat{P}\}$. Before we proceed, it is useful to consider what modifications would be required to treat the case where the excitation at $x_1 = 0$ is a specified velocity amplitude $V_0 = \hat{P}_0/(\rho_1 c_1)$. This condition would replace the first of Eq. (5), which would be implemented by setting $\Lambda_{1,1} = 1$ and $\Gamma_{1,1} = 0$, and $F_1 = \hat{P}_0/(\rho_1 c_1)$.

To determine the requested acoustic impedances, the solution for $\{\hat{P}\}$ is substituted into Eq. (4) to determine the port pressures. The resulting $\{P\}$ is substituted into Eq. (3) to find the port velocities. The acoustic impedance at each port then is obtained by dividing each element of the computed $\{P\}$ by the corresponding element of the computed $\{V\}$.

The frequency interval for the evaluation must extend beyond the third resonance, but we do not know what that frequency is. We use the solution to Example 3.4 as a guideline to select the frequency range. It was found that the third resonance peak for the case of a pressure excitation occurs close to 530 Hz. We shall cut off calculations at 40% above that frequency and then adjust the interval upward if the computations do not display three resonances. The parameters for the evaluation are as follows: lengths $L_1 = L_2 = 0.40$, $L_3 = 0.25$ m, diameters $D_1 = D_2 = 0.03$, $D_3 = 0.05$ mm, $A_n = \pi (D_n/2)^2$, $\rho_n = 1.2$ kg/m^3, $c_n = 340$ m/s, $\hat{P}_0 = 50/i$ Pa.

The result for the pressure at the junction is described in Fig. 2. The pressure magnitude is plotted on a logarithmic scale in order to view the lower range more clearly. The most striking

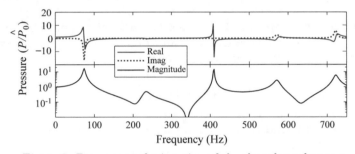

Figure 2. Pressure at the junction of the three branches as a ratio to the complex amplitude of the pressure excitation.

difference from the midpoint pressure in Example 3.4 is the increased number of resonances and greater irregularity of the frequency at which they occur. Resonances may be identified by the occurrence of peaks, as well as by frequency intervals in which the phase angle changes drastically. The upper graph indicates that the ratio of the junction pressure P_b to $\hat{P}_0$ is real at low frequencies, so a resonance corresponds to $P_b/\hat{P}_0$ being purely imaginary. Inspection of the lower graph shows that there are two strong resonances at 74 and 410 Hz, two weaker ones at 570 and 722 Hz, and a very weak one at 234 Hz. A comparison of these results with those for a single waveguide having $L = 800$ mm suggests that adding branches increases the number of resonances that will occur in a specific frequency interval. At the same time, it appears that it might make some resonance peaks lower and wider (decreased Q factor).

The acoustic impedances at the junction are depicted in Fig. 3. Some of the peaks in these plots occur close to the pressure resonances, but the singularity of $Z_{ac}^{(3)}$ does not show up elsewhere.

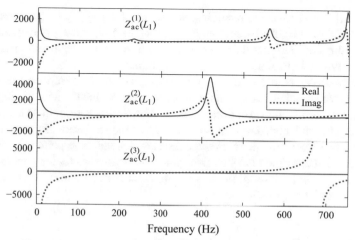

Figure 3. Acoustic impedances at the junction of the three
branches of the waveguide.

A partial explanation for these features lies in the mass flow continuity relation in
Eq. (1). In terms of acoustic impedance, this equation states that

$$\frac{A_1}{Z_{ac}^{(1)}(L_1)} = \frac{A_2}{Z_{ac}^{(2)}(0)} + \frac{A_3}{Z_{ac}^{(3)}(0)} \tag{6}$$

Suppose that at a certain frequency one of these impedances is much larger than the
other two. The reciprocal of that large impedance will be a small contributor to this
relation. This explains why the peaks in each impedance are not manifested in the
others.

The other useful aspect is suggested by the fact that the natural frequencies of
a waveguide whose ends are rigid are f_n (Hz) $= 0.5nc/L$. In the frequency interval of the graphs, these values are 0 and 425 Hz for segments 1 and 2, and 0 and
680 Hz for segment 3. The former are very close to the frequencies of the peaks in
$\left| Z_{ac}^{(2)}(0) \right|$, while the latter are precisely the frequencies at which $Z_{ac}^{(3)}(0)$ is singular. This correspondence is not coincidental. When segments 2 and 3 are isolated,
each becomes a uniform waveguide that is terminated with a known impedance. The
acoustic impedance at the opposite end, that is, at the junction, in such a system, is
given in Eq. (3.2.24). That result did not depend on the nature of the excitation, so it
may be used here. Evaluation of that expressions at $x = 0$ leads to

$$Z_{ac}^{(n)}(x) = \rho_0 c \left(\frac{e^{ikL_n} + R_{L,n}e^{-ikL_n}}{e^{ikL_n} - R_{L,n}e^{-ikL_n}} \right) \tag{7}$$

Because segment 3 truly has a rigid closure, singular values of $Z_{ac}^{(3)}(0)$ occur when
the denominator is zero, which corresponds to $e^{2ik_3 L_3} = 1$. These are the natural
frequencies of a single waveguide with a rigid closure. In the case of branch 2, $R_{L,2}$

is sufficiently close to one to result in peak values of $Z_{ac}^{(2)}(0)$ at frequencies that are close to those for a rigid closure. The fact that $R_{L,2}$ is not actually one explains why the peak values of $\left|Z_{ac}^{(2)}\right|$ are finite and do not occur exactly at the natural frequencies the branch would have if its termination was rigid.

We now arrive at the critical point. Because Eq. (7) describes the acoustic impedances of branches 2 and 3, it follows from Eq. (6) that the acoustic impedance of branch 1 at the junction also does not depend on the excitation. In turn, this means that the drive point impedance is independent of the excitation, just as it was in Example 3.4. As was true there, frequencies at which the drive point impedance has a minimum will correspond to situations where a pressure excitation generates maximum pressures in the overall field, while maxima of that quantity will correspond to relatively quiet responses.

3.5 Lumped Parameter Models

Several features of the analysis of connected waveguides have analogs in the theory of alternating current circuits. Perhaps the most obvious are the pressure continuity and conservation of mass equations for a junction, which resemble the statements that there is a unique voltage at a wire junction and that the current entering a junction must equal the current exiting it. Many acoustics textbooks make this analogy a central theme. We have not done so for several reasons, not the least of which is that unless one is interested in electrical transmission lines covering a long range or a very high frequency, voltage is taken to change only when the current flows through a discrete element like a resistor. In contrast, because the speed of sound is comparatively low, pressure fluctuations in a waveguide are an essential feature. However, there are some situations where a waveguide's behavior can be approximated by a single parameter. As with electrical circuits, these situations generally are restricted to situations where their largest dimension L and the frequency are sufficiently small that they fit the specification that $kL \ll 1$. In essence, these models collapse a two-port segment to a single port. The segment in this situation is referred to as a *lumped element* because its properties have been collected to a single location.

3.5.1 Approximations for Short Branches

We begin by considering an isolated branch of length L, with the particle velocity at $x = 0$ taken to be known as V_0. The pressure induced by this known velocity is given in Eq. (3.2.21). Our interest is acoustic impedances, so let us use Eq. (3.2.11) to eliminate R. The result is

$$P = \rho_0 c V_0 \left(\frac{Z_L \cos(k(L-x)) + i\rho_0 c \sin(k(L-x))}{iZ_L \sin(kL) + \rho_0 c \cos(kL)} \right)$$

$$V = V_0 \left(\frac{iZ_L \sin(k(L-x)) + \rho_0 c \cos(k(L-x))}{iZ_L \sin(kL) + \rho_0 c \cos(kL)} \right) \tag{3.5.1}$$

If $x = L$ were a termination, then Z_L is a property of the wall at that termination. Our interest is the situation where $x = L$ is a junction with other segments. In that case, the acoustic impedance at the end, $Z_{ac}(L)$, gives the ratio of pressure to velocity at that location. It follows that replacing Z_L with $Z_{ac}(L)$ in Eq. (3.5.1) will produce an expression for the pressure field in a short branch. We use it to evaluate the pressure at $x = 0$ and then divide the result by the particle velocity at that location, which is V_0. The result is a general relation between acoustic impedances at the ends of a branch,

$$Z_{ac}(0) = \rho_0 c \left[\frac{Z_{ac}(L) + i\rho_0 c \tan(kL)}{\rho_0 c + iZ_{ac}(L)\tan(kL)} \right] \tag{3.5.2}$$

Our interest is the behavior when $kL \ll 1$, which leads to the approximation $\tan(kL) \approx kL$. If $x = L$ is rigid, then $Z_{ac}(L)$ is infinite. In that case, the preceding reduces to

$$Z_{ac}(0) = -i\frac{\rho_0 c}{kL} = -i\frac{\rho_0 c^2}{\omega L} \tag{3.5.3}$$

The most important aspect of this limiting form is that it indicates that the reactance is negative and therefore spring-like. This quality stems from the compressibility of the fluid it contains.

The acoustic impedance derived above is a special case of the general behavior for any small cavity. In Fig. 3.13, an arbitrarily shaped cavity whose volume is $\mathcal{V}$ is filled with fluid. Displacement u of the piston, whose face has area $\mathcal{A}$, is positive into the cavity.

Our interest lies the situation where the size of the cavity is small compared to an acoustic wavelength. This means that kL is very small, where L is the largest dimension of the cavity. Under this restriction, the pressure everywhere within the cavity is the same function $p(t)$. Thus, the pressure acting on the piston is p. To relate this pressure to the motion of the cavity, we recall Eq. (2.3.13), which features

Fig. 3.13 A small cavity compressed by a piston that oscillates at a low frequency

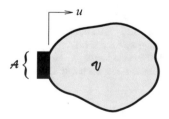

the bulk modulus $K = \rho_0 c^2$. That equation describes the pressure within a cavity when the volume is changed slowly. The volume displaced by the piston is $\mathcal{A}u$, which is the amount by which the cavity's volume decreases. The condensation is $(-\mathcal{A}u)/\mathcal{V}$, so the corresponding pressure is

$$p = \rho_0 c^2 \frac{\mathcal{A}u}{\mathcal{V}} \qquad (3.5.4)$$

In the frequency domain, $U = V_0/(i\omega)$, so the preceding proportionality of pressure and displacement leads to

$$\boxed{Z_{\text{cav}} = \frac{P}{V_0} = -i\rho_0 c^2 \frac{\mathcal{A}}{\mathcal{V}\omega}} \qquad (3.5.5)$$

This is fully consistent with Eq. (3.5.3), because the cavity there consists of a waveguide whose cross-sectional area $\mathcal{A}$ everywhere, so its volume is $\mathcal{A}L$.

Now let us examine the opposite case, which is the behavior for $kL \ll 1$ when the order of magnitude of $Z_{\text{ac}}(L)$ is no larger than $\rho_0 c$. Both terms in the numerator of Eq. (3.5.2) might have comparable magnitude, but the second term in the denominator is negligible. Consequently, we find

$$\boxed{Z_{\text{ac}}(0) = i\rho_0 ckL + Z_{\text{ac}}(L)} \qquad (3.5.6)$$

To explain this relation, we observe that smallness of kL means that the wavelength is large. Consequently, all particles in this segment have a common value V. This allows us to replace the acoustic impedances by P/V at each end. Doing so leads to $P_0 = P_L + (i\rho_0 ckL) V$. Now, consider Fig. 3.14, which is a free body diagram for this system of particles contained in the segment. The cross- sectional area is $\mathcal{A}$, so the total mass is $\rho_0 \mathcal{A}L$. Newton's second law for this system states that $\rho_0 \mathcal{A}L\dot{v} = p(0,t)\mathcal{A} - p(L,t)\mathcal{A}$. The frequency-domain version reduces to $(i\rho_0 ckL) V = P_0 - P_L$. This is the same as the pressure relation derived from Eq. (3.5.6). Hence, this relation is a statement that when a short segment opens to a region of low acoustic impedance, the pressure drop is an inertial effect due to acceleration of the particles.

This *open-port model* helps explain an apparent paradox. It was stated previously that if the end of a waveguide is open to a large body of fluid that is undisturbed,

Fig. 3.14 Free body diagram for a short waveguide having a pressure-release termination

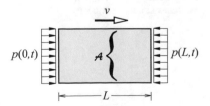

the pressure at that end fits the pressure-release model because the total pressure there must match the ambient pressure in the outside fluid. We will see in Chap. 4 that power flow is proportional to the product of pressure and particle velocity. This means that the power flow past an end that is pressure-release is zero, even though the particle velocity there is not zero. However, we know from experience that we can hear sound radiate from the open end of a tube. (Consider any wind instrument.) The radiated field obviously has acoustical energy, but the pressure-release model says that no energy is transported from the tube to the surroundings. Something is wrong!

The explanation lies in Fig. 3.15, which depicts the particle velocity in a circular waveguide at an open end. Because we are considering plane waves, the velocity is uniformly distributed over the cross section. The wall at the opening is taken to be planar and to extend far from the waveguide. The same situation is encountered when a vibrating piston is flush mounted on a large wall. The wall in that case is said to be an *infinite baffle*. Determination of the pressure field that is generated by this motion is the topic of Chap. 8. The case where the piston face is circular is especially amenable to analysis. This is the situation depicted in the figure.

Fig. 3.15 Particle velocity distribution at the opening of a waveguide in a wall

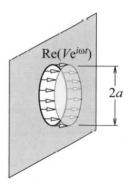

Section 8.5 shows that evaluation of an integral that extends over the face of a circular piston gives the pressure at any location. Our interest for a plane wave representation is the average pressure on the face, which is found by integrating the position-dependent pressure over the face of the piston to obtain the resultant force and then dividing that force by the area. The integration is described in the analysis of Eq. (8.5.4), which finds that the average pressure is

$$P_{av} = \frac{1}{\pi a^2} \iint_{A} P \, dA = \rho_0 c V \chi \, (ka) \tag{3.5.7}$$

where χ is a complex coefficient that depends on ka. (The parameter ka may be considered to be either a nondimensional frequency or nondimensional radius.) It is the *radiation impedance* of a piston,[11]

$$\chi(ka) = \left(1 - \frac{J_1(2ka)}{ka}\right) + i\left(\frac{\mathbf{H}_1(2ka)}{ka}\right) \tag{3.5.8}$$

In this expression, $J_1(2ka)$ and $\mathbf{H}_1(2ka)$ are, respectively, first-order Bessel and Struve functions evaluated at $2ka$. Bessel functions are encountered in a variety of topics involving the wave equation in polar or cylindrical coordinate, including acoustic radiation (Chap. 7) and nonplanar signals in waveguides (Chap. 9). They are in-line functions for most standard mathematical software. Struve functions occur less commonly and are seldom available for in-line evaluation. (Bessel and Struve functions are two of many types of special functions that arise as solutions of fundamental differential equations. An excellent general reference for special functions is the handbook edited by Abramowitz and Stegun.[12] We will refer to it frequently in the later chapters.) Fig. 3.16 shows a graph of $\chi(ka)$.

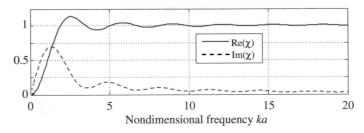

Fig. 3.16 Plot of the frequency dependence of the specific acoustic impedance χ for a vibrating circular piston of radius a

Division of the complex average pressure P_{av} by the complex velocity V yields the average plane wave acoustic impedance,

$$Z_{\text{piston}} = \rho_0 c \chi(ka) \tag{3.5.9}$$

Although this impedance was derived for a piston, it is equally valid for any uniform velocity distribution over a circular patch in an infinite baffle, specifically the open end of a circular tube in a large wall. With this, we may resolve the apparent paradox. The sound that emanates from the open end is acoustic radiation that is generated by the piston-like velocity distribution at the opening. The corollary is that rather than using a pressure-release model at any open end, we should treat an open termination as though the impedance there is Z_{piston} in Eq. (3.5.9). This correction is generally applicable, regardless of the length L and nondimensional frequency ka.

---

[11]Most references multiply the numerator and denominator of the fraction in Eq. (3.5.8) by two, so that only $2ka$ appears in the equation. Correspondingly, χ is considered to be a function of $2ka$. Apparently, this representation is done for historical reasons.

[12]M.I. Abramowitz and I.A. Stegun, *Handbook of Mathematical Functions*, Dover, (1965).

Using Z_{piston} at an open end might lead to drastically different behavior than that which is obtained from the pressure-release model. For example, $\chi(ka) \to 1$ as ka increases beyond 10. Thus, at very high frequencies, Z_{piston} matches the fluid's characteristic impedance, which means that a plane wave in the waveguide is not reflected at the end.

The pressure-release model, for which $R_L \approx -1$, can be justified if $ka \ll 1$. (At 440 Hz, $ka = 0.1$ corresponds to $a = 12.3$ mm in air and 53.5 mm in water.) When $ka \ll 1$, the limiting form is

$$\chi(ka) = \frac{1}{2}(ka)^2 + i\frac{8}{3\pi}ka \qquad (3.5.10)$$

The smallness of ka means that the real part of $\chi(ka)$, which gives resistance, is much smaller than the imaginary (reactive) part. We therefore will neglect the resistive part. The reactance is positive, so it is an inertance.

Suppose we imagine that the open end of a waveguide is extended from $x = L$ to $x = L + \ell$ where the pressure actually is zero. If $k\ell \ll 1$, then Eq. (3.5.6) with $x = L$ to the left and $Z_{\text{ac}}(L + \ell) = 0$ indicates that

$$Z_{\text{ac}}(L) = i\rho_0 ck\ell \qquad (3.5.11)$$

Let us select ℓ such that this expression for the acoustic impedance matches the low-frequency approximation of $Z_{\text{piston}}(ka)$, so that

$$i\rho_0 ck\ell = \rho_0 c\left(i\frac{8}{3\pi}ka\right) \implies \ell = \frac{8}{3\pi}a \qquad (3.5.12)$$

An interpretation of this result is that when ka is very small, the velocity at the open end causes a slug of fluid in the outside region surrounding the end to oscillate back and forth. This slug effectively extends the length of the waveguide by a distance $8a/(3\pi)$. This addition is referred to as an *end correction*. The idea is that the sound field properties for a pressure-release termination at $x = L + \ell$ may be used to describe the field in $0 < x < L$.

In a strict sense, there are other limitations to the validity of the end correction in Eq. (3.5.12) beyond the basic limitation to $ka \ll 1$. It was derived from solution of the acoustic field radiated by a circular cross-sectional piston flush mounted on a rigid wall. What if the cross section is not circular? What if there is no wall to justify using the model of a piston in a baffle. What if there is a flare at the opening, like any musical horn? In each case, there is a different end correction. Some may be predicted analytically, but those analyses are even more challenging than the one leading to $\chi(ka)$ in Eq. (3.5.10). If an approximate model is adequate, the opening, whose area is $\mathcal{A}$, may be treated as though it were a circle in a wide wall. The equivalent radius a is defined such that the areas are equal, $\mathcal{A} = \pi a^2$. The justification for this approximation is that smallness of ka implies that the fluid behavior is well predicted by incompressible theory, in which c is infinite. The net force acting on a vibrating surface in incompressible theory is proportional to the total volume that is displaced. Finding an equivalent circle gives that displacement. Of course, if ka is extremely small, the end correction may be ignored. In the opposite situation where ka is not small, the relation $Z_L = \rho_0 c \chi (\kappa a)$, whose value is obtained from Eq. (3.5.8) with a as the average radius, $a = (\mathcal{A}/\pi)^{1/2}$, may be used as an approximation. The alternative of using Eqs. (3.5.7) and (3.5.8) is always valid.

3.5.2 Helmholtz Resonator

A *Helmholtz resonator* combines a closed cavity and a short throat. Figure 3.17 provides a conceptual picture and associated free body diagram. The volume of the cavity is $\mathcal{V}$, and the cross-sectional area of the throat is $\mathcal{A}$. All dimensions are taken to be small compared to the wavelength, so we may consider the fluid particles in the throat to translate in unison and the pressure within the cavity to be the same everywhere. The extent of the mass in the throat that has been isolated is L', which is the actual length L plus corrections at the open end on the left and the entry to the cavity at the right. Thus, if we use the standard end correction for a circular cross section, $L' = L + (16/3\pi) a$. The forces act on the sides of the isolated mass equilibrate, while the forces acting in the direction of movement are the resultants of the cavity pressure P_{cav} and of the pressure $P_0^{(1)}$ at the open end. The mass of the moving particles is $\rho_1 \mathcal{A} L'$, where the density is designated as ρ_1 to allow for the possibility that the fluid in the resonator is different from ρ_0 for the waveguide.

Fig. 3.17 A Helmholtz resonator, in which a throat of cross-sectional area $\mathcal{A}$ is the passage to a cavity whose volume is $\mathcal{V}$

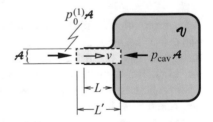

The equation of motion for the particles in the throat is

$$P_0^{(1)} \mathcal{A} - P_{\text{cav}} \mathcal{A} = \rho_1 \mathcal{A} L' (i\omega V) \tag{3.5.13}$$

The pressure within the cavity is $P_{\text{cav}} = Z_{\text{cav}} V$, where Z_{cav} is the cavity impedance in Eq. (3.5.5). Thus, this equation of motion gives

$$P_0^{(1)} = (\rho_1 L' i\omega V + Z_{\text{cav}}) V = i\rho_1 L' \left(\omega - \frac{\mathcal{A}c^2}{L' \mathcal{V} \omega}\right) V \tag{3.5.14}$$

The coefficient of V is the acoustic impedance at the entry point. The relation usually is written as

$$P_0^{(1)} = Z_H V, \quad Z_H = i\rho_1 L'\omega \left(1 - \frac{\omega_H^2}{\omega^2}\right) \tag{3.5.15}$$

where the frequency parameter is

$$\omega_H = c\left(\frac{A}{L'\mathcal{V}}\right)^{1/2} \tag{3.5.16}$$

This derived expression for Z_{II} is real, so it is reactive. In reality, dissipation effects, mostly concentrated in the boundary layer of the neck, add a resistive part. This resistive effect usually is negligible compared to the reactive part, except at ω_H, where it is the only effect that opposes the velocity in the short segment.

If we ignore the resistive part of Z_H, we see that a small pressure applied to the open end at frequency ω_H will generate a large particle velocity in the throat. This phenomenon is called a *Helmholtz resonance*, and ω_H is the *Helmholtz resonance frequency*. Explaining the behavior of this system was one of many pioneering acoustics contributions made by Herman von Helmholtz (1821–1891). One of his applications was to use a series of different size resonators to perform spectral analysis of signals. The cause of the resonances is that the compliance of the cavity acts in series with the inertance of the throat, much like a spring-mass system. The natural frequency of a Helmholtz resonator is ω_H. At this frequency, the pressure required to overcome the inertia of the throat exactly cancels the pressure required to overcome the compliance of the cavity, so little pressure is required to move the fluid.

Most of us have experienced a Helmholtz resonance, either by blowing across the open end of a beverage bottle, or else by partially opening one window or the sunroof of a moving automobile. In both cases, turbulence and vortex shedding associated with the air flowing over the leading and trailing edges of the opening generates a fluctuating pressure. If a spectral decomposition using FFTs indicates that the fluctuating pressure has significant components in the range of ω_H, the resulting particle velocity at the opening will be very large. In the case of the beverage bottle, we observe the resonance as a loud sound coming from the bottle because the large velocity at the throat is like a baffled piston. The interior sound field generated by flow across an opening in a moving vehicle is the subject of the next example.

A Helmholtz resonator can inhibit propagation of sound. To see why, suppose the open end of a Helmholtz resonator is attached somewhere along a waveguide, thereby creating a junction and splitting the waveguide. Segment 1 is upstream of the junction, segment 2 is downstream, and segment 3 is the resonator. The respective acoustic impedances at this junction are $Z_{ac}^{(1)}(L_1)$, $Z_{ac}^{(2)}(0)$, and Z_H. When the frequency is ω_H, we have $Z_H = 0$. According to Eqs. (3.4.43) and (3.4.44), a zero acoustic impedance for either downstream segment leads to $R_L^{(1)} = -1$ and $P_{junction} = 0$. A slightly less drastic change results if a small resistive part is incorporated into Z_H. In that case, $R_L^{(1)}$ will differ little from -1, and $P_{junction}$ will be very small relative to what it would be without the addition of the resonator. In any event, we can assert that little sound will be propagated downstream at the Helmholtz resonance frequency. In essence, the Helmholtz resonator acts like a stop filter for signals at ω_H, and it greatly reduces the pressure that is propagated downstream at frequencies that are close to ω_H. If a noise source such as a piece of machinery induces an unacceptably large narrow band noise at a certain frequency, transmission of this noise through a network one can be reduced by inserting a Helmholtz resonator that is tuned to the objectionable frequency. Multiple resonators tuned to different frequencies can widen the frequency band that is reduced. The effectiveness of this concept is examined in the second of the following examples.

EXAMPLE 3.11 Consider a model of the passenger compartment in an automobile whose sunroof is open. The compartment is considered to be a cavity in which the pressure does not depend on position. The opening in the roof is taken to be the throat. Develop an expression for the ratio of the pressure within the compartment to the pressure in the flow across the opening. Evaluate the frequency dependence of this transfer function based on the opening being 500 mm perpendicularly to the flow and the roof having been opened by 100 mm. A representative value for the roof thickness is 20 mm, and the volume of the passenger compartment is 5 m^3.

Significance

The nice feature of this example is that the by-product of demonstrating the use of lumped element impedances is an explanation of a common annoying phenomenon.

Solution

Our model considers the passenger compartment to be the cavity of a Helmholtz resonator and the opening in the roof to be the throat. We wish to relate the pressure P_{cav} in the cavity to the pressure at the exterior entrance to the throat, which we designate as P_{ext}. The variable that relates both pressures is the particle velocity in the throat. From Eq. (3.5.15), we know that

$$V = \frac{P_{ext}}{Z_H}$$

Because V also is the particle velocity at the entrance to the cavity, Eq. (3.5.5) tells us that

$$P_{cav} = Z_{cav} V = \frac{Z_{cav}}{Z_H} P_{ext}$$

Thus, Z_{cav}/Z_H is the requested transfer function. Substitution of the impedances in Eqs. (3.5.5) and (3.5.15) followed by a minor rearrangement of terms leads to the transmission factor being

$$Tr = \frac{Z_{\text{cav}}}{Z_{\text{H}}} = -\frac{\mathcal{A}c^2}{L'\mathcal{V}}\frac{1}{\left(\omega^2 - \omega_{\text{H}}^2\right)} \equiv \frac{1}{\left(1 - \omega^2/\omega_{\text{H}}^2\right)}$$

This is the same as the frequency response of a one-degree-of-freedom oscillator. Indeed, there is a true resonance at ω_{H}. Inclusion of resistance in the opening or dissipation within the vehicle compartment due to the walls not being rigid will prevent the amplitude from growing without bound.

The area of the rectangular opening is $\mathcal{A} = 0.05$ m^2, and the thickness of the roof is the length $L = 20$ mm. We do not know the actual end correction for a rectangular opening, so let us use the one for a circular opening. The radius of a circle whose area matches that of the opening is $a = 126$ mm. The roof constitutes a large wall on both sides of the opening, so we will apply the end correction in Eq. (3.5.12) to both sides of the throat, which leads to $L' = L + 2(8a/(3\pi)) = 234$ mm. (This calculation is typical of the usage of the lumped mass model to describe the passage of sound through any thin open slot. The contributions of the end corrections typically will be the major contributor to L'). The volume of the passenger compartment is stated to be $\mathcal{V} = 5$ m^3. We use Eq. (3.5.15) to evaluate ω_{H} corresponding to these parameters; the result is 11.2 Hz. This is infrasound, which is felt throughout the interior and is extremely annoying. Note that the wavelength of a plane wave at 20 Hz is 30.4 m, depending on ambient conditions. Thus, the requirement that the extent of the Helmholtz resonator be much less than an acoustic wavelength, on which the theory is based, is met by the passenger compartment.

Figure 1 is a graph of transmission factor as a function of frequency. The extreme amplification at ω_{H} is evident. To get an idea of its magnitude, we compute $Tr = 10.3$ at $\omega = 0.95\omega_{\text{H}}$ and $Tr = -9.8$ at $\omega = 1.05\omega_{\text{H}}$. In other words, in the frequency band from 5% below to 5% above the Helmholtz resonance, the transmission factor is at least 20 dB. Close the sunroof!

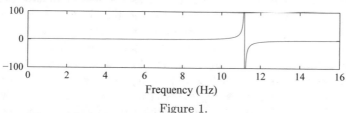

Figure 1.

EXAMPLE 3.12 The waveguide in the sketch has a Helmholtz resonator at each of its one-third points. A harmonically vibrating transducer at the left end induces a particle velocity whose amplitude is $\hat{V}_0$. If the resonators were not present, this system would be a uniform waveguide, with resonances at the natural frequencies of a waveguide having one end open and the other rigid. The resonator to the left has been tuned to the fundamental natural frequency of the uniform waveguide, and the resonator to the right has been tuned to the second natural frequency. All cross sections are circular, and the diameter of the throat of both resonators is 1 mm. Measurements of Z_H for each resonator indicate that both have a nonzero resistive part that is $0.001\rho_1 L\omega$. Determine the amplification factor V/V_0 at the open end on the right, as well as the drive point impedance. Compare both properties to the result that would be obtained if there were no Helmholtz resonators.

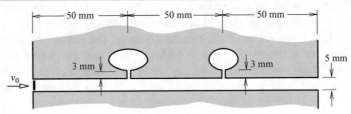

Figure 1.

Significance

The development of the model equation will demonstrate how discrete elements are incorporated into the algorithmic procedure for analyzing a network. The velocity amplification factor is important because the acoustic pressure that radiates into the surroundings at an open end is proportional to the velocity in the waveguide at that location. The results will assist us to recognize that using a Helmholtz resonator as a stop filter at ω_H has other effects in the response.

Solution

We begin by developing the model equations, after which we will set the system parameters. The formulation using junction pressures must be modified to account for the single port nature of lumped elements. The development is somewhat easier if the segments that are modeled as two-port elements are numbered first. Thus, the main waveguide segments are defined to be 1, 2, and 3, proceeding from left to right. After them comes the left resonator, which is designated as branch 4 and then the right resonator, which is branch 5. In a left to right progression, the left end is junction a, the locations of the resonators are junctions b and c, and the right end is junction d. The x_n coordinate for each two-port branch increases in the downstream direction, to the right. The Helmholtz resonators have a single port. Hence, the junction pressure and port pressure vectors are

$$\{\hat{P}\} = [P_a \quad P_b \quad P_c \quad P_d]^T$$
$$\{P\} = \left[P_0^{(1)} \quad P_L^{(1)} \quad P_0^{(2)} \quad P_L^{(2)} \quad P_0^{(3)} \quad P_L^{(3)} \quad P_H^{(4)} \quad P_H^{(5)}\right]^T \tag{1}$$

The port velocity vector is defined similarly to $\{P\}$. Note that the prime in the subscript of $P_{L'}^{(3)}$ is intended to indicate that the pressure is evaluated at the extended location that includes the end correction. This means that we will need to evaluate the velocity at the actual exit, $x_3 = L_3 = 50$ mm, as a separate calculation. The relations between the port pressures and those at the junctions are

$$P_0^{(1)} = P_a, \quad P_L^{(1)} = P_0^{(2)} = P_H^{(4)} = P_b$$
$$P_L^{(2)} = P_0^{(3)} = P_H^{(5)} = P_c, \quad P_{L'}^{(5)} = P_d \tag{2}$$

The nonzero elements of the 8×4 connectivity matrix corresponding to this mapping are $S_{1,1} = S_{2,2} = S_{3,2} = S_{4,3} = S_{5,3} = S_{6,4} = S_{7,2} = S_{8,3} = 1$. The density is the same in all branches, so we may use conservation of volume velocity, rather than of mass, to formulate the junction conditions. These are

$$A_1 V_L^{(1)} = A_2 V_0^{(2)} + A_4 V_H^{(4)}$$
$$A_2 V_L^{(2)} = A_3 V_0^{(3)} + A_5 V_H^{(5)} \tag{3}$$

where A_4 and A_5 are the throat areas. Additional conditions to be satisfied are the velocity at the drive point and an open-end condition at the right end,

$$V_0^{(1)} = \hat{V}_0, \quad P_{L'}^{(3)} = 0 \tag{4}$$

We fit Eqs. (3) and (4) to the standard form $[\Lambda]\{V\} + [\Gamma]\{P\} = \{F\}$. There are four equations, and $\{P\}$ and $\{V\}$ consist of eight variables, so the matrix sizes are 4×8 for $[\Lambda]$ and $[\Gamma]$, and 4×1 for $\{F\}$. The nonzero elements are $\Lambda_{1,2} = A_1$, $\Lambda_{1,3} = -A_2$, $\Lambda_{1,7} = -A_4$, $\Lambda_{2,4} = A_2$, $\Lambda_{2,5} = -A_3$, $\Lambda_{2,8} = -A_5$, $\Lambda_{3,1} = 1$, $\Gamma_{4,6} = 1$, $F_3 = V_0$.

The last step is assembly of the mobility matrix $[D]$. The relation between port velocities and pressures for the two-port elements (branches 1, 2, and 3) are described by the two-port mobility matrix $[D_n]$. These form the first three blocks on the diagonal of $[D]$. For each Helmholtz resonator, we have established that $P_H^{(n)} = Z_H^{(n)} V_H^{(n)}$, so the mobility is the reciprocal of $Z_H^{(n)}$. These reciprocals are entered on the diagonal of $[D]$ at the appropriate row. Thus, the mobility matrix is

$$\{V\} = [D]\{P\}$$
$$[D] = \begin{bmatrix} [D_1] & [0] & [0] & [0] & [0] \\ [0] & [D_2] & [0] & [0] & [0] \\ [0] & [0] & [D_3] & [0] & [0] \\ [0] & [0] & [0] & 1/Z_H^{(4)} & 0 \\ [0] & [0] & [0] & [0] & 1/Z_H^{(5)} \end{bmatrix} \tag{5}$$

All matrices required to form Eq. (3.4.30) have been defined. The last step is to assign values to the various parameters. When we add the given resistance to Z_H in Eq. (3.5.15), we find that

$$Z_H = \rho L' \omega \left[i \left(1 - \frac{\omega_H^2}{\omega^2} \right) + 0.001 \frac{L}{L'} \right] \tag{6}$$

The value of L' includes the end correction, which requires that a be small compared to λ. The wavelength at 2 kHz is 170 mm, and $2a = 1$ mm for the throat of the resonators, so this criterion is met for the resonators. The end correction is applied to both ends, as well as the opening at the end of branch 3. Thus, we set $L' = 3 + (2)(8)(0.5) / (3\pi) = 3.85$ mm for each resonator, and $L_3' = 50 + 8(5/2)/(3\pi) = 52.12$ mm. The other parameters for the evaluation are $L_1 = L_2 = 50$ mm, $\rho_0 = 1.2$ kg/m^3, and we well set $V_0 = 1$ to obtain all results as proportionalities relative to V_0.

Equation (3.4.30) is solved for $\{\hat{P}\}$ at each frequency in the band from zero to 2500 Hz. The corresponding port pressures $\{P\}$ at each frequency are found by evaluating $[S]\{\hat{P}\}$. The only particle velocity to be computed is at the exit, where $x_3 = L$. This is not the particle velocity $V_{L'}^{(3)}$, which is the velocity at $x_3 = L_3'$. An expression for $V^{(3)}(L_3)$ may be found by forming the velocity in branch 3 as the result of superposition of oppositely traveling waves whose amplitudes are given by Eq. (3.4.12), which leads to

$$
\begin{aligned}
V^{(3)}(L_3) &= \frac{1}{\rho_0 c} \left(B_1^{(3)} e^{-ikL_3} - B_2^{(3)} e^{ikL_3} \right) \\
&= \frac{1}{2i\rho_0 c \sin\left(kL_3'\right)} \left(P_0^{(3)} \cos\left(k\left(L_3' - L_3\right)\right) - P_L^{(3)} \cos\left(kL_3\right) \right)
\end{aligned}
\tag{7}
$$

It is requested to compare the results obtained from the model equations to those for a similar uniform waveguide. We will apply the end correction here also, so that $L' = 0.15$ m $+ 8a/(3\pi)$. Expressions for the pressure and particle velocity may be found by setting $R_L = -1$ in Eq. (3.2.21), which leads to

$$
\begin{aligned}
P &= i\rho c V_0 \frac{\sin\left(k\left(L' - x\right)\right)}{\cos\left(kL'\right)} \\
V &= V_0 \frac{\cos\left(k\left(L' - x\right)\right)}{\cos\left(kL'\right)}
\end{aligned}
\tag{8}
$$

The resonant frequencies are those for which $\cos\left(kL'\right) = 0$, which correspond to $\omega_n = 0.5\pi c/L'$, $1.5\pi c/L'$, ... These evaluate to 551 Hz, which is assigned to the left resonator and 1653 Hz for the right. The quantities we need to compute are V at the actual right end, $x = L$, and the drive point impedance, which is $P(0)/V_0$.

Before we examine the results, it is appropriate to consider how one can verify the model equation for an intricate system. We may verify that the port pressures at a common junction actually are equal and that the corresponding port velocities actually satisfy conservation of mass at that junction. A different type of check assigns a set of values to the system properties that have the effect of converting the system to one whose behavior we know. In the present system, we know that if the impedance of the resonators was extremely large, the velocity in each resonator's throat would be very small. Thus, we can carry out an evaluation with Z_H artificially multiplied by 10^6. This calculation was done for the present system, and the result was found to be indistinguishable from those obtained from Eq. (8).

The data for the velocity at the open end of the waveguide is plotted in Fig. 2 as magnitude and phase angle in the range $-180° \leq \theta < -180°$. The uniform waveguide shows resonant peaks at the expected frequencies, and the 180° phase shift as each frequency is passed verifies that these are resonances. The resonators act as expected. They have replaced the resonances of the uniform waveguide with nulls. However, this is achieved by replacing one resonant peak with two on either side of the nominal one. Anyone who has studied mechanical vibrations will recognize this phenomenon as being similar to the effect achieved with a tuned vibration absorber. If it is desired that frequency response be smooth, so that V at the exit does not vary greatly with frequency, then the resistive part of Z_H must be increased greatly. Here, repeating the calculation with $\text{Re}\,(Z_H)$ multiplied by 1000 would remove the nulls at the natural frequencies of the uniform waveguide because Z_H would no longer be zero at those frequencies. It also would lower all resonant peaks because the dissipation associated with this resistance would be effective at all frequencies. However, achieving such a large resistance would not be trivial.

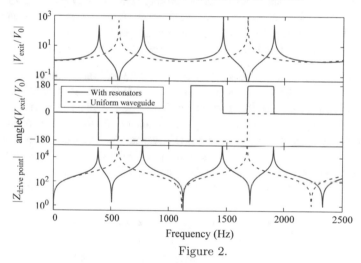

Figure 2.

3.6 Closure

At this juncture, we have developed considerable expertise to investigate and understand acoustic phenomena. Propagation and reflection of waves are basic issues, and frequency-domain analysis usually is the most tractable manner in which we may account for the variability of surface properties. Dissipation effects typically are of major importance only for propagation over long ranges or at very high frequencies, but in those situations, its role is primary. Despite the wealth of knowledge we have developed in this and the previous chapter, we are limited to situations where the signal is planar. All waves have been taken to propagate in the direction of one of the

Cartesian coordinate axes, so we cannot consider combinations of plane waves that he propagate in arbitrary directions. Furthermore, some basic issues such as energy and power transport arose, but were not studied. In the next chapter, we begin to remove these limitations.

3.7 Homework Exercises

Exercise 3.1 Nano-sized accelerometers are mixed with oil ($\rho_0 = 910$ kg/m^3, $c = 1460$ m/s) in order to measure the acceleration of the fluid as a wave passes through. The waveguide is terminated at the far end in a manner that eliminates reflections. In a specific case, the steady-state acceleration at $x = 2$ m is found to be $\dot{v} = 50\sin(3000t)$ m/s^2. The origin, $x = 0$, is at the end where the wave is initiated. Determine (a) the pressure and particle velocity waveforms at $x = 2$ m, (b) the pressure and particle velocity waveforms at $x = 0$, (c) the sound pressure level referenced to 1 μPa at $x = 0$, (d) the time-averaged intensity at $x = 0$, and (e) explain why the properties found in Parts (c) and (d) do not depend on the location.

Exercise 3.2 The sketch shows a piston that separates oil and water in a waveguide. The waveguide is cylindrical, with a diameter of 160 mm. Both ends are distant and terminated in a manner that prevents reflections. The piston's mass is 4 kg. An alternating electromagnetic field causes the piston to vibrate. The displacement of the piston is observed to be $u = 0.09\cos(1200t)$ mm/s. The fluid properties of the oil are $\rho_0 = 910$ kg/m^3, $c = 1460$ m/s, and $\rho_0 = 1005$ kg/m^3, $c = 1485$ m/s for the water. Determine the electromagnetic force acting on the piston.

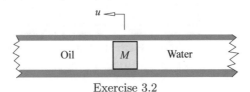

Exercise 3.2

Exercise 3.3 A pair of plane waves at frequency ω propagate in opposite directions within a tube. It is known that they combine to form a standing wave with the maximum pressure amplitude occurring at $x = 0$ and the pressure being zero at $x = L$. What properties of the complex amplitude of each wave can be deduced from this information?

Exercise 3.4 Two plane harmonic waves at the same frequency propagate to the left and right within a tube. The speed of sound at ambient conditions is 1480 m/s, and the ambient density is 1005 kg/m^3. A laser velocimeter is used to measure the particle velocity at two locations, $x = 0$ and $x = 40$ mm. The waveforms are shown below. (a) Express each wave in the form $p_j(x, t) = \mathrm{Re}\left\{A_j \exp\left[i(\pm kx - \omega t)\right]\right\}$. Give the values of k, ω, A_1, and A_2. (b) Plot the pressure waveform at $x = 20$ mm.

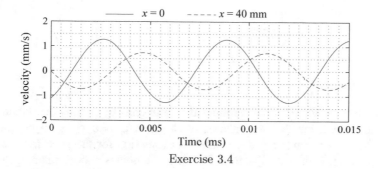

Exercise 3.4

Exercise 3.5 The pressure distribution in the graph describes a harmonic plane wave that propagates in the direction of increasing x. The profile corresponds to $t = 1.44\pi/\omega$, where ω is the frequency. The fluid is air. Use this data to determine frequence of the pressure and the pressure waveform at $x = 0$.

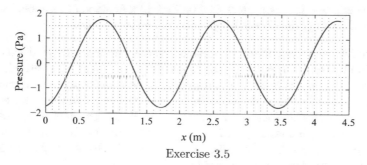

Exercise 3.5

Exercise 3.6 A piston at $x = 0$ whose velocity is $v = v_0 \sin(\omega t)$ generates a plane wave propagating in the $+x$ direction. The pressure at a specified location $x = \xi$ may be written as $p = P_1 \sin[\omega(t - \tau)]$, where τ is the smallest positive value of t at which $p = 0$. (a) Given a value for ξ, determine the most general expression for τ. (b) Consider a set of measurements taken at $x = \xi$ for two closely spaced frequencies, ω_1 and ω_2. This yields measured values $\tau = \tau_1$ and $\tau = \tau_2$, respectively. Derive an expression for c in terms τ_1, τ_2, and ξ. (c) Explain why the frequencies in part (b) must be close in order to use the scheme as the basis for an experiment to measure the speed of sound.

Exercise 3.7 A tube containing water ($\rho_0 = 1000$ kg/m^3, $c = 1480$ m/s) is terminated by a barrier at $x = 0$ consisting of a locally reacting material whose specific impedance is $\zeta = 0.6$. A 1.6 kHz source far to the right has generated a steady-state plane wave traveling in the negative x direction. The source is sufficiently distant, and the observation time is sufficiently early, to justify ignoring reflections at the location of the source. The pressure at the barrier due to the combination of the incident and reflected waves is $p = 100 \sin(\omega t)$ Pa. (a) Determine the functional dependence of the reflected steady-state wave on distance x and time t. (b) Determine the amplitude and phase angle relative to a pure sine of the *total* acoustic pressure as a function of time at $x = 2$ m.

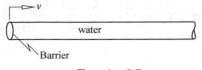

Exercise 3.7

Exercise 3.8 A waveguide is terminated at both ends by pistons that vibrate with equal amplitude, but different phases, such that $v(x = 0, t) = v_0 \sin(\omega t)$ and $v(x = L, t) = v_0 \sin(\omega t - \Delta)$. Derive an expression for the pressure as a function of x and t. Is there a value of Δ for which the field is a wave that propagates in the direction of increasing x? Is there a value of Δ for which the field is a standing wave?

Exercise 3.9 One end of a waveguide is driven by a harmonically varying pressure source whose complex amplitude is P_0. The other end is open to the atmosphere. Derive an expression for the pressure amplitude as a function of location. From that expression, identify the natural frequencies and mode functions.

Exercise 3.10 A waveguide may be driven harmonically at $x = 0$ by either a velocity source $\text{Re}(V_0 \exp(i\omega t))$ or a pressure source $\text{Re}(P_0 \exp(i\omega t))$ that varies harmonically. The end $x = L$ is rigid. (a) Determine the nondimensional pressure amplitudes $|p/(\rho_0 c V_0)|$ and $|p/P_0|$ at the midpoint as a function of $\omega L/c$ for each excitation. (b) Graph each relation for the case where V_0 and P_0 are constants. (c) Consider the situation where V_0 is constant, but P_0 is frequency dependent. Graph the dependence of $|P_0/(\rho_0 c |_0|)|$ on $\omega L/c$ for which the pressure and velocity excitations produce the same pressure amplitude at the midpoint.

Exercise 3.11 Beth and Abby are playing "jump rope" with Leah. Abby holds one end stationary, while Beth shakes the other end in the horizontal y and vertical direction z, such that the respective transverse displacements are $w_y = B \cos(\omega t)$ and $w_z = 2B \sin(\omega t)$. The rope, whose length is 3 m, carries a tensile force of 20 N, and its mass per unit length is 0.5 kg/m. The wave equations for the horizontal and vertical displacements are uncoupled, but they share the same phase speed. (a) Derive an expression for the displacement components of the rope at its midpoint. (b) Draw the path of the midpoint. (c) Identify the lowest frequency interval in which the vertical displacement of the midpoint is greater than $20B$.

Exercise 3.12 The particle velocity at $x = 0$ of a waveguide is $\text{Re}(V_0 \exp(i\omega t))$. The pressure waveforms that are measured at $x = L/3$ and $x = 2L/3$ are plotted below. The fluid is air at standard conditions, and the length of the waveguide is $L = 2.35$ m. Determine the specific impedance at $x = L$.

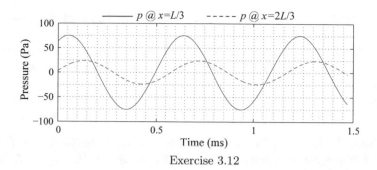

Exercise 3.12

Exercise 3.13 The velocity at $x = 0$ in a waveguide is the sum of three harmonics, $v = B \sin(\omega t) + 2B \sin(2\omega t) + 3B \sin(3\omega t)$. The other end is pressure-release. The fluid is air at standard atmospheric condition, the waveguide's length is 1.2 m, and $B = 4$ mm/s. Determine the RMS pressure at the midpoint as a function of frequency. The frequency range of interest is $\omega < 2\pi c/L$.

Exercise 3.14 The pressure at $x = 0$ in a waveguide is Re $(P_0 \exp(i\omega t))$. The other end, $x = L$, is made rigid by applying an axial force to a piston that fills the cross section, but would slide freely if not for the force. Derive an expression for this force, and evaluate it as a function of frequency for $\omega < 3\pi c/L$. The waveguide is 400 mm long and has a diameter of 20 mm, and the fluid is water.

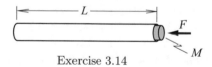

Exercise 3.14

Exercise 3.15 The waveguide depicted below is driven by a velocity source Re $(V_0 \exp(i\omega t))$ at one end. The other end is closed with piston M that is supported by spring K. In the absence of fluid loading, the natural frequency of this system would be $\omega_{nat} = (K/M)^{1/2}$. Evaluate $|p/(\rho_0 c V_0)|$ and $|v/V_0|$ at the face of the piston as a function of the nondimensional frequency kL. Three mass ratios are of interest, $\sigma = 0.01$, 1, and 100. The natural frequency parameter for each case is $\omega_{nat} = \pi c/L$. Identify the natural frequencies in each case.

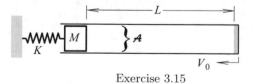

Exercise 3.15

Exercise 3.16 It is desired to test the procedure by which an impedance tube is used to identify the impedance of a material. To that end a set of data for a waveguide that is terminated by a locally reacting material should be generated, then processed to identify the impedance. The system for this test is a waveguide filled with water, with $v = 0.02 \cos(10000\pi - 0.4)$ m/s at $x = L$. The length of the waveguide is 1.4 m, and the specific impedance at the far end is $\zeta_L = 0.3 - 0.5i$. (a) One of the factors affecting the accuracy of the identification is the spacing Δx between sample locations. In any event, it must be less than a half-wavelength, so it must be that $dx = (\pi/k)/\beta$, where $\beta > 1$. Compare the data sets for $\beta = 2$ and $\beta = 2.24 \approx \sqrt{5}$ to explain why choosing β to be a small integer is not desirable. (b) Process the data set for $\beta = 2.24$ to identify ζ_L. Compare the identified and actual values.

Exercise 3.17 One end of a waveguide is open to the atmosphere, and the other is closed with a material whose specific impedance is $\zeta = 0.2i$ at any frequency. Determine the lowest three natural frequencies and corresponding mode functions.

Exercise 3.18 The impedance at $x = L$ of a waveguide is $Z_L = \rho_0 c \sigma i$. The velocity at the end $x = 0$ is controllable. (a) For the case where $\sigma = 2$ and $v = 0$ at $x = 0$, determine the lowest three natural frequencies and describe the corresponding mode functions. (b) Consider the range $-10 < \sigma < 10$. Graph the dependence of the nondimensional fundamental frequency $k_1 L$ on σ. For what range of σ is $x = L$ well approximated as pressure-release? For what range of σ is $x = L$ well approximated as rigid?

Exercise 3.19 A source situated on the ground emits a harmonic acoustic wave that spreads spherically. The pressure at radial distance r from the source is $p = \text{Re}[(S/r) \exp(-ik'r + \omega t)]$, where S is the source strength. The air may be considered to be homogeneous at 20°C and 20% relative humidity. Frequencies of interest are 100 Hz, 1 kHz, and 10 kHz. The sound pressure level at $r = 1$ m for each frequency is 130 dB. Determine the sound pressure level at $r = 10$, 100, and 1000 m for each frequency, and compare the result to what would be obtained if dissipation were negligible.

Exercise 3.20 The graph describes the pressure waveform in a plane wave at two locations separated by 80 mm. Determine the propagation wavenumber and absorption coefficient.

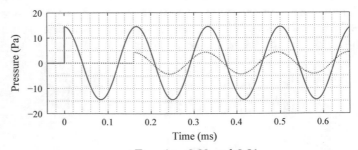

Exercises 3.20 and 3.21

Exercise 3.21 The graph describes the pressure waveform at two locations separated by 80 mm. Viscosity is the dominant source of dissipation. Determine the propagation wavenumber and the value of $\lambda + 2\mu$?

Exercise 3.22 Consider the tube with porous walls described in Exercise 2.5. Derive the dispersion equation for a plane harmonic wave. What is the primary effect of a small amount of porosity?

Exercise 3.23 In an ocean propagation experiment, the sound pressure is measured at a distance of 10 km from the source. The signal is harmonic, and it may be assumed to propagate as a planar wave. The sound pressure level is measured to be 200 dB//1 μPa. The ocean properties may be considered to be constant along the propagation path, with a salinity of 35% and a temperature of 20°, which are the conditions associated with Eq. (3.3.33). The sound speed is 1505 m/s. Determine the sound pressure level at the source for frequencies of 10 Hz, 1 and 100 kHz.

Exercise 3.24 The square wave in Example 3.7 is generated by an underwater projector. The maximum pressure at the source is 40 kPa. The ocean conditions are those for Eq. (3.3.34), for which the sound speed is 1505 m/s. It may be assumed that the wave radiated from the projector propagates as a plane wave. Determine the waveform that would be measured at $x = 10$ km. Frequencies of interest are 5 kHz and 500 Hz.

Exercise 3.25 A loudspeaker inside a very long narrow tunnel generates a periodic signal whose fundamental frequency is 250 Hz. The pressure it generates at $x = 0$ is $p = A \sin (\omega t) + (A/2) \sin (2\omega t) + (A/4) \sin (4\omega t)$. The sound pressure level at that location is 105 dB//20 μPa. The signal propagates as a plane wave. The atmospheric conditions are those for Eq. (3.3.33), for which the sound speed is 343 m/s. Plot the pressure waveform at $x = 100$ m, and compare it to the waveform that would result if dissipation were negligible. Then, repeat the evaluation for the case where the fundamental frequency is 2.5 kHz.

Exercise 3.26 In order quantify the effect of wall friction, consider two alternative situations. In the first, a pressure source at $x = 0$, $P (0) = \text{Re} (P_0 \exp (i\omega t))$, generates a wave in a very long cylindrical waveguide whose far end inhibits reflections. In the second, the same source at $x = 0$ generates the field within a waveguide whose end $x = L$ is rigidly terminated. It is desired to examine the ratio $|P (L) / P_0|$ in each case as a function of the cross-sectional diameter in the range $1 < D < 100$ mm. Carry out the analysis for frequencies of 10 Hz and 1 kHz.

Exercise 3.27 Solve Example 3.9 for the case where the pulse interval equals the time required for a pressure wave to travel from the left end of segment 1 to the junction. Perform the analysis for $R = -0.6$ and $T = 0.4$. Estimate the time after which the pressure in the left segment will not exceed $0.25\rho_0 c V_0$.

Exercise 3.28 A loose membrane stretched across the junction of the tubes allows the larger one to be filled with oil ($\rho_0 = 910$ kg/m^3, $c = 1460$ m/s), while the right one is filled with water. The pressure at the left end is $1000 \cos (\omega t)$ Pa. The termination at the right end is actively controlled to eliminate reflections at the end. The result is that $p = \rho_0 c v$ at that end if positive v is defined to be the right. Determine the time-averaged pressure and intensity at the junction of the wide and narrow sections. The frequency is 7.5 kHz.

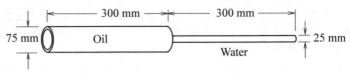

Exercise 3.28 and 3.29

Exercise 3.29 Solve Exercise 3.28 for the situation where the right end is open to a large vat of water.

Exercise 3.30 The right end of the forked waveguide is driven by a constant amplitude velocity source, with the result that the particle velocity there is $V_0 \sin (95t)$. The other ends are open to the atmosphere. Determine the particle velocity at both ends of the fork.

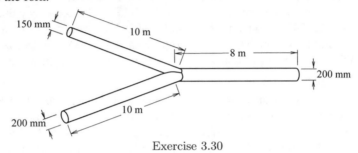

Exercise 3.30

Exercise 3.31 The pressure at the left end of the straight tube is known to be 200 $\cos(\omega t)$ Pa and the right end is open to the atmosphere. The frequency is 35 Hz. Determine the particle velocity at the right end and the pressure at the midpoint of the semicircular segment.

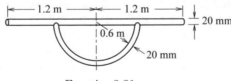

Exercise 3.31

Exercise 3.32 The sketch depicts a telescoping waveguide. Its construction allows the length L to change without altering the internal diameters, which are multiples of D. The right end is open to the atmosphere, which is at standard conditions. The excitation is a constant velocity source $V_0 \sin (\omega t)$ at the left end. Determine the particle velocity at the right end as a function of L ranging from 10 mm to 1 m. The frequency is 100 Hz.

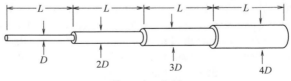

Exercise 3.32

Exercise 3.33 A 2-m-long waveguide has a 50 mm × 50 mm square cross section. It is open to the atmosphere at one end, and the other end is driven by a velocity source for which $v = v_0 \cos(\omega t)$, with $v_0 = 150$ mm/s and $\omega = 8000$ rad/s. In order to determine the pressure signal that would be radiated from the open end, the open end should be treated as a termination whose local impedance is described in Fig. 3.16. Determine the pressure and particle at the open end, first by assuming that dissipation is negligible and then by assuming that the dominant dissipation mechanism is tube wall friction.

Exercise 3.34 The piston-spring assembly that terminates the waveguide in the right sketch must reproduce the properties of the Helmholtz resonator in the left sketch. What are the values of M and K that achieve this objective?

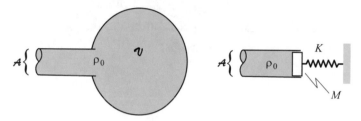

Exercise 3.34

Exercise 3.35 The long tube, whose cross section is square, is terminated at its right end by the hollow cube. The left end is driven by a source that sustains a constant 110 dB sound pressure level at that end, independent of frequency. Determine the amplitude of the particle velocity at the left end as a function of frequency in the range from 1 to 500 Hz. Use this result to identify the natural frequencies, and compare them to those if the tube was open at the right end. For this analysis, consider the cube to constitute a small volume lacking inertia.

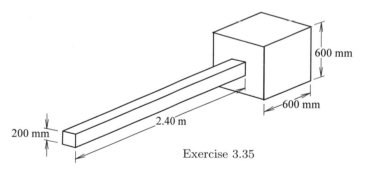

Exercise 3.35

Exercise 3.36 Modify the analysis in Exercise 3.35 to include the inertia of particles at the junction of the tube and the cubic cavity.

Exercise 3.37 The waveguide network depicted below is the same as the one in Exercise 3.31, except that a Helmholtz resonator has been inserted at the midpoint of the straight tube. The velocity at the left end is $v_0 \cos(\omega t)$. The natural frequency of the resonator has been tuned to the fundamental natural frequency of a straight tube rigid/pressure-release waveguide whose length is $L' = 2.4$ m, that is, $\omega_H = (\pi/2)\, c/L'$. Determine $|P|/(\rho_0 c v_0)$ at the left end and $|V|/v_0$ at the right end for cases in which the frequency is $\omega = 1.1\omega_H$ and $\omega = 1.001\omega_H$.

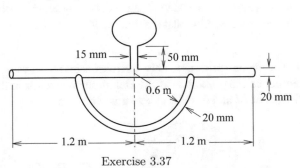

Exercise 3.37

Chapter 4
Principles and Equations for Multidimensional Phenomena

We will not analyze new phenomena in this chapter, nor will we develop new analytical techniques. Rather, our objective is to create the foundation for study of a variety of phenomena that do not fit the planar wave model. One possible difference is that the *wavefront*, which is the locus of points at which the acoustic disturbance was generated at a common time, is not planar. This is a feature of a spherical wave. Another possible difference from the plane wave model is that the disturbance at any instant might depend on position transverse to the direction in which the signal propagates, in addition to being a function of the propagation distance. It might be that this transverse dependence is best described in terms of Cartesian coordinates, which will be used to analyze planar waves that propagate in an arbitrary direction. However, spherical coordinates might be more appropriate, which is the case for acoustic radiation from spherical transducers, whereas cylindrical coordinates are crucial to the study of acoustic radiation from a piston in a wall and propagation of higher order modes in a circular waveguide.

In addition to these three commonly encountered coordinate systems, there are numerous less common ones that might be particularly suitable for a specific situation. Therefore, it is vital that the equations representing fundamental physical principles be posed in a manner that is independent of the specific coordinate system. This objective is met by invoking some fundamental concepts from vector calculus.

The first part of the development will rederive the conservation of mass and the momentum-impulse equations, and then the linear wave equation. These derivations will parallel those by which the equations governing planar waves were described. The second part of the chapter will develop principles and theorems regarding work, energy, and power.

Electronic supplementary material The online version of this chapter (DOI 10.1007/978-3-319-56844-7_4) contains supplementary material, which is available to authorized users.

© Springer International Publishing AG 2018, corrected publication 2021
J.H. Ginsberg, *Acoustics—A Textbook for Engineers and Physicists, Volume I*,
DOI 10.1007/978-3-319-56844-7_4

4.1 Fundamental Equations for an Ideal Gas

4.1.1 Continuity Equation

The continuity equation for one-dimensional planar waves was derived in Sect. 2.1.1 by examining a control volume whose position is fixed. There the boundary across which particles flowed was composed of planes of constant x because there was no movement in any other direction. Now, we must allow for movement in any and all directions. This means that the control volume should be some amorphous shape $\mathcal{V}$ whose bounding surface is denoted as $\mathcal{S}$. The vector extending from the origin of some coordinate system to an arbitrary point in $\mathcal{V}$ is the position vector $\bar{\xi}$, while the position vector to a point on $\mathcal{S}$ is denoted as $\bar{\xi}_S$. If Cartesian coordinates are appropriate to the system of interest, then position vectors would have components $x, y, and z$. However, it would be equally valid to describe the position in terms of any other set of coordinates if doing so expedited describing the configuration of a system.

Figure 4.1a depicts a control volume $\mathcal{V}$ and a differential surface patch dS. We consider $\mathcal{V}$ to be stationary, so particles can flow across $\mathcal{S}$ in any direction. To describe this flow, we refer to Fig. 4.1b. Because of its small size, dS may be considered to be flat, and therefore coincide with the local tangent plane. The direction, that is, perpendicular to this tangent plane pointing out of $\mathcal{V}$ is denoted as $\bar{n}$, whose value depends on the point $\bar{\xi}_S$ at which the tangent plane is situated. The central limit theorem of calculus indicates that all points on dS have the same velocity $\bar{v}\left(\bar{\xi}_S, t\right)$, which means that these points experience the same displacement $\bar{v}\left(\bar{\xi}_S, t\right) dt$ in an infinitesimal time interval from t to $t + dt$. The result is that points that were on dS shift to a new position. The height of this displaced patch above dS is $h = n\left(\bar{\xi}_S\right) \cdot \bar{v}\left(\bar{\xi}_S, t\right) dt$, so its volume is $h\,dS$. This is the region that is transported out of $\mathcal{V}$ in the interval dt. Hence, the mass that flows out of $\mathcal{V}$ in this interval is $\rho\left(\bar{\xi}_S, t\right) h\,dS$.

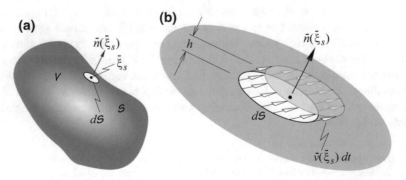

Fig. 4.1 Control volume for evaluating mass flow. **a** Basic configuration. **b** Evaluation of the flow across an infinitesimal surface patch dS

The total mass flow out of $\mathcal{V}$ is found by integrating the contribution of each surface patch. Conservation of mass requires that whatever flows out of $\mathcal{V}$ in any interval equal the decrease of mass that is contained within the domain. The latter quantity is obtained by integrating the differential mass $\rho\left(\bar{\xi}, t\right) d\mathcal{V}$, so it must be that

$$\iint_{S} \rho\left(x_x, t\right) \left[\bar{n}\left(\bar{\xi}_s\right) \cdot \bar{v}\left(\bar{\xi}_s, t\right) dt\right] d\mathcal{S} = \iiint_{\mathcal{V}} \rho\left(x, t\right) d\mathcal{V} - \iiint_{\mathcal{V}} \rho\left(x, t+t\right) d\mathcal{V}$$

(4.1.1)

Division of this relation by dt leads to

$$\iiint_{\mathcal{V}} \frac{\partial}{\partial t} \rho\left(x, t\right) d\mathcal{V} + \iint_{S} \rho\left(x_S, t\right) \bar{n}\left(\bar{\xi}_S\right) \cdot \bar{v}\left(\bar{\xi}_S, t\right) d\mathcal{S} = 0 \qquad (4.1.2)$$

The divergence theorem, also known as Gauss' theorem, converts the volume integral of the divergence of an arbitrary vector $\bar{U}$ to a surface integral, according to

$$\boxed{\iiint_{\mathcal{V}} \nabla \cdot \bar{U} d\mathcal{V} = \iint_{S} \bar{n}\left(\bar{\xi}_S\right) \cdot \bar{U}\left(\bar{\xi}_S\right) d\mathcal{S}} \qquad (4.1.3)$$

The surface integral in the mass equation fits this theorem with $\bar{U} = \rho\bar{v}$. Replacement of it with the volume integral leads to

$$\iiint_{\mathcal{V}} \left\{\frac{\partial}{\partial t} \rho\left(x, t\right) + \nabla \cdot \left[\rho\left(x, t\right) \bar{v}\left(\bar{\xi}, t\right)\right]\right\} d\mathcal{S} = 0 \qquad (4.1.4)$$

This condition must be met regardless of how $\mathcal{V}$ is defined—it could be the entire acoustical field, or any shaped subdomain. Its applicability is not restricted by the time at which the integral is formed. Under such conditions, the integral will vanish only if its integrand is zero. Thus, we are led to the derivative form of the general equation for conservation of mass,

$$\boxed{\frac{\partial \rho}{\partial t} + \nabla \cdot (\rho\bar{v}) = 0} \qquad (4.1.5)$$

Satisfaction of this equation assures that a velocity field will not open gaps in which there is no fluid. Hence, it also is known as the *continuity equation*. In the case of a simple planar wave that is propagating in the x direction, the only nonzero velocity component is v_x, so that $\nabla \cdot (\rho\bar{v}) = \partial\left(\rho v_x\right)/\partial x$. In this case, Eq. (4.1.5) reduces to Eq. (2.1.6) for plane waves, which is a partial confirmation of the general form.

4.1.2 Momentum Equation

The second fundamental principle that must be satisfied is Newton's second law. Any domain of the fluid contains an innumerable number of particles, so we have adopted a continuum model as a way of simplifying a description of the state of the fluid. Nevertheless, the governing laws are those that apply to a collection of particles. A basic theorem in classical dynamics is that such a system is governed by the momentum-impulse principle. The integral form holds that the change in the total momentum of a system of particles during a time interval equals the impulse of all forces exerted by agents that are external to that system. The derivative form of this principle is more useful for the derivation of field equations. It states that the rate at which the total momentum of a system changes must equal the resultant of the external force system. Figure 4.2 shows a control volume that moves with a collection of particles such that there is no mass flow across its boundary. The normal vector pointing out of the integration domain at point $\bar{\xi}_S$ is $\bar{n}\left(\bar{\xi}_S, t\right)$, with the dependence on time being a consequence of the motion of $\mathcal{V}$.

The total pressure will be denoted by the symbol $\hat{p}$ to distinguish it from the acoustic pressure p, which is the fluctuation relative to the ambient pressure p_0. The traction force associated with a positive pressure pushes the surface patch $d\mathcal{S}$ inward, which is opposite the sense of $\bar{n}$. Thus, the differential contribution to the resultant force attributable to the pressure acting on the patch of surface area $d\mathcal{S}$ is $-\hat{p}\left(\bar{\xi}_S, t\right)\bar{n}\left(\bar{\xi}_S, t\right) d\mathcal{S}$. The resultant force is obtained by adding the contribution of each patch, which means that it is obtained as a surface integral,

$$\sum \bar{F} = -\iint_{\mathcal{S}} \hat{p}\left(\bar{\xi}_S, t\right)\bar{n}\left(\bar{\xi}_S, t\right) d\mathcal{S} \qquad (4.1.6)$$

Fig. 4.2 Control volume that contains a specific set of particles

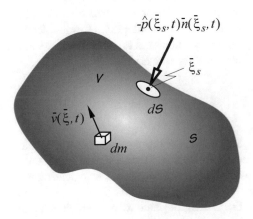

The velocity of the differential element of mass dm depicted in Fig. 4.2 is $\bar{v}\left(\bar{\xi}, t\right)$. It clearly is a function of t, but its dependence on $\bar{\xi}$ is a subtle aspect of kinematics. One way in which the motion of a particle may be described is to know the spatial coordinates of that particle as a function of time. This is the description, that is, typically used in particle and rigid body dynamics. In the study of the motion of deformable continua, this is said to be a *Lagrangian description* of a motion. It is not suitable for acoustical studies. For example, we might wish to make a measurement with a microphone at a specific position. We cannot, and probably would not want to, follow a specific particle with the microphone. Instead, we consider the properties of the particle that happens to be at a specific field point $\bar{\xi}$ at a specific instant t. At a later instant, that particle will have moved to a different location, and a different particle will be at $\bar{\xi}$. Thus, the velocity of the particle depends on the current value of time and the particle's current location, which is also a function of time. In other words, the functional dependence is $\bar{v}\left(\bar{\xi}\left(t\right), t\right)$. This is the *Eulerian description* of motion.

The momentum of this mass element is $\bar{v}\, dm$. The mass is invariant, so the rate at which the momentum of this element changes is $\left(d\bar{v}/dt\right)dm$, in other words mass times acceleration, as in Newton's second law. One reason we consider a specific collection of particles is that forces exerted between particles cancel when the individual contributions are added. The only force system that remains is the pressure, whose resultant is described in Eq. (4.1.6). Addition of the contribution of each element $dm = \rho\left(\bar{\xi}, t\right)d\mathcal{V}$ is a process of integration, so the principle of momentum and impulse for a system of particles requires that

$$\iiint_{\mathcal{V}} \left[\frac{d}{dt} v\left(\bar{\xi}, t\right)\right] \rho\left(\bar{\xi}, t\right) d\mathcal{V} = -\iint_{\mathcal{S}} \hat{p}\left(\bar{\xi}_S, t\right)\bar{n}\left(\bar{\xi}_S, t\right) d\mathcal{S} \qquad (4.1.7)$$

Application of the divergence theorem converts this to

$$\iiint_{\mathcal{V}} \rho\left(\bar{\xi}, t\right)\frac{d}{dt} v\left(\bar{\xi}, t\right) d\mathcal{V} = -\iiint_{\mathcal{V}} \nabla\hat{p}\left(\bar{\xi}, t\right) d\mathcal{V} \qquad (4.1.8)$$

The integrands on either side of the equality sign must match because the equality must hold regardless of how $\mathcal{V}$ is defined and what instant is considered. As was true for the continuity equation, these conditions require that the integrands are equal,

$$\rho\left(\bar{\xi}.t\right)\frac{d}{dt} \bar{v}\left(\bar{\xi}, t\right) = -\nabla p\left(\bar{\xi}, t\right) \qquad (4.1.9)$$

The fact that we are using an Eulerian description, in which $\bar{\xi}$ is the instantaneous position of a particle, slightly complicates the evaluation of acceleration. To account for this aspect of the time dependence of $\bar{v}$, we use the chain rule. Suppose that position is specified by the Cartesian coordinates $x, y, and z$, so that $\bar{v}$ depends on these coordinates and time. Then

$$\frac{d}{dt}\bar{v}\left(\bar{\xi}, t\right) = \frac{\partial}{\partial t}\bar{v}\left(\bar{\xi}, t\right) + \frac{dx}{dt}\frac{\partial}{\partial x}\bar{v}\left(\bar{\xi}, t\right) + \frac{dy}{dt}\frac{\partial}{\partial y}\bar{v}\left(\bar{\xi}, t\right) + \frac{dz}{dt}\frac{\partial}{\partial z}\bar{v}\left(\bar{\xi}, t\right) \quad (4.1.10)$$

It is possible to express these operations in a form that does not rely on a coordinate system. To recognize this, we observe that x, y, $and z$ are the coordinates of a particle, so dx/dt, dy/dt, and dz/dt are the velocity components of the particle. Furthermore, $\partial/\partial x$, $\partial/\partial y$, and $\partial/\partial z$ are the components of the gradient. Thus, an operator may be defined as

$$\boxed{\bar{v} \cdot \nabla \equiv \frac{dx}{dt}\frac{\partial}{\partial x} + \frac{dx}{dt}\frac{\partial}{\partial x} + \frac{dx}{dt}\frac{\partial}{\partial x}} \quad (4.1.11)$$

It will be noted that $\bar{v} \cdot \nabla$ is a scalar operator, so $\bar{v} \cdot \nabla \bar{v}$ is a vector. Some individuals are confused by this definition because they say that $\bar{v} \cdot \nabla$ should be equal to $\nabla \cdot \bar{v}$, so the velocity in the operator should be differentiated. At a basic level, $\bar{v} \cdot \nabla$ may be considered a shorthand way of writing the operations on the right side, with no special meaning beyond that. However, the usage of a gradient, rather than explicitly writing derivatives with respect to each component, does not merely make the notation compact. Rather, the vector notation facilitates usage of other types of coordinate systems. An alternative perspective evaluates $\bar{v} \cdot \nabla \bar{v}$ in a right to left sequence, so $\nabla \bar{v}$ is evaluated first. This quantity is a second rank dyadic, or equivalently a second rank tensor. The subtlety as to whether the operations should be performed left to right or right to left does not affect the final result.

With this understanding as to the meaning of operations, the final form of the momentum-impulse equation is

$$\boxed{\rho\left(\frac{\partial \bar{v}}{\partial t} + \bar{v} \cdot \nabla \bar{v}\right) = -\nabla \hat{p}} \quad (4.1.12)$$

Several names have been used to describe the total derivative of a quantity that depends on time and a variable position, but the most frequent and descriptive term is a *material derivative*. This derivative also is embedded in the continuity equation, Eq. (4.1.5), which may be written as

$$\frac{\partial \rho}{\partial t} + \bar{v} \cdot \nabla \rho + \rho \nabla \cdot \bar{v} \equiv \frac{d\rho}{dt} + \rho \nabla \cdot \bar{v} = 0$$

This is the form that would have been derived if a control volume that moves with a set of particles had been used to enforce mass conservation.

EXAMPLE 4.1 Riemann in 1860 derived a solution for the acoustic pressure and particle velocity in a nonlinear plane acoustic wave. An important parameter introduced in this solution is a generalization of the speed of sound from c in linear theory to a parameter $\tilde{c}$ that is defined to be

$$\tilde{c} = \left(\frac{dp}{d\rho}\right)^{1/2}$$

This definition resembles the definition for c, except that it is evaluated at the current, rather than the ambient, state. The Riemann solution is

$$p = F\left(x - (\tilde{c} + v)\,t\right), \qquad \frac{dv}{dp} = \frac{1}{\rho\tilde{c}}$$

where $F(\,)$ is an arbitrary function. The solution may be interpreted to state that the phase speed of a specific value of p at a specific x and t is $\tilde{c}$ plus the particle velocity v at that x and t. This representation fully prescribes the pressure p and particle velocity v. One way of recognizing this is to recall that the equation of state gives p as a function of ρ, which may be inverted to find ρ as a function of p. Substitution of that function into the second equation would convert it to an ordinary differential equation whose solution would be v as a function of p. Substitution of that relation into the argument of F leads to a single equation for p in which the argument of F depends on p. This solution and its implications are addressed in Chap. 13. Here, the task is to verify that the Riemann solution satisfies the fundamental equations. Specifically, we must prove that it satisfies the continuity and momentum equations.

Significance

The nature of this solution is fundamentally different from the solution for a linear plane wave. The operations that we shall perform demonstrate the level of ingenuity that is required to analyze nonlinear acoustic effects. The converse view is that it demonstrates why it is so important that the governing equations are linearizable.

Solution

A possible approach would follow the problem statement by solving the equation of state for ρ as a function of p, which would be used to describe v and $\tilde{c}$ as functions of p. The result would be an equation that says p is an arbitrary function whose argument depends on p, as well as the values of x and t. This cannot be done because the equation of state is not specified. A viable approach eliminates ρ and v. The equation of state relates p and ρ; solution of the second differential equation would give v as a function of p. This sequence of dependencies with p treated as a function of x and t is used in conjunction with the chain rule for differentiation to convert the derivatives of ρ and v in the continuity equation. The result is

$$\frac{d\rho}{dp}\frac{\partial p}{\partial t} + \left(\frac{d\rho}{dp}v + \rho\frac{dv}{dp}\right)\frac{\partial p}{\partial x} = 0 \qquad (1)$$

The particle velocity for a plane wave may be taken to be $\bar{v} = v\bar{e}_x$, so that $\bar{v} \cdot \nabla\bar{v} = v\partial v/\partial x$, which reduces the momentum equation to

$$\rho\frac{dv}{dp}\left(\frac{\partial p}{\partial t} + v\frac{\partial p}{\partial x}\right) = -\frac{\partial p}{\partial x} \tag{2}$$

According to the Riemann solution, $dv/dp \equiv 1/(dp/dv) = 1/(\rho\tilde{c})$ and $d\rho/dp \equiv 1/(dp/d\rho) = 1/\tilde{c}^2$. Substitution of these expressions transforms the continuity and momentum equations to the same equation,

$$\frac{1}{\tilde{c}}\frac{\partial p}{\partial t} + \left(\frac{v}{\tilde{c}} + 1\right)\frac{\partial p}{\partial x} = 0 \tag{3}$$

The task now is to evaluate the derivatives of the given expression for p. We must perform this operation without knowing the F function. How can we do so? The answer is to differentiate the solution with respect to t and x anyway. The derivatives of the right side of the expression are obtained by differentiating the function with respect to its argument, which we will denote simply as F', then differentiating the argument $x - (\tilde{c} + v)t$. Here the chain rule is used to differentiate $\tilde{c}$ and v, which have been taken to be functions of p. Thus,

$$\frac{\partial p}{\partial x} = \left(1 - \frac{\partial}{\partial x}(\tilde{c} + v)t\right)F' = \left(1 - \frac{\partial p}{\partial x}\frac{d}{dp}(\tilde{c} + v)t\right)F'$$

$$\frac{\partial p}{\partial t} = \left[-\left(\frac{\partial}{\partial t}(\tilde{c} + v)\right)t - (\tilde{c} + v)\right]F' = -\left[\frac{\partial p}{\partial t}\frac{d}{dp}(\tilde{c} + v)t + (\tilde{c} + v)\right]F' \tag{4}$$

The derivatives we seek appear on both sides of the equality. Collecting these terms leads to

$$\frac{\partial p}{\partial x} = \frac{F'}{1 + \dfrac{d}{dp}(\tilde{c} + v)tF'}$$

$$\frac{\partial p}{\partial t} = -\frac{(\tilde{c} + v)F'}{1 + \dfrac{d}{dp}(\tilde{c} + v)tF'} \tag{5}$$

Substitution of these expressions into Eq. (2) causes the left side to vanish identically. This proves that the Riemann solution is correct.

Even though a plane wave is the simplest type of nonlinear signal we might encounter, it contains many interesting features, as we will see in Chap. 13. At the same time, it is reasonable to suspect that analysis of other types of nonlinear signals would require a level of effort well beyond what suffices for the study of linear acoustics. Indeed, it is fortunate that linear acoustics is adequate for most applications.

4.2 Linearization

The momentum-impulse principle is represented by a vector equation, so each of its nontrivial components is a scalar differential equation. Enforcement of continuity yields another scalar differential equation. The unknowns are the pressure and density, and the nontrivial components of $\bar{v}$, whose number is the same as the number of scalar impulse-momentum equations. Thus, one more equation is needed before we may proceed further. As was the case for plane waves, the additional relation is the equation of state. It is convenient to restate it here from Chap. 2. A minor rearrangement of terms from Eq. (2.3.12) gives

$$p = \hat{p} - p_0 = c^2 \rho' \left[1 + \frac{1}{2} \frac{B}{A} \left(\frac{\rho'}{\rho_0} \right) + \cdots \right], \quad A = K = \rho_0 c^2 \qquad (4.2.1)$$

where $\rho' \equiv \rho - \rho_0$ is the density perturbation. A linearized equation of state results if the bracketed terms are approximated as one. Thus, the equation of state is taken to be

$$\boxed{p = c^2 (\rho - \rho_0)} \qquad (4.2.2)$$

Section 2.3 suggested that it is reasonable in most cases to use this approximation when the sound pressure levels are below 140 dB//20μPa in air and 250 dB//1μPa in water.

The next task is to linearize the continuity and momentum equations. The acoustic pressure is small compared to the bulk modulus $\rho_0 c^2$; the density perturbation ρ' is small compared to ρ_0. In a plane wave $v/c = p/ (\rho_0 c^2)$, which suggests that the particle velocity in general will be small compared to c. Smallness here means that any terms in the continuity and momentum equations that are a product of any combination of p, $(\rho - \rho_0)$, $\bar{v}$ and their derivatives will be small compared to terms that are linear in these variables. However, rather than deciding how to do this in an ad hoc manner, let us carry out this operation based on nondimensionalization of the governing equations. Each variable is multiplied and divided by its reference value, which is $\rho_0 c^2$ for pressure, ρ_0 for density, and c for velocity. A reference scale for time is selected to be $1/\omega$, where ω is a representative frequency, such as the upper limit in a spectral analysis of the pressure. Also, the gradient operator has units of inverse length, so we use $k = \omega/c$ as its reference quantity. Thus, we rewrite the continuity equation as

$$\left\{ \omega \rho_0 \frac{\partial \left((\rho_0 + \rho') / \rho_0 \right)}{\partial (\omega t)} + k \frac{\nabla}{k} \cdot \left[\rho_0 \left(\frac{\rho_0 + \rho'}{\rho_0} \right) c \left(\frac{\bar{v}}{c} \right) \right] \right\}$$

$$\equiv \rho_0 \omega \left\{ \frac{\partial (\rho'/\rho_0)}{\partial (\omega t)} + \frac{\nabla}{k} \cdot \left[\left(1 + \frac{\rho'}{\rho_0} \right) \frac{\bar{v}}{c} \right] \right\} = 0 \qquad (4.2.3)$$

where the restriction to conditions in which the ambient state is homogeneous allows us to set $\partial \rho_0 / \partial t = 0$. The only nonlinear term is $(\rho'/\rho_0)\,(\bar{v}/c)$. This term may be dropped because we have taken $|\rho'|\,/\rho_0 \ll 1$. The result is the linearized continuity equation,

$$\frac{\partial \rho'}{\partial t} + \rho_0 \nabla \cdot \bar{v} = 0 \qquad (4.2.4)$$

A similar analysis may be applied to the momentum equation. Because $\nabla p_0 \equiv 0$ for a homogeneous ambient state, the total pressure may be replaced by the acoustic pressure perturbation. Multiplication and division of each variable in Eq. (4.1.12) by its reference quantity give

$$\rho_0 \omega c \left(1 + \frac{\rho'}{\rho_0}\right) \left[\frac{\partial}{\partial (\omega t)}\left(\frac{\bar{v}}{c}\right) + \frac{\bar{v}}{c} \cdot \frac{\nabla}{k}\left(\frac{\bar{v}}{c}\right)\right] = -\rho_0 \omega c \frac{\nabla}{k}\left(\frac{p}{\rho_0 c^2}\right) \qquad (4.2.5)$$

The second term inside the brackets is quadratic in $\bar{v}/c$. Because $|\bar{v}|\,/c \ll 1$, it is small compared to the first term, which is linear in $\bar{v}/c$. Furthermore, multiplication of the small bracketed term by ρ'/ρ_0 will lead to a product of small quantities, so ρ'/ρ_0 is dropped from the term in parentheses. What remains is the linearized momentum equation,

$$\rho_0 \frac{\partial \bar{v}}{\partial t} = -\nabla p \qquad (4.2.6)$$

This is *Euler's equation* in the time domain. Its primary use will be to formulate the boundary condition that an acoustic pressure field $p\left(\bar{\xi}, t\right)$ must satisfy when it encounters a solid boundary.

The basic equations have been linearized, so we proceed to the derivation of a field equation in which the acoustic pressure is the sole independent variable. The equation of state is used to eliminate the density perturbation from the linearized continuity equation, which becomes

$$\frac{1}{c^2}\frac{\partial p}{\partial t} + \rho_0 \nabla \cdot \bar{v} = 0 \qquad (4.2.7)$$

The linearized momentum equation contains a time derivative of $\bar{v}$, whereas the continuity equation features the divergence of $\bar{v}$. Thus, we take the divergence of the former equation and the time derivative of the latter, which leads to

$$\rho_0 \nabla \cdot \left(\frac{\partial \bar{v}}{\partial t}\right) = -\nabla \cdot (\nabla p)$$
$$\frac{1}{c^2}\frac{\partial^2 p}{\partial t^2} + \rho_0 \frac{\partial}{\partial t}(\nabla \cdot \bar{v}) = 0 \qquad (4.2.8)$$

The order in which the divergence and time derivatives are taken has no affect on the result, and the operator $\nabla \cdot \nabla$ is defined to be the Laplacian operator ∇^2. Consequently, subtraction of the second of the preceding equations from the first leaves

$$\nabla^2 p - \frac{1}{c^2}\frac{\partial^2 p}{\partial t^2} = 0 \qquad (4.2.9)$$

This is the general version of the *wave equation*. In the special case where p depends on only one Cartesian coordinate, which may be taken to be x, then $\nabla^2 p = \partial^2 p/\partial x^2$, so the general wave equation reduces to the one-dimensional version.

The overall approach for any system will be much like that which we followed for planar waves. We will seek a general solution of the wave equation then fit it to the boundary conditions. The difference is that now the solution might be posed in terms of any set of spatial coordinates, and it might depend on more than one of those coordinates. The choice to use Cartesian, spherical, or cylindrical coordinates will usually be set by which description best fits the shape of the boundary. For example, suppose we know the velocity on the surface of a sphere of radius a. Satisfying a boundary condition that says that the particle velocity must match the specified value at $r = a$, where r is the radial distance in a spherical coordinate system, is much more conducive to analysis than matching the velocity at $\left(x^2 + y^2 + z^2\right)^{1/2} = a$.

The frequency-domain equations may be obtained as a special case of those for an arbitrary time dependence. The derivation begins with the representation of pressure and particle velocity as complex exponentials at frequency ω,

$$\begin{aligned} p\left(\bar{\xi}, t\right) &= \mathrm{Re}\left(P\left(\bar{\xi}\right)e^{i\omega t}\right)\\ \bar{v}\left(\bar{\xi}, t\right) &= \mathrm{Re}\left(\bar{V}\left(\bar{\xi}\right)e^{i\omega t}\right) \end{aligned} \qquad (4.2.10)$$

The complex pressure amplitude P is a scalar function of position. In contrast, the complex velocity amplitude is a vector function, which merely means that each of its (scalar) components may be a function of position. Differentiation with respect to time multiplies a complex amplitude by a factor $i\omega$, while a gradient acts on the position dependent complex amplitude. Thus, the wave equation reduces to the general form of the Helmholtz equation,

$$\nabla^2 P + k^2 P = 0 \qquad (4.2.11)$$

A similar treatment of the continuity equation, Eq. (4.2.7), leads to

$$P = i\frac{\rho_0 c}{k}\nabla \cdot \bar{V} \qquad (4.2.12)$$

and Euler's equation in the frequency domain reduces to

$$i\omega\rho_0 V = -\nabla P \qquad (4.2.13)$$

The former relation is most useful if we know the velocity field and wish to determine the pressure, whereas the latter is usually used to derive a boundary condition for the pressure at a boundary whose velocity is known. If we were to take the divergence, $\nabla\cdot$, of Euler's equation and use that result to eliminate $\nabla\cdot\bar{V}$ from the continuity equation, we would obtain $P = -\nabla^2 P/k^2$. In other words, we would derive the Helmholtz equation without the intermediate steps that led to the time-domain wave equation.

4.3 Plane Waves in Three Dimensions

A key feature of field equations that are linear is the ability to use the principle of superposition to adapt known solutions to describe a different configuration. The solution to several significant problems consists of a superposition of simple plane waves that propagate in directions that are not aligned with a coordinate axis. Let us consider one such wave. Figure 4.3 shows a unit vector $\bar{e}$ that points in the direction of propagation.

The position $\bar{\xi}$ is an arbitrary field point, whereas $\bar{\xi}_0$ is the reference position at which we know the pressure waveform, that is, we know that $p\left(\bar{\xi}_0, t\right) = F\left(t\right)$. The definition of a plane wave that propagates in the direction $\bar{e}$ is that at any instant t the same pressure is observed on any plane whose normal is $\bar{e}$. These planes constitute *wavefronts*. The corollary of this property is that the distance from $\bar{\xi}_0$ to $\bar{\xi}$ measured parallel to $\bar{e}$ is the propagation distance from the reference location; this is the distance s in Fig. 4.3. It is the difference of the projections of $\bar{\xi}$ and $\bar{\xi}_0$ onto $\bar{e}$, which can be computed as a scalar product, so we have

$$s = \bar{\xi}\cdot\bar{e} - \bar{\xi}_0\cdot\bar{e} \equiv \left(\bar{\xi} - \bar{\xi}_0\right)\cdot\bar{e} \tag{4.3.1}$$

Figure 4.3 describes $\bar{e}$ in terms of its projections onto the coordinate axes. These are its components parallel to the unit vectors $\bar{e}_x$, $\bar{e}_y$, and $\bar{e}_z$ pointing outward from

Fig. 4.3 Direction and position vectors for describing a simple plane wave

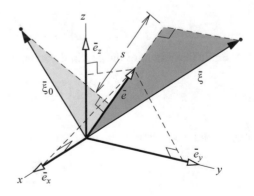

the origin along each coordinate axis. Thus, we can describe $\bar{e}$ in component form as

$$\bar{e} = (\bar{e} \cdot \bar{e}_x) \, \bar{e}_x + (\bar{e} \cdot \bar{e}_y) \, \bar{e}_y + (\bar{e} \cdot \bar{e}_z) \, \bar{e}_z \qquad (4.3.2)$$

The dot product of two unit vectors is the cosine of the angle between them. Let γ_x, γ_y, and γ_z be the angles between $\bar{e}$ and the coordinate axes. Then the propagation direction may be described as

$$\bar{e} = \cos(\gamma_x) \, \bar{e}_x + \cos(\gamma_y) \, \bar{e}_y + \cos(\gamma_z) \, \bar{e}_z \qquad (4.3.3)$$

Because $\bar{e}$ is a unit vector, it must be that

$$\cos(\gamma_x)^2 + \cos(\gamma_y)^2 + \cos(\gamma_z)^2 = 1 \qquad (4.3.4)$$

The descriptions of the propagation distance and the direction of propagation are needed to describe a plane wave in both the time domain and frequency domain. We begin by considering the former.

4.3.1 Simple Plane Wave in the Time Domain

The d'Alembert solution for a one-dimensional plane wave propagating in the direction of increasing x depends on the combination $t - x/c$, so a wave propagating in an arbitrary direction in space must depend only on $t - s/c$. The arbitrary function $F(t)$ is the time dependence of the pressure at the reference location, so the d'Alembert solution tells us that

$$\boxed{p\left(\bar{\xi}, t\right) = \rho_0 c v\left(\bar{\xi}, t\right) = F\left(t - \frac{\left(\bar{\xi} - \bar{\xi}_0\right) \cdot \bar{e}}{c}\right)} \qquad (4.3.5)$$

An alternative description uses the *phase variable* θ, which is the argument of the function,

$$\Theta = t - \frac{\left(\bar{\xi} - \bar{\xi}_0\right) \cdot \bar{e}}{c}, \quad p\left(\bar{\xi}, t\right) = \rho_0 c v\left(\bar{\xi}, t\right) = F(\Theta) \qquad (4.3.6)$$

The value of this form is that it emphasizes that if a specific phase is observed at $\bar{\xi}_0$ at time t', it will be observed at a selected field point $\bar{\xi}$ when $t = t' + \left(\bar{\xi} - \bar{\xi}_0\right) \cdot \bar{e}$. Because $\bar{\xi} = \bar{\xi}_0$ when a specific phase passes the reference location, this representation leads to the alternative interpretation of $\bar{\xi}_0$ as the location of the zero-phase signal when $t = 0$.

That Eq. (4.3.5) is a solution of the wave equation is apparent from the fact that we could define a coordinate system whose x-axis is parallel to $\bar{e}$, with origin at the point where $\bar{\xi}_0$ projects onto $\bar{e}$. However, as practice in working with the wave equation, let us verify explicitly that the above form is a solution. One approach explicitly describes the argument of F in terms of the coordinates and time. Thus, we write

$$p = F(\Theta)$$

$$\Theta = t - \frac{(\bar{\xi} - \bar{\xi}_0) \cdot \bar{e}}{c}$$

$$= t - \frac{(x - x_0)\cos\gamma_x + (y - y_0)\cos\gamma_y + (z - z_0)\cos\gamma_z}{c} \qquad (4.3.7)$$

The Laplacian operator in Cartesian coordinates is $\nabla^2 = \partial^2/\partial x^2 + \partial^2/\partial y^2 + \partial^2/\partial z^2$. The derivatives are with respect to the coordinates of $\bar{\xi}$, with the coordinates of $\bar{\xi}_0$ held constant. Because Θ is the only variable that explicitly depends on x, y, and z, this calls for the chain rule. For example,

$$\frac{\partial p}{\partial x} = \frac{\partial \Theta}{\partial x} F'(\Theta) = -\frac{\cos(\gamma_x)}{c} F'(\Theta)$$

$$\frac{\partial^2 p}{\partial x^2} = \frac{\partial}{\partial x}\left(\frac{\partial p}{\partial x}\right) = \frac{\partial \Theta}{\partial x}\left[-\frac{\cos(\gamma_x)}{c} F''(\Theta)\right] = \left(\frac{\cos(\gamma_x)}{c}\right)^2 F''(\Theta) \qquad (4.3.8)$$

As always, a prime denotes differentiation of a function with respect to its argument, so $F'(\Theta)$ is shorthand for $dF/d\Theta$. Because t appears by itself in Θ, we have $\partial^2 p/\partial t^2 = F''(\Theta)$. Hence, substitution of Eq. (4.3.7) into the wave equation reduces it to

$$\frac{\partial^2 p}{\partial x^2} + \frac{\partial^2 p}{\partial y^2} + \frac{\partial^2 p}{\partial z^2} - \frac{1}{c^2}\frac{\partial^2 p}{\partial t^2} = \left[\left(\frac{\cos(\gamma_x)}{c}\right)^2 + \left(\frac{\cos(\gamma_y)}{c}\right)^2\right.$$

$$\left. + \left(\frac{\cos(\gamma_z)}{c}\right)^2 - \frac{1}{c^2}\right] F''(\Theta) \equiv 0 \qquad (4.3.9)$$

where the result is identically zero as a consequence of Eq. (4.3.4).

An alternative proof that Eq. (4.3.5) satisfies the wave equation follows a vectorial approach. We shall pursue it here because it introduces operations that will simplify the study of more complicated problems. Let us leave Θ in its vectorial form in Eq. (4.3.7). The gradient of p is found by the chain rule to be

$$\nabla p = (\nabla\Theta) F'(\Theta) = -\nabla\left(\frac{(\bar{\xi} - \bar{\xi}_0) \cdot \bar{e}}{c}\right) F'(\Theta) \qquad (4.3.10)$$

All variables other than $\bar{\xi}$ are constant, so $\nabla(\bar{\xi}_0 \cdot \bar{e}) \equiv 0$. The other gradient term may be evaluated in component form according to

$$\nabla \left(\bar{\xi} \cdot \bar{e} \right) = \left(\frac{\partial}{\partial x} \bar{e}_x + \frac{\partial}{\partial y} \bar{e}_y + \frac{\partial}{\partial z} \bar{e}_z \right) \left(x \cos \gamma_x + y \cos \gamma_y + z \cos \gamma_z \right)$$
$$= \left(\cos \gamma_x \right) \bar{e}_x + \left(\cos \gamma_y \right) \bar{e}_y + \left(\cos \gamma_z \right) \bar{e}_z \qquad (4.3.11)$$

$$\boxed{\nabla \left(\bar{\xi} \cdot \bar{e} \right) = \bar{e}, \quad \nabla \Theta = -\frac{1}{c} \bar{e}}$$

It follows from this identity that

$$\nabla p = -\frac{1}{c} F' \left(\Theta \right) \bar{e}$$

$$(4.3.12)$$

$$\nabla^2 p \equiv \nabla \cdot \nabla p = -\frac{1}{c} \left(\nabla \Theta \right) F'' \left(\Theta \right) \cdot \bar{e} = \frac{1}{c^2} F'' \left(\Theta \right) \bar{e} \cdot \bar{e} = \frac{1}{c^2} F'' \left(\Theta \right)$$

The time derivative term in the wave equation was already found to be $\partial^2 p / \partial t^2 = F'' \left(\Theta \right)$. Substitution of this expression and the preceding for $\nabla^2 p$ makes it evident that $p = F \left(\Theta \right)$ satisfies the wave equation.

The identity in Eq. (4.3.11) will arise in several contexts, so let us consider it from a different perspective. In general, the gradient of any scalar σ is oriented in the direction of most rapid change of σ when the location changes. Because $\bar{\xi} \cdot \bar{e}$ is the distance s measured along $\bar{e}$ from the origin, the change of position that leads to the greatest change of s is a shift parallel to $\bar{e}$.

EXAMPLE 4.2 A plane wave propagating in the atmosphere in the horizontal plane at 30° north of east is measured by a string of pressure sensors running along a line from west to east, which is defined to be the x-axis. The y-axis is northward. A comparison of sensor readings taken at $t = 0$ indicates that the pressure distribution along the x-axis is a single cycle of a sine function beginning at $x = -2$ m and ending at $x = 8$ m. The amplitude of this sine function is 200 Pa. Determine the waveform observed at the location $x = 20$ m, $y = 30$ m.

Significance

This example is the conceptual extension of the evaluation of spatial profiles and waveforms for waves that propagate parallel to a coordinate axis. Several aspects of the vectorial description of the phase variable will be exploited to perform this change of perspective.

Solution

We begin by using the given information to construct a general description of the wave. Specifically, we know the direction of propagation is a unit vector at 30° northward from the positive x-axis, so that

$$\bar{e} = \cos \left(30° \right) \bar{e}_x + \sin \left(30° \right) \bar{e}_y = 0.866 \bar{e}_x + 0.5 \bar{e}_y$$

A reference location is not specified, so we may take it to be the origin, $\bar{\xi}_0 = \bar{0}$. The corresponding description of the pressure field is

$$p = F(\Theta), \quad \Theta = t - \frac{0.866x + 0.5y}{c} \tag{1}$$

The function F is determined by matching the general description to the observed profile, which we will denote as $f(x)$. The measured initial pressure dependence along the x-axis is

$$p(x, y = 0, t = 0) \equiv f(x) = 200 \sin\left(\frac{2\pi}{10}(x+2)\right)[h(x+2) - h(x-8)] \text{ Pa} \tag{2}$$

The factor $2\pi/10$ describes the fact that a cycle occupies $10\,\text{m}$, the factor $x + 2$ corresponds to the sine function starting at $x = -2\,\text{m}$, and the combination of step functions zeros out the function for $x < -2$ and $x > 8$.

The description of p in Eq. (2) must be consistent with the general form in Eq. (1) evaluated along $y = 0$ when $t = 0$. The corresponding phase variable is $\Theta(x,y=0, t = 0) = -0.866x/c$. We use this attribute to replace x in Eq. (2),

$$F\left(\Theta|_{y=0,t=0}\right) = f\left(-1.1547c\Theta|_{y=0,t=0}\right)$$

The description in Eq. (1) states that p everywhere is solely a function of Θ. This means that preceding expression is valid for any Θ. In other words

$$p = F(\Theta) = f(-1.1547c\Theta) \tag{3}$$

where the $f()$ function is defined in Eq. (2).

The specified field point is $x = 20\,\text{m}$, $y = 30\,\text{m}$. The phase variable at this location is

$$\Theta|_{x=20,y=30} = t - \frac{32.32}{c} \tag{4}$$

To evaluate the waveform at the field point, we substitute this phase variable into Eq. (3) to obtain

$$p(x = 20, y = 30, t) = f\left(-1.1547c\left(\Theta|_{x=20,y=30}\right)\right) = f(37.32 - 1.1547ct) \tag{5}$$

This result is evaluated over a time interval by replacing x in Eq. (2) with the argument in Eq. (5).

Evaluation of Eq. (5) must be carried out over the full duration in which the waveform is nonzero.range of t, then making adjustments if a full cycle of a sine is not seen. The interval also can be identified mathematically by examining the combination of step functions in Eq. (2). For the function argument in Eq. (3), this factor is $h(-1.1547c\Theta + 2) - h(-1.1547c\Theta - 8)$. Thus, p is nonzero in the interval

$$-8 < 1.1547c\Theta < 2$$

Substitution of the phase variable in Eq. (4) gives

$$-8 < 1.1547c \left(t - \frac{32.321}{c} \right) < 2 \implies 25.392 < ct < 34.052$$

A third method for selecting the interval uses physical reasoning. We know that at $t = 0$ the pressure extends along the x-axis from $x = -2$ to $x = 8$. The distance in the direction of propagation for points on the x-axis is $s = x\bar{e}_x \cdot \bar{e}$, where $\bar{e}$ is described in Eq. (1). Thus, the pulse initially is situated in the range $-1.732 < s < 6.928$ m. The field point is situated at $s = (20\bar{e}_x + 30\bar{e}_y) \cdot \bar{e} = 32.321$ m. The time the pulse arrives at the field point is the time for the leading edge to travel from $s = 6.928$ m to $s = 32.321$ m, and the pulse ends when the trailing edge travels from $s = -1.732$ to $s = 32.321$ m. Thus, the pressure is nonzero for $32.321 - 9.629 < ct < 32.321 + 1.732$. This is the same as the interval identified mathematically.

The pulse received at the field point $x = 20$ m, $y = 30$ m corresponding to $c = 340$ m/s is graphed in Fig. 1. Note that the waveform starts in the negative phase. This is because the initial profile features a negative pressure at the farthest x, so this phase is the first to arrive at the designated field point. If the waveform continued, the frequency would be the coefficient of t in the representation of p obtained from Eqs. (2) and (3), $\omega = (2\pi/10)(1.1547c)$. The corresponding period is $2\pi/\omega = 25.5$ ms, which matches the interval in the graph during which p is nonzero

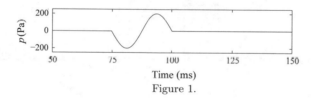

Figure 1.

4.3.2 Trace Velocity

Suppose there is a simple planar wave propagating in direction $\bar{e}$ that has the property that the pressure is zero ahead of a wavefront. This is said to be the *leading wavefront* or *leading edge*. It is convenient to define $t = 0$ as the instant when the leading edge passes $\bar{\xi}_0$. The waveform observed at the reference location $\bar{\xi}_0$ may be represented as

$$p\left(\bar{\xi}_0, t\right) = f(t) h(t) \tag{4.3.13}$$

where $f(t)$ is an arbitrary function and $h(t)$ is the step function, as always. Because Eq. (4.3.7) gives $\Theta = t$ at $\bar{\xi}_0$, it must be that

$$p\left(\bar{\xi}, t\right) = f(\Theta) h(\Theta) \tag{4.3.14}$$

Consider what is observed at the four positions depicted in Fig. 4.4: the reference position, the origin A, position B on the x-axis, and position C in the y-axis. (For the sake of clarity in depicting the signal, $\bar{e}$ is taken to lie in the xy plane.) A wavefront containing each point is depicted in the figure. There are two ways in which the picture may be interpreted. One is that it shows several wavefronts contemporaneously at a specific instant t. In that interpretation each wavefront corresponds to a different value of the phase, specifically

$$\Theta_0 = t, \quad \Theta_A = t + \frac{\bar{\xi}_0 \cdot \bar{e}}{c} = \Theta_0 + \frac{\bar{\xi}_0 \cdot \bar{e}}{c}$$

$$\Theta_B = t - \frac{\left(\xi_B - \bar{\xi}_0\right) \cdot \bar{e}}{c} = \Theta_A - \frac{s_B \sin\psi}{c} \qquad (4.3.15)$$

$$\Theta_C = t - \frac{\left(\xi_C - \bar{\xi}_0\right) \cdot \bar{e}}{c} = \Theta_A - \frac{s_C \cos\psi}{c}$$

Because the value of t is the same for all wavefronts, this viewpoint is equivalent to the way we construct a spatial profile.

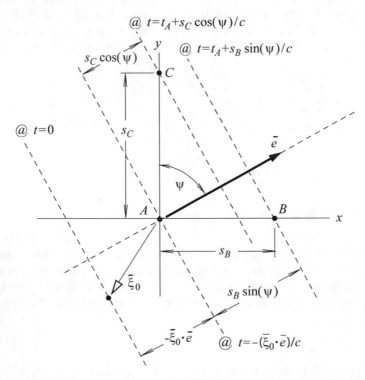

Fig. 4.4 Passage of a the $\Theta = 0$ wavefront of a simple plane wave past various observation points. The dashed lines mark the position of the wavefront at various instants. The pressure ahead of the wavefront is zero

A different view of Fig. 4.4 is that it marks the position of a wavefront of constant phase at a succession of instants. A phase of particular interest is $\Theta = 0$, because it separates the region that has not yet seen the signal from the region that has been insonified. Setting $\Theta = 0$ leads to the times being $t_0 = 0$, $t_A = -\bar{\xi}_0 \cdot \bar{e}/c$, $t_B = (\bar{\xi}_B - \bar{\xi}_0) \cdot \bar{e}/c = t_A + (s_B \sin \psi)/c$, and $t_C = (\bar{\xi}_C - \bar{\xi}_0) \cdot \bar{e}/c = t_A + s_C (\cos \psi)/c$. Note that $\bar{\xi}_0 \cdot \bar{e}$ is negative in the figure, so the timings are a sequence of increasing t. These instants are consistent with the requirement that the time required for the wavefront to travel from the origin to points B or C is the distance in the direction of propagation divided by the speed of sound. Now consider an experiment in which microphones are placed at each location and connected to a data acquisition system at a central receiving station. If we ignore the extremely small difference in time required for an electronic signal to propagate, we may consider each signal to be measured in real time. Thus, at time t_A, the signal from microphone A begins to grow, but the others are quiet. Then at $t_B = t_A + s_B \sin(\psi)/c$ and $t_C = t_A + s_C \cos(\psi)$ the signals from microphones B and C begin to grow. If we divide the distance from the origin to points B and C by the respective travel times, we obtain apparent propagation speeds of $c/\sin(\psi)$ from A to C and $c/\cos(\psi)$ from A to C, both of which are greater than the speed of sound. How this observation is interpreted depends on what the observer knows about acoustics. The ancients might think that sound travels much faster than c. On the other hand, we know that this observation is merely the geometric consequence of the leading wavefront traveling obliquely to a coordinate axis. The apparent propagation velocity along these lines is called the *trace velocity* c_{tr}. The fact that it is greater than c has no physical significance because it does not imply that the disturbance actually travels at that rate. In fact, if the wave travels parallel to the y-axis, so that $\bar{e} = \bar{e}_y$ and $\psi = 0$, the trace velocity along the x-axis is infinite, which is merely a different way of saying that the signal arrives simultaneously at all points on its leading wavefront.

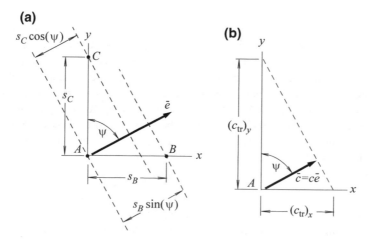

Fig. 4.5 Geometric constructions associated with propagation of a plane wave. **a** Propagation distance for points on the coordinate axes. **b** Trace velocity along the coordinate axes

The trace velocity figures prominently whenever we consider how waves are modified by incidence on a boundary. There tends to be confusion between the projections that are required to determine the propagation time and the trace velocity. The propagation time is evaluated by projecting the position vector onto the direction of propagation, as in Fig. 4.5a, then dividing that distance by the speed of sound. The trace velocity is constructed in Fig. 4.5b by creating a sound velocity vector $c\bar{e}$ in the direction of propagation, then seeking the location on each axis at which the line perpendicular to that vector intersects each coordinate axis. The trace velocities are denoted as $(c_{tr})_x$ and $(c_{tr})_y$ to emphasize that trace velocity is associated with directions, not points.

EXAMPLE 4.3 Hydrophones just below the surface of a large body of water record sound signals. Coordinate axes locating them are x positive eastward and y positive northward. The (x, y) coordinates of three such devices are $A : (-50, 0)$, $B : (0, 80)$, and $C : (-20, -60)$ m. The times at which a plane wave pulse is received at each location are $t_A = 14.934$ ms, $t_B = 4.778$ ms, $t_C = -20.182$ ms. Determine (a) the direction in which this wave is propagating, (b) the speed of sound, (c) the time at which the wave will arrive at the origin, (d) the time at which it will arrive at point D, whose coordinates are $(-200, 100)$ m, and (e) the trace velocities of the wave in the east-west and north-south directions.

Significance

A by-product of the examination of the relationship between arrival time and trace velocity will be better understanding of how to track waves that propagate in an arbitrary direction.

Solution

We could attempt a solution that uses diagrams of wavefronts intersecting each observation point. However, an analysis based on the phase variable will be easier to implement, and it is readily extended to cases where the measurement locations are not situated in a common plane. The given data describes when the sound begins, so it tracks the leading edge of the signal. It is not necessary to do so, but we will say that the phase at the leading edge is $\Theta = 0$. The reference location $\bar{\xi}_0$ is not specified, so we will leave $\bar{\xi}_0 \cdot \bar{e}$, which is a scalar, as an unknown. Thus, the propagation of the zero-phase signal is described by

$$c\Theta = 0 = ct - \bar{\xi} \cdot \bar{e} + \bar{\xi}_0 \cdot \bar{e}$$

The propagation direction $\bar{e}$ may be defined by the angle ψ measured counterclockwise from the eastward x direction, $\bar{e} = \cos(\psi)\,\bar{e}_x + \sin(\psi)\,\bar{e}_y$. Then, the time and position coordinates for which $\Theta = 0$ are related by

$$ct - x\cos\psi - y\sin\psi + \bar{\xi}_0 \cdot \bar{e} = 0 \qquad (1)$$

The unknowns to be determined are c, ψ, and $\bar{\xi}_0 \cdot \bar{e}$. (There is no need to determine the actual value of $\bar{\xi}_0$ because the only concern is propagation between the given locations.)

Let (x_α, y_α), $\alpha = A$, B, or C denote the coordinates of each hydrophone, and let t_α be the arrival time at hydrophone α. Then Eq. (1) states that

$$
\begin{aligned}
x_A \cos (\psi) + y_A \sin (\psi) &= ct_A + \bar{\xi}_0 \cdot \bar{e} \\
x_B \cos (\psi) + y_B \sin (\psi) &= ct_B + \bar{\xi}_0 \cdot \bar{e} \\
x_C \cos (\psi) + y_C \sin (\psi) &= ct_C + \bar{\xi}_0 \cdot \bar{e}
\end{aligned}
\tag{2}
$$

An algebraic solution is possible. The first step is to eliminate $\bar{\xi}_0 \cdot \bar{e}$ by forming the differences of the equations. This leads to

$$
\begin{aligned}
(x_B - x_A) \cos (\psi) + (y_B - y_A) \sin (\psi) &= c\,(t_B - t_A) \\
(x_C - x_A) \cos (\psi) + (y_C - y_A) \sin (\psi) &= c\,(t_C - t_A)
\end{aligned}
\tag{3}
$$

Division of each equation by the respective elapsed time yields a pair of equations for c. Equating them leads to

$$
\frac{(x_B - x_A) \cos (\psi) + (y_B - y_A) \sin (\psi)}{t_B - t_A} = \frac{(x_C - x_A) \cos (\psi) + (y_C - y_A) \sin (\psi)}{t_C - t_A}
$$

Rearrangement of the terms leads to

$$
\left[\frac{y_B - y_A}{t_B - t_A} - \frac{y_C - y_A}{t_C - t_A} \right] \sin \psi = \left[\frac{x_C - x_A}{t_C - t_A} - \frac{x_B - x_A}{t_B - t_A} \right] \cos \psi
\tag{4}
$$

For the given values of the coordinates and arrival times, the angle obtained from the arctangent function is $\psi = -23.000°$. The sound speed obtained from both of Eq. (3) is $c = -1454\,\text{m/s}$. However, $\psi = 157°$, which is $180°$ greater than the first value, also satisfies Eq. (4). It leads to $c = +1453$ m/s. Both alternatives are the same physical result. This is so because the two propagation directions $\bar{e}$ are reversed, which cancel the opposite signs for c. We shall use the second solution, so we have found that

$$
\psi = 157°, \quad c = 1454 \text{ m/s}
$$

Substitution of these results converts each of Eq. (2) to an equation for $\bar{\xi}_0 \cdot \bar{e}$. The value obtained from each equation should be the same. It is

$$
\bar{\xi}_0 \cdot \bar{e} = 24.31 \text{ m}
$$

According to Eq. (1), the arrival time at the origin is

$$
t_0 = -\frac{\bar{\xi}_0 \cdot \bar{e}}{c} = -16.72 \text{ ms}
$$

The arrival time at point D is found by substitution of the computed values of $\bar{\xi}_0 \cdot \bar{e}$ and c, and the point's coordinates into Eq. (1),

$$t_D = \frac{-\bar{\xi}_0 \cdot \bar{e} + x_D \cos\psi + y_D \sin\psi}{c} = 136.77 \text{ ms}$$

The trace velocities may be evaluated as the speed required to travel from the origin to points A and B, which are on the respective axes, in the elapsed times $t_A - t_0$ and $t_B - t_0$. The alternative is to use the property that $c\bar{e}$ is the projection of the trace velocity along each axis, as depicted in Fig. 4.5. The latter leads to

$$(c_{\text{tr}})_x = \frac{c}{\cos(\psi)} = -1580 \text{ m/s}$$

$$(c_{\text{tr}})_y = \frac{c}{\sin(\psi)} = 3721 \text{ m/s}$$

where the negative sign for $(c_{\text{tr}})_x$ indicates that the wave is propagating backward relative to the x-axis.

4.3.3 Simple Plane Wave in the Frequency Domain

As was true for plane waves that propagate parallel to a coordinate axis, a frequency-domain description substantially expedites analysis of a plane wave that propagates in an arbitrary direction. The complex amplitude of the pressure at the reference location $\bar{\xi}_0$ is designated as B. The generalized d'Alembert solution, Eq. (4.3.5) still applies, but the function must be harmonic at frequency ω, so it must be that

$$\begin{aligned} p &= \text{Re}\left[B \exp\left(i\omega\left(t - \frac{(\bar{\xi} - \bar{\xi}_0) \cdot \bar{e}}{c}\right)\right)\right] \\ &= \text{Re}\left[B \exp\left(i\left(\omega t - k\left(\bar{\xi} - \bar{\xi}_0\right) \cdot \bar{e}\right)\right)\right] \end{aligned} \qquad (4.3.16)$$

When a signal consists of a superposition of many plane waves having different frequencies that propagate in a variety of directions, it is useful to associate the wavenumber with the propagation direction of each wave. This is done by defining a *wavenumber vector*, which is

$$\bar{k} = k\bar{e} \qquad (4.3.17)$$

In terms of this parameter, the complex amplitude of a single harmonic plane wave

$$P = Be^{-i\bar{k}\cdot(\bar{\xi} - \bar{\xi}_0)} \qquad (4.3.18)$$

In most cases, we can define the origin of the coordinate system to be the reference position, so we set $\bar{\xi}_0 = \bar{0}$. Then the frequency domain description reduces to the form we usually employ,

$$\boxed{P = Be^{-i\bar{k}\cdot\bar{\xi}}} \tag{4.3.19}$$

Let us consider what Euler's equation tells us. With the aid of Eq. (4.3.11), we find that

$$\nabla P = Be^{-i\bar{k}\cdot\bar{x}}\nabla\left(-i\bar{k}\cdot\bar{\xi}\right) = -ik\nabla\left(\bar{e}\cdot\bar{\xi}\right)Be^{-i\bar{k}\cdot\bar{\xi}} = -ik\bar{e}P \tag{4.3.20}$$

Substitution of this relation into Eq. (4.2.13) leads to

$$\bar{V} = \frac{P}{\rho_0 c}\bar{e} \tag{4.3.21}$$

This is merely the vectorial generalization of the one-dimensional relation. It states that the particle velocity in a simple plane wave is in the propagation direction, with a magnitude that is the pressure divided by the characteristic impedance of the fluid.

Another use of Eq. (4.3.11) is to confirm that Eq. (4.3.19) satisfies the Helmholtz equation. Using that relation and Eq. (4.3.20) leads to

$$\begin{aligned}\nabla^2 P + k^2 &= \nabla \cdot \nabla P + k^2 P = \nabla \cdot (-ik\bar{e}P) + k^2 P \\ &= -ik\bar{e} \cdot \nabla P + k^2 P = 0\end{aligned} \tag{4.3.22}$$

The frequency-domain equivalent of the trace velocity results from an examination of the pressure in a specific direction. For points on the x-axis, the position is $\bar{\xi} = x\bar{e}_x$, so that

$$P(x) = Be^{-ikx\bar{e}\cdot\bar{e}_x} \tag{4.3.23}$$

For the arrangement in Fig. 4.4, $e \cdot \bar{e}_x = \sin(\psi)$, so that the wavenumber along the x-axis is $k \sin\psi$. If this was the true wavenumber, the corresponding phase velocity would be $\omega/(k \sin\psi) \equiv c/\sin\psi$, parallel to the x-axis. This is the trace velocity along the x-axis. Thus, the trace velocity is manifested in the frequency domain as a wavenumber that is a component of the wavenumber vector. The generalization describes the pressure along a line in an arbitrary direction $\bar{e}_s$. Let s be distance from the origin along this line, such that $\bar{\xi} = s\bar{e}_s$. Then the pressure dependence and trace wavenumber along this line is

$$\boxed{P(s) = Be^{-ik_{tr}s}, \quad k_{tr} = \bar{k}\cdot\bar{e}_s} \tag{4.3.24}$$

This expression states that the trace wavenumber in direction $\bar{e}_s$ of a plane wave is the component of the wavenumber parallel to $\bar{e}_s$. Some individuals confuse this construction with that of the trace velocity along $\bar{e}_s$ because the propagation speed c is the component of the trace velocity parallel to $\bar{e}$.

EXAMPLE 4.4 A simplified model of the sound radiated by a circular piston in a wall considers the pressure to be a plane wave in the vicinity of the piston. This plane wave is confined to a cylinder that extends outward from the piston face, which is the shaded area in the sketch, with zero pressure outside the cylindrical region. (This model will be examined in more detail in Example 4.7.) In the present situation, both pistons are driven at the same frequency ω. The pressure amplitude at the face of piston A is 20 kPa and the signal at that location is a pure cosine. The pressure amplitude at piston B is 10 kPa, but the phase angle at that location is not known. It is desired that the pressure amplitude at the intersection C of the centerline of each field be 24 kPa. Determine the phase lag or lead for the pressure at piston B required to attain this condition. Then determine the corresponding particle velocity at the intersection. The fluid is water and the frequency is 10 kHz.

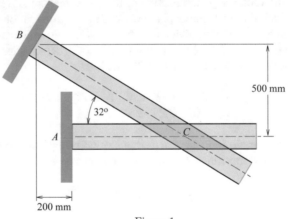

Figure 1.

Significance

The phenomenon to be explored is the interference of two plane waves at the same frequency. Two aspects that are highlighted are the dependence of phase angle on the propagation distance and the vectorial nature of the particle velocity.

Solution

The field is the superposition of the contribution of each plane wave. The waves have the same frequency, so the contributions add coherently. If the plane waves are in-phase at point C, then the pressure amplitude there would be 30 kPa, while it would be 10 kPa if the waves are 180° out-of-phase. It follows that we must ascertain the phase of each signal at point C.

A coordinate system serves to locate points. The center of piston A is a convenient origin, and defining the x-axis to be the centerline for piston A is consistent with the way the configuration is described. The positions of the significant points are

$$\bar{\xi}_A = \bar{0}, \quad \bar{\xi}_B = -0.2\bar{e}_x + 0.5\bar{e}_y \text{ m}$$
$$\bar{\xi}_C = [0.5\cot(32°) - 0.2]\,\bar{e}_x = 0.6002\bar{e}_x \text{ m}$$

The complex pressure amplitude for piston A is a pure cosine at the piston face, and its propagation direction is $\bar{e}_x$, so we know that

$$P_A = 20000e^{-ikx} \text{ Pa}$$

We will use $c = 1480$ m/s, so that $k = 2\pi(10000)/1480 = 42.454$ m^{-1}. Thus, the complex amplitude at point C due to the signal radiated by piston A is

$$P_A\left(\bar{\xi}_C\right) = 20000e^{-ik(0.6002)} = 18809.8 - 6796.5i \tag{1}$$

It is given that $|P_B| = 10000$ Pa at the piston, but we do not know the phase lag at that location. Therefore, we only know that

$$P_B\left(\bar{\xi}_B\right) = 10000e^{i\Delta}, \quad \Delta = k\bar{\xi}_B \cdot \bar{e}_B$$

The propagation direction is $\bar{e}_B = \cos(32°)\,\bar{e}_x - \sin(32°)\,\bar{e}_y$. The complex pressure amplitude at point C corresponding to these parameters is

$$P_B\left(\bar{\xi}_C\right) = 10000Be^{-ik(\bar{\xi}_C - \bar{\xi}_B)\cdot\bar{e}_B} = 10000e^{i\Delta}e^{-ik(0.8002\bar{e}_x - 0.50\bar{e}_y)\cdot\bar{e}_B}$$
$$= e^{i\Delta}(-7083.1 - 7059.1i) \tag{2}$$

Hence, the total pressure at $\bar{\xi}_C$ is

$$P = (-7083.1 - 7059.1i)e^{i\Delta} + (18809.8 - 6796.5i) \tag{3}$$

We must find the values of Δ for which $|P| = 24000$. We could evaluate $|P|$ for a 2π range of closely spaced Δ and compare the values to 24000. An alternative is to use Eq. (3) to form $|P|^2 \equiv PP^*$ and equate it to $2.400^2(10^8)$ Pa^2. This operation yields an equation whose roots are the value of Δ,

$$|P/1000|^2 - 24^2 = -170.51\cos\Delta + 361.84\sin\Delta + 500 = 0$$

We could pursue an algebraic solution, but the value of Δ is readily found by the application of numerical methods. There are two roots,

$$\Delta = 36.17° \text{ or } -165.73°$$

That there are two possible phase angles should not be surprising because a harmonic signal will equal a specified value at two instants in a period.

The complex particle velocity at point C may be found by vectorially adding the contribution of each wave according to

$$\bar{V} = \frac{1}{\rho_0 c}\left[P_A\left(\bar{\xi}_C\right)\bar{e}_x + P_B\left(\bar{\xi}_C\right)\bar{e}_B\right] \tag{5}$$

Two values of Δ lead to two possible pressure and particle velocities,

$$\Delta = 36.18° \implies \quad P = 17260 - 16676i \text{ Pa}$$
$$\bar{V} = (0.0118 - 0.0103i)\, \bar{e}_x + (0.0006 + 0.0035i)\, \bar{e}_y \text{ m/s}$$
$$\Delta = -165.72° \implies P = 23933 + 1791i \text{ Pa}$$
$$\bar{V} = (0.0156 + 0.0003i)\, \bar{e}_x + (-0.0018 - 0.0031i)\, \bar{e}_y \text{ m/s}$$

The complex nature of the velocity components obscures the meaning of the resultant vector. A description of the vector as a function of time is preferable. The data in Fig. 2 was constructed by synthesizing the x and y components of $\mathrm{Re}\left(\bar{V}\exp(i\omega t)\right)$, as well $\mathrm{Re}\left(P\exp(i\omega t)\right)$, over an interval of one period. The angle of $\bar{v}(t)$ is measured counterclockwise from the x-axis. The discontinuity of this quantity is due to limiting it to $-\pi < \Delta \le \pi$. By definition, $|\bar{v}| \ge 0$, but it closely follows $|p|$. The reason for this behavior is that the propagation direction $\bar{e}_B$ is close to $\bar{e}_x$ and $\bar{e}_A$ is $\bar{e}_x$, so the particle velocity is close to $(P_A + P_B)/(\rho_0 c)$ in the $\bar{e}_x$ direction. Confirmation of this observation may be found in the graph of angle$(\bar{v})$, which shows that the angle is very close to zero or 180° for most of a period. The transitions occur in the intervals where the pressure and particle velocity are small.

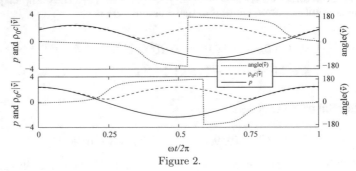

Figure 2.

Perhaps the most important aspects of these results is their fundamental nature. Interference effects between waves that propagate in different directions are more intricate and harder to anticipate than they are for one-dimensional waves. Understanding these effects often requires a more sophisticated perspective.

4.4 Velocity Potential

Suppose as the result of an analysis, we have determined the dependence of the pressure on position and time. How can we find the corresponding velocity field? The linearized momentum equation relates $\bar{v}$ and p. To use this relation, we would take the gradient of the pressure field, then integrate that result to obtain the velocity. The linearized continuity equation, Eq. (4.2.7), would not be suitable for this purpose because it is a single differential equation, but the velocity vector has three components. Now consider the alternative scenario, in which we have determined the velocity field as a function of position and time. The continuity equation could be used to determine the pressure. Doing so would entail evaluating the divergence of the velocity, then integrating it to find p. The momentum equation would be less desirable because it would require integrating the gradient of the pressure, which has three components. Although each approach is feasible, some aspects of an analysis would be simplified if the basic state variables were determined solely by differentiation. This attribute can be attained in the time domain by introducing a scalar *velocity potential* $\phi\left(\bar{\xi}, t\right)$. It is defined such that the particle velocity is the gradient of the potential,[1]

$$\boxed{\bar{v} = \nabla\phi}$$ (4.4.1)

A description of the pressure corresponding a known velocity potential may be found from the linearized momentum equation, Eq. (4.2.6), which becomes

$$\rho_0\frac{\partial}{\partial t}\nabla\phi = -\nabla p$$ (4.4.2)

The order in which the time derivative and gradient are performed will not alter the result, so the preceding states that the gradient of "something" must equal the gradient of "something else."It follows that the difference of these "things" can at the most be a function of time, in other words, $p = -\rho_0\partial\phi/\partial t + F(t)$. If there is some location at which the conditions are ambient, then p and $\bar{v}$ are zero there. Consequently, $F(t)$ is zero there, and therefore everywhere. The consequence is that

$$\boxed{p = -\rho_0\frac{\partial\phi}{\partial t}}$$ (4.4.3)

The field equation governing ϕ is found from the continuity equation. For a linearized theory, we set $\rho = \rho_0 + p/c^2$, so this equation is

$$\frac{1}{c^2}\frac{\partial p}{\partial t} + \rho_0\nabla\cdot\bar{v} = \frac{1}{c^2}\frac{\partial p}{\partial t} + \rho_0\nabla^2\phi = 0$$ (4.4.4)

Substitution of p from Eq. (4.4.3) leads to

$$\boxed{\nabla^2\phi - \frac{1}{c^2}\frac{\partial^2\phi}{\partial t^2} = 0}$$ (4.4.5)

This, of course, is the linear wave equation. It follows that an analysis that seeks a solution for the velocity potential would follow the same approach as one that uses pressure as the fundamental state variable. The difference is that boundary and initial

---

[1]Any velocity field can be represented in terms of a scalar potential function ϕ and a solenoidal vector field $\bar{\chi}$ according to $\bar{v} = \nabla\phi + \nabla\times\bar{\chi}$. The circulation is $\nabla\times\bar{v} \equiv \nabla\times(\nabla\times\bar{\chi})$. The implication of neglecting viscosity is that it is not possible to induce circulation. Hence, the velocity potential may be used to represent the velocity field of any inviscid fluid, even in an analysis that includes nonlinear effects.

conditions would lead directly to conditions on derivatives of ϕ. This applies equally to a frequency-domain analysis, in which case we would set

$$\phi = \text{Re}\left(\Phi\left(\bar{\xi}\right) e^{i\omega t}\right)$$ (4.4.6)

The complex amplitude is governed by the Helmholtz equation,

$$\nabla^2 \Phi + k^2 \Phi = 0$$ (4.4.7)

The complex velocity and pressure amplitudes are related to Φ by

$$\bar{V} = \nabla \Phi, \quad P = -i\omega \rho_0 \Phi$$ (4.4.8)

As was mentioned earlier, formulation of an analysis in terms of the velocity potential does not alter the basic solution procedure, but it might make some operations less awkward.

EXAMPLE 4.5 The system in Example 3.3 is a semi-infinite waveguide that is closed at $x = 0$ by a piston whose movement is opposed by a spring and a dashpot. Formulate the equations governing the velocity potential when the force F driving the piston is an arbitrary function of time.

Significance

The notion here is that being able to compare a formulation using ϕ to one that uses p will enable you to decide which you prefer. Letting the excitation force be an arbitrary function of time suggests that we work in the time domain, which will give a clearer picture of how derivatives of the potential arise.

Solution

The piston's displacement is u, which is positive when the piston moves in the positive x direction. The equation of motion for the piston is

$$M\frac{d^2u}{dt^2} = -Ku - D\frac{du}{dt} + F - p|_{x=0}\,\mathcal{A}$$ (1)

The pressure field within the waveguide is a forward propagating wave because the radiation/causality condition requires that no part of the signal propagates from infinity. Thus, the basic forms are

$$\phi = f\left(t - \frac{x}{c}\right)$$

$$v = \frac{\partial \phi}{\partial x} = -\frac{1}{c}f'\left(t - \frac{x}{c}\right)$$ (2)

$$p = -\rho_0 \frac{\partial \phi}{\partial t} = -\rho_0 f'\left(t - \frac{x}{c}\right)$$

where f' denotes the derivative of the function with respect to $t - x/c$. We substitute the representations in Eq. (2) into (1), which leads to

$$M\frac{d^2u}{dt^2} + D\frac{du}{dt} + Ku - \rho_0 A f'(t) = F \tag{3}$$

It also is necessary to satisfy continuity at the face of the piston, which requires that $du/dt = v$ at $x = 0$, so that

$$\frac{du}{dt} = -\frac{1}{c}f'(t) \tag{4}$$

Substitution of this expression converts Eq. (3) to an ordinary differential equation for u,

$$M\frac{d^2u}{dt^2} + (D + \rho_0 c A)\frac{du}{dt} + Ku = F \tag{5}$$

After Eq. (5) has been solved for $u(t)$ corresponding to a specified $F(t)$, the pressure is given by the combination of Eqs. (1) and (4) to be

$$p = \rho_0 c \left.\frac{du}{dt}\right|_{t-x/c} \tag{6}$$

Note that this analysis does not require integrations to relate particle velocity and pressure.

For the sake of comparison, let us now review the analysis using pressure as the fundamental variable. The general solution is

$$p = g(t - x/c) \tag{7}$$

The equation of motion, Eq. (1), becomes

$$M\frac{d^2u}{dt^2} + D\frac{du}{dt} + Ku + Ag(t) = F \tag{8}$$

Euler's equation governs acceleration, so continuity at the piston face requires

$$\rho_0 \frac{d^2u}{dt^2} = -\left.\frac{\partial p}{\partial x}\right|_{x=0} = \frac{1}{c}g'(t) \tag{9}$$

This relation may be integrated. It gives $g(t) = \rho_0 c \,(du/dt)$, which is the standard relation between pressure and particle velocity in a plane wave. Substitution of this relation into Eq. (8) gives the same differential equation as Eq. (5).

We see that the differences between the two formulations are not profound. In most situations, we may use either at our discretion. However, sometimes a basic concept is formulated in terms of velocity potential. An example is the nonlinear wave equation, which is derived in Chap. 13.

4.5 Energy Concepts and Principles

For the most part, an analytical determination of an acoustic response will entail finding a solution of a field equation that also satisfies the relevant boundary conditions. In classical mechanics, the principle of work and energy provides a supplement that is sometimes useful as a shorter way of arriving at a solution, but it also is useful for interpreting responses. In a similar way, the first law of thermodynamics for fluids may be used to explain a variety of phenomena, and in some cases anticipate the general form of a solution.

4.5.1 Energy and Power

Before we can formulate the first law of thermodynamics, we must describe the acoustical energies. We have adopted a continuum model, so a mass element is the fluid contained in a differential element of volume dV at an arbitrary instant. The mass of the particles in this element is ρdV, where ρ is the density at the current position and time. The infinitesimal size of the element means that all particles in it have the same velocity. Thus, the kinetic energy per unit volume of this element is

$$\boxed{\mathcal{T} = \frac{1}{2}\rho\bar{v}\cdot\bar{v}}$$

(4.5.1)

Description of the potential energy requires a little more effort. Consider a hypothetical experiment in which we enclose the fluid in a rectangular box whose walls are perfectly insulated, so that thermal energy cannot be transferred into or out of the fluid from the exterior. The walls of this box can move, thereby compressing the fluid, but fluid cannot leak out. If the walls move extremely slowly, the work done to move the walls in opposition to the pressure of the fluid will be stored solely as potential energy. (The fluid is ideal, so no work is done to overcome viscous stresses.)

To quantify this concept, suppose the walls are parallel to the planes of a Cartesian coordinate system, with sides ℓ_1, ℓ_2, and ℓ_3 in the x, y, and z directions, respectively. The brick of fluid confined within the box is shown in Fig. 4.6. The fluid is in thermodynamic, as well as static, equilibrium, so the *absolute* pressure everywhere is $\hat{p}$. The forces depicted in the figure are the pressure acting on each face. Each force is balanced by an opposite force on the opposite face of the brick.

Starting from the state in Fig. 4.6, the walls at $x_1 = \ell_1$, $x_2 = \ell_2$, and $x_3 = \ell_3$ are moved inward very slowly, thereby decreasing the sides by an infinitesimal amount

Fig. 4.6 A brick of fluid that
is compressed by pressure

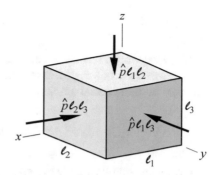

$\delta\ell_j$. (The use of δ to denote a differential change is prompted by the fact that we
are considering a hypothetical situation.) The decrease in the volume of the brick
causes the pressure to increase by an infinitesimal amount δp. The particles remain
at rest because the walls move very slowly. Furthermore, the insulated walls do not
permit heat to be transferred out. Hence, it must be that whatever work is done by
the pressure resultants is stored in the fluid as potential energy.

The work done by the pressure resultants is infinitesimal because the displacement
is infinitesimal. Therefore, we can ignore the change in each face area and use the
starting pressure $\hat{p}$ to evaluate the work. For example, the work done by the resultant
action on the face $x = \ell_1$ when it displaces inward is $\left(\hat{p}\ell_2\ell_3\right)\delta\ell_1$. Similar expressions
apply for the other faces. Also, the faces that coincide with the coordinate planes are
stationary, so the pressure resultants acting on them do not work. Adding the work
done by each pressure resultant gives

$$\delta\mathcal{W} = \hat{p}\left(\ell_2\ell_3\delta\ell_1 + \ell_1\ell_3\delta\ell_2 + \ell_1\ell_2\delta\ell_3\right) \tag{4.5.2}$$

The original volume of the brick is $\mathcal{V} = \ell_1\ell_2\ell_3$ and the volume after the walls
move is $\mathcal{V} + \delta\mathcal{V} = (\ell_1 - \delta\ell_1)(\ell_2 - \delta\ell_2)(\ell_3 - \delta\ell_3)$. Products of "$\delta$" quantities
are second order differentials. Dropping them leads to $\delta\mathcal{V} = (\mathcal{V} + \delta\mathcal{V}) - \mathcal{V} =$
$-(\ell_2\ell_3\delta\ell_1 + \ell_1\ell_3\delta\ell_2 + \ell_1\ell_2\delta\ell_3)$. Thus, the differential work may be written as

$$\delta\mathcal{W} = \hat{p}\left(-\delta\mathcal{V}\right) \tag{4.5.3}$$

The volume change is related to the density change by conservation of mass. The
inward movement of the walls results in changing the density from ρ to $\rho + \delta\rho$. The
mass m of the brick is constant, so it must be that

$$m = (\rho + \delta\rho)(\mathcal{V} + \delta\mathcal{V}) = \rho\mathcal{V} \implies \delta\mathcal{V} = -\frac{\mathcal{V}}{\rho}\delta\subset = -\frac{m}{\rho^2}\delta\subset \tag{4.5.4}$$

Note that $\mathcal{V}$ was replaced by m/ρ because the volume is not constant, whereas the
same mass always is contained in $\mathcal{V}$. We use the preceding to eliminate $\delta\mathcal{V}$ from the
expression for $\delta\mathcal{W}$, which leads to

$$\delta \mathcal{W} = \frac{\hat{p}\delta\rho}{\rho^2}m \tag{4.5.5}$$

The total work $\mathcal{W}$ is the accumulation of the work done in each incremental step as the size of the box is decreased, so $\mathcal{W}$ is found by integrating $\delta\mathcal{W}$ as the density changes from the initial ambient value to the final value ρ, that is,

$$\mathcal{W} = m \int_{\rho_0}^{\rho} \frac{\hat{p}(\eta)}{\eta^2} d\eta \tag{4.5.6}$$

where η is a dummy variable representing density. The preceding is the work required to compress a region of fluid to its current state. Evaluation of this integral requires specification of $\hat{p}(\rho)$, which is the equation of state.

It was noted when we began the derivation that the work described above is stored as potential energy. Furthermore, the constant mass is the product of the density and volume in the current state. Hence, dividing Eq. (4.5.6) by the volume yields the *potential energy* per unit volume, $\mathcal{U}$, of an ideal fluid.

$$\boxed{\mathcal{U} = \rho \int_{\rho_0}^{\rho} \frac{\hat{p}(\eta)}{\eta^2} d\eta} \tag{4.5.7}$$

Note that the factor ρ multiplying the integral is the same as the upper limit of the integral, that is, the volume in the disturbed state. Because the kinetic and potential energy are volume specific quantities, their SI units are joules per cubic meter.

Now that we know how to evaluate the mechanical energies, we may examine how these energies are transported. We consider a control volume $\mathcal{V}$ that is fixed in space. Heat flow across its surface $\mathcal{S}$ is assumed to be negligible because heat conduction between particles is neglected in our idealization of acoustical processes. When this assumption is combined with the assumption that viscosity is unimportant, the first law of thermodynamics reduces to the work-energy principle for a system of particles. We will use the time derivative form of this principle, which asserts that the rate of change of the mechanical energy of all particles contained in $\mathcal{V}$ equals the rate at which energy flows into V across $\mathcal{S}$ plus the power $\mathcal{P}_{\text{in}}$ input to $\mathcal{V}$ by all external forces. The mechanical energy is the sum of the potential and kinetic energies, so it must be that

$$\boxed{\frac{\partial}{\partial t}(\mathcal{T} + \mathcal{U})_{\text{inside } \mathcal{V}} = \frac{\partial}{\partial t}(\mathcal{T} + \mathcal{U})_{\text{across } \mathcal{S}} + \mathcal{P}_{\text{in}}} \tag{4.5.8}$$

Because this relation describes the time rate of change of energy, it is a statement of power balance.

The forces external to the control volume are the resultant of the pressure distribution. Consider an infinitesimal surface patch $d\mathcal{S}$ at position $\bar{\xi}_S$. Its infinitesimal size means that the pressure is constant over it, so the surface traction is the product of the pressure at that location and the surface area. The normal direction at that location is

$\bar{n}\left(\bar{\xi}_S\right)$, which is defined to point out of $\mathcal{V}$. Therefore the traction force acting on the surface patch is $-\hat{p}\left(\bar{\xi}_S, t\right)\bar{n}\left(\bar{\xi}_S, t\right)dS$, where $\hat{p}$ is the absolute pressure, not merely the disturbance from the ambient value. The power input to the control volume by this force is the product of the velocity of the point where it is applied and the component of this force in the direction of that velocity. The total quantity is a surface integral,

$$\mathcal{P}_{\text{in}} = -\iint_S \hat{p}\left(\bar{\xi}_S, t\right)\bar{n}\left(\bar{\xi}_S, t\right)dS \cdot \bar{v}\left(\bar{\xi}_S, t\right) \tag{4.5.9}$$

The total mechanical energy contained in $\mathcal{V}$ is the integral over that domain of the energy per unit volume,

$$\left(\mathcal{T} + \mathcal{U}\right)_{\text{inside } \mathcal{V}} = \iiint_{\mathcal{V}} \left[\mathcal{T}\left(\bar{\xi}, t\right) + \mathcal{U}\left(\bar{\xi}, t\right)\right]d\mathcal{V} \tag{4.5.10}$$

To describe the rate at which energy flows into $\mathcal{V}$, we refer to Fig. 4.1. The volume of fluid that departs from the control volume across an infinitesimal surface patch in a time interval dt is shown there to be $\left[\bar{n}\left(\bar{\xi}_s\right) \cdot \bar{v}\left(\bar{\xi}_s, t\right)dt\right]dS$. The infinitesimal extent of this volume means that the energy per unit volume throughout it is $\mathcal{T}\left(\bar{\xi}_s, t\right) + \mathcal{U}\left(\bar{\xi}_s, t\right)$. An outward flow corresponds to a loss from $\mathcal{V}$, rather than a gain, which is accounted for by a negative sign. Thus, the rate at which energy flows into the control volume is

$$\frac{\partial}{\partial t}\left(\mathcal{T} + \mathcal{U}\right)_S = -\iint_S \left[\mathcal{T}\left(\bar{\xi}_s, t\right) + \mathcal{U}\left(\bar{\xi}_s, t\right)\right]\left[\bar{n}\left(\bar{\xi}_s\right) \cdot \bar{v}\left(\bar{\xi}_s, t\right)dt\right]dS \tag{4.5.11}$$

The result of using these descriptions to fill in the terms in the power balance equation, Eq. (4.5.8), is

$$\frac{\partial}{\partial t}\iiint_{\mathcal{V}} \left[\mathcal{T}\left(\bar{\xi}, t\right) + \mathcal{U}\left(\bar{\xi}, t\right)\right]d\mathcal{V} = -\iint_S \left[\mathcal{T}\left(\bar{\xi}_s, t\right) + \mathcal{U}\left(\bar{\xi}_s, t\right) + \hat{p}\left(\bar{\xi}_s, t\right)\right]$$
$$\bar{v}\left(\bar{\xi}_s, t\right) \cdot \bar{n}\left(\bar{\xi}_s, t\right)dS \tag{4.5.12}$$

The control volume is stationary, which means that the integration limits for the term on the left side are constant. Therefore, the time derivative may be brought inside the integral. The divergence theorem converts the term in the right side to an integral over $\mathcal{V}$. These operations transform the power balance equation to

$$\iiint_{\mathcal{V}} \frac{\partial}{\partial t}[\mathcal{T}(\bar{\xi}, t) + \mathcal{U}(\bar{\xi}, t)]d\mathcal{V} = -\iiint_{\mathcal{V}} \nabla \cdot \{\bar{v}(\bar{\xi}, t)[\mathcal{T}(\bar{\xi}, t) + \mathcal{U}(\bar{\xi}, t) + \hat{p}(\bar{\xi}, t)]\}d\mathcal{V} \tag{4.5.13}$$

We now invoke the argument that because V is defined arbitrarily and the relation must be true for any V, the integrands must match. Thus, the first law of thermodynamics reduces to a partial differential equation

$$\frac{\partial}{\partial t} (T + U) + \nabla \cdot [(T + U)\, \bar{v}] + \nabla \cdot (\hat{p}\bar{v}) = 0 \qquad (4.5.14)$$

where T and U are given by Eqs. (4.5.1) and (4.5.7).

In its present form, this equation appears to be too complicated to be of much use, but that is because of its generality. The sole assumptions contained in it are that whatever process it describes occurs adiabatically and that the fluid is inviscid. Consequently, there is no limit to the magnitude of the pressure and particle velocity.

4.5.2 Linearization

Our primary interest is with situations where linearized approximations of the governing equations are sufficiently accurate. Thus, we let $p = \hat{p} - p_0$ and neglect terms in the power balance equation that are higher than quadratic products of p and/or v. The linearized equation of state is

$$\hat{p} - p_0 = c^2 (\rho - \rho_0) \qquad (4.5.15)$$

We also restrict the fluid to be homogeneous in its undisturbed state, so that p_0 and ρ_0 are constants. The potential energy per unit volume in Eq. (4.5.7) in that case becomes

$$U = \rho \int_{\rho_0}^{\rho} \frac{\left[p_0 + c^2 (\eta - \rho_0) \right]}{\eta^2}\, d\eta \qquad (4.5.16)$$

It is convenient to switch to the density perturbation $\rho' = \rho - \rho_0$ and correspondingly to change the integration variable such that $\eta' = \eta - \rho_0$. By doing so, we may make use of the fact that the magnitude of ρ'/ρ_0 and η'/ρ_0 is much less than one. This permits Taylor series expansions that simplify the integrals. The operations are

$$
\begin{aligned}
U &= \frac{\rho_0 + \rho'}{\rho_0^2} \int_0^{\rho'} \frac{(p_0 + c^2\eta')}{(1 + \eta'/\rho_0)^2}\, d\eta' \\
&\approx \frac{\rho_0 + \rho'}{\rho_0^2} \int_0^{\rho'} (p_0 + c^2\eta') \left(1 - 2\frac{\eta'}{\rho_0} \right) d\eta' \\
&= \frac{\rho_0 + \rho'}{\rho_0^2} \int_0^{\rho'} \left(p_0 + c^2\eta' - 2\frac{p_0}{\rho_0}\eta' \right) d\eta'
\end{aligned}
\qquad (4.5.17)
$$

The terms that have been dropped in each approximation would lead to terms in U that would be cubic or higher in ρ'. When additional terms of this nature are dropped after the integral is evaluated, what remains is

$$\mathcal{U} = \frac{p_0}{\rho_0}\rho' + \frac{c^2}{2\rho_0}\left(\rho'\right)^2 \qquad (4.5.18)$$

To understand this expression, consider the analogous case of a spring that carries a prestress tensile force F_0 prior to elongation. Let x be the displacement from this initial position. Then the force in that position will be $F_0 + kx$, and the work done to increase the displacement by dx will be $(F_0 + kx)\,dx$. The total work to attain an elongation Δ is the integral of this infinitesimal work, which is $F_0\Delta + k\Delta^2/2$. The first term is the work to overcome the initial force, whereas the second is the work done to overcome elastic restoring force. In the same way, the first term Eq. (4.5.18) is the work done to compress the fluid in opposition to the ambient pressure, while the second term is the incremental work required to increase the pressure.

Linearization also simplifies $\mathcal{T}$, because $\bar{v} \cdot \bar{v}$ already is a quadratic term. Thus, the current density may be replaced by the ambient value. Being able to drop terms higher than quadratic also reduces the second term in Eq. (4.5.14) to

$$(\mathcal{T} + \mathcal{U})\,\bar{v} \approx \frac{p_0}{\rho_0}\left(\rho'\bar{v}\right) \qquad (4.5.19)$$

In combination with constancy of p_0 and ρ_0 for a homogeneous fluid, substitution of Eqs. (4.5.18) and (4.5.19) into the power balance equation gives

$$\frac{\partial}{\partial t}\left[\frac{1}{2}\rho_0\bar{v}\cdot\bar{v} + \frac{c^2}{2\rho_0}\left(\rho'\right)^2\right] + \frac{p_0}{\rho_0}\frac{\partial}{\partial t}\rho' + \frac{p_0}{\rho_0}\nabla\cdot\left(\rho'\bar{v}\right) + \nabla\cdot\left(p\bar{v}\right) = 0 \quad (4.5.20)$$

The last step entails replacing ρ' with $\rho - \rho_0$

$$\frac{\partial}{\partial t}\left[\frac{1}{2}\rho_0\bar{v}\cdot\bar{v} + \frac{c^2}{2\rho_0}(\rho - \rho_0)^2\right] + \frac{p_0}{\rho_0}\left[\frac{\partial}{\partial t}\rho + \nabla\cdot\left(\rho\bar{v}\right)\right] + \nabla\cdot\left[(p - p_0)\,\bar{v}\right] = 0$$
$$(4.5.21)$$

The second bracketed term is the left side of the (nonlinear) continuity equation, Eq. (4.1.5), so it vanishes. The final form is obtained by using Eq. (4.5.15) to eliminate the density. The resulting power balance equation is

$$\boxed{\frac{\partial E}{\partial t} + \nabla \cdot \bar{I} = 0} \qquad (4.5.22)$$

The quantity E is the *acoustical energy* per unit volume. It consists of the quadratic terms in the linearized mechanical energy,

$$\boxed{E = \frac{1}{2}\rho_0\bar{v}\cdot\bar{v} + \frac{c^2}{2\rho_0 c^2}\left(p'\right)^2} \qquad (4.5.23)$$

We previously encountered the product of p' and v, which is the *intensity*, in Chap. 1 when we considered time averages of harmonically varying quantities. For one-dimensional waves, the vectorial nature of the intensity is manifested by its sign, but the basic fact that it is a vector with several components is one the most common mistakes made by beginning students. The intensity is denoted as $\bar{I}$,

$$\boxed{\bar{I} = p\bar{v}} \tag{4.5.24}$$

The direction in which $\bar{I}$ is oriented, which is the same as the direction of the particle velocity, is the direction in which power is flowing, and its magnitude is the power flow per unit area crossing a plane whose normal coincides with $\bar{I}$. Thus, it would be equally correct to call $\bar{I}$ the energy flux. Intensity units are watts per square meter, which is equivalent to pascal-meters per second.

Equation (4.5.23) is a local law that describes energy transport at a point. There are situations in which we are interested in what occurs over a finite region of fluid. Such a relation is obtained by integrating the differential equation over the domain $\mathcal{V}$, followed by application of the divergence theorem. The result is

$$\boxed{\frac{\partial}{\partial t}\iiint_{\mathcal{V}} E\, d\mathcal{V} = -\iint_{S} \bar{I} \cdot \bar{n}\, dS} \tag{4.5.25}$$

As before, $\bar{n}$ is the normal direction pointing outward from the surface. This is the linearized statement of the work-energy principle for an adiabatic process in a ideal fluid: The rate at which the acoustic energy in any domain increases equals the power that flows into it.

Although the linearized power balance equation might appear to be an addition to the basic continuity and momentum equations, that is not the case. Rather, those basic equations are embedded in Eq. (4.5.23). To demonstrate this attribute, we begin by taking the dot product of the linearized time-domain Euler equation with the particle velocity, to which we apply an identity for the divergence of a product. Specifically, we write

$$\rho_0 \bar{v} \cdot \frac{\partial \bar{v}}{\partial t} = -\bar{v} \cdot \nabla p \equiv -\nabla \cdot (p\bar{v}) + p\nabla \cdot \bar{v} \tag{4.5.26}$$

The last term on the right side is simplified by using the continuity equation, Eq. (4.2.7), to describe $\nabla \cdot \bar{v}$. Thus, we find that

$$\rho_0 \bar{v} \cdot \frac{\partial \bar{v}}{\partial t} = -\nabla \cdot (p\bar{v}) + p\left(-\frac{1}{\rho_0 c^2}\frac{\partial p}{\partial t}\right) \tag{4.5.27}$$

The identity for the derivative of a dot product converts this relation to

$$\frac{\partial}{\partial t}\left(\frac{1}{2}\rho_0 \bar{v} \cdot \bar{v} + \frac{1}{2\rho_0 c^2}p^2\right) = -\nabla \cdot (p\bar{v}) \tag{4.5.28}$$

This equation is the same as Eq. (4.5.23).

A corollary of the fact that a linearized power transfer equation may be derived from the momentum and continuity equations is that it provides no new information. Indeed, whereas it is a single equation relating pressure and particle velocity, the momentum equation consists of as many as three component equations that are supplemented by the scalar continuity equation. Their combination in the form of the wave equation consists of a single differential equation for pressure.

Nevertheless, there the power transfer offers some interesting and useful insights. For example, the d'Alembert solution for a simple plane wave in the time domain, for which $p = \rho_0 c v$, corresponds to

$$T = \frac{1}{2}\rho_0 \left[\frac{1}{\rho_0 c} F \left(t - \frac{(\bar{\xi} - \bar{\xi}_0) \cdot \bar{e}}{c} \right) \right]^2$$

$$E = \frac{1}{2\rho_0 c^2} \left[F \left(t - \frac{(\bar{\xi} - \bar{\xi}_0) \cdot \bar{e}}{c} \right) \right]^2 + T = 2T \qquad (4.5.29)$$

$$\bar{I} = \frac{1}{\rho_0 c} \left[F \left(t - \frac{(\bar{\xi} - \bar{\xi}_0) \cdot \bar{e}}{c} \right) \right]^2 \bar{e} = E c \bar{e}$$

These expressions tell us that at any instant and location the total acoustical energy per unit volume E in a plane wave is equally divided between potential and kinetic energy. We also see that the intensity is $\bar{I} = E c \bar{e}$. To interpret this relation, we align the x-axis with $\bar{e}$, so the yz plane is perpendicular to the direction of propagation. The magnitude of $\bar{I}$ is the power per unit area that is transported across this plane in a unit time interval, so the energy per unit area crossing the plane in an interval from t to $t + dt$ is $|\bar{I}| dt = E(c\,dt)$. At the beginning of this interval, this energy was contained in a region extending dx behind the plane, so the energy per unit area contained in that region was $E dx$. Equating this energy to the energy that was transported gives $dx = c dt$. From this, we deduce that the energy per unit volume is transported at the speed of sound in the direction that the plane wave is propagating. Of course, this is not surprising given that specific values of the pressure and particle velocity propagate at $c\bar{e}$, and the mechanical energy depends solely on these state variables.

4.5.3 Power Sources

Although the insights regarding the energy properties of a planar wave are interesting, they come after the basic analysis. However, consideration of acoustic energy sometimes can assist us to formulate a solution. The means for doing so is derived from the integral form of the power balance principle, Eq. (4.5.25). Integration of that expression over an arbitrary time interval $t_1 < t < t_2$ gives

$$E|_{t_2} - E|_{t_1} = -\int_{t_1}^{t_2} \left[\iint_S \bar{I}\left(\bar{\xi}_S, t\right) \cdot \bar{n}\left(\bar{\xi}_S\right) dS \right] dt \qquad (4.5.30)$$

This relation merely states that the change of acoustical energy for particles in any domain equals the work done by the pressure field acting on the domain's surface.

Let us apply this principle to the case where the field varies harmonically. At steady state, all state variables are replicated over any interval that is the period, $T = 2\pi/\omega$, so there is no difference between the acoustical energy from the beginning to the end of that interval. Consequently, the net work done at the boundary S during a period must be zero. It is more meaningful to convert this to a statement regarding averages over a period, so we divide by the period T to obtain

$$\boxed{\iint_S \bar{I}_{\text{av}}\left(\bar{\xi}_S\right) \cdot \bar{n}\left(\bar{\xi}_S\right) dS = 0} \qquad (4.5.31)$$

where the time-averaged intensity at any location is

$$\boxed{\bar{I}_{\text{av}}\left(\bar{\xi}\right) = \frac{1}{T}\int_{t_1}^{t_1+T} p\left(\bar{\xi}, t\right) \bar{v}\left(\bar{\xi}, t\right) dt = \frac{1}{2}\operatorname{Re}\left(P\left(\bar{\xi}\right)\bar{V}^*\left(\bar{\xi}\right)\right)} \qquad (4.5.32)$$

This equation tells us that in any situation where a harmonic signal propagates through an ideal fluid, the time-averaged power flow into or out of any region is zero.

Care should be taken to not misinterpret this statement. Instantaneously, the intensity on any surface S enclosing a region of the fluid need not be zero. Usually, power flows out of a region over part of a period and into the region for the remainder of the period. Another misconception to avoid is to think that the average power flow is zero everywhere on S. All that Eq. (4.5.31) tells us is that the total for the entire domain is zero; it is quite common that power will be input to a region over some portion of S, while it is radiated away from the remainder of S.

This notion leads us to the concept of an *acoustic source*. Suppose that an ideal fluid surrounds an inner domain that contains an energetic device such as a hydrophone. This device has a surface velocity in the normal direction, which we shall denote as $v_e(\bar{\xi}_e, t)$. Presumably, the nature of this transducer is such that v_e is imposed. It is conventional to take v_e to be positive if it points out of the transducer, and therefore into the fluid.

The acoustic domain $\mathcal{V}$ in this scenario consists of the region between the surface S_e that is the boundary of the transducer and some outer boundary, which is denoted S_o. The surface S then consists of the union of S_e and S_o. This is the situation depicted in Fig. 4.7.

The normal vector $\bar{n}$ on S_o and S_e is defined to point out of the fluid domain. Continuity at the interface S_e of the fluid and the transducer requires that the normal velocity of each medium be equal. Because v_e is taken to be positive if it is into the

Fig. 4.7 Acoustic domain bounded by surface $\mathcal{S}_O$ and surrounding an energetic device whose surface is $\mathcal{S}_e$

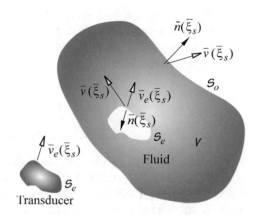

fluid, the continuity condition is

$$\boxed{\bar{v} \cdot \bar{n} = -v_e, \ \bar{\xi}_s \in \mathcal{S}_e} \tag{4.5.33}$$

The time averaged intensity for a harmonic signal is $\mathrm{Re}(P\bar{V}^*)/2$. The component of this quantity normal to $\mathcal{S}_e$ therefore is

$$\bar{I}_{av} \cdot \bar{n} = -\frac{1}{2} \mathrm{Re}\left(PV_e^*\right) \tag{4.5.34}$$

The presence of a negative sign in the preceding expression tells us that power flows in the direction of $-\bar{n}$. In other words, it exits the transducer, which is equivalent to saying that it flows into the fluid. Correspondingly, the total power input to the fluid domain by the transducer is

$$\boxed{\mathcal{P}_{in} = \iint\limits_{\mathcal{S}_e} \bar{I}_{av}\left(\bar{\xi}_s\right) \cdot \bar{n}_e\left(\bar{\xi}_s\right) d\mathcal{S} = \iint\limits_{\mathcal{S}_e} \frac{1}{2} \mathrm{Re}\left(P\left(\bar{\xi}_s\right)\bar{V}_e^*\left(\bar{\xi}_s\right)\right) \cdot \bar{n}_e\left(\bar{\xi}_s\right) d\mathcal{S}} \tag{4.5.35}$$

To apply this expression in the context of Eq. (4.5.31), it is necessary to recall that $\mathcal{S}$ is the union of $\mathcal{S}_o$ and $\mathcal{S}_e$. Thus, it must be that

$$\iint\limits_{\mathcal{S}_o} \bar{I}_{av}\left(\bar{\xi}_s, t\right) \cdot \bar{n}\left(\bar{\xi}_s\right) d\mathcal{S} = \mathcal{P}_{in} \tag{4.5.36}$$

The term on the left is the acoustic power that is radiated from domain $\mathcal{V}$. In recognition of the place from which this power originates, the transducer within $\mathcal{S}_e$ is said to be an acoustic source.

A slight generalization considers the possibility that S_o surrounds multiple transducers whose surfaces are $(S_e)_n$. The surface S is the union of S_o and the surfaces of the transducers, so the power radiating from S_o must equal that which is transferred into V by all transducers,

$$\iint\limits_{S_o} \bar{I}_{av}\left(\bar{\xi}_s, t\right) \cdot \bar{n}\left(\bar{\xi}_s\right) dS = \sum_n (\mathcal{P}_e)_n \qquad (4.5.37)$$

One use of this relation is to evaluate the power radiated by transducers and vibrating structures. Suppose there are one or more devices that generate sound. If their shape is irregular, it will be cumbersome to describe how the average intensity and normal direction vary along the surfaces $(S_e)_n$. This difficulty may be circumvented by considering a regularly shaped region V whose surface S_o surrounds all of the radiating bodies. Probably the simplest such region is a sphere. In fact, considering such a domain leads to several general insights. Let r be the radius of the sphere. Position on this surface is defined by the angle ψ measured from a fixed axis to a radial line and angle θ that defines how far to rotate this radial line about the fixed axis, as depicted in Fig. 4.8. Thus, both P and $\bar{V}$ on S_o can be taken to be functions of r, ψ, and θ.

Suppose that r is extremely large. Then, just as the Earth appears to be flat to someone on its surface, the sphere appears locally to be flat. This suggests that the relationship between pressure and particle velocity should tend to be $P = \rho_0 c V$, with the particle velocity being in the outward radial direction $\bar{e}_r$. Correspondingly, it is reasonable to anticipate that the complex exponential for a true plane wave figures prominently in P, of course, with r replacing x. However, we also recognize that there may be additional dependence on r because the signal is not truly planar.

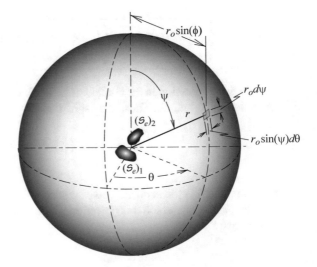

Fig. 4.8 Coordinates for a sphere that surrounds two energetic sources

Another consideration is that the sources will generally send out different signals in different directions. Consequently, the strength of the locally planar wave can be expected to depend on the location on the surface of the sphere. These expectations are captured by the conjecture that the signal is well approximated by

$$P = f(\psi, \theta)\, g(r)\, e^{-ikr}, \quad \bar{V} = \frac{P}{\rho_0 c} \bar{e}_r \tag{4.5.38}$$

where $f(\psi, \theta)$ and $g(r)$ are unspecified. Correspondingly, the average intensity on S_o is expected to be

$$\bar{I}_{av} = \frac{1}{2\rho_0 c} |f(\psi, \theta)|^2 |g(r)|^2 \tag{4.5.39}$$

Because $\bar{e}_r$ is the outward normal on the sphere, the power $\mathcal{P}$ radiated by sources within $\mathcal{V}$ is described by

$$\mathcal{P} = \sum_n (\mathcal{P}_e)_n = \int_0^\pi \int_{-\pi}^\pi \frac{1}{2\rho_0 c} |f(\psi, \theta)|^2 |g(r)|^2 (r \sin\psi d\theta)(r d\psi) \tag{4.5.40}$$

Evaluation of the surface integral removes any dependence on ψ and θ. Let F denote the value of this integral, so we find that

$$\mathcal{P} = \frac{F}{2\rho_0 c} |g(r)|^2 r^2 \tag{4.5.41}$$

The radiated power must be the same for any value of r, provided that the sphere encloses all sources. It follows that $|g(r)|$ must vary inversely with r according to

$$|g(r)| = \left(\frac{2\rho_0 c \mathcal{P}}{F}\right)^{1/2} \frac{1}{r} \tag{4.5.42}$$

This line of reasoning leads to the conjecture that the complex pressure amplitude should tend to

$$\boxed{P = \left(\frac{2\rho_0 c \mathcal{P}}{F}\right)^{1/2} f(\psi, \theta) \frac{e^{-ikr}}{r}} \tag{4.5.43}$$

This expression describes the *farfield* of a set of acoustic sources. Whenever we see a signal whose pressure depends inversely on the distance from a fixed origin, we say that it exhibits *spherical spreading*. We will encounter this behavior in a variety of contexts in Chaps. 6 and 7.

EXAMPLE 4.6 The pressures at the origin $x = y = 0$ in a pair of plane waves are given by

$$p_1 = 10\cos(\omega t), \quad p_2 = 20\cos(\omega t + \pi/3)\ \text{Pa}$$

The propagation direction of each wave is depicted in the figure. Determine and graph the complex pressure amplitude, the mean-squared pressure, and the time-averaged intensity as functions of distance along line AB. The fluid is air at standard atmospheric conditions and the frequency is 200 Hz.

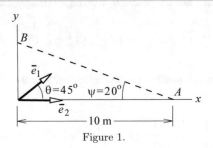

Figure 1.

Significance

It cannot be emphasized too frequently that intensity is a vector quantity. This example uses the properties of plane waves to reinforce recognition of this basic fact. It also shows that the trace of a wavenumber vector is useful for explaining phenomena.

Solution

The total pressure signal is a scalar sum of the contribution in each plane wave. It is useful for the interpretation of the results to perform the operations algebraically. Thus, the propagation directions are

$$\bar{e}_1 = \cos\theta \bar{e}_x + \sin\theta \bar{e}_y, \quad \bar{e}_2 = \bar{e}_y$$

The reference location for both waves is the origin, $\bar{\xi}_0 = \bar{0}$. At that location we have

$$P_1\left(\bar{\xi}_0\right) = B_1 = 10, \quad P_2\left(\bar{\xi}_0\right) = B_2 = 20e^{i\pi/3}$$

The individual simple plane waves at an arbitrary location $\bar{\xi}$ are

$$P_1\left(\bar{\xi}\right) = B_1 e^{-ik\bar{\xi}\cdot\bar{e}_1}, \quad P_2\left(\bar{\xi}\right) = B_2 e^{-ik\bar{\xi}\cdot\bar{e}_2} \tag{1}$$

The particle velocity associated with each wave is

$$\bar{V}_1\left(\bar{\xi}\right) = \frac{1}{\rho_0 c} P_1\left(\bar{\xi}\right)\bar{e}_1, \quad \bar{V}_2\left(\bar{\xi}\right) = \frac{1}{\rho_0 c} P_2\left(\bar{\xi}\right)\bar{e}_2 \tag{2}$$

To describe how quantities vary along the straight line AB let s be distance from point A along the line, so that the coordinates of a point are functions of s given by

$$x(s) = x_A - s\cos\psi, \quad y(s) = s\sin\psi, \quad \bar{\xi} = x(s)\bar{e}_x + y(s)\bar{e}_u$$

Application of the identity for the cosine of the sum of two angles reveals that the dependence on s of the total pressure is

$$P(s) = P_1(s) + P_2(s), \quad P_1(s) = B_1 e^{-ikx_A} e^{iks\cos(\psi+\theta)}, \quad P_2(s) = B_2 e^{-ikx_A} e^{iks\cos\psi}$$
(3)

Embedded in these expressions is the wavenumber vector for each simple wave. Let $\bar{e}_s$ be the unit vector from A to B. The component of $k\bar{e}_1$ parallel to $\bar{e}_s$ is $-k\cos(\theta+\psi)$, where the negative sign results from $\bar{e}_1$ projecting in the direction of $-\bar{e}_s$. Similarly, the component of $k\bar{e}_2$ parallel to $\bar{e}_s$ is $-k\cos\psi$. Thus, each term in the pressure corresponds to propagation in the direction of $-\bar{e}_s$ at the corresponding trace wavenumber. These wavenumbers are not equal, so it follows that an interference pattern is set up, in which the two terms add at some locations and cancel at others. We could describe this dependence by a pair of curves in which the abscissa is s and the ordinate is $\text{Re}(P(s))$ and $\text{Im}(P(s))$. Figure 2 shows an alternative description. At any point along line AB, the complex value of $P(s)$ is represented as a vector in the complex plane. A longer vector at a point indicates that $|P|$ is larger there, and the angle of that vector relative to the x-axis is $\arg(P(s))$. [This plot was created with the MATLAB "quiver" function whose arguments are two column vectors of $x(s)$ and $y(s)$ coordinates, followed by two column vectors of $\text{Re}(P(s))$ and $\text{Im}(P(s))$.] To avoid confusion with the intensity, it must be emphasized that this vectorial description of pressure is solely an aid to visualizing the influence of the real and imaginary parts of P.

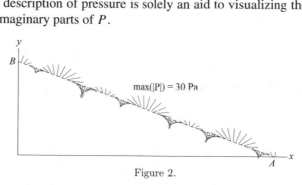

Figure 2.

Working algebraically helped us to understand the behavior of $P(s)$. Therefore, let us follow a similar process to form $(p^2)_{av}$ and $\bar{I}_{av}$. We begin with the former by writing

$$\begin{aligned}
(p^2)_{av} &= \frac{1}{2}[P_1(s) + P_2(s)][P_1^*(s) + P_2^*(s)] \\
&= \frac{1}{2}|P_1(s)|^2 + \frac{1}{2}|P_2(s)|^2 + \text{Re}(P_1(s)P_2^*(s)) \\
&= \frac{1}{2}|B_1|^2 + \frac{1}{2}|B_2|^2 + \text{Re}(B_1 B_2^* e^{iks[\cos(\psi+\theta)-\cos\psi]})
\end{aligned}$$
(5)

This expression shows that the mean-squared pressure fluctuates about a mean that is the sum of the mean-squared values of the simple waves. The wavenumber of the fluctuating part is the difference of the trace wavenumbers along $\bar{e}_s$ of the individual waves. The description of $\left(p^2\right)_{\mathrm{av}}$ in Fig. 3 uses a line perpendicular to $\bar{e}_s$ to convey the value.

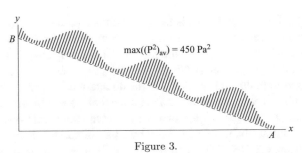

Figure 3.

The analysis of intensity proceeds similarly. We use Eq. (3) to describe the contribution of the individual waves to the pressure and particle velocity,

$$
\begin{aligned}
\bar{I}_{\mathrm{av}} &= \frac{1}{2\rho_0 c}\,\mathrm{Re}\left\{[P_1\,(s) + P_2\,(s)]\left[P_1^*\,(s)\,\bar{e}_1 + P_2^*\,(s)\,\bar{e}_2\right]\right\} \\
&= \frac{1}{2\rho_0 c}\,\mathrm{Re}\left[|P_1\,(s)|^2\,\bar{e}_1 + |P_2\,(s)|^2\,\bar{e}_2 + P_1\,(s)\,P_2^*\,(s)\,\bar{e}_2 + P_1^*\,(s)\,P_2\,(s)\,\bar{e}_1\right] \\
&= \frac{1}{2\rho_0 c}\,|B_1|^2\,\bar{e}_1 + \frac{1}{2\rho_0 c}\,|B_2|^2\,\bar{e}_2 + \frac{1}{2\rho_0 c}\,\mathrm{Re}\{B_1 B_2^* e^{iks[\cos(\psi+\theta)-\cos\psi]}\bar{e}_1 \\
&\quad + B_1 B_2^* e^{-iks[\cos(\psi+\theta)-\cos\psi]}\bar{e}_2\}
\end{aligned}
$$

(6)

The two leading terms are the average intensity if either plane wave was the only one present. The other terms are the effect of interference between the waves. As was the case for the mean-squared pressure, the phase of this effect depends on the difference of the trace wavenumbers of the wave in the $\bar{e}_s$ direction. The contribution in the direction of $\bar{e}_2$ is oppositely phased from that $\bar{e}_1$ contribution. Figure 4 describes $\bar{I}_{\mathrm{av}}$ as a real vector distribution along line AB. The isolated contributions are $|P_1|^2/(2\rho_0 c) = 0.0867\,(\bar{e}_x + \bar{e}_y)$ and $|P_2|^2/(2\rho_0 c) = 0.4902\bar{e}_x$ W/m^2. Interference results in a spatially periodic variation along line AB.

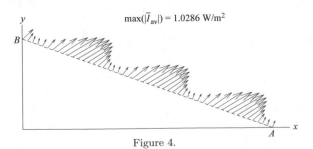

Figure 4.

Perhaps the most important lesson to be gained from this example is that although we can use superposition to construct a pressure field, the same is not true for an analysis of power transport. A simple explanation is that if there are two signals of any kind, the power transport is the result of each signal's pressure pushing a particle at the total velocity, which contains a contribution from the other signal.

EXAMPLE 4.7 A harmonically vibrating circular piston whose diameter is $2a$ is embedded in a large wall. If the frequency is sufficiently large, such that $ka \gg 2\pi$, the sound field is said to be a sound beam because it decreases greatly with increasing distance off the center line of the piston. An extremely crude model of the field considers it to be a simple plane wave in the *nearfield,* which is a cylindrical region whose diameter is $2a$. This region extends outward from the wall to a distance $R_0 = 0.5ka^2$ called the *Rayleigh length.* Beyond this distance, the field is considered to spread spherically with constant amplitude along the spherical wavefront. The spherical wave is contained inside a cone whose apex is centered on the piston and whose angle is such that the cone intersects the nearfield cylindrical domain at the Rayleigh distance. Outside the region that is the union of the cylinder and cone, the pressure is considered to be zero. This model is depicted in the figure. Derive an expression for the power radiated by the piston by considering the pressure distribution along the wall. Then compare that power to the power radiated by the spherically spreading wave and from that evaluation determine the farfield pressure signal.

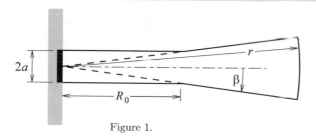

Figure 1.

Significance

The study of sound beams is an important topic in acoustics, because they occur in a wide variety of applications, including speaker systems. This example serves as their conceptual introduction, while at the same time extending the evaluation of radiated power to situations where it is not possible to fully surround a transducer with a sphere.

Solution

The plane wave in the nearfield model propagates in the direction of the centerline of the piston. We designate it as the axis x for a coordinate system whose origin is at the center of the piston. We encountered a piston in a wall in the previous chapter, where we explored the end correction for a waveguide whose end is open. We saw there that the specific acoustic impedance χ for a circular piston is close to one when $ka > 2\pi$. Considering the nearfield to be a plane wave whose complex velocity is $V = P/(\rho_0 c)$ is consistent with that result. Thus, the nearfield is taken to be described by

$$\bar{V} = V_0 e^{-ikx} h(a - R)\, \bar{e}_x, \quad P = \rho_0 c e^{-ikx} h(a - R),$$

where R is distance perpendicular to the x-axis in any sense. Thus, $h(a - R)$ zeros the pressure outside the cylinder.

We may evaluate the radiated power by integrating the average intensity over the entire wall. The intensity is zero off the face of the piston, and it is constant on the face at $\bar{I}_{av} = 0.5\rho_0 c\, |V_0|^2\, \bar{\iota}_1$. Furthermore, $\bar{I}_{av}$ is parallel to the outward normal to the piston. The face area of the piston is πa^2, so the radiated power is

$$\mathcal{P} = \frac{\pi}{2} a^2 \rho_0 c\, |V_0|^2$$

The sound beam model adopted here considers the pressure at fixed radial distance r to be constant inside the cone, and the particle velocity in that region is taken to be in the radial direction. This direction is described by a unit vector $\bar{e}_r$ that extends from the origin to a point in the fluid. The angle between $\bar{e}_r$ and $\bar{e}_z$ is denoted as ψ. Thus, the model represents the farfield signal as

$$p = \frac{A}{r} e^{-ikr} h(\beta - \psi), \quad \bar{V} = \frac{P}{\rho_0 c} \bar{e}_r$$

where β is the semi-vertex angle of the cone. From the given figure, we find that

$$\beta = \tan^{-1}\left(\frac{a}{R_0}\right) = \tan^{-1}\left(\frac{2}{ka}\right)$$

The value of ka is large, so it is reasonable to set $\beta = 2/(ka) \ll 1$. The average intensity inside the cone is

$$\bar{I}_{av} = \frac{1}{2\rho_0 c} \frac{|A|^2}{r^2} h(\beta - \psi)\, \bar{e}_r$$

which is independent of the angular coordinate θ measured around the x-axis.

We cannot surround the piston with a full sphere because the acoustic domain is an infinite half-space bounded by the wall. However, we can surround the piston with a hemisphere. The range of polar angles for this surface is $0 \le \psi \le \pi/2$, while the angle around this axis occupies a full circle, $-\pi \le \theta \le \pi$. Thus the radiated power may be evaluated from the far field intensity according to

$$P = \int_0^\pi \int_{-\pi}^\pi \bar{I}_{av} \, (r \sin \psi d\theta) \, (rd\psi)$$

This expression is valid regardless of what $\bar{I}_{av}$ is for the piston. However, our model takes it to be independent of θ and to be zero for $\psi > \beta$, so the above expression reduces to

$$P = 2\pi \int_0^\beta \bar{I}_{av} \left(r^2 \sin \psi\right) d\psi = \frac{\pi}{\rho_0 c} |A|^2 [1 - \cos (\beta)]$$
$$\approx \frac{\pi}{\rho_0 c} |A|^2 \left(\frac{\beta^2}{2}\right) = \frac{2\pi}{\rho_0 c \, (ka)^2} |A|^2$$

The power leaving the wall cannot be dissipated within the region of the cylinder and cone, so it must equal the power radiating from the spherical surface. Thus, we find that

$$\frac{2\pi}{\rho_0 c \, (ka)^2} |A|^2 = \frac{\pi}{2} a^2 \rho_0 c \, |V_0|^2 \implies |A| = \frac{1}{2} ka^2 \rho_0 c \, |V_0| = R_0 \rho_0 c \, |V_0|$$

Correspondingly, the pressure in the farfield is given by

$$P = \rho_0 c \, |V_0| \frac{R_0}{r} e^{-ikr}$$

This result is reasonably good for locations that are very close to the z-axis, but the actual behavior is that the pressure falls off to zero at some value of ψ that is close to β, beyond which it might grow to a lesser value than the axial pressure. Depending on ka, there might be several such regions off-axis where the pressure is nonzero. They are called *side lobes*. Despite these details, the picture of a sound beam developed here certainly would provide an adequate description for most nonacousticians.

4.6 Closure

The wave equation for time-domain studies, or the Helmholtz equation for frequency-domain investigations, combined with the corresponding version of the momentum equation, are the basic tools that we will employ to explore a variety of systems and phenomena. Our analysis of plane waves that propagate in an arbitrary direction served to illustrate the application of these basic tools. We will encounter plane wave behavior in a variety of more general situations. Considerations involving intensity and energy transport tend to be used as diagnostics, particularly for noise control issues. However, as we saw with the farfield of the signal radiated by acoustic sources, these concepts sometimes are quite useful to anticipate and understand at a fundamental level solutions that are obtained by mathematical analysis, as well as to verify analyses, especially numerical simulations.

4.7 Homework Exercises

Exercise 4.1 Consider a pressure field that depends on the x and y coordinates in the horizontal plane according to $p = A \sin(\omega t - \alpha_1 x - \beta_1 y) \cos(\omega t - \alpha_2 x - \beta_2 y)$, where ω, α_1, α_2, β_1, and β_2 are constants. (a) Represent this signal as plane waves. Specify the frequency, phase speed, and propagation of each. (b) What conditions must the coefficients satisfy in order that the expression for p satisfies the Helmholtz equation?

Exercise 4.2 The pressure field within a two-dimensional waveguide with rigid walls is the subject of the next chapter. That study shows that the field may be described as $p = A \sin(\omega t - \kappa x - \mu y) + A \sin(\omega t - \kappa x + \mu y)$. (a) For what value(s) of μ will the y-component of particle velocity be zero at $y = 0$ and $y = H$? (b) What value(s) of κ correspond to the value(s) of μ identified in Part (a)?

Exercise 4.3 Two plane harmonic waves, both at frequency ω, propagate in different directions $\bar{e}_1$ and $\bar{e}_2$ in the horizontal plane, as shown below. The pressure signal measured at the origin for each wave is $p_1(x = 0, t) = A \cos(\omega t)$, $p_2(x = 0, t) = B \sin(\omega t)$. (a) Derive expressions giving the time dependence at an arbitrary location $\bar{x}$ of the total pressure, and of the particle velocity components parallel to the x and y. In these expressions, describe the position of $\bar{x}$ in terms of polar coordinates (R, ϕ). (b) Evaluate the expressions in part (a) for the case $\phi = 45°$, $R = 3.75\pi c/\omega$, $\theta_1 = 36.87°$, $\theta_2 = 53.13°$, and $B = 2A$. Give the amplitude and phase angle of the pressure and each velocity component relative to a pure cosine function.

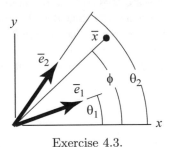

Exercise 4.3.

Exercise 4.4 Two microphones in the horizontal plane are used to monitor the passage of a plane wave. Microphone A is located at the origin and microphone B is distances x_B to the east and y_B to the south of the origin. The speed of sound is a known value c. The leading edge of pulse passes the origin at $t = 0$, and its arrival time at microphone B is $t_B > 0$. Derive an expression for the direction of propagation of the wave and prove that there are two possible directions.

Exercise 4.5 A line of microphones is placed along the east-west x-axis. At $t = 0$, the pressure recorded by these microphones is the triangular distribution in the graph. It is known that this signal is a simple plane wave propagating in direction $\bar{e}$, which is 30° west of north, and $c = 340$ m/s. (a) Draw a graph of the pressure as a function of

time that would be measured by the microphone at $x = y = 0$. (b) Draw a graph of
the pressure as a function of time that would be measured at $x = 400\,\mathrm{m}$, $y = 200\,\mathrm{m}$.

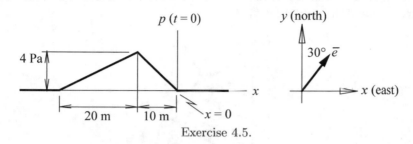

Exercise 4.5.

Exercise 4.6 In an experiment, microphones were aligned in arrays along the east
and south axes. The experiment measured a planar acoustic wave propagating in the
horizontal plane at a speed of 340 m/s in an unknown direction $\bar{e}$ whose eastward
component is positive. The measurements along the east axis were taken completely,
but most of the records for those along the south axis were lost. At $t = 0$, the spatial
distribution of pressure along the east axis is known to be as shown, with the leading
edge at position A and the trailing edge at position B. However, only the position of
the trailing edge of the wave along the south axis is known. (a) Determine the angle θ
describing the direction of propagation of this wave relative to the easterly direction.
(b) For the initial instant, $t = 0$, determine the distance s along the south axis from
the origin to the leading edge of the wave. (c) At what time t_1 will the trailing edge of
the wave arrive at position A? (d) At what time t_2 will the leading edge of the wave
arrive at $x = 750$ m, $y = 600\,\mathrm{m}$?

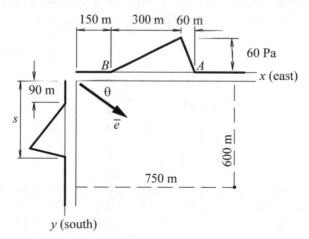

Exercise 4.6.

Exercise 4.7 Consider a plane wave propagating in the $x - y$ plane in the direction
$\bar{e}$ shown. The waveform has a duration τ in the shape of a single sine lobe. (a) For the
instant when the wavefront arrives at the origin, determine the pressure distribution

that would be detected by a set of sensors arranged along the straight line C_1. Sketch the results as a function of arclength s_1 measured along C_1. (b) Repeat the analysis in Part (a) for the circle C_2. Sketch the results as a function of arclength s_2 measured along c_2 from its intersection with the x-axis.

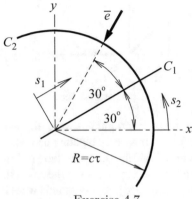

Exercise 4.7.

Exercise 4.8 Two plane waves propagate in the horizontal plane in directions $\bar{e}_1$ and $\bar{e}_2$. The steady-state square waveform in the sketch is the same for wave 1 at point A and wave 2 at point B. (a) Identify locations along the x-axis at which the individual pressure waveforms are in-phase. Describe the waveforms of the total pressure and particle velocity components at those locations. (b) Identify locations along the x - axis at which the pressure waveforms are 180° out-of-phase. Describe the waveforms of the total pressure and particle velocity components at those locations.

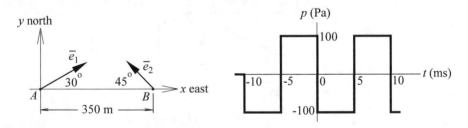

Exercise 4.8.

Exercise 4.9 Plane waves at the same frequency ω travel in opposite directions within a waveguide. The pressure is $p = \text{Re}\left[B\left(\exp i\left(\omega t - kx\right)\right) + C\left(\exp i\left(\omega t + kx\right)\right)\right]$. (a) Derive expressions for the instantaneous intensity $I(x, t)$ and the time-averaged intensity $I_{av}(x)$. Are there any combinations of the complex amplitudes B and C for which $I(x, t)$ or $\bar{I}_{av}$ do not depend on x? (b) Consider the case where $B = 2 - 3i$ and $C = -5 - 2i$. Evaluate $p(x, t)$ and $v(x, t)$ for $0 < kx < 4\pi$ at $\omega t = 0, 0.25\pi$, and 0.5π. Graph pv vs. x at each instant and compare each graph to the graphs of $I(x, t)$ and $\bar{I}_{av}$ according to part (a).

Exercise 4.10 Two plane waves at the same frequency propagate in directions $\bar{e}_1$ and $\bar{e}_2$. The respective complex amplitudes at the origin are A_1 and A_2. Derive an expression for the time-averaged intensity.

Exercise 4.11 The acoustic signal in a fluid consists of the superposition of a plane wave propagating in the positive x direction, and another plane wave propagating in the negative y direction. At the origin of xyz, the signals are $P_x \cos(\omega_1 t)$ and $P_y \sin(\omega_2 t)$, respectively. (a) Derive an expression for the time-averaged intensity at an arbitrary (x, y) location for the case where ω_2 is significantly different from ω_1. (b) Derive an expression for the instantaneous intensity at arbitrary (x, y, t) for the case where $\omega_2 = \omega_1$.

Exercise 4.12 A pair of plane harmonic waves at 2 kHz underwater propagate in orthogonal directions. At the origin the first wave, whose propagation direction is $\bar{e}_1 = -\bar{e}_x$, is known to be $500 \cos(\omega t)$ Pa. The signal measured at field point D is $300 \sin(\omega t)$ Pa. The polar coordinates locating this point are $r = 50$ m, $\theta = 20°$. (a) Determine the pressure at the origin for the second wave, whose propagation direction is $\bar{e}_2 = \bar{e}_y$. (b) Determine the time-averaged intensity at point D corresponding to the second wave being the result in part (a).

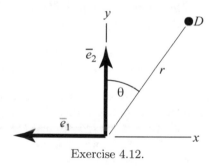

Exercise 4.12.

Exercise 4.13 It is desired to use the velocity potential $v_x = \partial\phi/\partial x$ to analyze one-dimensional waves in a viscous fluid. The effect of viscosity was modeled by the addition of a term to Eq. (3.3.4), but it does not alter the continuity equation or the equation of state. (a) Derive an equation for p in terms of ϕ and no other variable. (b) Derive a field equation governing ϕ.

Exercise 4.14 A circular cylinder whose diameter is 600 mm is filled with air. The cylindrical wall and the left end are rigid and immobile. The other end is closed by a piston. Application of force F moves the piston inward, thereby compressing the air. The walls, end, and piston are thermally insulated, so no heat flows out of the cylindrical cavity. Prior to the compression process, the air is at standard atmospheric conditions, $L = 3$ m, and F is zero. The force is increased gradually, causing the piston to move slowly inward, thereby increasing the internal pressure by 2 kPa. Determine the distance the piston moves inward in this process, the work done by F, and the average value of F. How does this average value compare to the value of F required to hold the piston in place in the compressed state?

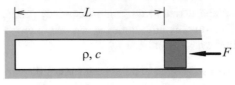

Exercise 4.14.

Exercise 4.15 A person on the ground speaks continuously at a fairly constant level. The ground is rigid and flat. Consequently, the pressure amplitude anywhere on a virtual hemisphere centered at the speaker may be taken to be inversely proportional to the radius r of the hemisphere. A friend of the speaker who is 50 m away estimates that the sound pressure level she hears is 65 dB. Estimate the acoustic power output by the speaker.

Exercise 4.16 The field surrounding a vibrating object at a long range is thoroughly probed. A certain direction is defined to be the polar axis for a set of spherical coordinates. At a distance of 30 m from the object in the axial direction, the sound pressure level is measured to be 95 dB at 400 Hz. Also, data processing has led to the conclusion that the pressure amplitude at any fixed (large) radial distance varies as $\cos \psi$ and is independent of the azimuthal angle θ. (This is a dipole field, which is a topic covered in Chap. 6.) Determine the power radiated by this object.

Exercise 4.17 Perform the analysis in Exercise 4.16 in the situation where the pressure amplitude at a fixed radial distance is proportional to $\sin(\psi)^2 \sin \theta \cos \theta$.

Exercise 4.18 A useful model for acoustic studies is an infinitely long cylinder that radiates sound due to a vibration pattern on its surface that is length-wise periodic. If L is the axial distance z over which the pattern repeats, then the pressure resulting from a time harmonic vibration must have the form $p = \mathrm{Re}\,(f\,(R)\,\Theta\,(z)\,g\,(z))\exp(i\omega t))$, where R is the distance from the z-axis to a field point, θ is the circumferential angle, and z is the axial distance. Periodicity requires that $g\,(z+L) = g\,(z)$, but $g\,(z)$ otherwise is arbitrary. For simplicity, let the function $\Theta\,(\theta) = 1$, corresponding to an axisymmetric situation. Derive an expression for the time-averaged power that flows across any axial interval L of a virtual concentric cylinder whose radius is much greater than that of the vibrating cylinder. From that result, deduce how $|f\,(r)|$ must depend on R when R is very large

Chapter 5
Interface Phenomena for Planar Waves

Our attention now turns to situations where a plane wave propagates obliquely relative to a planar interface with another medium. The consequence will be a wave that is somewhat different from the one encountered in a one-dimensional waveguide. A wave might also be transferred into the receiving medium. The analyses will employ Cartesian coordinates aligned with the boundary to describe the multidimensional phenomena. Each topic is important for a variety of applications, such as noise control for wall design. Each topic represents an advance toward our goal of fully understanding how acoustical waves behave.

The chapter opens by employing the concept of trace velocity developed in Chap. 4 to identify a general feature of the way sound waves are induced by vibrating machines and structures. This is followed by a time-domain study of the reflection of plane waves from a purely resistive boundary. As was the case for the one-dimensional situation in Chap. 2, this type of impedance is not realistic. A frequency-domain formulation makes it possible to generalize the reflective properties of the boundary, as well as to handle a type of incidence that causes difficulty for a time-domain analysis. After that, we will explore transmission phenomena that arise when a plane wave arrives at the interface between two media. Those developments will be extended to treat transmission through multiple layers composed of various media. There is much similarity between these topics and our earlier study of acoustic transmission lines.

5.1 Radiation Due to Surface Waves

Situations in which an acoustic wave interacts with a boundary may be classified as to whether the phenomena are associated with radiation or scattering. In a *radiation* problem, a known movement of the boundary induces an acoustic signal in the adjacent fluid. In a *scattering* problem, the input is a known acoustic signal, for which the boundary represents an obstacle. The encounter with the boundary modifies the acoustic signal, and it might also result in the transmission of a disturbance into

Electronic supplementary material The online version of this chapter
(DOI 10.1007/978-3-319-56844-7_5) contains supplementary material, which is
available to authorized users.

or through the boundary. In this chapter, the boundary consists of a plane whose extent is infinite. We begin by considering a general radiation problem, in which a known displacement wave propagates along the plane. Such a situation occurs with a seismic wave in the ground or ocean floor, and it also has relevance to the interaction of submerged structures and fluids. Waves running along a surface also are encountered in room acoustics when one is interested in sound generated by a vibrating wall, as well as in noise generation by vibrating machines.

5.1.1   Basic Analysis

Figure 5.1 depicts a situation where a wave propagates in the x direction at phase speed c_w along a solid surface. A variety of such waves can occur in an elastic solid. They differ in the relation between the phase speed and the wavelength. This relation is irrelevant to our current concern, which is to determine the acoustic signal in the fluid that is above the surface when the surface wave is known. The fluid domain is taken to extend to infinite distance above and along the surface. Such a domain is said to be a *half-space*.

Fig. 5.1 A displacement wave that propagates at phase speed c_f along a plane that bounds a fluid, and the acoustic wave it generates

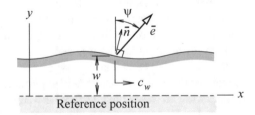

Our interest is restricted to situations that are describable by the laws of linear acoustics. This means that the surface displacement w in Fig. 5.1 is limited to being extremely small. Visually, the surface would appear to always lie in the xz-plane. Mathematically, we take the continuity boundary condition to apply at $y = 0$. Another consequence of smallness of w is that the normal to the surface, which is $\bar{n}$ in Fig. 5.1, is essentially parallel to the y direction. Thus, we set $\bar{n} = \bar{e}_y$. As a result of these approximations, the surface displacement may be set to $w\bar{e}_y$ at $y = 0$. As shown in the figure, positive w is into the fluid. This displacement is a wave that propagates in the x direction, with no variation in the z direction. The phase speed of this wave is c_w, so the phase variable is $t - x/c_w$. Such a wave is described by

$$w = F\left(t - \frac{x}{c_w}\right) \tag{5.1.1}$$

According to the continuity condition, the surface and the fluid particles that contact it move in unison in the normal direction. The motion of a fluid is described

by its velocity, and $\bar{e}_y$ is the normal direction, so the continuity condition requires
that

$$\bar{v} \cdot \bar{e}_y\big|_{y=0} = \frac{\partial w}{\partial t} = F'\left(t - \frac{x}{c_w}\right) \tag{5.1.2}$$

Time-Domain Perspective

We found in Sect. 4.3.2 that a plane wave at an arbitrary orientation seems to propagate along a plane at the trace velocity, which is greater than the speed of sound. It is reasonable to conjecture that the converse will be true. Hence, let us begin by considering the case where $c_w > c$. Such a surface disturbance is said to be *supersonic*. The propagation direction $\bar{e}$ of the plane wave is described in terms of the angle ψ in Fig. 5.1, which is measured from the normal to the surface. Thus, we will try an ansatz for the pressure given by

$$p = f\left(t - \frac{\bar{x} \cdot \bar{e}}{c}\right) \equiv f\left(t - \frac{x \sin \psi + y \cos \psi}{c}\right) \tag{5.1.3}$$

The particle velocity in this plane wave is

$$\bar{v} = \frac{p}{\rho_0 c} \bar{e} \tag{5.1.4}$$

Substitution of this description into the boundary continuity condition leads to

$$\frac{1}{\rho_0 c} f\left(t - \frac{x \sin \psi}{c}\right) \cos \psi = F'\left(t - \frac{x}{c_w}\right) \tag{5.1.5}$$

This equality must hold for all x and t. The trace velocity of the term to the left is $(c/\sin \psi)\,\bar{e}_x$, while the trace velocity of the term on the right is $c_w \bar{e}_x$. Equating these velocities leads to

$$\sin \psi = \frac{c}{c_w}, \quad c_w > c \tag{5.1.6}$$

This relation is a special version of *Snell's law*, which governs the transmission of a plane wave at an interface between different media. When this condition holds, equating both sides of the continuity condition leads to

$$f(t) = \frac{\rho_0 c}{\cos \psi} F'(t) \tag{5.1.7}$$

Thus, we have established that the pressure field is

$$p = \frac{\rho_0 c}{\cos \psi} F'\left(t - \frac{x \sin \psi + y \cos \psi}{c}\right) \tag{5.1.8}$$

This solution is general from the standpoint of the form of the surface displacement. But it also has a fundamental limitation, which is recognizable when we consider slowing the phase speed of the surface wave. The ultimate result is that ψ approaches 90° as c_w approaches c. The consequence is that the magnitude of the pressure becomes infinite, which obviously is not possible. The reason for this dilemma is apparent when we consider the implication of ψ being 90°, in which case $\bar{e} = \bar{e}_x$. Thus, the limit gives a simple plane wave that travels parallel to the surface. The particle velocity in such a wave is solely in the x direction, which is inconsistent with the requirement that the particle velocity at the surface must have a y component that matches w.

This is not the only limitation of Eq. (5.1.8). What happens if $c_w < c$? This is the case of a *subsonic* surface wave. The value of ψ obtained from Eq. (5.1.6) is complex if $c/c_w > 1$, but a complex angle conflicts with the requirement that $p(x, t)$ be real. The physical reason for this situation is that the trace velocity of a simple plane wave is greater than or equal to c, so it cannot match a subsonic phase speed c_w. We must conclude that the pressure in the case of a subsonic surface wave cannot be a plane wave, but if it is not, what is it? Fortunately, an analysis in the frequency domain can be carried out. Although this will limit all waves to being harmonic, the case where $w(x, t)$ is an arbitrary function, as in Eq. (5.1.1), can be treated through Fourier series or Fourier transform techniques.

Frequency-Domain Formulation

It will assist our analysis of the subsonic case if we first use the frequency-domain tools to analyze the case of a supersonic surface wave. Thus, the surface displacement is now represented as

$$\boxed{w = \mathrm{Re}\left(W e^{i\omega(t - x/c_w)}\right)} \tag{5.1.9}$$

The continuity condition, Eq. (5.1.2), requires that

$$\bar{v} \cdot \bar{e}_y\big|_{y=0} = \mathrm{Re}\left(i\omega W e^{i\omega(t - x/c_w)}\right) \tag{5.1.10}$$

The reasoning that led us previously to consider a plane wave in the time domain is equally applicable here, so we begin with

$$p = \mathrm{Re}\left(B e^{i\left(\omega t - k\bar{\xi}\cdot\bar{e}\right)}\right), \quad \bar{e} = (\sin\psi)\,\bar{e}_x + (\cos\psi)\,\bar{e}_y \tag{5.1.11}$$

where the coefficient B may be complex and $k = \omega/c$.

A frequency-domain analysis focuses on the coefficient of $e^{i\omega t}$ in all variables. For the pressure this coefficient is

$$P = B e^{-ik\bar{\xi}\cdot\bar{e}} = B e^{-ik(x\sin\psi + y\cos\psi)} \tag{5.1.12}$$

We know that this form is a solution of the Helmholtz equation, so the only condition that remains is continuity. The particle velocity in a simple plane wave is $p/(\rho_0 c)$ in the $\bar{e}$ direction, so the continuity equation requires that

$$\frac{P}{\rho_0 c} \bar{e} \cdot \bar{e}_y \bigg|_{y=0} = \frac{B}{\rho_0 c} (\cos \psi) \, e^{-ikx \sin \psi} = i\omega W e^{-i\omega x/c_w} \tag{5.1.13}$$

This condition must be satisfied at all x, which can only occur if the trace wavenumbers on each side match, that is, $k \sin \psi$ must equal ω/c_w. Therefore, it must be that

$$\boxed{\psi = \sin^{-1}\left(\frac{c}{c_w}\right)} \tag{5.1.14}$$

Now that its exponentials match, Eq. (5.1.13) leads to

$$B = i\omega \frac{\rho_0 c}{\cos \psi} W \tag{5.1.15}$$

Substitution of this coefficient into Eq. (5.1.12) leads to

$$\boxed{\begin{aligned} P &= \rho_0 c \frac{i\omega W}{\cos \psi} e^{-ik(x \sin \psi + y \cos \psi)} \\ \bar{V} &= \frac{i\omega W}{\cos \psi} e^{-ik(x \sin \psi + y \cos \psi)} \left[(\sin \psi)\, \bar{e}_x + (\cos \psi)\, \bar{e}_y\right] \end{aligned}} \tag{5.1.16}$$

Equation (5.1.14) is the same as Eq. (5.1.6), and the fact that $i\omega W$ is the complex amplitude of dw/dt means that Eq. (5.1.15) is equivalent to Eq. (5.1.8). It follows that the same difficulties arise if c_w approaches c from above, and if $c_w < c$. The analysis of the case where c_w exactly equals c would be quite difficult, because it requires consideration of nonideal effects. We shall exclude it from consideration. (Doing so is not much of a limitation, because an exact equality is extremely unlikely.) Thus, the preceding expression for B is taken to apply for any situation in which $c_w > c$.

The frequency-domain solution for the subsonic case is quite accessible. A simple plane wave cannot exist because the trace of the wavenumber vector $k\bar{e}$ parallel to the surface will be smaller than the surface wavenumber ω/c_w. Nevertheless, these wavenumbers must match. Thus, let us try a solution that has the required dependence parallel to the surface, and an unspecified dependence on y. Such a form is

$$P = F(y)^{-i\omega x/c_w} \tag{5.1.17}$$

This expression must be a solution of the Helmholtz equation. Matching the coefficients of the complex exponentials in that equation leads to

$$\frac{d^2 F}{dy^2} + \left(k^2 - \frac{\omega^2}{c_w^2}\right) F = 0 \tag{5.1.18}$$

If $c < c_w$, the coefficient of F is a real positive number, so the solution is a pair of complex exponentials. If $c > c_w$, so that $k < \omega/c_w$, then the coefficient of F is real and negative. It is convenient to represent this value as $-k^2\mu^2$, that is,

$$\mu = \left(\frac{c^2}{c_w^2} - 1\right)^{1/2} > 0 \tag{5.1.19}$$

Then, the general solution for F is

$$F = B_1 e^{-k\mu y} + B_2 e^{+k\mu y} \tag{5.1.20}$$

In the supersonic case, the sole amplitude coefficient was B, which was determined by satisfying continuity at the surface. Here, we have two coefficients. What should we do?

To answer this question, we turn to a fundamental requirement that may be traced to Sommerfeld.[1] Basically it is a dual requirement. The first part, which we encountered already, states that radiation from a body into a domain of infinite extent cannot result in a signal returning to the body from infinity. The second part states that the amplitude of the signal radiated from a body cannot vary in a manner that will lead to radiated power increasing with increasing distance. The latter is a property we established in Sect. 4.5.3, but Sommerfeld's statement applies in the time domain also. Together, these requirements are the *Sommerfeld radiation condition*.

In the present situation, this condition requires $B_2 = 0$ in Eq. (5.1.20), because otherwise P will grow without bound as y increases. Thus, the pressure field for a subsonic surface wave must fit

$$P = B e^{-i\omega x/c_w} e^{-k\mu y}, \quad c_w < c \tag{5.1.21}$$

This description states that the field radiated by a subsonic surface wave propagates parallel to the surface at the trace wavenumber and it decays exponentially with increasing distance from the surface. Such behavior is said to be *evanescent,* because the wave vanishes at long distance. Clearly, this is not a simple plane wave.

It still remains to satisfy the continuity condition. We cannot invoke the simple plane wave property that $p = \rho_0 c v$. However, it is always correct is to use Euler's equation, which in the present situation gives

$$\begin{aligned}\bar{V} &= -\frac{1}{i\omega\rho_0}\left(\frac{\partial P}{\partial x}\bar{e}_x + \frac{\partial P}{\partial y}\bar{e}_y\right)\\ &= \frac{1}{\rho_0 c}\left(\frac{c}{c_w}\bar{e}_x - i\mu\bar{e}_y\right)B e^{-i\omega x/c_w}e^{-k\mu y}\end{aligned} \tag{5.1.22}$$

[1]A. Sommerfeld, "A Green's Function of the Oscillation Equation," *Jahresber. Dtsch. Math. Ver.* **21**, 309–353 (1912).

When we use this expression to form Eq. (5.1.2), we find that

$$B = -\frac{\rho_0 c \omega}{\mu} W \tag{5.1.23}$$

As c_w approaches c from below, the value of μ approaches zero, causing a singularity in B. Thus, the evanescent solution also is not valid if c_w is exactly equal to c.

Before we examine the implications of this solution, it is useful to identify a single representation that describes the response for any c_w/c. Consider what would happen if we were to use mathematical software to evaluate Eq. (5.1.13) with $c_w = c/2$. The result of evaluating $\psi = \sin^{-1}(2)$ would be either $\psi = 1.5708 - 1.3170i$ or the complex conjugate value, $\psi = 1.5708 + 1.3170i$. (The value that is returned depends on the way in which the software implements a branch cut, which is required for a unique definition of some complex functions.)

Based on this behavior of the arcsine function, let us examine the sine and cosine functions when $\psi = \pi/2 \pm i\Psi$, $\Psi > 0$. The definitions of these functions in the theory of complex variables give

$$\sin \psi = \frac{e^{i\pi/2}e^{\mp\Psi} - e^{-i\pi/2}e^{\pm\Psi}}{2i} \equiv \cosh \Psi$$

$$\cos \psi = \frac{e^{i\pi/2}e^{\mp\Psi} + e^{-i\pi/2}e^{\pm\Psi}}{2} \equiv \mp i \sinh \Psi \tag{5.1.24}$$

where sinh and cosh are the hyperbolic sine and hyperbolic cosine functions. Note that the preceding identities have been written to preserve the order of the alternate signs in the expression for ψ. The pressure will evanesce if $\exp(-iky\cos\psi)$ is a decaying exponential, which results if $\cos\psi$ is a negative imaginary value. Hence, we should set $\psi = \pi/2 + \iota\Psi$. Equation (5.1.24) then give

$$\cos \psi = -i \sinh \Psi = -i \left[(\cosh \Psi)^2 - 1\right]^{1/2} = -i \left(\frac{c^2}{c_w^2} - 1\right)^{1/2} \equiv -i\mu \tag{5.1.25}$$

It follows that Eq. (5.1.12) may be used to describe the pressure field in both the subsonic and supersonic cases, provided that we select the branch cut of the arcsine function correctly. The test for computer routines is:

$$\text{If Im}\left(\sin^{-1}(r)\right) > 0 \text{ when } r > 1, \text{ then set } \psi = \sin^{-1}\left(\frac{c}{c_w}\right)$$

$$\text{If Im}\left(\sin^{-1}(r)\right) < 0 \text{ when } r > 1, \text{ then set } \psi = \text{conj}\left(\sin^{-1}\left(\frac{c}{c_w}\right)\right) \tag{5.1.26}$$

5.1.2   Interpretation

The acoustic signals in the supersonic and subsonic cases are fundamentally different. Most obviously, a supersonic surface wave leads to an acoustic signal that will propagate to long distances from the surface, whereas the acoustic signal in the subsonic case becomes negligibly small. Energy cannot be dissipated in any domain that consists solely of an ideal fluid, so it must be that in the first case power flows from the surface into the fluid, whereas it does not in the second.

To explore this, we form the time-averaged intensity, which is $\mathrm{Re}(P\bar{V}^*)/2$. For the supersonic case, the signal is a plane wave, so $\rho_0 c \bar{V} = P\bar{e}$. Equation (5.1.12) gives

$$\bar{I}_{av} = \frac{PP^*}{2\rho_0 c}\bar{e} = \frac{BB^*}{2\rho_0 c}\left[(\sin\psi)\,\bar{e}_x + (\cos\psi)\,\bar{e}_y\right] \qquad (5.1.27)$$

We use Eq. (5.1.15) to eliminate B, followed by Eq. (5.1.14) to eliminate ψ. The result is

$$\bar{I}_{av} = \frac{1}{2}\rho_0 c\omega^2\,|W|^2\,\frac{(c/c_w)\,\bar{e}_x + \left[1 - (c/c_w)^2\right]^{1/2}\,\bar{e}_y}{1 - (c/c_w)^2},\qquad c_w > c \qquad (5.1.28)$$

The intensity in the subsonic case results from combining Eqs. (5.1.21) and (5.1.22) and then applying Eqs. (5.1.23) and (5.1.19). Doing so yields

$$\bar{I}_{av} = \frac{1}{2}\rho_0 c\omega^2\,|W|^2\,\frac{c/c_w}{(c^2/c_w^2 - 1)}\,\exp\left(-2k\left(\frac{c^2}{c_w^2} - 1\right)^{1/2}y\right)\bar{e}_x,\quad c_w < c \qquad (5.1.29)$$

There is no $\bar{e}_y$ component in this expression because the particle velocity in that direction is 90° degrees out-of-phase from the pressure. Consequently, the time average of the y component of intensity is zero.

The solid boundary is an acoustic source. The power per unit surface area it inputs to the fluid is the component of the surface intensity in the direction of $\bar{e}_z$. In other words, $d\mathcal{P}/d\mathcal{A} = \bar{I}_{av} \cdot \bar{e}_y$. The time-averaged intensity in both the subsonic and supersonic cases does not depend on x. At the surface, where $y = 0$, this quantity is

$$\frac{d\mathcal{P}}{d\mathcal{A}} = \bar{I}_{av}\cdot\bar{e}_y = \begin{cases} \dfrac{\rho_0 c\omega^2}{2\,(\cos\psi)^2}\,|W|^2\,L\cos\psi \ \text{if}\ c_w > c \\[2mm] 0 \ \text{if}\ c_w < c \end{cases} \qquad (5.1.30)$$

In other words, a supersonic wave on the surface radiates power into the fluid, but a subsonic wave does not.

In the subsonic case, the acoustic energy (potential plus kinetic) is confined to the vicinity of the surface. We say that it is *trapped* there. These relations describe time-averaged quantities. In any case, power flows into the fluid for part of a cycle and out of the fluid into the surface for the other part. The difference is that the average is zero in the subsonic case. This is a primary reason why a submarine whose engines produce enormous amounts of power and has a high acoustic pressure near the hull can radiate less acoustic energy than an incandescent light bulb.

This behavior has important general implications for noise control. Suppose one wishes to reduce the noise that is radiated by a vibrating object. One approach is to attempt to reduce the vibration amplitude, but that might be problematic. A different approach modifies the object by attaching closely spaced ribs, such as I-beams, that run perpendicularly to the direction in which the surface wave propagates. The spacing between the ribs should be less than the acoustic wavelength at the frequency where the vibration amplitude is largest. By presenting a set of obstacles to the surface wave, the ribs reduce the surface wavelength to less than $2\pi/k$, which is the property of a subsonic surface wave.

Another difference between the acoustic response in the subsonic and supersonic cases is the particle displacement in the fluid. Displacement is the change in a particle's position relative to its original location, so we denote it as $\Delta\bar{\xi}$. Because displacement is the time integral of velocity, it is $\bar{V}/i\omega$ in the frequency domain. The velocity is described by Eq. (5.1.16) from which we find that the time-domain representation of displacement is

$$\Delta\bar{\xi} = \text{Re}\left(\frac{\bar{V}}{i\omega}e^{i\omega t}\right) = \text{Re}\left\{\frac{W}{\cos\psi}e^{i(\omega t - kx\sin\psi - ky\cos\psi)}\left[(\sin\psi)\,\bar{e}_x + (\cos\psi)\,\bar{e}_y\right]\right\}$$
(5.1.31)

In all cases $\sin\psi = c/c_w$ is real. In the case of a supersonic wave, $\cos\psi$ is real, so the displacement components are in-phase. Consequently, a particle oscillates back and forth along a straight path parallel to the direction of propagation. The amplitude of this displacement is $|W|/\cos\psi$. For a subsonic wave, $\cos\psi = -i\mu$, so the displacement components are 90° out-of-phase. The nature of the displacement is easier to recognize if we select $t = 0$ such that W is real. Then, the displacement is described by

$$\Delta\bar{\xi} = \text{Re}\left\{We^{i(\omega t - kx(c/c_w))}e^{-\mu y}\left(i\frac{c/c_w}{\mu}\bar{e}_x + \bar{e}_y\right)\right\}$$

$$= \frac{W}{(c^2 - c_w^2)^{1/2}}e^{-\mu y}\left[-c\sin\left(\omega t - \frac{\omega}{c_w}x\right)\bar{e}_x\right.$$
(5.1.32)

$$\left. + (c^2 - c_w^2)^{1/2}\cos\left(\omega t - \frac{\omega}{c_w}x\right)\bar{e}_y\right]$$

The path in this case is defined by

$$\frac{(\Delta\bar{\xi}\cdot\bar{e}_x)^2}{c^2} + \frac{(\Delta\bar{\xi}\cdot\bar{e}_y)^2}{c^2 - c_w^2} = \frac{W^2}{c^2 - c_w^2}e^{-2\mu y}$$
(5.1.33)

This is the equation of an ellipse. Because $c^2 > c^2 - c_w^2$, the semi-major diameter of the ellipse is $a_w = Wc/(c^2 - c_w^2)^{1/2}e^{-\mu y}$ parallel to the x-axis, and the semi-minor axis is $b_w = We^{-i\mu y}$ parallel to the y-axis. The ellipse shrinks rapidly with increasing y and it becomes more prolate (that is, flattened) as c_w approaches c.

EXAMPLE 5.1   A wave in the floor of a large tank filled with an unspecified liquid creates a wave in the fluid. The instrumentation used to measure the dynamic response consists of two hydrophones that are located at $\bar{\xi}_A = 0.1\bar{e}_y$, $\bar{\xi}_B = 1\bar{e}_x + 0.2\bar{e}_y$ m, where the origin is located at the bottom of the tank. The hydrophone measurements are shown below. In addition, an accelerometer mounted on the floor indicates that the acceleration magnitude is $19.15\,\text{m/s}^2$. From these measurements determine the density and sound speed of the liquid, and the amplitude and phase speed of the ground displacement.

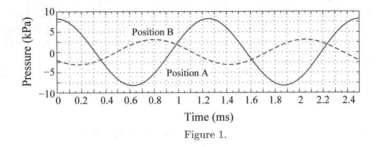

Figure 1.

Significance

The solution will require that we work backward from the observed response to the basic analytical results, which will reinforce our understanding of the various relations that were derived.

Solution

Our strategy will be to extract features from the given time-pressure plots and then match those features to the derived relations. A different pressure amplitude is observed at each elevation, so this must be the subsonic case. The period of both signals is 1.25 ms, which corresponds to a frequency of 800 Hz, so $\omega = 5026$ rad/s. A visual inspection of the plot indicates that the amplitude of the pressures is slightly greater than 8 kPa for $x = 0$, $y = 0.1$ m, and approximately 3 kPa for $x = 1$, $y = 0.6$ m. We will use the more accurate values of 8.23 and 3.10 kPa, which are found from a scan of the actual data. The phase lag of the signals also is relevant. The signal at $x = 0$, $y = 0.1$ seems to be a pure cosine, and the signal at $x = 1$, $y = 0.6$ has its first peak at $t = 0.8$ ms; both observations are confirmed by scans of the data. Thus, the observed signals are

$$p(0, 0.1, t) = 8.23 \left(10^3\right) \cos(\omega t), \quad p(1, 0.6, t) = 3.10 \left(10^3\right) \cos(\omega (t - 0.0008))$$

The corresponding complex pressure amplitudes are

$$P(0, 0.1) = 8.23 \left(10^3\right), \quad P(1, 0.6) = 3.10 \left(10^3\right) e^{-i0.0008\omega} = 3.10 \left(10^3\right) e^{-4.021i}$$

$$\tag{1}$$

Equation (5.1.21) gives the form of the pressure in the subsonic case. Equating that expression to the observed signals leads to

$$P\,(0, 0.1) = Be^{-k\mu(0.1)} = 8.23\,(10^3)$$
$$P\,(1, 0.6) = Be^{-i\omega(1.0)/c_w}e^{-k\mu(0.6)} = 3.10\,(10^3)\,e^{-4.021i} \tag{2}$$

The ratio of these equations is

$$\frac{P\,(0, 0.1)}{P\,(1, 0.6)} = e^{0.5k\mu}e^{i(1.0)\omega/c_w} = \frac{8.23}{3.10}e^{4.021i} \tag{3}$$

The phase angles on either side must match, so

$$\frac{(1.0)\,\omega}{c_w} = 4.021 \implies c_w = 1250\,\text{m/s}$$

Matching the magnitudes in Eq. (3) gives

$$e^{0.5k\mu} = \frac{8.23}{3.10} \implies k\mu = \frac{1}{0.5}\ln\left(\frac{8.23}{3.10}\right) = 1.9528 \tag{4}$$

With $k\mu$ and c_w known, the value of B may be found from either of Eq. (2). The first gives

$$B = 8.23e^{0.1k\mu} = 10005$$

Because we know $k\mu$, c_w and ω, the value of c may be extracted from the definition of μ, Eq. (5.1.19). Setting $k = \omega/c$ there leads to

$$k\mu = 1.9528 = \frac{\omega}{c}\left(\frac{c^2}{c_w^2} - 1\right)^{1/2} \implies c = \left(\frac{1}{c_w^2} - \frac{1.9528^2}{\omega^2}\right)^{-1/2} = 1430\,\text{m/s}$$

The displacement amplitude W and density ρ_0 appears solely in the expression for B, Eq. (5.1.23). The value of μ also appears there. It may be computed from its definition, or from Eq. (4). The former gives

$$\mu = \left[\left(\frac{c}{c_w}\right)^2 - 1\right]^{1/2} = 0.5556$$

Substitution into Eq. (5.1.23) of the parameters that are known at this stage gives

$$B = 10005 = -\frac{\rho_0\,(1430)\,(5026)}{0.5556}W \tag{5}$$

Because ρ_0 is a positive real value, it must be that W is negative. The magnitude of the ground acceleration in general is $\omega^2\,|W|$, so matching it to the given value gives

$$|W| = \frac{19.15}{\omega^2} = 75.8 \ \mu\text{m}$$

Then Eq. (5) gives

$$\rho_0 = \left| \frac{B\mu}{c\omega} \right| = 1020 \ \text{kg/m}^3$$

An interesting question, which is the subject of a Exercise 5.3, is how would the analysis have been altered if the pressure amplitudes at both locations were the same?

5.2 Reflection from a Surface Having a Local Impedance

The remainder of this chapter addresses a variety of problems in which a plane wave in a half-space encounters a planar boundary. In the first case, a plane wave arrives at a solid surface at an arbitrary angle. This is a problem of the reflection of a plane wave at *oblique incidence* relative to the surface. The special case of normal incidence should be the same as the one-dimensional wave in previous chapters.

An important aspect of our investigation is that it considers the surface's behavior to fit the special model of a *local impedance*. "Impedance" tells us that the ratio of the surface pressure to the normal velocity at any location is known, while "local" indicates that the pressure at any location depends only on the velocity at that location, and *vice versa*. Such a surface is said to be *locally reacting*. We dealt with such a surface previously in the context of one-dimensional waves. The model herein is the same, with the addition of an assumption that there is no variation of the impedance property along the surface. A mattress is an example of a system for which the model might be useful. Older construction used interlinked springs. If one were to lie down on one side of the mattress, the entire mattress would deflect. In contrast, a newer type of mattress uses a molded foamy polymer. It deflects little away from the region where one lies. A local impedance model could be used for the latter construction. Indeed, foamy polymeric materials are often used for acoustical treatments, and they often are modeled as having a local impedance.

In the case of one-dimensional waves, we were only able to analyze reflections in the time domain if the surface had a purely resistive impedance. Analysis of general impedance conditions was performed only in the frequency domain. It therefore should not be surprising that the following time-domain analysis is limited to a purely resistive local impedance. After that, we will use a frequency domain formulation to study reflection at a surface having arbitrary impedance properties.

5.2.1 Reflection from a Time-Domain Perspective

Consider a plane wave traveling in an arbitrary direction relative to the reflecting surface. It is said to be the *incident wave*, so its parameters will be denoted with a subscript "I". The propagation direction is described by the unit vector $\bar{e}_I$. The *angle of incidence* is ψ_I measured between $\bar{e}_I$ and the normal to the surface. An arbitrary point on the surface is selected as the origin for xyz axes. As shown in Fig. 5.2, the y-axis is defined to be normal to the surface, and the x-axis is defined to lie in the plane formed by $\bar{e}_I$ and $\bar{e}_y$.

Fig. 5.2 Propagation directions for a plane wave incident on a locally reacting planar surface and for the reflected wave that results

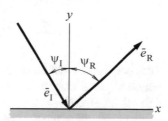

The reflecting surface is taken to have a resistive impedance Z everywhere, and the inward direction for the surface is $-\bar{e}_y$, so the boundary condition at the surface is

$$p = Z\left(-\bar{v}\cdot\bar{e}_y\right) @ y = 0 \tag{5.2.1}$$

It is convenient to let the origin be the reference location for the incident wave and to let $F(t)$ be the pressure waveform that is observed at this location. The distance in the $\bar{e}_I$ direction from the origin to a field point $\bar{\xi}$ is $\bar{\xi}\cdot\bar{e}_I$. A plane wave propagating in an arbitrary direction was described in Eq. (4.3.5), so the incident wave is

$$p_I = F_I\left(\Theta_I\right), \quad \Theta_I = t - \frac{\bar{\xi}\cdot\bar{e}_I}{c}$$
$$\bar{e}_I = (\sin\psi_I)\,\bar{e}_x - (\cos\psi_I)\,\bar{e}_y \tag{5.2.2}$$

Evaluation of the Signal

Clearly, p_I cannot satisfy the boundary condition by itself. The additional signal is the *reflected wave*. Because the incident wave and the surface properties are independent of the z coordinate, the reflected wave must share that property. Given that, it is reasonable to conjecture that it too is a plane wave, with a propagation direction $\bar{e}_R$ that is coplanar with the plane containing $\bar{e}_I$ and the y-axis. The angle ψ_R between the y-axis and $\bar{e}_R$ in Fig. 5.2 is the *angle of reflection*. Thus, we will attempt to construct a solution based on

$$p = p_I + p_R$$
$$p_R = F_R\left(\Theta_R\right), \quad \Theta_R = t - \frac{\bar{\xi}\cdot\bar{e}_R}{c} \tag{5.2.3}$$
$$\bar{e}_R = \sin\left(\psi_R\right)\bar{e}_x + \cos\left(\psi_R\right)\bar{e}_y$$

The particle velocity in each plane wave is the pressure divided by the characteristic impedance in the respective direction of propagation,

$$\bar{v}_I = \frac{P_I}{\rho_0 c}\bar{e}_I, \quad \bar{v}_R = \frac{P_R}{\rho_0 c}\bar{e}_R \tag{5.2.4}$$

The incident and reflected waves individually satisfy the wave equation, so we proceed to satisfy the impedance boundary conditions. Substitution of Eqs. (5.2.2), (5.2.3), and (5.2.4) into Eq. (5.2.1) gives

$$F_I\left(\Theta_I|_{y=0}\right) + F_R\left(\Theta_R|_{y=0}\right)$$
$$= \frac{Z}{\rho_0 c}\left[F_I\left(\Theta_I|_{y=0}\right)\cos\psi_I - F_R\left(\Theta_R|_{y=0}\right)\cos\psi_R\right] \tag{5.2.5}$$

This condition must be satisfied at every location on the surface at every instant, which requires that the argument of F_R be the same as the argument of F_I. Their definitions give $\Theta_I = t - x\sin\psi_I$ and $\Theta_R = t - x\sin\psi_R$ at $y = 0$. Therefore, the angle of reflection must equal the angle of incidence,

$$\boxed{\psi_I = \psi_R} \tag{5.2.6}$$

When this condition is met, satisfaction of Eq. (5.2.5) requires that F_R be proportional to F_I, according to

$$F_R\left(\Theta_I|_{y=0}\right) = R F_I\left(\Theta_R|_{y=0}\right) \tag{5.2.7}$$

where

$$\boxed{R = \frac{Z - Z_0}{Z + Z_0}, \quad Z_0 = \frac{\rho_0 c}{\cos\psi_I}} \tag{5.2.8}$$

As was true for the one-dimensional case, R is the reflection coefficient.[2] The complete representation then is

$$\boxed{\begin{aligned} p &= F_I\left(\Theta_I\right) + R F_I\left(\Theta_R\right) \\ \Theta_I &= t - \frac{\bar{\xi}\cdot\bar{e}_I}{c} \equiv t - \frac{x\sin\psi_I - y\cos\psi_I}{c} \\ \Theta_R &= t - \frac{\bar{\xi}\cdot\bar{e}_R}{c} \equiv t - \frac{x\sin\psi_I + y\cos\psi_I}{c} \end{aligned}} \tag{5.2.9}$$

The quantitative properties of R are like those discussed in regard to Eq. (3.2.11) for normal incidence, except that the effective fluid impedance is $\rho_0 c/\cos\psi_I$. This dependence arises because the normal component of particle velocity in the incident wave decreases as $\cos\psi_I$, whereas the surface pressure in that wave is not dependent on ψ_I. The consequence is that the effective fluid impedance increases with increasing ψ_I. From the perspective of the fluid, the surface appears to become softer. An interesting corollary is that if Z is greater than the fluid's characteristic impedance, then the reflection coefficient will vanish at $\psi_I = \cos^{-1}(\rho_0 c/Z)$. Normal incidence,

---

[2]In many texts, one will find that the expression for R is the form obtained when the numerator and denominator of Eq. (5.2.8) are multiplied by $\cos\psi_I$. The form used here is chosen for consistency with the next section, wherein the interaction between two fluids is investigated.

for which $\psi_{\mathrm{I}} = 0$, gives the same value of R as we previously encountered for one-dimensional waves.

The limit where $\psi_{\mathrm{I}} \to \pi/2$, which leads to $R \to -1$ regardless of the magnitude of Z, is called *grazing incidence*. This tendency contains a subtlety because the solution is not valid if ψ_{I} actually is $90°$. One reason why this is so is that the general solution reduces to $p = 0$ everywhere if $\psi_{\mathrm{I}} = 90°$ and $R = -1$. A different reason results from considering a plane wave that actually propagates parallel to the surface. The particle velocity in such a plane wave is parallel to the surface. Unless Z is infinite, this is inconsistent with the condition of a compliant surface, which requires that a surface pressure induce a velocity normal to the surface. This feature is comparable to the reason why a plane wave can only exist in a waveguide whose walls are rigid.

Wavefronts and Rays

It might seem to be problematic that neither the reflecting surface nor the waves can truly have infinite extent. This is not an issue if the incident wave has a finite duration. To understand why this is so, suppose that $F_{\mathrm{I}}(t)$ is a pulse that is nonzero only for $0 \le t \le \tau$. Thus, the trailing edge of both the incident and reflected waves at any instant is at distance $c\tau$ behind the respective leading edge. Figure 5.3 depicts the region where the pressure in each pulsed wave is nonzero at an arbitrary instant. It can be seen that the *insonified region* on the surface extends a distance $c\tau/\sin\psi_{\mathrm{I}}$ in the x direction. The reflection analysis is valid if the surface extends beyond this region.

At any instant, the signal at all points on a *wavefront* departed from the source at the same time. Thus, the edges of the shaded regions in Fig. 5.3 are the leading and trailing wavefronts of the incident and reflected pulses at some instant t. At time $t + \tau$, the trailing edges will be at the locations where the leading edges were at

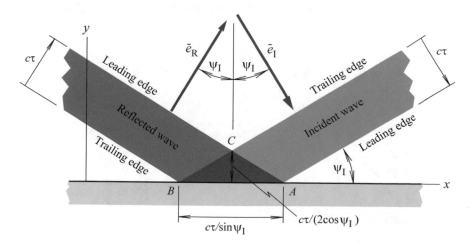

Fig. 5.3 Regions in which pressure due to a pulsed incident and its reflection are nonzero

time t. This means that the shaded regions will have shifted by distance $c\tau/\sin\psi$ in the direction of increasing x, which is consistent with the trace velocity along the surface being $c/\sin\psi_{\mathrm{I}}$.

Another aspect of a pulsed signal is displayed in Fig. 5.3. A trigonometric evaluation based on triangle ABC being isosceles reveals that the region where the reflected and incident signals overlap extends to a height of $c\tau/(2\cos\psi_{\mathrm{I}})$ above the surface. It follows that points that are farther above the surface than $c\tau/(2\cos\psi_{\mathrm{I}})$ will see the incident wave fully pass before the reflected wave arrives. In contrast, at points that are closer to the surface than $c\tau/(2\cos\psi_{\mathrm{I}})$, there will be some interval during which both signals are present.

There is another way in which a signal may be described. Suppose we were to track the instantaneous position of a signal that departs from a specific location on the source at a specific instant. The locus of such positions as time elapses is a *ray*. In media whose properties are time dependent, as well as in some nonlinear situations, signals leaving a specific source point at different instants might follow different rays. The situation for a plane wave is much simpler because the rays are straight lines parallel to the propagation direction $\bar{e}_{\mathrm{I}}$. If we were to move backward along the incident ray that intersects the arbitrary point $\bar{\xi}$ in Fig. 5.4a, we would eventually arrive at the source point from which this ray was emitted.

The reflected ray that intersects point $\bar{\xi}$ is constructed by passing a line parallel to $\bar{e}_{\mathrm{R}}$ through $\bar{\xi}$ back to the surface. This locates the surface point $\bar{\xi}_s$ from which the reflected ray originated. This ray and the incident ray that generated it constitute the *reflected path*, and the incident ray through $\bar{\xi}$ forms the *direct path*.

A diagram of the rays that arrive at a field point can be used to understand the signal that will be observed at a specific field point $\bar{\xi}$. Figure 5.4b depicts direct and reflected paths for a field point $\bar{\xi}$ whose distance above the surface is y. Lines perpendicular to $\bar{e}_{\mathrm{I}}$ are wavefronts of the incident wave; the dashed one in the figure is the leading wavefront of the incident pulse at the instant it arrives at the field point. The pulse in the incident wave along the direct path just touches the field point at the instant

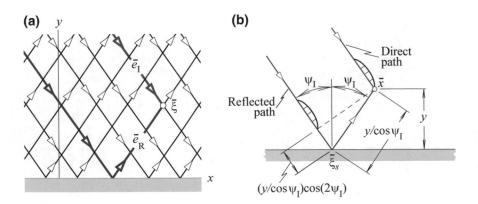

Fig. 5.4 Rays of the incident and reflected waves that arrive at an arbitrary field point $\bar{x}$

described by the figure, but the profile along the reflected path has not yet arrived at the surface. A geometrical analysis would show that the additional distance the wave must travel along the reflected path before it arrives at $\bar{\xi}$ is $(y/\cos\psi_1)(1+\cos 2\psi_1) \equiv 2y\cos\psi_1$. The signal propagates along the rays at the speed of sound, from which it follows that the time required for the reflected signal to arrive is $\Delta t = 2\,(y/c)\cos\psi_1$. If $\Delta t < \tau$, there will be an interval in which the pulses in the incident and reflected waves overlap. Otherwise, they will be observed as distinct events.

This is not a new observation, in that we identified it by examining the region in Fig. 5.3 that is insonified by both pulses. Indeed, we could have identified the time between arrival of the incident and reflected pulses by considering Fig. 5.3. To do so we would observe that at a height y above the reflection plane the distance in the x direction from the leading wavefront of the incident signal to the leading wavefront of the reflected signal is $2y\cot\psi_1$. The trace velocity is $c/\sin\psi_1$, so the time required for the reflected wave to arrive at a field is $\Delta t = 2y\cot\psi_1/(c/\sin\psi_1)$, which is the same as the result derived by considering the properties of rays.

The alternative analyses of the time delay for arrival of the reflected wave exemplify a general aspect. Both wavefronts and rays are useful for viewing wave propagation. Both perspectives can be useful, and they often are used jointly.

EXAMPLE 5.2 A simple model for the shock wave radiated by a supersonic aircraft is the N-wave pulse of an incident wave. It is permissible to approximate this signal as a plane wave at an angle of incidence of $60°$ with a duration of $T = 29.4$ ms. The reflecting surface has a purely resistive impedance of 500 Rayl. Evaluate the waveforms of pressure and particle velocity received at points at several elevations above the origin.

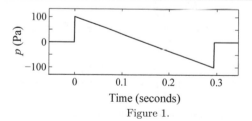

Figure 1.

Significance

The effect of the angle of incidence on the arrival time and the relationship between particle velocity and pressure, as well as the evaluation of waveforms, are the aspects that are emphasized by this example.

Solution

The problem statement is not specific in regard to where the waveforms should be evaluated, so our selection will be such that we obtain a good picture of how distance to the reflecting surface affects the signal that is received. According to Fig. 5.3, at

$y > cT/(2\cos\psi_1)$ the incident signal will pass completely before the reflected pulse
has arrived. Below this elevation, the waveforms overlap over an interval. Thus, we
will compute the waveforms at $x = 0$, $y = (0.4n)cT/(2\cos\psi_1)$, $n = 0, 1, 2, 3$.

For an angle of incidence of $60°$ and standard atmospheric conditions, the reflection coefficient corresponding to $Z = 500$ Rayl is $R = -0.240$. The surface is said
to be hard when $Z > \rho_0 c$, but this terminology stems from the analysis of normal
incidence. At this angle of incidence, the surface appears to be softer than the fluid.
The incident waveform is not too complicated, so we could graphically construct
the combination of incident and reflected signals. However, several such constructions are called for, so we shall use mathematical software. The pressure is given by
Eq. (5.2.9), with

$$\bar{e}_I = 0.8660\bar{e}_x - 0.50\bar{e}_y, \quad \bar{e}_R = 0.8660\bar{e}_x + 0.50\bar{e}_y$$
$$\Theta_I = t - \frac{0.8660x - 0.50y}{c}. \quad \Theta_R = t - \frac{0.8660x + 0.50y}{c}$$
$$F_I(t) = 100\left(1 - \frac{2t}{T}\right)[h(t) - h(t-\tau)]$$

The total pressure and particle velocity are the superposition of the contributions of
both plane waves,

$$P = F_I(\Theta_I) + RF_I(\Theta_R), \quad \bar{v} = \frac{1}{\rho_0 c}[F_I(\Theta_I)\bar{e}_I + RF_I(\Theta_R)\bar{e}_R]$$

We will display the velocity as graphs of the x and y components, which are

$$\bar{v}\cdot\bar{e}_x = \frac{0.8660}{\rho_0 c}[F_I(\Theta_I) + RF_I(\Theta_R)]$$

$$\bar{v}\cdot\bar{e}_y = \frac{0.50}{\rho_0 c}[-F_I(\Theta_I) + RF_I(\Theta_R)]$$

We must determine the time interval to be depicted by the plots. At the origin
$\Theta_I = \Theta_R = t$. The waveform at any location is found by replacing t with Θ_I or Θ_R
in F_I. It follows that the interval of the incident pulse at any location corresponds to
$0 < \Theta_I < T$. For a point on the y-axis, where $x = 0$, this translates to $-0.50y/c <
t < -0.50y/c + T$. The same reasoning leads to recognition that the interval over
which $F_R(\Theta_R)$ is nonzero at points on the y-axis is $0.50y/c < t < 0.50y/c + T$.
The largest value of y for an evaluation has been set at $0.6cT/\cos(\psi_1)$, so the
range of time we will consider is $-0.3T/\cos(\psi_1) < t < 0.3T/\cos(\psi_1) + T$. The
computerized evaluation of the pressure and particle velocity components according
to the preceding expressions is done by defining a column vector of t values in the
designated range and then evaluating the corresponding Θ_I and Θ_R values for the
location of interest. A direct substitution gives the response variables.

According to Fig. 2, at the highest elevation, $y = 120$ m, the entire incident pulse
elapses before the reflected wave arrives. The x component of particle velocity
reverses phase in the reflection, whereas the y component does not, because R is

negative. At $y = 80$ m, the reflected wave arrives when the incident wave is in its negative phase, whereas at $y = 40$ m the reflected wave arrives while the incident wave is in its positive phase. Whether a particle velocity is enhanced or lessened by arrival of the reflected wave depends on the sign of the components of $\bar{e}_R$. At $y = 0$, the reflected wave departs when the incident wave arrives, so the individual waveforms overlap. The maxima of the pressure and x component of particle velocity at this location are lessened relative to the incident wave, while the y component is enhanced.

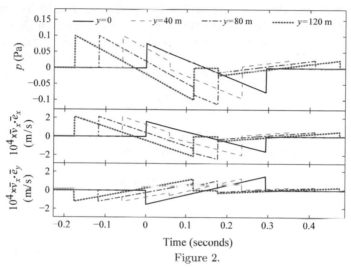

Figure 2.

5.2.2 Reflection from a Frequency-Domain Perspective

Phenomenologically, the primary difference between frequency- and time-domain analyses of reflection is a consequence of the persistent nature of a harmonic wave, as compared to the abbreviated nature of a plane wave pulse. The result is that interference between the incident and reflected waves plays an important role. Consistent with our general approach, we will analyze reflection of a harmonic plane wave as though we had not already performed a time-domain analysis.

Field Properties

Figure 5.2, which depicts the propagation directions of the incident and reflective waves, provides the starting point for the development. Thus, the harmonic plane wave incident on the surface propagates in direction $\bar{e}_I$ at angle of incidence ψ_I and the reflected wave's direction is $\bar{e}_R$ at angle of reflection ψ_R. It remains true that there is no variation in the direction perpendicular to the plane formed by $\bar{e}_I$ and $\bar{e}_y$, so we take $\bar{e}_R$ to also lie in this plane. The surface has a local impedance Z. Whereas we required that Z be real and constant to perform the time-domain analysis, it now may be complex and frequency dependent. However, we still require that it be the same value at all locations on the surface.

The governing equations are linear, which permits us to construct a solution by superposing simple plane waves. Thus, the complex amplitudes of the pressure field are taken to be

$$P = P_I + P_R$$
$$P_I = B_I e^{-ik\bar{e}_I \cdot \bar{\xi}}, \quad P_R = B_R e^{-ik\bar{e}_R \cdot \bar{\xi}} \tag{5.2.10}$$

where the unit vectors are as described in Eqs. (5.2.2) and (5.2.3). The factors B_I and B_R represent the complex amplitude of the respective waves at the origin. The particle velocity is the superposition of the contribution of each plane wave, so that

$$\bar{V} = \bar{V}_I + \bar{V}_R$$
$$\bar{V}_I = \frac{P_I}{\rho_0 c} \bar{e}_I, \quad \bar{V}_R = \frac{P_R}{\rho_0 c} \bar{e}_R \tag{5.2.11}$$

A particle velocity that is inward to the surface is in the negative $\bar{e}_y$ direction, so the local impedance boundary condition requires that

$$P = -Z\bar{V} \cdot \bar{e}_y \ @ \ y = 0 \tag{5.2.12}$$

Substitution of the assumed forms leads to

$$B_I e^{-ikx \sin \psi_I} + B_R e^{-ikx \sin \psi_R} = -\frac{Z}{\rho_0 c} \left(-B_I \cos \psi_I e^{-ikx \sin \psi_I} + B_R \cos \psi_R e^{-ikx \sin \psi_R} \right) \tag{5.2.13}$$

All complex exponentials in this expression must have the same argument because the equality must be satisfied at all x. For this reason, the angle of reflection must equal the angle of incidence,

$$\boxed{\psi_R = \psi_I} \tag{5.2.14}$$

Because the complex exponentials are the same, the collected coefficients on each side of the equal sign also must be equal. The solution of that equality is

$$\boxed{B_R = R B_I, \quad R = \frac{Z - Z_0}{Z + Z_0}, \quad Z_0 = \frac{\rho_0 c}{\cos \psi_1}} \tag{5.2.15}$$

If Z is complex, then the reflection coefficient R will also be complex. For real Z, this expression is the same as that derived in the time-domain analysis.

The pressure field corresponding to the preceding is

$$\boxed{p(x, y, t) = \text{Re}\left[B_I e^{i(\omega t - kx \sin(\psi_1))} \left(e^{iky \cos(\psi_1)} + R e^{-iky \cos(\psi_1)} \right) \right]} \tag{5.2.16}$$

Thus, the surface pressure is

$$p(x, 0, t) = \text{Re}\left[B_{\text{I}}(1 + R)e^{i(\omega t - kx \sin(\psi_1))}\right]$$
(5.2.17)

Equality of the angle of reflection and the angle of incidence sets both trace wavenumbers to $k \sin \psi_1$. Equivalently, both plane waves have identical trace wavelengths, $2\pi/(k/\cos(\psi_1))$. This quality is depicted in Fig. 5.5. Another way of stating the same property is to say that the trace velocity of both waves is $c/\sin \psi_1$.

At the surface, the total pressure amplitude is $|1 + R|$ times the amplitude of the incident wave. If $R = 1$, corresponding to a rigid surface, the surface pressure will be twice the incident. The best way to understand the broader implications of this observation is to examine the mean-squared pressure throughout the field.

Time Averages

An evaluation of time-averaged quantities begins with the representation of the field as a sum of incident and reflected waves. Thus, for the mean-squared pressure we write

$$\langle p^2 \rangle_{\text{av}} = \frac{1}{2}(P_{\text{I}} + P_{\text{R}})(P_{\text{I}}^* + P_{\text{R}}^*)$$
$$= \frac{1}{2}\left(|P_{\text{I}}|^2 + |P_{\text{R}}|^2\right) + \text{Re}\left(P_{\text{I}}P_{\text{R}}^*\right)$$
(5.2.18)

while the time-averaged intensity is described by

$$\bar{I}_{\text{av}} = \frac{1}{2\rho_0 c}\text{Re}\left((P_{\text{I}} + P_{\text{R}})(P_{\text{I}}^* \bar{e}_{\text{I}} + P_{\text{R}}^* \bar{e}_{\text{R}})\right)$$
$$= \frac{1}{2\rho_0 c}\left[|P_{\text{I}}|^2 \bar{e}_{\text{I}} + |P_{\text{R}}|^2 \bar{e}_{\text{R}} + \text{Re}\left(P_{\text{R}}P_{\text{I}}^* \bar{e}_{\text{I}} + P_{\text{I}}P_{\text{R}}^* \bar{e}_{\text{R}}\right)\right]$$
(5.2.19)

Fig. 5.5 Trace velocity and trace wavelength for reflection of a harmonic wave

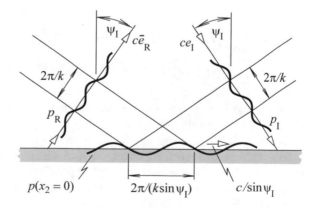

The first two terms in $\left(p^2\right)_{\text{av}}$ and $\bar{I}_{\text{av}}$ are the values that would be obtained if each wave existed without the other. The other terms represent interference between the waves. In the case of $\bar{I}_{\text{av}}$, the third term corresponds to the power required to propagate the incident wave in opposition to the pressure field in the reflected wave, while the fourth term is the corresponding effect for the reflected wave. These interference effects first arose in Example 4.6.

Now that we have a general understanding of averages, let us examine their spatial dependence by substituting the complex amplitudes in Eq. (5.2.10) with $B_{\text{R}} = R B_{\text{I}}$. The result for the mean-squared pressure is

$$\boxed{\left(p^2\right)_{\text{av}} = \frac{1}{2} |B_{\text{I}}|^2 \left[1 + |R|^2 + 2\,\text{Re}\left(R e^{2iky\cos\psi_1}\right)\right]} \qquad (5.2.20)$$

This expression shows that there is a sinusoidal variation of $\left(p^2\right)_{\text{av}}$ in the direction perpendicular to the surface. The wavelength of this variation is $2\pi/\left(k\cos\psi_1\right)$. We encountered a similar form in Eq. (3.2.52) for an impedance tube. The only difference is that the wave here is at oblique incidence, so the wavenumber perpendicular to the surface is $k\cos\psi_1$. The preceding expression also shows that the mean-squared pressure does not depend on distance parallel to the surface. This property is a consequence of the infinite extent of the reflecting surface and the fact that its properties do not vary spatially.

A very hard surface is one for which $R \approx 1$. The mean-squared pressure on the surface in that case is approximately $2|B_{\text{I}}|^2$, whereas it is $|B_{\text{I}}|^2/2$ in the absence of the reflected wave. This result is not dependent on the angle of incidence, and it is independent of frequency. Thus, although it might be comforting to try to sleep in an airplane by resting your head on the cabin's wall, you would hear less noise if you did not do so.

An evaluation of power flow will shed insight into the role of the wall. To begin this analysis, we substitute the explicit descriptions of P_{I} and P_{R} into the preceding expression for $\bar{I}_{\text{av}}$, which leads to

$$\begin{aligned}
\bar{I}_{\text{av}} &= \frac{|B_{\text{I}}|^2}{2\rho_0 c} \left[\bar{e}_{\text{I}} + |R|^2\,\bar{e}_{\text{R}} + \text{Re}\left(R e^{-ik(\bar{e}_{\text{R}}-\bar{e}_{\text{I}})\cdot\bar{\xi}}\bar{e}_{\text{I}} + R^* e^{ik(\bar{e}_{\text{R}}-\bar{e}_{\text{I}})\cdot\bar{\xi}}\bar{e}_{\text{R}}\right)\right] \\
&= \frac{|B_{\text{I}}|^2}{2\rho_0 c} \left[\bar{e}_{\text{I}} + |R|^2\,\bar{e}_{\text{R}} + \text{Re}\left(2R^* \sin\psi_1\, e^{2iky\cos\psi_1}\right)\bar{e}_x\right]
\end{aligned} \qquad (5.2.21)$$

Note that the $\bar{e}_y$ component of the interference term vanishes because it is the difference of two quantities that are complex conjugates, which is purely imaginary. Equation (5.2.21) indicates that in addition to the intensity associated with the individual incident and reflected waves, each of which is independent of position and frequency, interference leads to power flow parallel to the surface. The amount varies sinusoidally with distance from the surface, in a manner similar to $\left(p^2\right)_{\text{av}}$. Consider a virtual wall perpendicular to the surface. The power flow past this wall would be obtained by integrating $\bar{I}_{\text{av}} \cdot \bar{e}_x$ along the wall. Because the contributions of the isolated incident and reflected waves to the intensity are independent of y, the integral of these contributions would be proportional to the height of the wall. In contrast, the contribution of the interference term to the power flow across this would fluctuate sinusoidally with the height of the wall.

An important quantity is the power that is transported into the boundary. We isolate a surface patch of length L in the x direction and unit width in the z direction region. The normal to the interface pointing out of the fluid domain, that is, into the boundary, is $\bar{n} = -\bar{e}_y$, so the interaction term in the time-averaged intensity does not contribute to the power transported across the horizontal surface. The value of $\bar{I}_{av}$ is independent of the x coordinate, so the power flowing from the fluid across the reflecting surface is

$$\mathcal{P}_{surf} = \bar{I}_{av} \cdot (-\bar{e}_y) L = \frac{|B_I|^2}{2\rho_0 c} \left(1 - |R|^2\right) L \cos \psi_1 \qquad (5.2.22)$$

A pictorial explanation of this result appears in Fig. 5.6, which shows the bundle of rays of the incident wave that arrive at the surface patch. The bundle representing reflection of these rays also appears there. There is no variation of each plane wave perpendicularly to its rays, and a trigonometric construction would show that the cross-sectional area of both bundles is $(L \cos \psi_1) (1)$. Thus, the power arriving at the interface in the incident wave and the power emanating from the boundary in the reflected wave are

$$\mathcal{P}_I = \frac{|P_I|^2}{2\rho_0 c} (L \cos \psi_1), \quad \mathcal{P}_R = \frac{|P_R|^2}{2\rho_0 c} (L \cos \psi_1) \qquad (5.2.23)$$

At any location, $|P_I| = |B_I|$ and $|P_R| - |R| \, |B_I|$. Thus, Eq. (5.2.22) states that

$$\boxed{\mathcal{P}_{surf} = \mathcal{P}_I - \mathcal{P}_R} \qquad (5.2.24)$$

Fig. 5.6 Incident and reflected rays of plane waves on a $L \times 1$ patch of the interface

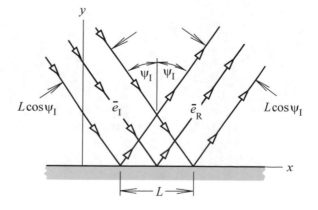

This is a direct corollary of the energy theorem: The power transported across the interface is the difference between the power that arrives in the incident wave and the power that emanates in the reflected wave.

The *absorption coefficient* α is the ratio of the power per unit surface area that flows into the surface to the power contained in the incident wave, that is,

$$\alpha = 1 - |R|^2 \tag{5.2.25}$$

(The terminology here is standard, but it might cause confusion. The term "absorption coefficient" and the symbol α were used in Chap. 3 to describe the attenuation of waves as they propagate. In fact, the two parameters seldom arise in the same context, so we will retain the standard.) Because R depends on Z and ψ_1, and Z might depend on ω, the absorption coefficient of a specific material is a function of the frequency and angle of incidence. A manufacturer of materials for acoustical liners should provide a thorough description of the absorption coefficient, or its complement, $|R|^2$, which is called the *energy reflection coefficient*. An implicit aspect of the material specification is the fluid to which it applies. The difference of characteristic impedance between gases and liquids is enormous. Thus, a material that might have desirable acoustical properties for noise control in a room is likely to be quite unsuitable for coating a submarine.

EXAMPLE 5.3 Acoustical tiles are a common noise control treatment for rooms. Consider a tile that consists of 1.5 mm diameter holes through a 10-mm-thick panel that may be considered to be rigid. The openings of the holes constitute 10% of the surface area of a tile. The tiles are suspended 200 mm below the ceiling, which also may be considered to be rigid. Thus, the set of holes in the tile may be considered to be the neck of a Helmholtz resonator, and the region above the tiles may be considered to be the cavity. The relation between the pressure at an opening and the particle velocity in a hole is given by Eq. (3.5.60), which is modified by the addition of a resistive term resulting from wall friction. The result is that

$$Z_H = i \rho_0 L' \omega \left[1 - i\eta \left(\frac{\omega_H}{\omega} \right)^{1/2} - \frac{\omega_H^2}{\omega^2} \right]$$

where ω_H is the Helmholtz resonance frequency and the loss factor $\eta = 0.5$ for the tile. Determine the reflection coefficient for the tile as a function of the angle of incidence, and also determine the corresponding absorption coefficient. Carry out these evaluations for frequencies that are $\omega = 0.5\omega_H$, ω_H, and $2\omega_H$.

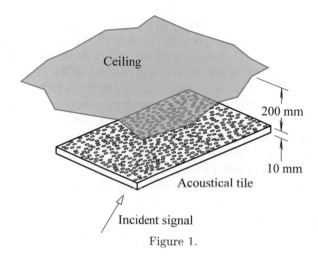

200 mm

10 mm

Acoustical tile

Incident signal

Figure 1.

Significance

In addition to exploring a common method for controlling noise with a locally reacting material, this example will enable us to see quantitative trends that underlie the
various formulas.

Solution

The given impedance Z_H is the ratio of pressure to particle velocity at the inlet to a
hole. An audible signal has a wavelength that is much greater than a hole's diameter.
Consequently, the effective particle velocity is an average of zero for 90% of the
tile's surface and V_H for the 10% coverage of the holes. In other words, the average
normal velocity is $\bar{V} \cdot \bar{e}_y = 0.1 V_H$ at $y = 0$. Correspondingly, the local impedance
seen by the incident acoustic wave is

$$Z = \frac{P(y=0)}{\bar{V} \cdot \bar{e}_y (y=0)} = 10 Z_H = 10 i \rho_0 L' \omega \left[1 - i \eta \left(\frac{\omega_H}{\omega} \right)^{1/2} - \frac{\omega_H^2}{\omega^2} \right]$$

The Helmholtz resonance frequency is given by Eq. (3.5.60),

$$\omega_H = c \left(\frac{\mathcal{A}}{L' \mathcal{V}} \right)^{1/2} \tag{5.2.26}$$

The parameters are determined by considering a 1 m square section of tile. The
cavity volume is set by the 0.2 m gap between the tile and the ceiling, so $\mathcal{V} = 1\ \text{m}^2$
$\times$ 0.2 m. The neck area $\mathcal{A}$ is the 10% of a 1 m^2 patch that is open to the cavity,
so $\mathcal{A} = 0.1\ \text{m}^2$. For the neck length L', we apply the low-frequency end correction
in Eq. (3.5.57) to both ends of a hole. Hence, the effective length is the sum of the
tile's thickness, which is 10 mm, and $8/(3\pi)$ of the hole's 0.75 mm radius, doubled
to account for each end. This gives $L' = 0.011273$ m. The Helmholtz resonance
frequency for these parameters is 2264 rad/s.

Figure 2 displays Re (R) and Im(R) as a function of the angle of incidence. At $0.5\omega_{\mathrm{H}}$, the specific impedance is $\zeta = 0.2654 - 1.1262i$. Thus, it is mainly compliant, which is the nature of a Helmholtz resonator below its resonant frequency. The reflection factor at normal incidence is close to $R = -i$. At this frequency, the value of R changes little until ψ_1 is greater than 40°, where it progressively tends to $R = -1$ for grazing incidence. At $\omega = \omega_{\mathrm{H}}$, the specific impedance is purely resistive, $\zeta = 0.375$. The value of R is real for all angles of incidence, gradually transitioning from $R = -0.45$ at normal incidence to $R = -1$ at grazing incidence. At $\omega = 2\omega_{\mathrm{H}}$, the impedance is $\zeta = 0.5309 + 1.1262i$, which is an inertance combined with a smaller resistance. The angular dependence of Re (R) in this case is like that for $\omega = 0.5\omega_{\mathrm{H}}$, while Im (R) here behaves like $-$ Im (R) at the lower frequency.

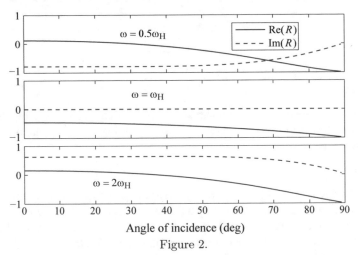

Figure 2.

The greatest power transfer into the surface occurs when $|R|$ is smallest. Figure 3 indicates that the ceiling tile absorbs the most when $\omega = \omega_{\mathrm{H}}$, which is consistent with the results for R. Recall that when the frequency is ω_{H}, the reactive part of the impedance is zero. In this situation, the magnitude of R will depend on how close the resistive part of Z matches the fluid's characteristic impedance.

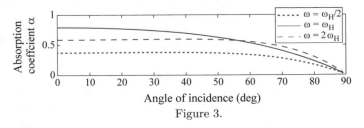

Figure 3.

5.3 Transmission and Reflection at an Interface Between Fluids

The concern here is adjacent half-spaces filled with different fluids. A plane acoustic wave in one fluid is incident at the planar interface, which induces an acoustic wave in the second fluid. This situation is readily obtained if one considers a gas above the free surface of a liquid. It also can be obtained with two liquids if they are immiscible. In both cases, the interface between the fluids will be a horizontal plane. It is possible for the interface to have a different orientation or to fill each half-space with a different gas. However, such configurations require that the media be separated by an impermeable sheet that has no rigidity.

The situation is different if the medium receiving the incident wave is an elastic solid. Several types of waves may exist in a solid. One such wave is said to be dilatational because it expands and compresses the material through which it propagates, much like acoustic waves in a fluid. However, a solid can sustain shear stresses, as well as direct stresses. Not surprisingly, a shear wave also may propagate through a solid. In this wave, there is neither compression nor expansion—it is unlike anything encountered in an ideal fluid.

The analysis of waves in a solid is a generalization of the approach we will follow here, but the details are beyond the present scope. In a few situations, we will treat a solid as though shear waves do not exist. This is acceptable for normal incidence. However, when a dilatational wave is obliquely incident at an interface, a shear wave is generated in order to satisfy continuity conditions on stresses. Ignoring the existence of shear waves allows us to consider a broader range of material properties. In that case, the most we can hope for is that the analysis yields results that are qualitatively correct.

5.3.1 Time-Domain Analysis

A time-domain analysis will provide a basic understanding of the underlying processes, and it also will disclose situations where a frequency-domain analysis must be pursued. We define the xz-plane to coincide with the interface at which the two half-spaces meet. The situation in fluid 1 is like the case of reflection from a locally reacting surface, in that it contains incident and reflected plane waves. Trace matching along the interface will lead to the conclusion that the angle of reflection equals the angle of incidence, so we will denote both angles as ψ_1 to associate them with that fluid. Equally important is the wave that is induced in medium 2. It is reasonable to begin with the assumption that this wave also is planar; we say that it is the *transmitted wave,* which propagates at the *angle of transmission* ψ_2 relative to the normal to the interface. Figure 5.7 is a picture of the rays for each wave.

Figure 5.8 describes the field when a pulse is incident at a planar interface. The wavefronts of the incident, reflected, and transmitted waves are perpendicular to the respective propagation directions. The distance from the leading to trailing wavefront

Fig. 5.7 Rays of the
incident, reflected, and
transmitted wave at a planar
interface between two fluids

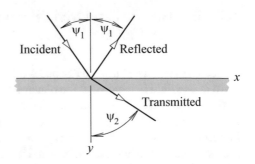

in each wave is the speed of sound in the respective fluid multiplied by the pulse
duration τ. The insonified region, that is, the region in which the acoustic pressure is
not zero, is shaded. It should be noted that the y direction has been defined to point
into fluid 2. Doing so is merely a matter of convenience.

The picture in Fig. 5.8 is a snapshot of the insonified region as it translates to
the right at the trace velocity. It follows that each set of waves must have the same
trace velocity in the x direction. If this were not true, it would not be possible to
satisfy boundary conditions at the interface between the media. To evaluate the trace
distance between points A and B, we observe that the duration of the pulse is τ.
The distance between these wavefronts therefore is $c_1\tau$ for the incident and reflected
waves, and $c_2\tau$ for the transmitted wave. The trace distance is $c_1\tau/\sin\psi_1$ for the
incident and reflected waves, and it is $c_2\tau/\sin\psi_2$ for the transmitted wave. This
corresponds to trace velocities that are $c_1/\sin\psi_1$ and $c_2/\sin\psi_2$. The condition that

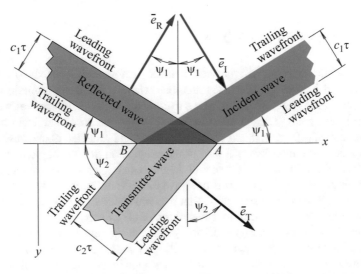

Fig. 5.8 Reflected and transmitted plane waves resulting from the incidence of a plane acoustic
pulse at the interface between two fluids

all trace velocities must be equal is met if

$$\boxed{\frac{1}{c_{tr}} = \frac{\sin \psi_2}{c_2} = \frac{\sin \psi_1}{c_1}} \tag{5.3.1}$$

This is *Snell's law*.

A question that arises at this juncture pertains to the relative size of c_1 and c_2. If c_2 is less than c_1, then ψ_2 will be smaller than ψ_1. In contrast, $c_2 > c_1$ leads to the possibility that Snell's law will give $\sin \psi_2 > 1$, which occurs if $\psi_1 > \sin^{-1}(c_1/c_2)$. The value of ψ_2 in this case is complex. The angle of incidence for which $\psi_2 = 90°$ is called the *critical angle of incidence*,

$$\boxed{\psi_{cr} = \sin^{-1}(c_1/c_2)} \tag{5.3.2}$$

The fact that Snell's law might lead to a value of $\sin \psi_2$ that is greater than one suggests the acoustic field will be like the one generated by a subsonic surface wave. It is for this reason that our attempt to construct a time-domain solution is limited to cases of regular incidence, which occurs if $0 \le \psi_1 < 90°$ for $c_2 < c_1$ and $0 \le \psi_1 < \psi_{cr}$ for $c_2 > c_1$. The frequency-domain analysis we will pursue will not be subject to this limitation.

Snell's law prescribes the directions in which the reflected and transmitted waves propagate, but it still remains to evaluate their waveforms. In mathematical form, what we have established is

$$
\begin{aligned}
p_I &= F_I(\Theta_I), \quad \Theta_I = t - \frac{\bar{\xi} \cdot \bar{e}_I}{c_2} \\
p_R &= F_R(\Theta_R), \quad \Theta_R = t - \frac{\bar{\xi} \cdot \bar{e}_R}{c_2} \\
p_T &= F_T(\Theta_T), \quad \Theta_T = t - \frac{\bar{\xi} \cdot \bar{e}_T}{c_2}
\end{aligned}
\tag{5.3.3}
$$

with the propagation directions set at

$$
\begin{aligned}
\bar{e}_I &= (\sin \psi_1)\,\bar{e}_x + (\cos \psi_1)\,\bar{e}_y \\
\bar{e}_R &= (\sin \psi_1)\,\bar{e}_x - (\cos \psi_1)\,\bar{e}_y \\
\bar{e}_T &= (\sin \psi_2)\,\bar{e}_x + (\cos \psi_2)\,\bar{e}_y, \quad \psi_2 = \sin^{-1}\left(\frac{c_2}{c_1}\sin \psi_1\right)
\end{aligned}
\tag{5.3.4}
$$

The task now is to determine the functions F_R and F_T given F_I. As always, we assume that bubbles, which are the manifestation of *cavitation*, are not present. In that case, continuity conditions apply at the interface of the two fluids. The physical situation is somewhat like the conditions we encountered in Sect. 3.4.1, where we addressed the junction of waveguides. Both fluids must have identical normal particle velocity components at the interface. In addition, the pressure is transmitted across the interface. The coordinate y is zero at the interface, with $y < 0$ for fluid 1 and $y > 0$ for fluid 2. We will denote the position of an interface point to be $y = 0^-$ if it is in fluid 1, or $y = 0^+$ if it is in fluid 2. Thus, the boundary conditions to be satisfied are

$$p_T|_{y=0^+} = (p_I + p_R)|_{y=0^-}$$

$$\bar{e}_y \cdot \bar{e}_T \frac{p_T}{\rho_2 c_2}\bigg|_{y=0^+} = \left(\bar{e}_y \cdot \bar{e}_I \frac{p_I}{\rho_1 c_1} + \bar{e}_y \cdot \bar{e}_R \frac{p_R}{\rho_1 c_1}\right)\bigg|_{y=0^-} \qquad (5.3.5)$$

Snell's Law was derived by equating the phase variable of each wave at $y = 0$, so the boundary condition will be satisfied everywhere on the surface if it is satisfied at one point. Selecting the origin for this purpose is best because all phase variables are t at that location. The result of substitution of Eq. (5.3.3) into the boundary conditions with $\bar{\xi} = \bar{0}$ is

$$F_T(t) = F_I(t) + F_R(t)$$

$$\cos(\psi_2) \frac{F_T(t)}{\rho_2 c_2} = \cos(\psi_1) \frac{F_I(t)}{\rho_1 c_1} + \cos(\psi_1) \frac{F_R(t)}{\rho_1 c_1} \qquad (5.3.6)$$

The only way these conditions can be met for an arbitrary value of t is if the functions are proportional, that is,

$$\boxed{F_R(t) = R F_I(t), \quad F_T(t) = T F_I(t)} \qquad (5.3.7)$$

In this expression, R is the reflection coefficient as before, and T is the *transmission coefficient*. (It is permissible to use T here, as well as to denote a time period, because they occur in different contexts.) These coefficients are found from the boundary conditions to be

$$\boxed{R = \frac{Z_2 - Z_1}{Z_2 + Z_1}, \quad T = \frac{2Z_2}{Z_2 + Z_1}, \quad Z_j = \frac{\rho_j c_j}{\cos \psi_j}} \qquad (5.3.8)$$

The parameters Z_1 and Z_2 are *effective impedance factors*. The values of Z_j are real numbers if $\psi_1 < \psi_{cr}$. In that case, T and R are real. The form of R is like the reflection coefficient for a locally reacting surface. However, unlike a local impedance, the value of Z_2 depends on the angle of incidence and it is independent of frequency. As the angle of incidence increases, the normal velocity decreases proportionally to $\cos \psi_1$, which increases the ratio of pressure to normal velocity. The transmission coefficient has some interesting attributes. If $Z_2 > Z_1$, then $|T|$ will be greater than one, which means that the transmitted signal will be larger than the incident. The limiting case occurs when ψ_1 approaches ψ_{cr}. Then, ψ_2 approaches $90°$, which leads to $R \to 1$ and $T \to 2$. Although these values seem to violate the principle of energy conservation, they do not. The power flowing into fluid 2 is less than the power incident on the boundary because the particle velocity in fluid 2 is reduced when $Z_2 > Z_1$. We will explore this in greater detail when we examine the behavior from a frequency-domain perspective.

We could obtain a frequency-domain picture of the reflection by representing $F_1(\Theta)$ as a complex exponential. However, we shall pursue the frequency-domain analysis independently in the next section. Doing so will lead to results for cases where $\psi_1 > \psi_{cr}$.

EXAMPLE 5.4 The ocean bottom in a certain region is a sediment that is reasonably well described as a liquid. In order to identify its sound speed and density, a hydrophone sensor has been buried 30 m below the bottom at location B. Location A is a reference hydrophone in the water. The sound speed and density of the water were measured to be 1490 m/s and 1030 kg/m^3. In one experiment, a uniform pressure pulse whose duration is 10 ms propagates from a remote source and arrives at the ocean bottom at an angle of incidence of 20°. Arrival of the source signal at location A defines $t = 0$, and the maximum acoustic pressure measured there is 35 kPa. The pressure pulse arrives at location B when $t = 91.2$ ms, and the maximum pressure measured there is 50.4 Pa. Determine the density and sound speed of the sediment and the extent of the regions in both fluids that are insonified at $t = 91.2$ ms.

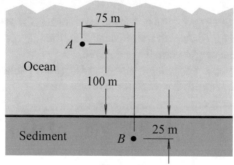

Figure 1.

Significance

This example applies the laws of reflection and transmission to an examination of rays and wavefronts. It also demonstrates the usage of the trace velocity as a fundamental analytical tool. A clearer picture of the transmission process will emerge.

Solution

The duration of the pulse is 10 ms, whereas the minimum time for a reflected wave at $\psi_R = 20°$ to travel from the bottom to point A is $(100/\sin\psi_R)/c = 74.4$ ms. This calculation tells us that the incident signal measured at position A ends before the reflected signal arrives there. Because $|R| \leq 1$, it must be that the maximum pressure observed at position A corresponds to the incident wave. The transmitted waveform is scaled relative to the incident one by the transmission factor, so the given information leads to

$$T = \frac{50.4}{35} = 1.44$$

We will obtain the requested information without explicit reference to the functional representations in Eq. (5.3.3). However, this is a good opportunity to demonstrate those equations. Let $f(t)$ be the pressure waveform observed at position A. The incident signal is stated to be a uniform pulse whose duration is $\tau = 0.01$ s and whose amplitude is 35 kPa, so

$$f(t) = 35\left(10^3\right)\left[h(t) - h(t - \tau)\right] \text{ Pa}$$

where $h(t)$ is the usual notation for a step function. The first of Eq. (5.3.3) states that

$$p_I\left(\bar{\xi}_A, t\right) = f(t) = F_I\left(t - \frac{\bar{\xi}_A \cdot \bar{e}_I}{c_1}\right)$$

In other words, the wave function is

$$F_I(t) = f\left(t + \frac{\bar{\xi}_A \cdot \bar{e}_I}{c_1}\right)$$

The interpretation of this expression is that the waveform at the origin is the waveform at point A shifted by $\bar{\xi}_A \cdot \bar{e}_I / c_1$. In the present situation, $\bar{\xi}_A \cdot \bar{e}_I$ will be negative, so what is observed at the origin actually is delayed.

At the origin, all phase variables equal t, so Eq. (5.3.7) states that the functions for the incident and transmitted waves at the origin are

$$P_R|_{\bar{\xi}=0} = Rf\left(t + \frac{\bar{\xi}_A \cdot \bar{e}_I}{c_1}\right), \quad P_T|_{\bar{\xi}=0} = Tf\left(t + \frac{\bar{\xi}_A \cdot \bar{e}_I}{c_1}\right)$$

At any location other than the origin, the waveform is delayed by $\bar{\xi} \cdot \bar{e}_R / c_1$ for the reflected wave and $\bar{\xi} \cdot \bar{e}_T / c_2$ for the transmitted wave, so that

$$P_R = Rf\left(t + \frac{\bar{\xi}_A \cdot \bar{e}_I}{c_1} - \frac{\bar{\xi} \cdot \bar{e}_R}{c_1}\right), \quad P_T = Tf\left(t + \frac{\bar{\xi}_A \cdot \bar{e}_I}{c_1} - \frac{\bar{\xi} \cdot \bar{e}_T}{c_2}\right)$$

At first glance, it might seem that the requested travel time is the time required for the incident signal to travel from position A to its intersection with the surface and then from that intersection to position B. This reasoning is not correct because it assumes that the incident ray that intersects position A will generate the transmitted ray that intersects position B. Figure 2 shows the incident ray through position A and the transmitted ray through position B in the situation where $\psi_2 > \psi_1$. Because the positions and ψ_1 are fixed, there is only one value of ψ_2 that will fulfill such a condition, but ψ_2 is set by Snell's law. Hence, we cannot assume that a single ray intersects both points.

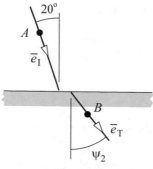

Figure 2.

Because consideration of ray paths does not seems to be helpful, let us consider wavefronts. Figure 3 is comparable to Fig. 5.8, except it does not show the reflected wave and it describes two instants: $t = 0$, when the leading edge of the incident wave arrives at position A, and $t = 0.0912$ s, when the leading edge of the transmitted wave arrives at position B.

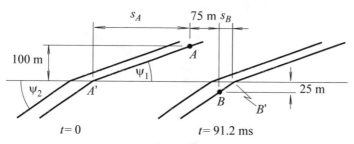

Figure 3.

In Fig. 3 points A' and B' are the locations where the leading edge containing A and B, respectively, intersects the interface. The intersection of any wavefront with the interface moves at the trace velocity, which is $c_{tr} = c_1 / \sin \psi_1$. Hence, the distance between these points, which is $s_A + 75 + s_B$, must equal c_{tr} times the elapsed time, 0.0912 s. Both s_A and s_B are sides of right triangles. The sum of those lengths and horizontal distance between A and B must equal the distance covered at the trace velocity in the time interval,

$$100 \cot \psi_1 + 75 + 25 \cot \psi_2 = \frac{c_1}{\sin \psi_1} (0.0912)$$

Both c_1 and ψ_1 are given, so this relation may be solved for the transmission angle, which is found to be

$$\psi_2 = 27.73°$$

The value of c_2 may then be found from Snell's law,

$$c_2 = c_1 \frac{\sin \psi_2}{\sin \psi_1} = 2027 \text{ m/s}$$

For reference, the critical angle is

$$\psi_{cr} = \sin^{-1}\left(\frac{c_1}{c_2}\right) = 47.32°$$

The region insonified by the incident and transmitted waves at $t = 91.2$ ms is described by Fig. 4. The distance between the leading and trailing edges of the incident and reflected waves is $c_1\tau$, while the matching distance for the transmitted wave is $c_2\tau$. The distance from point A to the location where the leading edges intersect the interface is found from the construction in Fig. 3.

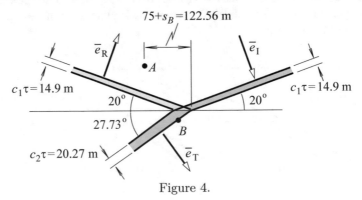

Figure 4.

It still remains to determine ρ_2. This property appears in the expression for the transmission coefficient. We have determined that $T = 1.44$, but T also is given by Eq. (5.3.8). We solve that expression for Z_2, which gives

$$Z_2 \equiv \frac{\rho_2 c_2}{\sin \psi_2} = \frac{T Z_1}{2 - T} = \left(\frac{T}{2 - T}\right)\frac{\rho_1 c_1}{\sin \psi_1}$$

All parameters appearing here other than ρ_2 are known. The solution is

$$\rho_2 = 2649 \text{ kg/m}^3$$

5.3.2   Frequency-Domain Analysis

A time-domain analysis of transmission in a case of supercritical incidence constitutes a forbidding task. This situation is readily addressed in the frequency domain, primarily because time dependence is not an issue. The result will be a single mathematical form that represents all cases, subcritical and supercritical. That solution will be analyzed to determine the power flow across the interface.

The General Solution

The fact that there is a plane wave in fluid 1 that reflects from a planar boundary suggests that there is much similarity between the cases where the interface is formed with another fluid and that in which the interface is a locally reacting material. In particular, the requirement that all waves have the same trace wavelength along the interface tells us that there is a reflected plane wave for which the angle of incidence equals the angle of reflection. The trace matching requirement also suggests that a plane wave is transmitted into fluid 2. It is reasonable to let the complex amplitudes of the reflected and transmitted waves be proportional to the incident wave's complex amplitude. Thus, we begin by assuming that the frequency-domain solution is described by

$$\boxed{\begin{array}{c} P_1 = P_I + P_R, \quad P_2 = P_T \\ P_I = B_I e^{-ik_1 \bar{e}_I \cdot \bar{\xi}}, \quad P_R = R B_I e^{-ik_1 \bar{e}_R \cdot \bar{\xi}}, \quad P_T = T B_I e^{-ik_2 \bar{e}_T \cdot \bar{\xi}} \end{array}} \qquad (5.3.9)$$

where R and T are the reflection and transmission coefficients. The propagation directions are taken to be

$$\begin{aligned} \bar{e}_I &= (\sin \psi_1) \, \bar{e}_x + (\cos \psi_1) \, \bar{e}_y \\ \bar{e}_R &= (\sin \psi_1) \, \bar{e}_x - (\cos \psi_1) \, \bar{e}_y \\ \bar{e}_T &= (\sin \psi_2) \, \bar{e}_x + (\cos \psi_2) \, \bar{e}_y \end{aligned} \qquad (5.3.10)$$

The frequency ω is the same for all waves, so the wavenumbers are

$$k_1 = \frac{\omega}{c_1}, \quad k_2 = \frac{\omega}{c_2} \qquad (5.3.11)$$

The corresponding particle velocities are

$$\bar{V}_1 = \bar{V}_I + \bar{V}_R, \quad \bar{V}_2 = \bar{V}_T$$

$$\bar{V}_I = \frac{P_I}{\rho_1 c_2} \bar{e}_I, \quad \bar{V}_R = \frac{P_R}{\rho_1 c_2} \bar{e}_R, \quad \bar{V}_T = \frac{P_T}{\rho_2 c_2} \bar{e}_T \qquad (5.3.12)$$

The boundary conditions enforce continuity of pressure and the normal component of particle velocity at the interface. The origin of a reference coordinate system is defined to be situated somewhere on the interface. The y direction is normal to the boundary pointing into fluid 2, as it was for the time-domain analysis. The x-axis lies in the propagation plane on the interface. Thus, equating the trace of each wave requires that $k_1 \bar{\xi} \cdot \bar{e}_I = k_1 \bar{\xi} \cdot \bar{e}_R = k_2 \bar{\xi} \cdot \bar{e}_T$ at $y = 0$, that is $k_1 (\sin \psi_1) x = k_2 (\sin \psi_2) x$. cancellation of the common factor ωx leads to Snell's Law, as previously stated in Eq. (5.3.1),

$$\frac{1}{c_{\mathrm{tr}}} = \frac{\sin \psi_1}{c_1} = \frac{\sin \psi_2}{c_2} \qquad (5.3.13)$$

It is helpful to restate the conditions for supercritical incidence. The angle of incidence may be any value from $\psi_1 = 0$ to $\psi_1 = \pi/2$. If c_2 is less than c_1, then the value of ψ_2 will always be less than ψ_1, except for normal incidence, in which case $\psi_2 = \psi_1 = 0$. If c_2 equals c_1, then $\psi_2 = \psi_1$. (This obviously would be the case where both fluids are the same, but it could happen that two fluids have the same sound speed but different densities.) In the more complicated case of $c_2 > c_1$, the value of ψ_2 will be greater than ψ_2, until it equals $\psi_2 = \pi/2$ at the critical incidence angle,

$$\psi_{\mathrm{cr}} = \sin^{-1} \left(\frac{c_1}{c_2} \right) \qquad (5.3.14)$$

If ψ_1 exceeds ψ_{cr}, then Snell's Law gives a complex value for ψ_2.

In the subcritical incidence case, ψ_2 is real. Because Snell's Law is derived by equating at $y = 0$ the complex exponentials in Eq. (5.3.9), they may be factored out of the boundary conditions. The complex amplitude B_{I} also is a common factor, so what remains is

$$T = 1 + R$$
$$\frac{T \cos \psi_2}{\rho_2 c_2} = \frac{(1 - R) \cos \psi_1}{\rho_1 c_1} \qquad (5.3.15)$$

The solution of these equations gives the same expression for the reflection and transmission coefficients as Eq. (5.3.8),

$$R = \frac{Z_2 - Z_1}{Z_2 + Z_1}, \quad T = \frac{2Z_2}{Z_2 + Z_1} \qquad (5.3.16)$$

where the effective impedance factors are $Z_j = \rho_j c_j / \cos \psi_j$, as in Eq. (5.3.8). After the values of ψ_2, R, and T have been determined, the response in both fluids may be evaluated according to Eqs. (5.3.9) and (5.3.12).

In subcritical incidence cases, the value of ψ_2 obtained from Snell's Law is real. In that case, the value of T and R depends on the relative impedance ratio $(\rho_2 c_2) / (\rho_1 c_1)$ and the value of ψ_1, but they do not depend on the frequency. Suppose we used a Fourier series to represent a temporally periodic incident pressure, or more generally used a DFT to represent a pulse. The Fourier coefficients of p_{I} at each harmonic frequency would be multiplied by the same R to obtain the Fourier coefficients of p_{R}. Similarly, the same T would multiply the Fourier coefficients of p_{I} to obtain the coefficients of p_{T}. Thus, we would find that $p_{\mathrm{R}} = R p_{\mathrm{I}}$ and $p_{\mathrm{T}} = T p_{\mathrm{I}}$, which we already know from the time-domain analysis.

A pictorial representation of the field in both fluids at subcritical incidence is provided by Fig. 5.9. The incident wave is taken to have a finite width in order to see the individual waves. The result is that the insonified region is finite and fixed,

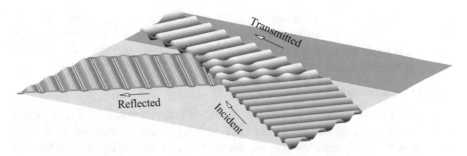

Fig. 5.9 Spatial profile at $t = 0$ of the field in both fluids as a result of incidence at the interface. The angle of incidence is subcritical at $\psi_1 = 50°$, $c_2/c_1 = \rho_2/\rho_1 = 1.2$. The incident plane wave has a finite width

with the waves passing through each. The bumps in the triangular region where the incident and reflected waves overlap are caused by interference between those waves. The angle of incidence is 50°, and the angle of transmission is 66.817°. The critical angle is 56.443°. The amplitude coefficients are $T = 1.4032$ and $R = 0.4032$.

The values of R and T exhibit some interesting trends. When c_1 is much greater than c_2, Snell's Law indicates that ψ_2 is very small, regardless of the value of ψ_1. If we approximate ψ_2 in this case as being essentially zero, we find that $Z_2 \approx \rho_2 c_2$ for any ψ_1. The concept of a local impedance is that it is independent of the angle of incidence. Hence, fluid 2 almost behaves as though it is a locally reacting material. Notwithstanding the omission of shear waves, such behavior is encountered in many polymeric materials used for acoustical treatments. If it also is true that $\rho_1 c_1 \gg \rho_2 c_2$, then $R \approx -1$ and $T \approx 0$, which means that fluid 2 appears to be pressure-release from the perspective of fluid 1. The converse attribute results when $c_2 \gg c_1$ and $\rho_2 c_2 \gg \rho_1 c_1$, provided that ψ_1 is sufficiently smaller than ψ_{cr}. This case leads to $R \approx +1$, and $T \approx 2$. From the perspective of fluid 1, fluid 2 appears to be rigid. The value of $T = 2$ corresponds to doubling of the pressure when an incident wave is reflected from a rigid boundary. Note that although the pressure amplitude transmitted into fluid 2 in this case is twice the incident value, the large value of $\rho_2 c_2$ means that the particle velocity is very small. This is comparable to the case of a very stiff elastic bar, which may carry a stress with very little deformation.

Both the pressure-release and rigid limits are encountered at the free surface of a large body of water. For a plane wave in air that is transmitted into water, at standard conditions we have $c_2/c_1 = 4.35$ and $\rho_2/\rho_1 = 833$. From this, we conclude that sound in air that is incident on a body of water will reflect as though the surface was rigid. Conversely, the surface will appear to be a pressure-release plane to sound in the water. Because $R = -1$ results in inversion of the waveform, a technique for inverting a waveform in an experiment is to generate it in water and aim it at the free surface.

Grazing incidence corresponds to ψ_1 approaching $\pi/2$. (Recall that ψ_1 actually being $\pi/2$ is excluded from consideration.) This is a case of regular incidence, $\psi_2 < \pi/2$, if $c_2 < c_1$. The value of Z_1 grows without bound with increasing ψ_1, leading to $R \to -1$ and $T \to 0$, regardless of the actual characteristic impedances. Grazing incidence when $c_2 > c_1$ means that ψ_1 exceeds the critical value. Before we can explore this case, we must first determine the waves generated when the angle of incidence exceed ψ_{cr}.

Supercritical Incidence

Snell's law states that $\sin \psi_2 > 1$ if $c_2 > c_1$ and ψ_1 is supercritical. This condition is similar to the situation we encountered in Sect. 5.1.1. With this in mind, let us redo the solution in the supercritical case by assuming that the transmitted wave evanesces in the y direction. We know that all waves must have the same trace wavenumber, $k_1 \sin \psi_1$. Hence, let us try the following ansatz for P_T,

$$P_T = T B_1 e^{-ik_1 x \sin \psi_1} e^{-k_2 \mu y} \qquad (5.3.17)$$

An expression for μ is found by making P_T satisfy the Helmholtz equation for fluid 2. Doing so leads to

$$-(k_1 \sin \psi_1)^2 + (k_2 \mu)^2 + k_2^2 = 0 \qquad (5.3.18)$$

Setting $k_1 = c_1 \omega$ and $k_2 = c_2 \omega$ leads to

$$\mu = + \left[\left(\frac{c_2}{c_1} \sin \psi_1 \right)^2 - 1 \right]^{1/2} \qquad (5.3.19)$$

Because fluid 2 extends infinitely in the positive y direction, the Sommerfeld condition dictates that the signal not grows without bound. This condition is met if $\text{Re}(\mu) > 0$, which is emphasized by explicitly displaying the plus sign.

An expression for T now may be derived by satisfying the boundary conditions. The particle velocity is found by applying Euler's equation to Eq. (5.3.17), which gives

$$\bar{V}_T = \frac{T B_1}{i\omega \rho_2} \left[(ik_1 \sin \psi_1) \bar{e}_x + k_2 \mu \bar{e}_y \right] e^{-ik_1 x \sin \psi_1} e^{-k_2 \mu y} \qquad (5.3.20)$$

The pressure and velocity boundary conditions at $y = 0$ now require that

$$T = 1 + R, \quad \frac{T\mu}{i\rho_2 c_2} = \frac{(1-R)\cos \psi_1}{\rho_1 c_1} \qquad (5.3.21)$$

These relations are equivalent to replacing $\cos \psi_2$ in the boundary conditions for subcritical incidence with $-i\mu$, from which it follows that Eq. (5.3.8) continues to describe the reflection and transmission coefficients, except that the impedance factors are

$$Z_1 = \frac{\rho_1 c_1}{\cos \psi_1}, \quad Z_2 = \frac{\rho_2 c_2}{\mu} i \text{ if } \psi_1 > \psi_{cr} \qquad (5.3.22)$$

A corollary of these relations is that $|R| = 1$. We will see in the next section that this property has important implications for power flow.

As was true for surface waves, there is no need to alter a computer program to evaluate the transmitted signal in the supercritical case. Rather, we may implement the procedure in Eq. (5.1.26). According to it, we first determine how our software implements the branch cut for the arcsine of a complex number. To obtain ψ_2 from Snell's law, we use the direct output of $\psi_2 = \arcsin((c_2/c_1) \sin \psi_2)$ or the complex conjugate of that output, according to which alternative gives $\text{Re}(\psi_2) \geq 0$. When this scheme is followed, we may use Eqs. (5.3.9), (5.3.12), and (5.3.16) to evaluate the transmitted field for any c_2/c_1.

At any angle of incidence, the time-domain description of the harmonic wave in fluid 2 is

$$p_T = \text{Re}\left[T B_I e^{i(\omega t - k_1 x \sin \psi_1 - k_2 y \cos \psi_2)}\right] \tag{5.3.23}$$

We have seen that this is a plane wave if ψ_2 is real, but what is it if ψ_2 is complex? When we set $\cos \psi_2 = -i\mu$, this signal becomes a wave that propagates in the x direction, with an amplitude that depends on the y coordinate, specifically

$$p_T = \text{Re}\left[\tilde{P}(y) e^{i(\omega t - k_1 x \sin \psi_1)}\right], \quad \tilde{P}(y) = T B_I e^{-i\mu y} \tag{5.3.24}$$

This pressure field is a plane wave, in the sense that all points at the same x have the same phase. However, it is not a simple wave because the amplitude is not constant on a wavefront. Such behavior is generically referred to as a *nonuniform plane wave*. The propagation velocity of this wave is $c_1/\sin \psi_1$ parallel to the interface. The definition of the supercritical case is that $\sin \psi_1 > c_1/c_2$, which means that $c_{tr} < c_2$. In other words, the disturbance moves along the interface supersonically from the viewpoint of fluid 1, but subsonically relative to fluid 2. In subcritical incidence, the trace speed exceeds the speed of sound in both fluids.

Figure 5.10 depicts supercritical transmission for the same conditions as those for Fig. 5.9, except that the angle of incidence is 57°. The other parameters are $\psi_2 = \pi/2 + 0.1132i$, $R = 0.9591 + 0.2832i$, $T = 1.9591 + 0.2832i$. Even though ψ_1 is only slightly larger than the critical value, which is 56.443°, the transmitted signal decays very rapidly with increasing distance from the interface. The variation parallel to the interface is sinusoidal, in order to match the signal at the interface with fluid 1. Qualitatively, the interference between the incident and reflected waves in fluid 1 is no different from the process for the subcritical case. This interference would occur throughout fluid 1 if the wavefront of the incident wave was infinitely long. The extent in the figure is finite, so the extent of the reflected and transmitted wavefronts is finite. As time elapses, the region that is insonified remains fixed.

Power Flow

We found that supercritical incidence leads to Z_1 being real and Z_2 being imaginary. Consequently, $|R| = 1$ for any $\psi_1 > \psi_{cr}$. This is an interesting result from the viewpoint of power flow across the interface. The waves in fluid 1 are P_I and P_R, whose

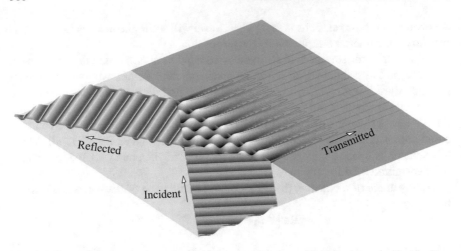

Fig. 5.10 Spatial profile at $t = 0$ of the field in both fluids as a result of incidence at the interface. The angle of incidence is supercritical at $\psi_1 = 57°$, $c_2/c_1 = \rho_2/\rho_1 = 1.2$. The incident plane wave has a finite width

descriptions in Eq. (5.3.9) are the same as Eq. (5.2.10) for reflection from a locally reacting surface. It follows that the power that flows out of fluid 1 over an $L \times 1$ surface patch is as given by Eq. (5.2.22).

$$\mathcal{P}_1 = \left(\bar{I}_1\right)_{\text{av}} \cdot \left(-\bar{e}_y\right) L = \frac{|B_1|^2}{2Z_1} \left(1 - |R|^2\right) L \qquad (5.3.25)$$

Hence, if $|R| = 1$, then the average power flow across the interface is zero. This might seem to conflict with the fact that T is not zero in the supercritical case, which means that a signal is transmitted into fluid 2. To examine why these properties are consistent, let us evaluate the time-averaged power crossing the $L \times 1$ surface patch from the perspective of fluid 2. In the subcritical case, the time-averaged intensity in the plane wave is

$$\left(\bar{I}_2\right)_{\text{av}} = \frac{|T|^2 |B_1|^2}{2\rho_2 c_2} \bar{e}_{\text{T}} \qquad (5.3.26)$$

The normal at the interface into fluid 2 is $\bar{e}_y$. Because $\bar{e}_{\text{T}} \cdot \bar{e}_y = \cos \psi_2$, the power flowing into fluid 2 over an area of L (1) is

$$\mathcal{P}_2 = \left(\bar{I}_2\right)_{\text{av}} \cdot \bar{e}_y L = \frac{|T|^2 |B_1|^2 L}{2Z_2} \qquad (5.3.27)$$

To analyze the power flow in the case of supercritical incidence , we evaluate Eqs. (5.3.17) and (5.3.20) at $y = 0$. The resulting time-averaged intensity in fluid 2 at the interface is

$$\left(\bar{I}_2\right)_{\text{av}}\big|_{y=0} = \frac{1}{2}\,\text{Re}\left(P_2 V_2^*\big|_{y=0}\right)$$

$$= \text{Re}\left[\left(T B_I e^{-ik_1 x \sin \psi_1}\right)\left(\frac{T B_I}{i\omega\rho_2}\left[(ik_1 \sin \psi_1)\,\bar{e}_x + k_2\mu\bar{e}_y\right]e^{-ik_1 x \sin \psi_1}\right)^*\right]$$

$$= \frac{|T|^2 |B_I|^2}{2\rho_2 c_2}\left(\frac{c_2}{c_1}\sin \psi_1\right)\bar{e}_x$$

(5.3.28)

The important feature here is that $\left(\bar{I}_2\right)_{\text{av}}$ is parallel to the interface, because the particle velocity normal to the interface is 90° out-of-phase relative to the pressure. The consequence is that the average power flow across the interface is zero, $\mathcal{P}_2 = 0$. At any instant, power might flow into fluid 2, or it might flow back into fluid 1. Only the average is zero.

Energy is not absorbed at the interface, so it must be that $\mathcal{P}_1 = \mathcal{P}_2$ for any incidence case, which is satisfied if

$$\boxed{\begin{aligned}\frac{1 - |R|^2}{Z_1} &= \frac{|T|^2}{Z_2} \text{ if } \psi_1 < \psi_{\text{cr}}\\ |R| &= 1 \text{ if } \psi_1 > \psi_{\text{cr}}\end{aligned}}$$

(5.3.29)

We already have seen that the supercritical condition is true. Substitution of Eq. (5.3.16) in the case of subcritical incidence would show that the first condition is satisfied by the results for R and T. Note that both Z_1 and Z_2 are real when $\psi_1 < \psi_{\text{cr}}$. Thus, $|T|^2 = T^2$ and $|R|^2 = R^2$. Equation (5.3.29) has not been adjusted to account for this property, because the form in which it appears applies when there are multiple fluids in layers, which is the topic of the next section.

A quantity of interest in this and other situations where a plane wave encounters a discontinuity is the *transmission loss*. It is based on comparing the normal components of the time-averaged intensity in the incident and transmitted waves. The preceding relations give

$$\frac{\left(\bar{I}_2\right)_{\text{av}} \cdot \bar{e}_y}{\left(\bar{I}_1\right)_{\text{av}} \cdot \bar{e}_y} = \frac{|T|^2 |B_I|^2 / Z_2}{|B_I|^2 / Z_1} = \begin{cases} \dfrac{4 Z_1 Z_2}{(Z_2 + Z_1)^2} \text{ if } \psi_1 < \psi_{\text{cr}} \\ 0 \text{ if } \psi_1 > \psi_{\text{cr}} \end{cases}$$

(5.3.30)

The transmission loss, denoted TL, describes this ratio on a decibel scale. The intensity ratio is less than one, so the logarithm of the ratio is negative. For this reason, the definition is

$$TL = -10 \log_{10}\left(\frac{\left(\bar{I}_2\right)_{\text{av}} \cdot \bar{e}_y}{\left(\bar{I}_1\right)_{\text{av}} \cdot \bar{e}_y}\right)$$

(5.3.31)

An important aspect is that the transmission loss only depends on the properties of the fluids, but not on which medium contains the transmitted wave. For example, the

388 5 Interface Phenomena for Planar Waves

transmission loss at an air-water interface for normal incidence is 61 dB, regardless of whether the incident wave is in the air or the water. This large number is prime justification for modeling the interface between these fluids as rigid or pressure-release.

EXAMPLE 5.5 A harmonic plane wave in the ocean is incident on the bottom. Hydrophones measure the signal at location A, which is 50 m above the bottom, and at location B, which is at the same elevation, but 200 m westward. The measured signal at location A (incident plus reflected) is used to set $t = 0$ such that $p\left(\bar{\xi}_A, t\right) = 500 \sin\left(\omega t\right)$ Pa for any value of ω. The corresponding pressure at location B is monitored to be $p\left(\bar{\xi}_B, t\right) = P_B \sin\left(\omega t - \chi\right)$ Pa. For the first measurement, the frequency is set at 250 Hz, which leads to $\chi = 0.541461$ rad. Then, the frequency is decreased very gradually until χ is reduced to zero. The frequency for this condition is 249.2839 Hz. The ocean properties are $c_1 = 1500$ m/s and $\rho_1 = 1030$ kg/m^3, both of which differ from the properties of fresh water as a result of the salinity. The bottom may be treated as a liquid whose sound speed is 1700 m/s and whose density is 1200 kg/m^3. For the case where the frequency is 250 Hz, determine the amplitude P_B, as well as the signal that would be measured at location C, which is buried 10 m below the ocean's floor at a location that is directly under point B.

Significance

This example will invoke many of the relations governing reflection and transmission, concurrently with affording us the opportunity to refresh our expertise relative to the usage of complex exponentials to describe wave propagation.

Solution

Rather than knowing the incident pressure, we are given information about the sum of the incident and reflected signals at two locations in fluid 1 (ocean water). We cannot determine the transmission coefficient without the knowledge of the angle of incidence, so we begin by considering the pressure in fluid 1. Let the origin be at the ocean floor below point A, so the locations are $\bar{\xi}_A = -50\bar{e}_y$, $\bar{\xi}_B = 200\bar{e}_x - 50\bar{e}_y$ m. (Recall that y is defined to point into the ocean bottom.) The representation of the signal at both locations is

$$
\begin{aligned}
P\left(\bar{\xi}_A\right) &= \frac{500}{i} = B_{\mathrm{I}}\left(e^{+ik_1(50)\cos\psi_1} + Re^{-ik_1(50)\cos\psi_1}\right) \\
P\left(\bar{\xi}_B\right) &= \frac{P_B}{i}e^{-i\chi} = B_{\mathrm{I}}e^{-ik_1(200)\sin\psi_1}\left(e^{+ik_1(50)\cos\psi_1} + Re^{-ik_1(50)\cos\psi_1}\right)
\end{aligned}
\tag{1}
$$

The primary parameter is ψ_1, so we form the ratio of these complex pressures, with the result that

$$
\frac{P_B}{500}e^{-i\chi} = e^{-ik_1(200)\sin\psi_1}
$$

The value of P_B is real, so matching both sides leads to

$$P_B = 500 \text{ Pa}, \quad 200 k_1 \sin \psi_1 = \chi \tag{2}$$

It might seem reasonable to use the given fact that $\chi = 0.541461$ rad at 250 Hz to determine ψ_1 directly from Eq. (2). Doing so could be wrong because the measured phase angle is limited to $-\pi \le \chi < \pi$. Therefore, the given value of χ could differ from the value of $200 k_1 \sin \psi_1$ by a multiple of 2π. We need to make use of the data for both frequencies. Let n be the number of cycles, so that at 250 Hz, the second of Eq. (2) gives

$$200 \left(\frac{500\pi}{c_1} \right) \sin \psi_1 - n \, (2\pi) = 0.541461$$

It is stated that gradually reducing ω reduces χ to zero, which implies that the number of cycles at the second frequency is also n. Thus, evaluation of Eq. (2) at ω_2 gives

$$200 \left(\frac{1566.297}{c_1} \right) \sin \psi_1 - n \, (2\pi) = 0$$

Taking the difference of these relations yields

$$\psi_1 = \sin^{-1} \left(\frac{0.541461 c_1}{200 \, (500\pi - 1566.297)} \right) = 64.50° \tag{3}$$

Note that extra precision for the value of the frequencies is required to obtain a reasonably precise value of ψ_1.

We may now evaluate the signal in fluid 2. The critical angle of incidence is

$$\psi_{cr} = \sin^{-1} \left(\frac{c_1}{c_2} \right) = 61.93°$$

Because $\psi_1 > \psi_{cr}$, this is a supercritical incidence case. We handle this situation by following the procedure in Eq. (5.1.26), which gives

$$\psi_2 = \frac{\pi}{2} + 0.21359i \tag{4}$$

so that

$$\cos \psi_2 = -0.21522i$$

The reflection and transmission coefficients are found from Eq. (5.3.16) to be

$$R = 0.7493 + 0.6622i, \quad T = 1.7493 + 0.6622i$$

The value of B_I is found by substituting ψ_1 from Eq. (3) into Eq. (1). The wavenumber at 250 Hz is $k_1 = 500\pi/c_1$, which gives

$$B_I = \frac{500}{i\left(e^{+ik_1(50)\cos\psi_1} + Re^{-ik_1(50)\cos\psi_1}\right)} = 90.16 + 238.1i = 254.7e^{i1.2089} \text{ Pa}$$

We then find from Eq. (5.3.9), with $k_2 = 500\pi/c_2$, that the transmitted pressure at $\bar{\xi}_C = 200\bar{e}_x - 10\bar{e}_y$ is

$$P_T\left(\bar{\xi}_C\right) = T B_I e^{-ik_2(200\sin\psi_2 + 10\cos\psi_2)} = 33.531 + 55.753i = 55.06e^{i1.0293} \text{ Pa}$$

The fact that the pressure amplitude at $\bar{\xi}_C$ is substantially smaller than the amplitude at the interface is a consequence of the exponential decay in the y direction.

5.4   Propagation Through Layered Media

The interfaces between the atmosphere and the ocean, and between the ocean and a sedimentary bottom are two of the few occasions where it is reasonable to consider the system to consist of two adjacent half-spaces. Even in these cases, it is necessary that the ocean be sufficiently deep to treat it as though its depth is infinite. Here, we will remove that limitation. The development has a wide range of applications, including the design of materials for noise and vibration control.

5.4.1   Basic Analysis of Three Fluids

We begin by considering a layer of finite width (or depth) H that is sandwiched between two fluids that extend infinitely above and below. The situation in the half-spaces is like the previous case of two fluids. Consideration of the interface between fluids 1 and 2 suggests that an incident plane wave in fluid 1 will generate a reflected wave in fluid 1 and a transmitted plane wave in fluid 2. This transmitted wave will be incident at the interface with fluid 3, which means that it will be transmitted into fluid 3 and also be reflected back into fluid 2. This reflected wave appears to be incident at the interface between fluids 1 and 2, which means that it will transmit a wave into fluid 1, and also be reflected back into fluid 2. The process continues *ad infinitum*. The net result of this combination is that there are incident and reflected waves in fluid 1, a transmitted wave in fluid 3, and waves that propagate back and forth in fluid 2. The overall picture is described in Fig. 5.11.

Like the analysis of two adjoined half-spaces, a time-domain analysis of this system would be limited to the case of subcritical incidence. For this reason, we will pursue only a frequency-domain study. Because the distinction between waves that

Fig. 5.11 Waves in two half-spaces separated by a finite thickness layer

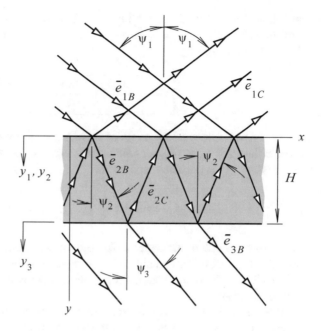

are incident, reflected, and transmitted is blurred by the multiple reflection process, it is useful to modify the notation. The complex amplitude of waves that propagate in the sense from fluid 1 to fluid 3 will be denoted as B_n, and amplitudes of waves that propagate in the opposite sense will be denoted as C_n, where n is the fluid (or layer) number. In this notation, $n = 1, 2$, or 3, and $C_3 \equiv 0$. The corresponding propagation directions are $\bar{e}_{nB}$ and $\bar{e}_{nC}$. Another modification is to use a local coordinate system for each layer, such that y_n measures distance within each layer. As shown in Fig. 5.11, $y_1 = y_2 = 0$ at the interface of fluids 1 and 2, and $y_2 = H$ and $y_3 = 0$ at the interface of fluids 2 and 3. In this scheme, the plane waves in layer n are described by

$$P_n = B_n e^{-ik_n \bar{\xi} \cdot \bar{e}_{nB}} + C_n e^{-ik_n \bar{\xi} \cdot \bar{e}_{nC}}$$

$$= \left(B_n e^{-iK_n y_n} + C_n e^{iK_n y_n} \right) e^{-ik_n x \sin \psi_n}, \quad C_3 = 0$$

$$\bar{V}_n = \frac{1}{\rho_n c_n} \left[B_n e^{-iK_n y_n} \left(\sin \psi_n \bar{e}_x + \cos \psi_n \bar{e}_y \right) \right.$$

$$\left. + C_n e^{iK_n y_n} \left(\sin \psi_n \bar{e}_x - \cos \psi_n \bar{e}_y \right) \right] e^{-ik_n x \sin \psi_n}$$

(5.4.1)

where K_n is the trace wavenumber in the direction that is normal to the interfaces,

$$K_n = \frac{\omega}{c_n} \cos \psi_n \tag{5.4.2}$$

It will be noted that both $\bar{e}_{nB}$ and $\bar{e}_{nC}$ are taken to be at angle ψ_n with respect to the y direction in order that both waves have the same trace wavenumber along the interface. The trace wavenumbers must match at the interface between fluids 1 and 2, and also at the interface between fluids 2 and 3, so there is only one trace wavenumber. Setting $k_{tr} = k_1 \sin \psi_1 = k_2 \sin \psi_2 = k_3 \sin \psi_3$ leads to an extension of Snell's law, specifically

$$\boxed{\frac{1}{c_{tr}} = \frac{\sin \psi_1}{c_1} = \frac{\sin \psi_2}{c_2} = \frac{\sin \psi_3}{c_3}} \tag{5.4.3}$$

There are two values of the angle of incidence that are critical. Evanescent waves arise in fluid 2 if $\psi_1 > (\psi_{cr})_2$, while evanescence of the transmitted wave in fluid 3 results if $\psi_1 > (\psi_{cr})_3$. These angles are

$$(\psi_{cr})_2 = \sin^{-1}\left(\frac{c_1}{c_2}\right), \quad (\psi_{cr})_3 = \sin^{-1}\left(\frac{c_1}{c_3}\right) \tag{5.4.4}$$

Each critical angle of incidence exists only if the sound speed in fluid n is greater than that in fluid 1. For example, if $c_3 < c_1 < c_2$, then evanescent waves will exist in layer 2 if $\psi_1 > (\psi_{cr})_2$, but a simple plane wave will be transmitted into fluid 3, regardless of ψ_1.

The complex amplitudes of pressure and normal velocity at each interface are obtained by setting y_n in Eq. (5.4.1) to the appropriate value. The continuity conditions constitute four algebraic equations,

$$P_1 = P_2 \ @ \ y_1 = y_2 = 0 \implies B_1 + C_1 = B_2 + C_2$$

$$\bar{V}_1 \cdot \bar{e}_y = \bar{V}_2 \cdot \bar{e}_y \ @ \ y_1 = y_2 = 0 \implies \frac{B_1 - C_1}{Z_1} = \frac{B_2 - C_2}{Z_2}$$

$$P_2 \ @ \ y_2 = H = P_3 \ @ \ y_3 = 0 \implies B_2 e^{-iK_2 H} + C_2 e^{iK_2 H} = B_3$$

$$\bar{V}_2 \cdot \bar{e}_y \ @ \ y_2 = H = \bar{V}_3 \cdot \bar{e}_y \ @ \ y_3 = 0 \implies \frac{B_2 e^{-iK_2 H} - C_2 e^{iK_2 H}}{Z_2} = \frac{B_3}{Z_3}$$

$$\tag{5.4.5}$$

where the Z_n factors are as previously defined, $Z_n = \rho_n c_n / \cos \psi_n$. There are two strategies we could pursue to solve the preceding equations. One solves the equations simultaneously based on considering B_1 for the incident wave to be known. This approach will be developed in the next section. Here, we will implement a progressive elimination scheme that sequentially solves two equations. There are three waves at the 2–3 interface, so we start by solving the second pair of equations for B_2 and C_2 in terms of B_3. This gives

$$B_2 = \left[\frac{1}{2}\left(1 + \frac{Z_2}{Z_3}\right) e^{iK_2 H}\right] B_3$$

$$C_2 = \left[\frac{1}{2}\left(1 - \frac{Z_2}{Z_3}\right) e^{-iK_2 H}\right] B_3 \tag{5.4.6}$$

These expressions are substituted into the continuity equations for the 1–2 interface, which then are solved for B_1 and C_1, with the result that

$$B_1 = \frac{1}{4}\left[\left(1+\frac{Z_1}{Z_2}\right)\left(1+\frac{Z_2}{Z_3}\right)e^{iK_2H} - \left(1-\frac{Z_1}{Z_2}\right)\left(1-\frac{Z_2}{Z_3}\right)e^{-iK_2H}\right]B_3$$

$$C_1 = \frac{1}{4}\left[\left(1-\frac{Z_1}{Z_2}\right)\left(1+\frac{Z_2}{Z_3}\right)e^{-iK_2H} + \left(1+\frac{Z_1}{Z_2}\right)\left(1-\frac{Z_2}{Z_3}\right)e^{-iK_2H}\right]B_3$$

$$(5.4.7)$$

Now recall that B_1 is the amplitude of the incident wave, while B_3 is the amplitude of the wave transmitted into fluid 3 and C_1 is the amplitude of the reflected wave in fluid 1. Thus, the transmission and reflection coefficients are

$$T = \frac{B_3}{B_1}, \quad R = \frac{C_1}{B_1} \equiv \frac{C_1}{B_3}\frac{B_3}{B_1} \tag{5.4.8}$$

After some manipulations and invocation of the definitions of the sine and cosine functions in terms of complex variables, these coefficients are found to be

$$T = \frac{2Z_2Z_3}{(Z_1Z_2+Z_2Z_3)\cos(K_2H) + i\left(Z_1Z_3+Z_2^2\right)\sin(K_2H)}$$

$$R = \frac{(Z_2Z_3-Z_1Z_2)\cos(K_2H) + i\left(Z_2^2-Z_1Z_3\right)\sin(K_2H)}{(Z_1Z_2+Z_2Z_3)\cos(K_2H) + i\left(Z_2^2+Z_1Z_3\right)\sin(K_2H)}$$

$$(5.4.9)$$

It is evident that both R and T are complex even if the angle of incidence is less than critical. The only exception occurs if K_2H is such that either $\cos(K_2H)$ or $\sin(K_2H)$ vanishes. We will examine the significance of this behavior in an example. Supercritical incidence occurs in fluid 2 if ψ_1 exceeds $(\psi_{cr})_2$. In this situation, we may evaluate the complex angle ψ_2 by following the procedure in Eq. (5.1.26). This leads to $\cos\psi_2 = -i\mu$, which may be used directly to evaluate the corresponding values of the sine and cosine of K_2H. The definition of these functions in terms of complex exponential will lead to hyperbolic functions, such that

$$\cos(K_2H) \equiv \cosh(\mu k_2H), \quad \sin(K_2H) \equiv -i\sinh(\mu k_2H) \tag{5.4.10}$$

The case where $\psi_1 > (\psi_{cr})_3$ and ψ_2 is real is somewhat different. Following the branch cut in Eq. (5.1.26) leads to Z_3 being purely imaginary. The other terms in Eq. (5.4.9) remain real. The consequence is that the numerator in Eq. (5.4.9) for R is the negative complex conjugate of the denominator, which may be proven by letting $Z_3 = i\alpha$. Thus, $|R| = 1$ when the angle of incidence is greater than the critical value for transmission into fluid 3.

The various features of R and T are exhibited in Fig. 5.12, where the properties of fluids 2 and 3 have been selected such that supercritical incidence is possible in either

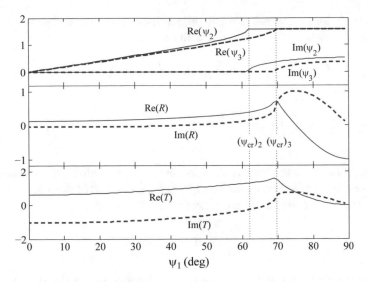

Fig. 5.12 Dependence of the propagation angles and the reflection and transmission coefficients on the angle of incidence. The system consists of two fluids in half-spaces separated by a finite fluid layer; $\rho_1 = 1000\,\text{kg/m}^3$, $c_1 = 1480\,\text{m/s}$, $\rho_2 = 1200\,\text{kg/m}^3$, $c_2 = 1680\,\text{m/s}$, $\rho_3 = 1400\,\text{kg/m}^3$, $c_3 = 1580\,\text{m/s}$, $k_2 H = \pi/3$

fluid, with $(\psi_{\text{cr}})_2 < (\psi_{\text{cr}})_3$. The graph of ψ_2 and ψ_3 as functions of ψ_1 shows that the imaginary part of each is zero if ψ_1 is below the respective critical value, after which the real part of the "angle" remains at $\pi/2$. The plots of R and T show that most of the incident signal passes into fluid 3 at small ψ_1, and that $R \to -1$ and $T \to 0$ in the limit of grazing incidence, as they did in the two-fluid case. Increasing ψ_1 beyond $(\psi_{\text{cr}})_3$ drastically alters the trends for both R and T, but the curve is continuous. A numerical evaluation of $|R|$ in this supercritical case would confirm that it is unity.

Power Flow

The fundamental difference between supercritical incidence in fluid 2 and fluid 3 may be understood by considering how power flows through the layers. A block of fluid 2 has been isolated in Fig. 5.13. Its upper and lower sides are $L \times 1$ patches on the interfaces. The fluid in the isolated control volume neither dissipates nor creates energy, so the time-averaged power flowing out of this region must equal that which flows into it. The signal in fluid 1 always consists of the incident and reflected waves, so we may use Eq. (5.3.25) to describe the time-averaged power $\mathcal{P}_1$ flowing into the region from fluid 1. Regardless of whether ψ_1 is greater than $(\psi_{\text{cr}})_2$, the x dependence of the signal in fluid 2 is solely a phase variation associated with the trace velocity along the interface. Thus, at a specified depth y, the time-averaged intensity on the left face, $\left(\bar{I}_{\text{left}}\right)_{\text{av}} \cdot \bar{e}_x$ in the figure always equals the corresponding value $\left(\bar{I}_{\text{right}}\right)_{\text{av}} \cdot \bar{e}_x$ on the right face. It follows that the time-averaged power $\mathcal{P}_{\text{left}}$ flowing in from the left always equals the power $\mathcal{P}_{\text{right}}$ flowing out on the right.

Fig. 5.13 Isolated region in the intermediate layer between two-fluid half-spaces used to evaluate power flow

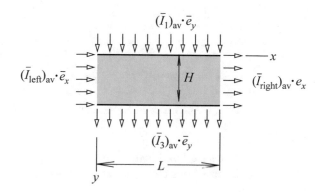

In terms of the transmission coefficient, the signal in fluid 3 has the same form as the signal in fluid 2 in the previous section. This means that Eq. (5.3.27), with the symbols altered to fit the present configuration, may be employed to describe P_3 when $\psi_1 < (\psi_{cr})_2$, and that $P_3 = 0$ when $\psi_1 > (\psi_{cr})_2$. (Recall that in the supercritical case, the evanescent wave's particle velocity in the y direction is 90° out-of-phase from the pressure, so the time-averaged intensity is directed in the x direction, parallel to the interface.) Conservation of energy requires that $P_1 + P_{\text{left}} = P_3 + P_{\text{right}}$, but $P_{\text{left}} = P_{\text{right}}$. Therefore, it must be that

$$\boxed{\begin{array}{l} \dfrac{1 - |R|^2}{Z_1} = \dfrac{|T|^2}{Z_3} \text{ if } \psi_1 < (\psi_{cr})_3 \\[2mm] |R| = 1 \text{ if } \psi_1 > (\psi_{cr})_3 \end{array}} \tag{5.4.11}$$

These conditions are like Eq. (5.3.29) for the case of a single interface between two fluids. The coefficients in Eq. (5.4.9) must satisfy it. This may be proven analytically with some effort or else tested computationally for a variety of parameters.

From a qualitative perspective, the signals in fluids 1 and 3 behave as though the layer containing fluid 2 was not present. That is, the incident wave generates a reflected plane wave in fluid 1, regardless of the angle of incidence. A plane wave is transmitted into fluid 3 if the angle of incidence is below the critical value for fluid 3, and supercritical incidence leads to an evanescent wave that travels at the trace velocity parallel to the interface. The power transmitted into fluid 3 over a patch of the interface is always the difference between the power in the incident wave over that area and the power that is reflected, with supercritical incidence reducing the transmitted power to zero. The sole effect of the presence of the layer containing fluid 2 is to alter the values of the transmission and reflection coefficients at a specific angle of incidence.

EXAMPLE 5.6 Consider a barrier in the air that covers a wide planar area. The barrier is filled with an unspecified gas between two thin plastic sheets that have negligible rigidity. Identify criteria for the properties of the gas and the thickness such that the magnitude of the transmission coefficient is minimized at a specified frequency Ω. Then, identify the bandwidth over which the sound pressure level in the transmitted signal is no more than 6 dB greater than the minimum value.

Significance

The design issues explored here have practical application. Resolving these issues will greatly improve our understanding of how the fluid properties affect acoustic transmission.

Solution

We will begin by examining what factors affect $|T|$, which will suggest criteria for the properties of the layer. Toward that end, Eq. (5.4.9) is specialized to describe the case where the same fluid fills both half-spaces. The result is

$$T = \frac{2Z_1 Z_2}{2Z_1 Z_2 \cos\left(K_2 H\right) + i\left(Z_1^2 + Z_2^2\right)\sin\left(\omega H/c_2\right)} \tag{1}$$

where K_2 has been replaced with ω/c_2 because $\psi_1 = 0$.

The first item we consider is the determination of the selection of H that will minimize $|T|$ if Z_1 and Z_2 are set. Fixing these values means that the numerator in Eq. (1) is fixed, so the optimum H will maximize the denominator's magnitude. A little manipulation shows that

$$|T| = \frac{2Z_1 Z_2}{\left[4Z_1^2 Z_2^2 + \left(Z_1^2 - Z_2^2\right)^2 \sin\left(\omega H/c_2\right)^2\right]^{1/2}} \tag{2}$$

We wish to minimize $|T|$ at $\omega = \Omega$, which results if the sine term equals one. Thus, we equate the argument $\Omega H/c_2$ of this term to an odd multiple of $\pi/2$. Many values of H are possible. We select the smallest one, which would be the most economical choice, thus

$$H = \frac{\pi c_2}{2\Omega} \tag{3}$$

This value is the optimum H for minimizing $|T|$, but it does not address the question of what would be the best properties for fluid 2. When the sine term in Eq. (2) is one, the transmission coefficient is

$$\min\left(|T|\right) = \frac{2Z_1 Z_2}{Z_1^2 + Z_2^2} = \frac{2r}{1 + r^2} \tag{4}$$

where r is the ratio of characteristic impedances,

$$r = \frac{Z_2|_{\psi_2=0}}{Z_1|_{\psi_1=0}} = \frac{\rho_2 c_2}{\rho_1 c_1}$$

The minimum value $|T|$ is very small if $r \approx 0$ or $r \gg 1$. Both alternatives correspond to fluid 2 being very different from fluid 1. Thus, if there is a very large *impedance mismatch* at the interfaces, there is little transmission into the lower half-space.

Selecting fluid 2 to give a large impedance mismatch, along with setting H according to Eq. (3), gives the smallest possible $|T|$ at Ω. However, the best selection for r might only reduce $|T|$ in a very small frequency interval. An exploration of frequency dependence may be pursued by substituting H from Eq. (3) and $Z_2 = rZ_1$ into Eq. (2). The result is an expression for the transmission coefficient at the optimal H as a function of ω / Ω and r,

$$|T(\omega/\Omega, r)| = \frac{2r}{\left[4r^2 + \left(1 - r^2\right)^2 \sin\left(\frac{\pi\omega}{2\Omega}\right)^2\right]^{1/2}} \tag{5}$$

An important aspect of this relation is that $|T(\omega/\Omega, 1/r)| = |T(\omega/\Omega, r)|$. This feature is exploited in Fig. 1, which shows the frequency dependence for impedance matches ranging from very small to extremely large. The worst case results if ω/Ω is an even integer, in which case $|T| = 1$. Our interest stems from the observations that the minimum $|T|$ occurs when ω/Ω is an odd integer, and that there is a frequency interval surrounding each of these frequencies within which $|T|$ is close to the minimum.

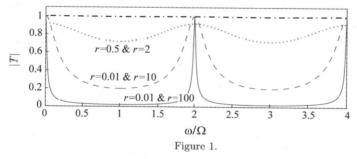

Figure 1.

The problem statement specifically asks for the bandwidth in which $\left(P_T^2\right)_{av}$ is no more that 6 dB greater than $\left(P_T^2\right)_{av}$ at frequency Ω. Because $|P_T|^2 = |T|^2 |P_1|^2$, this property is attained if $|T(\omega/\Omega, r)|^2 < 4|T(1, r)|^2$. We will examine the behavior in the vicinity of $\omega = \Omega$. The properties for $\omega = 3\Omega, 5\Omega, \ldots$ are the same. When we use Eq. (5) to describe the transmission ratio, we obtain an equation relating ω/Ω and r, specifically

$$\frac{4r^2}{4r^2 + \left(1 - r^2\right)^2 \sin\left(\frac{\pi\omega}{2\Omega}\right)^2} = 4\frac{4r^2}{\left(1 + r^2\right)^2}$$

This gives

$$\sin\left(\frac{\pi\omega}{2\Omega}\right)^2 = \frac{\left(1+r^2\right)^2 - 16r^2}{4\left(1-r\right)^2} \tag{6}$$

The denominator of this expression is positive, so the right side is negative if $16r^2 > \left(1 + r^2\right)^2$. The values of r at the limit of this equality are found by solving a quadratic equation for r^2. The result is that there are no real values of ω/Ω if $0.26795 < r < 3.73205$. To understand the significance of this condition consider Fig. 1 for $r = 0.5$. The value of $|T|$ for any frequency is less than twice the minimum value at $\omega/\Omega = 1$. Thus, the absence of real roots for ω/Ω means that there is no frequency at which $|T(\omega/\Omega, r)|$ is larger than twice $|T(1, r)|$. In other words, the 6 dB criterion is met for any frequency.

If $r < 0.26795$ or $r > 3.73205$, Eq. (6) yields two real values for ω/Ω. This comes about because $\sin(\theta) = \alpha$ is satisfied by $\theta = \sin^{-1}(\alpha)$ and $\theta = \pi - \sin^{-1}(\alpha)$. Thus, the limiting frequencies are

$$\frac{\omega_{\min}}{\Omega} = \frac{2}{\pi}\sin^{-1}\left[\left(\frac{\left(1+r^2\right)^2 - 16r^2}{4\left(1-r^2\right)^2}\right)^{1/2}\right] \tag{7}$$

$$\frac{\omega_{\max}}{\Omega} = 2 - \frac{\omega_{\min}}{\Omega}$$

In Fig. 2, the upper graph describes the smallest possible $|T|$ as a function of the impedance ratio. This is the value when H is set according to Eq. (3). The lower graph describes Eq. (7). The shaded band at a specific Z_2/Z_1 is the range of ω/Ω for which $|T|$ is less than or equal to twice the optimal value in the upper graph.

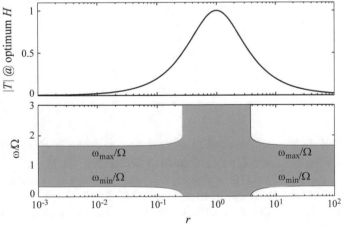

Figure 2.

Figure 2 is very useful for identifying the desirable range of impedances. The unlimited bandwidth in the interval surrounding $r = 1$ is not useful because the optimal value of $|T|$ is close to one in this range of impedances. In contrast, if $r < -0.1$ or $r > 10$, then the value of $|T|$ will not exceed 0.2 and the frequency interval within which $|T| < 0.4$ will be $0.3 < \omega/\Omega < 1.69$.

Before we leave this system, let us examine the pressure field in the finite layer, because doing so will relate the present developments to our prior work with one-dimensional waveguides. Equation (5.4.6) describes the amplitude coefficients for this layer, and $B_3 = T B_1$, so the pressure in this region is

$$
\begin{aligned}
P &= \frac{1}{2}\left[\left(1+\frac{Z_2}{Z_1}\right) e^{i(\omega/c_2)(H-y)} + \frac{1}{2}\left(1-\frac{Z_2}{Z_1}\right) e^{-i(\omega/c_2)(H-y)}\right] T B_1 \\
&= \left[\cos\left(\frac{\omega}{c_2}(H-y)\right) + i\frac{Z_2}{Z_1}\sin\left(\frac{\omega}{c_2}(H-y)\right)\right] T B_1
\end{aligned}
\tag{5.4.12}
$$

If Z_2/Z_1 is very small, then the first term is dominant. In that case, dP/dy is close to zero at $y = 0$, that is, the velocity is very small. This is like a rigid termination in a waveguide. The reflection coefficient at $y = H$ therefore is close to one, so there is little power transfer across the interface into the lower half-space. In contrast, if Z_2/Z_1 is large, the second term is dominant, so $y = H$ tends to a pressure-release condition. The reflection coefficient at $y = H$ then is close to minus one, so this case also exhibits little power flow into the lower half-space.

5.4.2 Multiple Layers

Multilayer composite materials are finding increasing use in a variety of applications. Such materials are solids and therefore do not fit the specification that the medium cannot sustain a shear stress. However, the case of normal incidence is correctly described by an analysis of fluid layers. This is so because one type of wave in an elastic medium is a dilatational wave, in which there is no shear stress. When a dilatational wave is normally incident on an interface with another solid, the result is the reflection and transmission of dilatational waves. In contrast, oblique incidence also generates shear waves. Thus, although it would be approximate, a study of transmission through multiple layers of fluid can provide some useful information for solid composite materials. The same model also has application to problems involving acoustical oceanography.

The basic setup in Fig. 5.14 is like that used to analyze transmission through a single layer. Let N be the number of layers having finite thickness, with half-spaces on either side. The half-space for the incident and reflected waves is numbered $n = 1$, and the half-space for the transmitted wave is $n = N + 2$. A local coordinate system y_n describes distance from the incident side of the nth layer, and x is measured in the same way in all layers, with the x-axis defined to lie in the plane that contains the incident wave's direction and the normal to the interfaces.

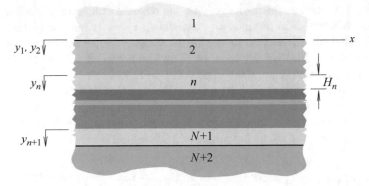

Fig. 5.14 Geometry and coordinate systems for propagation through multiple fluid layers

In each layer, $\bar{e}_{nB}$ and $\bar{e}_{nC}$ are the directions of the plane waves that propagate in the sense of increasing and decreasing y_n, respectively. The corresponding complex amplitudes are B_n and C_n. The incident wave amplitude is B_1, so the reflection coefficient is $R = C_1/B_1$ and the transmission coefficient is $T = B_{N+2}/B_1$. There is no backward wave in the lower half-space, so $C_{N+2} \equiv 0$. This is the arrangement described by the description of pressure and particle velocity in Eq. (5.4.1). The first interface is defined by $y_1 = y_2 = 0$, and all others are $y_n = H_n$, $y_{n+1} = 0$ for $n = 2, ..., N + 1$. The pressure and particle velocity in the y direction must vary continuously across every interface. Thus, it must be that

$$
\begin{aligned}
P_1 \ @\ (y_1 = 0) &= P_2 \ @\ (y_2 = 0) \\
\bar{V}_1 \cdot \bar{e}_y \ @\ (y_1 = 0) &= \bar{V}_2 \cdot \bar{e}_y \ @\ (y_2 = 0) \\
\left. \begin{aligned}
P_n \ @\ (y_n = H_n) &= P_{n+1} \ @\ (y_{n+1} = 0) \\
\bar{V}_n \cdot \bar{e}_y \ @\ (y_n = H_n) &= \bar{V}_{n+1} \cdot \bar{e}_y \ @\ (y_{n+1} = 0)
\end{aligned} \right\} & \quad n = 2, 3, ..., N + 1
\end{aligned}
$$

$$(5.4.13)$$

There is much similarity between these conditions and those for junctions in an acoustic transmission line, as described in Sect. 3.4.1. Thus, it should not be surprising that we will develop a sequential transfer function.

As always, the trace velocities parallel to an interface must be the same. Thus, Snell's law now requires that

$$
\boxed{\frac{1}{c_{\text{tr}}} = \frac{\sin \psi_n}{c_n}, \quad n = 1, ..., N + 2}
$$

$$(5.4.14)$$

Snell's law gives all waves the same complex exponential x dependence along the boundary. It is factored out. The result is that substitution of Eq. (5.4.1) leads to a set of simultaneous equations for the B_n and C_n coefficients. The equations for the first interface have a slightly different form because both y_1 and y_2 are zero there, and those for the last interface are different because C_{N+2} is zero in that half-space. Thus, the equations to solve are

$$B_1 + C_1 - (B_2 + C_2) = 0$$

$$\frac{B_1 - C_1}{Z_1} - \frac{B_2 - C_2}{Z_2} = 0$$

$$\left.\begin{array}{c} B_n e^{-iK_n H_n} + C_n e^{iK_n H_n} - (B_{n+1} + C_{n+1}) = 0 \\ \dfrac{B_n e^{-iK_n H_n} - C_n e^{iK_n H_n}}{Z_n} - \dfrac{B_{n+1} - C_{n+1}}{Z_{n+1}} = 0 \end{array}\right\} \quad n = 2, 3, ..., N \qquad (5.4.15)$$

$$B_{N+1} e^{-iK_{N+1} H_{N+1}} + C_{N+1} e^{iK_{N+1} H_{N+1}} - B_{N+2} = 0$$

$$\frac{B_{N+1} e^{-iK_{N+1} H_{N+1}} - C_{N+1} e^{iK_{N+1} H_{N+1}}}{Z_{N+1}} - \frac{B_{N+2}}{Z_{N+2}} = 0$$

where $Z_n = \rho_n c_n / \cos \psi_n$, as before.

The set of boundary conditions may be assembled into a block-diagonal matrix form by defining submatrices having two rows,

$$[A_1] = \begin{bmatrix} 1 & 1 & -1 & -1 \\ \dfrac{Z_2}{Z_1} & -\dfrac{Z_2}{Z_1} & -1 & 1 \end{bmatrix}$$

$$[A_n] = \begin{bmatrix} E_n & \dfrac{1}{E_n} & -1 & -1 \\ \dfrac{Z_{n+1}}{Z_n} E_n & -\dfrac{Z_{n+1}}{Z_n} \dfrac{1}{E_n} & -1 & 1 \end{bmatrix}, \quad n = 2, ..., N \qquad (5.4.16)$$

$$[A_{N+1}] = \begin{bmatrix} E_{N+1} & \dfrac{1}{E_{N+1}} & -1 \\ \dfrac{Z_{N+2}}{Z_{N+1}} E_{N+1} & -\dfrac{Z_{N+2}}{Z_{N+1}} \dfrac{1}{E_{N+1}} & -1 \end{bmatrix}$$

where E_n represents the complex exponential phasor,

$$E_n = e^{-iK_n H_n}, \quad K_n = \frac{\omega}{c_n} \cos \psi_n \qquad (5.4.17)$$

With these definitions, Eq. (5.4.15) may be written as

$$\begin{bmatrix} [A_1] & [0] & [0] & [0] & [0] \\ [0] & \ddots & [0] & [0] & [0] \\ [0] & [0] & [A_n] & [0] & [0] \\ [0] & [0] & [0] & \ddots & [0] \\ [0] & [0] & [0] & [0] & [A_{N+1}] \end{bmatrix} \begin{Bmatrix} B_1 \\ C_1 \\ \vdots \\ C_{N+1} \\ B_{N+2} \end{Bmatrix} = \{0\} \qquad (5.4.18)$$

Critical incidence occurs in any layer $n > 1$ whenever the angle of incidence ψ_1 is sufficiently large that Snell's law gives $\sin \psi_n > 1$. The critical angles are

$$(\psi_{cr})_n = \sin^{-1}\left(\frac{c_1}{c_n}\right) \qquad (5.4.19)$$

If ψ_1 exceeds the critical value for layer n, then Snell's law will give $\sin \psi_n > 1$. This may be handled by following Eq. (5.1.26) to evaluate ψ_n.

There are $N + 2$ layers, including the two half-spaces, and two plane waves propagate within each layer except the last. Thus, there are $2N + 3$ wave coefficients. The full set of continuity conditions represent $2(N + 1)$ equations, which means that they may be solved if we take one of the coefficients to be known. By definition, the reflected wave at the closest interface is $C_1 = RB_1$ and the wave transmitted into the far half-space is $B_{N+2} = TB_1$. Thus, setting $B_1 = 1$ would allow us to solve for the reflection and transmission coefficients.

If one seeks to establish how the transmission and reflection coefficients vary over a wide range of frequencies and angles of incidence, it will be necessary to solve Eq. (5.4.18) many times. If N is very large, such an evaluation will be numerically intensive. A important feature of the equations in this regard is their tightly banded structure, which permits the application of highly efficient numerical algorithms.

However, an even more efficient solution algorithm may be derived by sequentially solving the interface conditions to create a transfer function. The various $[A_n]$ matrices are partitioned into left and right submatrices, according to

$$
\begin{aligned}
[A_n]_{2\times4} &\equiv \big[[\alpha_n]_{2\times2} \;\; [\beta_n]_{2\times2}\big], \quad n = 1, 2, \; N \\
[A_{N+1}]_{2\times3} &\equiv \big[[\alpha_{N+1}]_{2\times2} \;\; \{\beta_{N+1}\}_{2\times1}\big]
\end{aligned}
\tag{5.4.20}
$$

We disregard for the moment the fact that the incident wave is known. This allows us to apply the same analysis to all interfaces other than the last.

The pair of equations for a generic interface may be solved for the complex amplitudes in the next layer in terms of the amplitudes in the preceding one. The equations have the form

$$
[\alpha_n] \begin{Bmatrix} B_n \\ C_n \end{Bmatrix} + [\beta_n] \begin{Bmatrix} B_{n+1} \\ C_{n+1} \end{Bmatrix} = \begin{Bmatrix} 0 \\ 0 \end{Bmatrix}, \quad n = 1, 2, ..., N
\tag{5.4.21}
$$

from which it follows that

$$
\begin{Bmatrix} B_{n+1} \\ C_{n+1} \end{Bmatrix} = [S_n] \begin{Bmatrix} B_n \\ C_n \end{Bmatrix}, \quad n = 1, 2, ..., N
$$
$$
[S_n] = -[\beta_n]^{-1}[\alpha_n]
\tag{5.4.22}
$$

We begin with the first interface and work forward. The first two steps are

$$
\begin{Bmatrix} B_2 \\ C_2 \end{Bmatrix} = [S_1] \begin{Bmatrix} B_1 \\ C_1 \end{Bmatrix}, \quad \begin{Bmatrix} B_3 \\ C_3 \end{Bmatrix} = [S_2] \begin{Bmatrix} B_2 \\ C_2 \end{Bmatrix} = [S_2][S_1] \begin{Bmatrix} B_1 \\ C_1 \end{Bmatrix}
\tag{5.4.23}
$$

Continuing this process leads to a general relation,

$$\begin{Bmatrix} B_{N+1} \\ C_{N+1} \end{Bmatrix} = [S_N][S_{n-1}]\cdots[S_1]\begin{Bmatrix} B_1 \\ C_1 \end{Bmatrix} \equiv \left[\prod_{j=1}^{N}[S_j]\right]\begin{Bmatrix} B_1 \\ C_1 \end{Bmatrix} \qquad (5.4.24)$$

The matrix form of the equations for the last interface is

$$[\alpha_{N+1}]\begin{Bmatrix} B_{N+1} \\ C_{N+1} \end{Bmatrix} + \{\beta_{N+1}\}\, B_{N+2} = \begin{Bmatrix} 0 \\ 0 \end{Bmatrix} \qquad (5.4.25)$$

Thus, the amplitudes in the incident and receiving half-spaces are related by

$$[S]\begin{Bmatrix} B_1 \\ C_1 \end{Bmatrix} + \{\beta_{N+1}\}\, B_{N+2} = \begin{Bmatrix} 0 \\ 0 \end{Bmatrix}$$
$$[S] = [\alpha_{N+1}]\left[\prod_{j=1}^{N}[S_j]\right] \qquad (5.4.26)$$

Our concern here is the situation where B_1 is the incident wave, C_1 is the reflected wave, and B_{N+1} is the transmitted wave. We set $B_1 = 1$, $C_1 = R$, and $B_{N+2} = T$, which leads to

$$\begin{Bmatrix} 1 \\ R \end{Bmatrix} = \{G\}\, T, \quad \{G\} = -[S]^{-1}\{\beta_{N+1}\} \qquad (5.4.27)$$

The scalar version of the preceding is

$$T = \frac{1}{G_1}, \quad R = G_2 T = \frac{G_2}{G_1} \qquad (5.4.28)$$

Thus, the overall procedure requires that the $[A_n]$ matrix for each of the interface be constructed and partitioned into left and right submatrices according to Eq. (5.4.20). The transfer matrix $[S]$ is created by sequentially multiplying the submatrices according to Eq. (5.4.26), and then, R and T may be found by forming $\{G\}$ according to Eq. (5.4.27). If the pressure in any intermediate layer is of interest, it may be computed according to Eq. (5.4.24). However, if only R and T are of interest, the intermediate matrices may be discarded after they are used.

Some problems do not fit exactly the specifications of finite layers bounded by half-spaces. For example, in the next example, the displacement at the first interface is specified and the last interface is locally reacting. Such situations may be handled by applying the transfer matrix formulation to satisfy the equations for the internal interfaces, $y_3 = 0$ to $y_{N+1} = 0$. This will yield expressions for B_{N+1} and C_{N+1} in terms of B_2 and C_2. Those relations may then be applied to enforce the interface equations at $y_2 = 0$ and $y_{N+1} = H_n$.

EXAMPLE 5.7   The motor shown below has an unbalanced rotor whose center of mass is situated at distance ε relative to the shaft. A rigid steel mounting plate is attached to the motor in order to apply a uniform pressure distribution to its support. A force analysis indicates that maintaining the eccentric distance ε as the rotor spins generates an outward force $\varepsilon\omega^2 m$ acting on the motor's bearings. Because this force is the reaction to the centripetal force acting on the rotating parts, it rotates at angular speed ω. Hence, the upward vertical force applied to the shaft by the rotor is $\varepsilon\omega^2 m \cos(\omega t)$. (The sideways force is carried by snubbers.) This force and the resultant of the pressure p_{plate} acting on the mounting plate accelerate the mass of the motor assembly. Thus, the equation of motion for the motor assembly is

$$p_{\text{plate}}\mathcal{A} + \varepsilon\omega^2 m \cos(\omega t) = -M\frac{dv_{\text{plate}}}{dt}$$

where M is the total mass of all parts of the motor, including the mounting plate and rotor, and $\mathcal{A}$ is the contact area between the pad and the plate. The sign of the acceleration term results from taking a downward velocity of the pad to be positive.

   If the motor were directly attached to the ground, the vibration of the ground would be excessive. A rubber isolation pad will be inserted below the mounting plate to reduce the vibration imparted to the ground. The pad consists of two layers of polymeric materials having equal depth H. In the study of vibrations, *insertion loss* is defined as the amplitude reduction in decibels achieved by inserting a system that absorbs vibration between two parts that would otherwise be joined. In other words, $IL = 20\log_{10}(V_0/V_{\text{ins}})$, where V_0 is the velocity without the inserted part. Consider the case where the polymers for the pad are silicone, $\rho = 1300\,\text{kg/m}^3$, $c = 900\,\text{m/s}$, above neoprene, $\rho = 1250\,\text{kg/m}^3$, $c = 1500\,\text{m/s}$, and $\mathcal{A} = 0.2\,\text{m}^2$. The ground is locally reacting, with $Z_{\text{gr}} = (0.1 - 3i)(10^6)$ Rayl. The motor's properties are $M = 120\,\text{kg}$ and $me = 0.01\,\text{kg-m}$. Determine the value of H that will maximize the insertion loss at 800 Hz. What are the vibration amplitudes of the mounting plate and of the ground in these conditions?

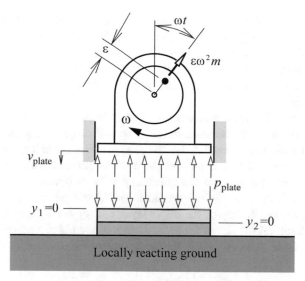

Figure 1.

Significance

This example will show how the basic formulation of multiple waves in layers may
be modified to address configurations that do not feature upper and lower fluid half-
spaces. The specific application to vibration control can be quite useful in some
situations.

Solution

We begin by introducing the frequency-domain ansatz to the motor's equation of
motion, and then solving for the pressure exerted on the plate,

$$P_{\text{plate}} = -\left(i\omega \frac{M}{A} V_{\text{plate}} + \frac{\varepsilon m}{A} \omega^2\right) \qquad (1)$$

In the reference condition, the motor assembly is mounted directly on the ground.
In that case, the plate and the ground have the same velocity, so that $V_{\text{plate}} = V_{\text{gr}}$ and
$P_{\text{plate}} = Z_{\text{gr}} V_{\text{gr}}$. Equating this relation to Eq. (1) yields the reference velocity,

$$\left(V_{\text{gr}}\right)_0 = -\left(\frac{\varepsilon m \omega^2}{Z_{\text{gr}} A + i\omega M}\right) \qquad (2)$$

If the imaginary part of Z_{gr} is much greater than the real part, that is, if Z_{gr}
is essentially reactive, then the maximum velocity will occur at the frequency that
minimizes the imaginary part of the denominator, so that

$$\omega_{\text{max}} \approx \frac{-\operatorname{Im}(Z_{\text{gr}})A}{M} \qquad (3)$$

This condition represents a resonance of the motor assembly coupled to the motor. It only occurs if $\text{Im}\left(Z_{gr}\right)$ is negative, that is, if the impedance is spring-like. The given value of Z_{gr} gives $\omega_{\max}/(2\pi) = 796\,\text{Hz}$, which is very close to the specified frequency. This suggests that the vibration amplitude will be quite large if the pad is not used. (In practice, Z_{gr} is likely to be frequency dependent. If it is, Eq. (3) is not a simple equation for $\omega_{\max}$.)

The analysis of the response when the pad is in place combines the continuity conditions at the interface between the silicone and neoprene layers with the local impedance condition at the bottom and the plate pressure at the top. These equations will be solved numerically to determine the wave amplitudes at a range of H values, from which particle velocities at the interface may be evaluated.

The coordinate systems we use are those introduced previously, specifically, x horizontal, $y_1 = 0$ at the top of the upper layer, and $y_2 = 0$ at the top of the lower layer. The velocity at the top is the same as that of the plate. The velocity $into$ the ground is $\bar{v}_2 \cdot \bar{e}_y$ at $y_2 = H$, so the frequency-domain equations for each interface are

$$P_1|_{y=0} = P_{\text{plate}}, \quad \bar{V}_1\big|_{y=0} \cdot \bar{e}_y = V_{\text{plate}}$$

$$P_1|_{y_1=H} = P_2|_{y_2=0}, \quad \bar{V}_1\big|_{y_1=H} \cdot \bar{e}_y = \bar{V}_2\big|_{y_2=0} \cdot \bar{e}_y$$

$$P_2|_{y_2=H} = Z_{gr}\,\bar{V}_2\big|_{y_2=H} \cdot \bar{e}_y$$

The pressure and particle velocity within each layer are described by Eq. (5.4.1). The propagation angles are $\psi_1 = \psi_2 = 0$ because the mounting plate creates a uniform pressure distribution, so there is no dependence on the x coordinate. Thus, substitution of the wave representations and Eq. (1) into the preceding equations gives

$$B_1 + C_1 = -\left(i\omega\frac{M}{\mathcal{A}}V_{\text{plate}} + \frac{\varepsilon m}{\mathcal{A}}\omega^2\right)$$

$$\frac{1}{\rho_1 c_1}(B_1 - C_1) = V_{\text{plate}}$$

$$E_1 B_1 + \frac{C_1}{E_1} - B_2 - C_2 = 0 \qquad\qquad (2)$$

$$\frac{1}{\rho_1 c_1}\left(E_1 B_1 - \frac{C_1}{E_1}\right) - \frac{1}{\rho_2 c_2}(B_2 - C_2) = 0$$

$$E_2 B_2 + \frac{C_2}{E_2} = \frac{Z_{gr}}{\rho_2 c_2}\left(E_2 B_2 - \frac{C_2}{E_2}\right)$$

where $E_n = \exp\left(-i\left(\omega/c_n\right)H\right)$.

Equation (2) is a set of five scalar equations for the B_n and C_n coefficients for a specified H. We use the second of that set to eliminate V_{plate} from the first equation and then solve for the wave amplitudes. After that, the complex velocity amplitude of the ground is found by evaluating the velocity of the second layer at its bottom,

$$V_{gr} = \bar{v}_2|_{y_2=H} \cdot \bar{e}_y = \frac{E_2 B_2 - C_2/E_2}{\rho_2 c_2}$$

The insertion loss is $20\log_{10}\left(\left|(V_{gr})_0\right|/\bar{V}_{gr}\right)$. The displacements may be found by using the fact that $\bar{U} = \bar{V}/(i\omega)$. Presumably, the displacement amplitudes are the desired quantities. The complex velocity is $i\omega$ times the complex displacement.

The insertion loss was computed for thicknesses ranging from a very small value to one meter. The result appears in Fig. 2. Three maxima, which correspond to the greatest reduction, occur in the one meter range of thicknesses. The one at $H = 0.86$ m gives the greatest reduction, 39.4 dB. However, that thickness is quite large. The maximum at $H = 0.16$ m gives a reduction of 36.7 dB, which is nearly as good, and represents a more economical use of materials. In the absence of the rubber pad, the displacement amplitude of the ground and of the plate would be $|V_{gr}|/\omega = 24.8$ mm, which obviously would be quite unacceptable. With the pad height at $H = 0.16$ m, the displacement amplitude of the plate is 0.772 mm, and that of the ground is 0.362 mm. Thus, the pad is extremely effective at reducing the motion of the ground and the motor at 800 Hz.

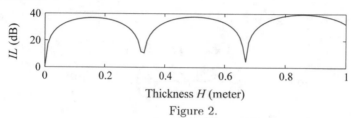

Figure 2.

EXAMPLE 5.8 Suppose the polymeric pad in the previous example is replaced by a multilayer composite. Describe how the transfer function formulation may be used to determine the insertion loss.

Significance

The versatility of the transfer function formulation is the highlight.

Solution

As always, we let N be the number of layers. Unlike the basic derivation, there is neither an upper or lower half-space. The base of the motor is $y_2 = 0$, and the locally reacting ground as $y_{N+1} = H_{N+1}$. The velocity of the plate is the same as that of the upper surface of the pad. Thus, Eq. (1) of the previous example requires that

$$\left.\left(AP_2 + i\omega M\bar{V}_2 \cdot \bar{e}_y\right)\right|_{y_2=0} = \varepsilon\omega^2 m \tag{1}$$

When we introduce the frequency-domain ansatz, and set the velocity of the ground equal to the velocity at the bottom of the pad, the local impedance condition is

$$\left.\left(P_{N+1} - Z_{gr}\bar{V}_{N+1} \cdot \bar{e}_y\right)\right|_{y_{N+1}=H_{N+1}} = 0 \tag{2}$$

Equation (5.4.1) describes the pressure and particle velocity in each layer in terms of the complex wave amplitudes. Substitution of those relations with all $\psi_n = 0$ converts Eqs. (1) and (2) to

$$\mathcal{A}\left(B_2 + C_2\right) + \frac{i\omega M}{\rho_2 c_2}\left(B_2 - C_2\right) = \varepsilon\omega^2 m \cos\left(\omega t\right)$$

$$\left(1 - \frac{Z_{gr}}{\rho_{N+1}c_{N+1}}\right)B_{N+1}e^{-i(\omega/c_{N+1})H_{N+1}} + \left(1 + \frac{Z_{gr}}{\rho_{N+1}c_{N+1}}\right)C_{N+1}e^{i(\omega/c_{N+1})H_{N+1}} = 0$$

(3)

Equation (5.4.24) for $n = 2$ to N enforces the pressure and continuity equations at all interfaces between adjacent layers. Thus, we have

$$\left\{\begin{matrix} B_{N+1} \\ C_{N+1} \end{matrix}\right\} = \left[\tilde{S}\right]\left\{\begin{matrix} B_2 \\ C_2 \end{matrix}\right\}, \quad \left[\tilde{S}\right] = [S_N]\left[S_{n-1}\right]\cdots[S_2]$$

(4)

The second of Eq. (3) gives

$$C_{N+1} = \alpha_{N+1}B_{N+1}$$

$$\alpha_{N+1} = -\left(\frac{\rho_{N+1}c_{N+1} - Z_{gr}}{\rho_{N+1}c_{N+1} + Z_{gr}}\right)e^{-2i(\omega/c_{N+1})H_{N+1}}$$

Substitution of this relation converts Eq. (4) to

$$\left\{\begin{matrix} B_2 \\ C_2 \end{matrix}\right\} = \{\Gamma\}\, B_{N+1}, \quad \{\Gamma\} = \left[\tilde{S}\right]^{-1}\left\{\begin{matrix} 1 \\ \alpha_{N+1} \end{matrix}\right\}$$

(5)

From this relation, we find that $B_2 = \Gamma_1 B_{N+1}$ and $C_2 = \Gamma_2 B_{N+1}$. Substitution of these relations into the first of Eq. (3) gives

$$\left[\left(\mathcal{A} + \frac{i\omega M}{\rho_2 c_2}\right)\Gamma_1 + \left(\mathcal{A} - \frac{i\omega M}{\rho_2 c_2}\right)\Gamma_2\right]B_{N+1} = \varepsilon\omega^2 m \cos\left(\omega t\right)$$

This expression may be solved for B_{N+1}. Then, the other wave amplitudes may be found from the intermediate transfer functions leading to Eq. (3).

5.5   Solid Barriers

A solid barrier that separates two large chambers is visually similar to a finite layer sandwiched between two half-spaces. However, there is a fundamental difference, in that the laws governing its motion are not those of fluid mechanics. Many common types of construction lead to equations of motion that have a common form. We will begin with a general discussion of this form and how it enters into an analysis of the sound that is transmitted through the barrier. After that, we will consider the specific relations for some common construction.

5.5.1 General Analysis

In Fig. 5.15, a compliant barrier divides a fluid into two half-spaces; the region through which the incident wave propagates is designated as layer 1. The fluids on both sides are taken to have the same physical properties, so $\rho_2 = \rho_1$ and $c_2 = c_1$. (It would not be difficult to modify the analysis to accommodate a situation where each half-space is composed of a different fluid.) It is helpful to refer to the side of the barrier that is exposed to the incident wave as the "front" of the barrier, so the other side is the "rear".

Incidence of a plane wave at the front of the barrier results in the generation of a reflected wave. In addition, the acoustic pressure applied to the front pushes the barrier, thereby inducing a transverse (that is normal) displacement field w, positive in the y direction. This displacement disturbs the fluid in the rear. This disturbance is a transmitted plane wave.

Our fundamental model assumes that the thickness h does not change when the barrier moves. Furthermore, a corollary of the restriction to a linearized analysis is that the displacement is small, so that the description in Fig. 5.15 greatly exaggerates the slope of the barrier. Smallness of w makes it permissible to consider the surfaces to always be at $y_1 = 0$ and $y_2 = 0$. It also allows us to consider the normal to the barrier to always be parallel to the y-axis. The combination of these attributes leads to recognition that the velocity of the front and back of the barrier is $\partial w\,(x, t)\,/\partial t$ in the $\bar{e}_y$ direction. (These restrictions are valid if no waves propagate in the y direction within the barrier. For elastic media, the restriction generally is valid if the thickness is substantially less than the trace acoustic wavelength, which usually is true.)

Fig. 5.15 A plane wave that is incident on a compliant wall and the waves that it generates

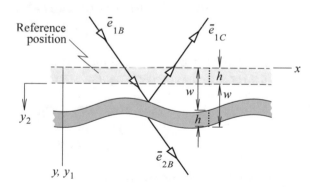

An analysis of the equations of motion for the barrier, based on the assumption that we know the pressure that acts on both sides, would lead to a field equation that governs the displacement field. Consistent with our use of linearized acoustics laws, we assume that this equation is linear. Based on the assumption that the incident, reflected, and transmitted waves propagate in a common plane, which we will verify, there is no dependence of the acoustic field in the z direction. Thus, the displacement field is solely a function of x and t. The consequence is that the barrier's differential equation of motion may be written as

$$\mathcal{L}(w) = p_1|_{y_1=0} - p_2|_{y_2=0} \tag{5.5.1}$$

where $\mathcal{L}(w)$ represents an arbitrary linear operator that might contain various terms that are derivatives with respect to x and t in unspecified combinations. It should be noted that the usage of local coordinate systems whose origins are situated on the front and back of the barrier will simplify evaluation of the pressures applied to the barrier.

The conditions imposed on the acoustic response are velocity continuity at both faces of the barrier. The normal velocity component of the barrier is $\partial w/\partial t$ in the y direction. The thickness of the barrier is considered to be invariant, so this velocity must match the normal particle velocity of the fluid on each face. Hence, the continuity conditions are

$$\bar{v}_1|_{y=0} \cdot \bar{e}_y = \bar{v}_2|_{y_2=0} \cdot \bar{e}_y = \frac{\partial w}{\partial t} \tag{5.5.2}$$

Satisfaction of velocity continuity is possible only if the barrier displacement is a surface wave that propagates in the x direction with a wavenumber that equals the trace of the acoustic waves. Thus, we shall attempt to form a solution in which each half-space has the same acoustical signal as in previous cases, while the wave in the barrier matches these waves in the x direction. It is useful at this stage to display the time dependence explicitly, so the ansatz is taken to be

$$\begin{aligned} p_n &= \mathrm{Re}\left(P_n(x, y)\, e^{i\omega t}\right) \\ w &= \mathrm{Re}\left(\hat{W}(x)\, e^{i\omega t}\right) \end{aligned} \tag{5.5.3}$$

In layer 1 there are the incident and reflected waves, and a transmitted wave propagates in layer 2, so the complex amplitudes are

$$\begin{aligned} P_1 &= B_1 e^{-ik_1\bar{e}_{1B}\cdot\bar{\xi}} + C_1 e^{-ik_1\bar{e}_{1C}\cdot\bar{\xi}} \\ P_2 &= B_2 e^{-ik_1\bar{e}_T\cdot\bar{\xi}} \\ \hat{W} &= W e^{-ik_{tr}x} \end{aligned} \tag{5.5.4}$$

where the propagation directions are

$$\begin{aligned} \bar{e}_I = \bar{e}_T &= (\sin\psi)\,\bar{e}_x + (\cos\psi)\,\bar{e}_y \\ \bar{e}_R &= (\sin\psi)\,\bar{e}_x - (\cos\psi)\,\bar{e}_y \end{aligned} \tag{5.5.5}$$

Satisfaction of the continuity conditions requires that the acoustic waves have the same trace wavenumber parallel to the surface of the barrier. Furthermore, designation of the wavenumber for the barrier as k_{tr} is a reminder that it must match the trace of the acoustic waves. The consequence is that Snell's law must hold, but both half-spaces contain the same fluid. Hence, the angles of reflection and transmission must equal the angle of incidence. Let ψ denote this angle, so that the trace wavenumber parallel to the barrier of the acoustic waves is

$$k_{\mathrm{tr}} = k_1 \sin \psi \tag{5.5.6}$$

We wish to determine C_1, B_2, and W corresponding to specified $B_1 \equiv P_1$. Two equations come from velocity continuity at the barrier. Substitution of Eqs. (5.5.3) and (5.5.4), followed cancellation of the trace exponentials, yields

$$\frac{B_1 - C_1}{Z_1} = \frac{B_2}{Z_1} = i\omega W \tag{5.5.7}$$

The third equation is derived from the barrier's equation of motion, Eq. (5.5.1). The linear nature of $\mathcal{L}(w)$ has an important consequence because a derivative of a complex exponential becomes an algebraic factor. This applies to derivatives with respect to x, as well as those with respect to t. Thus, substitution of the basic form of w merely leads to frequency-dependent coefficients multiplying the complex exponential propagator. The barrier's normal velocity amplitude is $i\omega W$. The collected coefficients multiplying $i\omega W$ is the *barrier impedance* Z_{B}, such that

$$\mathcal{L}(w) = \mathcal{L}\left(\mathrm{Re}\left(\hat{W}(x)e^{i\omega t}\right)\right) \equiv \mathrm{Re}\left[Z_{\mathrm{B}}(i\omega W)e^{-ik_1 x \sin \psi}e^{i\omega t}\right] \tag{5.5.8}$$

In the preceding, the factor $i\omega$ has been associated with W because the combination is the complex amplitude of the barrier's velocity, so that Z_{B} has dimensions of force divided by velocity. (The actual expression for Z_{B} for various types of configurations will be obtained by matching the actual result of the substitution to this standard form.) Thus, the equation of motion for the barrier reduces to

$$\boxed{i\omega Z_{\mathrm{B}} W e^{-ik_1 x \sin \psi} = P_1|_{y_1=0} - P_2|_{y_2=0}} \tag{5.5.9}$$

The pressures at the faces of the barrier are described by Eq. (5.5.3). Because the complex exponential on the left side matches those of P_1 at $y = 0$ and of P_2 at $y_2 = 0$, the coefficients of that factor must match. Thus, we find that

$$i\omega Z_{\mathrm{B}} W = B_1 + C_1 - B_2 \tag{5.5.10}$$

Equations (5.5.7) and (5.5.10) constitute three algebraic equations. To solve them, we set $C_1 = R B_1$ and $B_2 = T B_1$. The resulting reflection and transmission coefficients are

$$\boxed{R = \frac{Z_{\mathrm{B}}}{2Z_1 + Z_{\mathrm{B}}}, \quad T = \frac{2Z_1}{2Z_1 + Z_{\mathrm{B}}}} \tag{5.5.11}$$

The displacement amplitude of the barrier is

$$W = \frac{2}{i\omega\,(2Z_1 + Z_B)}\,B_1 \tag{5.5.12}$$

The appearance of a factor of two multiplying Z_1 is a consequence of the fluid on both faces of the barrier having equal effect in terms of opposing the motion of the barrier.

An interesting aspect of the intermediate description of velocity continuity, Eq. (5.5.7), is that it requires that

$$T = 1 - R \tag{5.5.13}$$

This relation should not be confused with conservation of energy. The important feature of power flow analysis for a barrier is that it is not an acoustical material. Thus, we must adopt the formulation in Chap. 4, wherein an acoustic source was inserted into the acoustical domain. Figure 5.16 shows a suitable domain. In it, blocks of fluid extending an arbitrary distance away from each face of the barrier are isolated from their surroundings, and from the barrier.

Fig. 5.16 Acoustic domain used to analyze power flow across a barrier

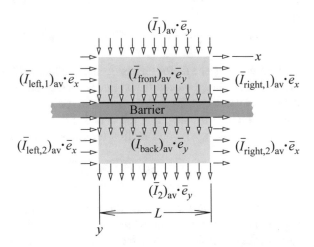

As was true for the analysis of power flow through many fluid layers, the x dependence of all variables is solely a phase difference, so the time-averaged intensities on the right and left sides are equal. Thus, conservation of energy in the acoustic domain requires that

$$(\mathcal{P}_1 - \mathcal{P}_{\text{front}}) + (\mathcal{P}_{\text{back}} - \mathcal{P}_2) = 0 \tag{5.5.14}$$

The first term, $\mathcal{P}_1$, is associated with the time-averaged intensity on the upper face, where the acoustic signal is the combination of incident and reflected waves. This term is described by Eq. (5.2.22). The power that flows out of the upper region across the front face of the barrier is $\mathcal{P}_{\text{front}}$. The normal velocity there is $\bar{V} \cdot \bar{e}_y = i\omega W$, so that

$$\mathcal{P}_{\text{front}} = \frac{1}{2} \int_0^L \text{Re}\left(P_1|_{y=0} \left(-i\omega W^*\right) \right) dx \tag{5.5.15}$$

Note that the velocity in this description is that of the barrier in order to expose the role of the barrier. An expression similar to the preceding using P_2 at $y_2 = 0$ describes $\mathcal{P}_{\text{back}}$, which is the power flow into the lower domain from the barrier. Finally, the power $\mathcal{P}_2$ flowing out of the lower region across the lower face is due to the transmitted wave.

Thus, conservation of energy requires that

$$\begin{aligned}
&\left(\frac{|B_1|^2}{2Z_1} \left(1 - |R|^2\right) L - \frac{1}{2} \int_0^L \text{Re}\left(P_1|_{y=0} \left(-i\omega W^*\right) \right) dx \right) \\
&+ \left(\frac{1}{2} \int_0^L \text{Re}\left(P_2|_{y_2=0} \left(-i\omega W^*\right) \right) dx - \frac{|B_1|^2}{2Z_1} |T|^2 L \right) = 0
\end{aligned} \tag{5.5.16}$$

Merging the integrals leads to $P_1|_{y=0} - P_2|_{y_2=0}$ being a factor in the integrand. According to Eq. (5.5.9), this term is proportional to W. Thus, the power balance equation reduces to

$$\frac{|B_1|^2}{2Z_1} |T|^2 = \frac{|B_1|^2}{2Z_1} \left(1 - |R|^2\right) - \frac{1}{2} \text{Re}\left(Z_B\right) \omega^2 |W|^2 \tag{5.5.17}$$

To interpret this relation, note that $|T|^2 + |R|^2 = 1$ only if $\text{Re}\left(Z_B\right) = 0$. This means that only if the barrier impedance is purely reactive will all the acoustical energy be reflected or transmitted. If $\text{Re}\left(Z_B\right) > 0$, then $|T|^2 + |R|^2 < 1$, which means that power is dissipated within the barrier. (This may be proven independently by setting $Z_B = (\alpha + \beta i) Z_1$ in Eq. (5.5.11). The property that $|T|^2 + |R|^2 = 1$ will be found to hold only if $\alpha = 0$.)

The transmission loss TL of a barrier is an extremely useful metric for noise control applications. It compares the intensity in the fluid on the transmitted and incident sides. The preceding expression for the normal component of intensity in the transmitted wave is the same as it is for an interface between liquids. The same fluid is on both sides of the barrier, so $Z_2 = Z_1$. Thus, the fundamental definition in Eq. (5.3.31) reduces to

$$TL = -10 \log_{10} \left(\frac{(\bar{I}_2)_{\text{av}} \cdot \bar{e}_y}{(\bar{I}_1)_{\text{av}} \cdot \bar{e}_y} \right) = -10 \log_{10} \left(|T|^2 \right) \tag{5.5.18}$$

5.5.2 Specific Barrier Models

Several systems are described by equations of motion that fit Eq. (5.5.1). Here, we examine a few in increasing order of the difficulty in their analysis.

Inertial Blanket

The simplest type of barrier is one whose stiffness is negligible. Stiffness is a measure of the elastic effect, so the implication here is that the barrier stores little potential energy when it is displaced. Thus, the primary impedance effect would be inertial. Many individuals say that such a barrier is a limp mass. A common object fitting this description is a heavy blanket. Another would be a sheet of loosely packed sand confined between thin sheets of plastic, but maintaining a constant thickness would be problematic. The true value of this model is that it characterizes a limiting behavior for other systems.

The xz-plane has been defined to coincide with the plane of the barrier, so an element of mass is $\rho_B h\, dx\, dz$, where ρ_B is the density of the barrier. The net force exerted on this element by the fluid is the difference between the pressures on the front and back of the barrier multiplied by the element of surface area. This force must accelerate the element, with positive displacement w taken to be in the normal sense at the back of the barrier. Newton's second law requires that

$$\rho_B h \frac{\partial^2 w}{\partial t^2} = p_1|_{y=0} - p_2|_{y_2=0} \qquad (5.5.19)$$

To identify Z_B for this construction, we substitute the frequency-domain description of each variable. The result is that

$$-\rho_B h \omega^2 W e^{-ik_1 x \sin \psi} = P_1|_{y=0} - P_2|_{y_2=0} \qquad (5.5.20)$$

A comparison of this expression Eq. (5.5.9) leads to the recognition that the barrier impedance is

$$\boxed{Z_B = i\omega \rho_B h} \qquad (5.5.21)$$

This impedance is a positive imaginary value, so it is an inertance. This means that energy is not dissipated within an inertial blanket. The value of Z_B occurs solely in the denominator of T in Eq. (5.5.11). Because Z_B increases with frequency, we conclude that an inertial blanket becomes increasingly effective in reducing sound transmission as the frequency increases.

Stretched Membrane

A prototype for another type of barrier is the head of a musical drum. It consists of a sheet of material that is stretched, so that it carries a tensile stress. Such a barrier is referred to as a *stretched membrane*. Let Γ denote the resultant traction (dimensions of force divided by length) obtained by integrating the tensile stress through the thickness of the membrane, and let $\rho_{mem}h$ denote the mass per unit surface area.

The equation of motion for this system is the wave equation in the coordinates that lie in the plane of the membrane, with the net pressure acting as the excitation.[3] Positive displacement w of the membrane is in the normal sense at the back. The corresponding field equation is

$$\rho_{\text{mem}}h\frac{\partial^2 w}{\partial t^2} - \Gamma\left(\frac{\partial^2 w}{\partial x^2} + \frac{\partial^2 w}{\partial z^2}\right) = P_1|_{y=0} - P_2|_{y_2=0} \tag{5.5.22}$$

Substitution of the basic ansatz in Eq. (5.5.3) into the preceding leads to

$$\left[-\rho_{\text{mem}}h\omega^2 + \Gamma\,(k_1\sin\psi)^2\right]We^{-ik_1 x\sin\psi} = P_1|_{y=0} - P_2|_{y_2=0} \tag{5.5.23}$$

A comparison of this expression to Eq. (5.5.9) reveals that the barrier impedance of a stretched membrane is

$$Z_{\text{B}} = i\omega\left[\rho_{\text{B}}h - \frac{\Gamma}{c_1^2}\,(\sin\psi)^2\right] \tag{5.5.24}$$

Consider now a different situation, in which the same membrane is placed in a vacuum. Both pressure terms in the right side of Eq. (5.5.22) then are not present. The remaining terms express the conditions for which a wave generated by initial conditions may propagate without further excitation. The phase speed of such a wave is denoted as c_{mem}, which we can determine by substituting $w = \text{Re}\left(We^{i\omega(t-x/c_{\text{memb}})}\right)$ into the equation. The result is

$$c_{\text{mem}} = \left(\frac{\Gamma}{\rho_{\text{mem}}h}\right)^{1/2} \tag{5.5.25}$$

Introduction of this parameter and the trace speed of the incident wave into Eq. (5.5.24) leads to

$$Z_{\text{B}} = i\omega\rho_{\text{mem}}h\left(1 - \frac{c_{\text{mem}}^2}{c_{\text{tr}}^2}\right) \equiv i\omega\rho_{\text{mem}}h\left[1 - (\sin\psi)^2\,\frac{c_{\text{mem}}^2}{c_1^2}\right] \tag{5.5.26}$$

This is a reactance. If $\sin\psi \ll c_1/c_{\text{mem}}$, then the membrane acts like an inertial blanket. In contrast, a compliant barrier impedance is obtained if $\sin\psi > c_1/c_{\text{mem}}$. If $\psi = \sin^{-1}(c_1/c_{\text{mem}})$, then Z_{B} vanishes, which leads to $T = 1$ and $R = 0$. We say that the membrane becomes *transparent* in this condition. Transparency occurs because no excitation is required to propagate a wave in the membrane at phase speed c_{memb}, but that is the trace speed of the incident wave in this condition. The membrane cannot oppose an applied pressure at this frequency.

---

[3] A comprehensive presentation of analytical solutions may be found in the recent text by C. Y. Yang and C. M Yang, *Structural Vibration: Exact Solutions for Strings, Membranes, Beams, and Plates*, CRC Press (2013).

Elastic Plate

An elastic plate has the same appearance as a membrane, in that each is flat and covers an area whose extent is large compared to its thickness. The difference is the manner in which they support forces that are applied transversely to their plane. A membrane is the planar analog of a stretched string, in that both use a tensile stress that extends through their thickness to resist transverse loads. In contrast, an elastic plate, like a beam, uses internal moments to resist transverse loads. In principle, a plate could be preloaded with tensile stresses, just as a beam could carry a preexisting tensile force, but in most applications no such preload is present. Because a plate extends in two directions, whereas a beam runs in only one, the theory of plates is much more complicated. In particular, whether closed-form analytical solutions for their dynamic displacement exist depends on how they are supported. Here, we consider the plate to extend without limit in a plane. In essence, our analysis will disclose the nature of the general solution, which constitutes the foundation for plates having finite size.

As mentioned above, an elastic plate and a beam are analogous, which means that the equations of motion for both systems contain fourth-order spatial derivatives of their displacement. For a plate, which extends in the x and z directions, the equation governing the transverse displacement w is

$$m_{pl}\frac{\partial^2 w}{\partial t^2} + D\left(\frac{\partial^4 w}{\partial x^4} + 2\frac{\partial^4 w}{\partial x^2 \partial z^2} + \frac{\partial^4 w}{\partial x^4}\right) = P_1|_{y=0} - P_2|_{y_2=0} \qquad (5.5.27)$$

The parameter D is the flexural rigidity, and m_{pl} is the mass per unit length. If a plate is composed of a single material whose thickness is h, then

$$D = \frac{Eh^3}{12\left(1 - \nu^2\right)}, \quad m_{pl} = \rho h$$

where E is Young's modulus, ν is Poisson's ration, and ρ is the material density. However, composite construction is increasingly being used, in which case D and m_{pl} are equivalent values. The former may be identified by a (static) strength of materials analysis, and the latter is the mass of the plate in a 1×1 area of the plate.

Our interest is the response to an incident plane wave, so we take w to be as given by Eq. (5.5.3) and therefore independent of z. The result of substituting these representation and then canceling the common complex exponential phasor and is

$$\left[D\left(k_1 \sin \psi\right)^4 (1 + i\eta) - m_{pl}\omega^2\right] W e^{-ik_1 x \sin \psi} = P_1|_{y=0} - P_2|_{y_2=0} \qquad (5.5.28)$$

Here, we have introduced the structural damping loss factor η to account for dissipation within the plate. Such effects tend to be quite small for metallic plates, but they can be substantial for composite constructions. It usually is best to measure it for the system of interest because dissipation effects are difficult to characterize fully. A constant value of η often is used to represent intricate systems; a value between

0.001 and 0.1 is typical. A comparison of the preceding relation to Eq. (5.5.9) indicates that the barrier impedance is

$$Z_B = i\omega \left[m_{pl} - \frac{D\omega^2}{c_1^4} (\sin\psi)^4 (1 + i\eta) \right] \tag{5.5.29}$$

Suppose dissipation is not significant, $\eta \approx 0$. In that case, Z_B is imaginary, which means that it is reactive. At very low frequencies $Z_B \approx i\omega m_{pl}$, which matches Eq. (5.5.21) for an inertial blanket. This same behavior occurs at normal incidence for any frequency. This is so because there is no variation in the x direction, so the plate does not bend. As ω is increased with ψ fixed, Im (Z_B) decreases, eventually becoming zero when $\omega^2 (\sin\psi)^4 = m_{pl} c_1^4 / D$. Transparency, $T = 1$, is an undesirable condition because the plate will not shield the back from the incident wave. The frequency at which this condition occurs varies inversely to $(\sin\psi)^2$, so $\psi = \pi/2$ corresponds to the minimum value. This is the *coincidence frequency*,

$$\omega_{co} = c_1^2 \left(\frac{m_{pl}}{D} \right)^{1/2} \tag{5.5.30}$$

Isolation of this combination of parameters in Eq. (5.5.29) leads to an alternative representation of Z_B, specifically

$$Z_B = i\omega m_{pl} \left[1 - \left(\frac{\omega}{\omega_{co}} \right)^2 (\sin\psi)^4 (1 + i\eta) \right] \tag{5.5.31}$$

The maximum positive value of Im (Z_B) is

$$\max \left(\text{Im} \left(Z_B \right) \right) = \frac{2}{3} i\omega m_{pl} \text{ at } \omega = \frac{\omega_{co}}{\sqrt{3} (\sin\psi)^2} \tag{5.5.32}$$

If $\omega < \omega_{co}$ and $\eta = 0$, then Z_B is purely an inertance and $|T| < 1$, regardless of ψ.

In the case where the plate is composed of a single material, the coincidence frequency may be rewritten in terms of the phase speed of extensional waves in a bar composed of that material, which is a fundamental material property. This phase speed is related to the physical properties by

$$c_{bar} = \left(\frac{E}{\rho_{pl}} \right)^{1/2} \tag{5.5.33}$$

Then $D/m_{pl} = c_{bar}^2 h^2 / \left(12 \left(1 - \nu^2 \right) \right)$, so that

$$\omega_{co} = \sqrt{12 \left(1 - \nu^2\right) \frac{c_1^2}{c_{bar}h}} \qquad (5.5.34)$$

If the frequency is much greater than the coincidence value, then $\mathrm{Im}\,(Z_B)$ is large negatively, corresponding to a compliance. This value is essentially proportional to ω^3, so it leads to $T \to 0$. This is highly desirable. However, the frequency at which this condition occurs goes up as the inverse square of $\sin \psi$. If ψ is very small, then the frequency where the plate is most effective will be very high. In most situations, the sound that is incident on a barrier might arrive from any and all angles, and the frequency might fall in a rather wide band. Designing the plate to have a specific value for ω_{co} might be effective in some parameter range, but the most important requirement usually is that the inertance be sufficiently high at low frequencies and angles of incidence. In other words, the first step usually is to design the plate based on criteria for an inertial blanket.

A different view of the significance of the coincidence frequency comes from the consideration of a wave in the plate that acts as the excitation of an acoustic field. In the absence of the fluid, the plate's displacement would be governed by the homogeneous terms in Eq. (5.5.27). The homogeneous solutions for the plate displacement describe how the plate responds to a remote disturbance if it is in a vacuum. The displacement consists of a wave that may propagate freely along the surface in any direction. The form of the solution for a wave that propagates in the direction of increasing x is

$$w = \mathrm{Re}\left(W e^{i\omega\left(t - x/c_{pl}\right)}\right) \qquad (5.5.35)$$

To determine the phase speed c_{pl}, we substitute this ansatz into the plate equation of motion, so that

$$D \frac{\omega^4}{c_{pl}^4} - m_{pl}\omega^2 = 0$$

This equation has four roots for c_{pl}. Two are purely imaginary, which correspond to evanescent waves that are relevant to the vibration of a finite-sized plate. The other two are positive and negative real values that represent waves traveling in the positive and negative x direction. The positive value is

$$c_{pl} = \left(\frac{D}{m_{pl}}\right)^{1/4} \omega^{1/2} \qquad (5.5.36)$$

This is a dispersion relation.

Now suppose an elastic plate is surrounded by a gas, which means that fluid loading is a weak effect. In such situations, it is reasonable to decouple the fields. That is, the pressure effects are ignored in an analysis of the plate displacement. Then, that displacement is used as the excitation for a separate analysis of the acoustic field.

(Decoupling of the analyses of the waves in the structure and the fluid usually is not valid if the fluid is a liquid because the pressures are substantial. The result is a significant modification of the dispersion equation.) In this scenario, the free wave in Eq. (5.5.35) describes how the plate would respond to a remote disturbance. This displacement represents a surface wave that induces a signal in the fluid half-spaces at the front and rear of the plate. This situation was the topic of Sect. 5.1. Thus, a simple plane wave radiates away from the plate if c_{pl} is greater than c_1 for the fluid. The result of using Eq. (5.5.30) to eliminate D/m_{pl} is that

$$c_{pl} > c_1 \implies \left(\frac{D}{m_{pl}}\right)^{1/4} \omega^{1/2} > c_1 \implies \omega > \omega_{co}$$

In other words, a surface wave that is induced in the plate will generate a simple plane wave in the fluid if the surface wave's frequency exceeds the coincidence frequency. The angle for the radiated wave is

$$\psi = \sin^{-1}\left(\frac{c_1}{c_{pl}}\right) \tag{5.5.37}$$

In contrast, if ω is less than ω_{co}, then a surface wave in the plate induces an evanescent field in the fluid. The acoustic energy in that case is trapped near the plate. This obviously is a desirable condition if one wishes to reduce the noise that radiates from a vibrating plate.

Porous Barrier

A *porous* material has openings that extend through the thickness, which means that fluid can flow from front to back or vice versa. We have already encountered a porous material, specifically the acoustical tile in Example 5.3. One of the concepts introduced there will be used here, specifically that the surface behavior is examined in an average sense. Doing so is valid if the spacing and distance between holes are much less than an acoustic wavelength. Because porosity allows fluid to flow through the barrier, the fluid velocity at the front and back of the barrier needs not be the same as that of barrier. We will retain the notation $\bar{V}_1$ for the front, $\bar{V}_2$ for the back, and $i\omega\hat{W}$ for the velocity of the barrier.

To replace the previous velocity continuity conditions, we return to some basic concepts. First, consider a control volume through the thickness of the barrier and extending slightly beyond each face into the fluid domain. We consider the limiting situation on which the blanket is very thin relative to a wavelength, so that the isolated region's volume is essentially zero. Conservation of mass then requires that whatever mass flows into the control volume from the front equals that which flows out at the back, which leads to

$$\bar{v}_1 \cdot \bar{e}_y\big|_{y=0} = \bar{v}_2 \cdot \bar{e}_y\big|_{y_2=0} \tag{5.5.38}$$

The other relation we need originates in steady fluid dynamics. The velocity of fluid particles is $\bar{v}_1 \cdot \bar{e}_y$ (or $\bar{v}_2 \cdot \bar{e}_y$) evaluated at $y_1 = 0$. Correspondingly, the velocity of particles relative to the barrier is $\bar{v}_1 \cdot \bar{e}_y - \dot{w}$. This velocity is taken to be proportional to the pressure differential through the pores. If the pressure in the front is greater than the pressure in the back, then particles will flow from front to back. The corresponding relation is

$$\bar{v}_1 \cdot \bar{e}_y\big|_{y_1=0} - \dot{w} = \frac{1}{h\Upsilon} \left(p_1\big|_{y_1=0} - p_2\big|_{y_2=0} \right) \qquad (5.5.39)$$

where Υ is the *flow resistivity*. The frequency-domain version of this relation is obtained by replacing each variable by its complex amplitude,

$$\bar{V}_1 \cdot \bar{e}_y\big|_{y_1=0} - i\omega\hat{W} = \frac{1}{h\Upsilon} \left(P_1\big|_{y_1=0} - P_2\big|_{y_2=0} \right) \qquad (5.5.40)$$

The value of Υ is assumed to be independent of frequency.

In addition to these conditions, it is necessary to satisfy the equation of motion for the barrier when it is subjected to pressures p_1 and p_2 on either face. A general form is provided by Eq. (5.5.9),

$$i\omega Z_B W e^{-ik_1 x \sin\psi} = P_1\big|_{y_1=0} - P_2\big|_{y_2=0} \qquad (5.5.41)$$

The value of Z_B depends on the barrier's inertial, elastic, and dissipative properties.

The next step entails converting the continuity, porosity, and barrier equations to algebraic equations for the amplitudes of the waves. The basic representations of the pressure and barrier displacement in Eq. (5.5.3) still apply. All trace wavenumbers must match, so it follows that

$$k_{tr} = k_1 \sin\psi_1, \quad \psi_2 = \sin^{-1}\left(\frac{c_2}{c_1} \sin\psi_1 \right) \qquad (5.5.42)$$

Because of this condition, all complex exponentials for propagation in the x direction are the same. Therefore, they may be canceled when Eq. (5.5.4) is substituted into the general equations. The result is three simultaneous equations,

$$\begin{aligned} B_1 - C_1 &= B_2 \\ \frac{B_1 - C_1}{Z_1} - i\omega W &= \frac{1}{\Upsilon}(B_1 + C_1 - B_2) \\ i\omega Z_B W &= B_1 + C_1 - B_2 \end{aligned} \qquad (5.5.43)$$

To solve these equations, we set $C_1 = RB_1$ and $B_2 = TB_1$. The value of B_1 factors out of the expressions for the reflection and transmission coefficients, which are found to be

$$R = \dfrac{Z_{\mathrm{B}}}{(2Z_1 + Z_{\mathrm{B}}) + \dfrac{2Z_{\mathrm{B}}Z_1}{\Upsilon}}$$

$$T = \dfrac{2Z_1}{(2Z_1 + Z_{\mathrm{B}}) + \dfrac{2Z_{\mathrm{B}}Z_1}{\Upsilon}}\left(1 + \dfrac{Z_{\mathrm{B}}}{\Upsilon}\right) \tag{5.5.44}$$

In some cases, the velocity of the barrier is of interest; it is

$$i\omega W = \dfrac{2}{(2Z_1 + Z_{\mathrm{B}}) + \dfrac{2Z_{\mathrm{B}}Z_1}{\Upsilon}}B_1 \tag{5.5.45}$$

These expressions have been written in a form that facilitates recognizing the effect of porosity. If the barrier is nearly impermeable, then Υ is very large. In the limit as $\Upsilon \to \infty$, the values of R and T approach Eq. (5.5.11) and W approaches Eq. (5.5.12). This is the behavior one would anticipate, because a very large value of Υ implies that the barrier moves in unison with the fluid. The opposite limit, $\Upsilon = 0$, leads to $R = 0$, $T = 1$, and $W = 0$. This too is the behavior that is anticipated, because zero flow resistance means that the barrier does not block the fluid's response.

EXAMPLE 5.9 A common type of wall construction uses sheetrock panels that are a composite plate of gypsum with paper on both surfaces to prevent fracture under large tensile stresses. As is shown in the sketch, the construction uses two sheetrock panels separated by wooden braces spaced at regular intervals. The question to explore is whether a signal whose angle of incidence is $60°$ is transmitted less with this construction than one that uses one sheetrock panel having twice the thickness. It may be assumed that the wooden braces do not affect the propagation of waves. Consider a frequency range from zero to twice the coincidence frequency of a single panel of sheetrock. The extensional wave phase speed of the gypsum that composes sheetrock is 4000 m/s and its density is 2320 kg/m^3.

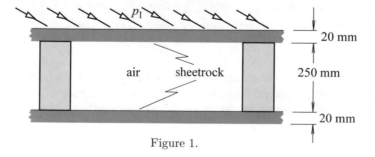

Figure 1.

Significance

The issues examined here have obvious practical applications. Equally important, the solution will synthesize the fundamental concepts by requiring that the basic formulation for a barrier be merged with that for layers.

Solution

The derived equations for a single barrier do not fit the configuration of the wall system. Thus, we will return to the general formulation of the equations governing a barrier. Fluid 1 is designated to be the upper half-space within which the incident wave propagates, and fluid 3 is the lower half-space into which the signal is transmitted. Fluid 2 is the air contained between the panels. The directions perpendicular to the panels in the downward sense are defined such that $y_1 = 0$ at the upper face of the top panel, $y_2 = 0$ at the lower face of the top panel, and $y_3 = 0$ at the bottom face of the lower panel.

All fluid regions are composed of air, so $c = 340 \, \text{m/s}$, $\rho = 1.2 \, \text{kg/m}^3$, and $k = \omega/c$ for all regions. Furthermore, the trace velocities in all regions must be the same, and they must equal the phase speed of the wave in each panel. Thus, the wavenumber parallel to the panel for all media is $k \sin \psi_1$.

Each panel behaves like an elastic plate, whose equation of motion has the form in Eq. (5.5.9), so that

$$
\begin{aligned}
i\omega Z_B W_U e^{-ikx \sin \psi_1} &= P_1|_{y_1=0} - P_2|_{y_2=0} \\
i\omega Z_B W_L e^{-ikx \sin \psi_1} &= P_2|_{y_2=H} - P_3|_{y_3=0}
\end{aligned}
\tag{1}
$$

where Z_B is described by Eq. (5.5.26). In addition, the normal velocities of a panel and a fluid particle must match at each interface, so it must be that

$$
\begin{aligned}
V_1 \cdot \bar{e}_y|_{y_1=0} &= V_2 \cdot \bar{e}_y|_{y_2=0} = i\omega W_U e^{-ikx \sin \psi_1} \\
V_2 \cdot \bar{e}_y|_{y_2=H} &= V_3 \cdot \bar{e}_y|_{y_3=0} = i\omega W_L e^{-ikx \sin \psi_1}
\end{aligned}
\tag{2}
$$

The waves within each half-space have the form in Eq. (5.5.3), and waves propagate in both directions in fluid 2, so we try a solution whose form is

$$
\begin{aligned}
P_1 &= \left(e^{-iky_1 \cos \psi} + R e^{+iky_1 \cos \psi} \right) e^{-ikx \sin \psi} \\
P_2 &= \left(B_2 e^{-iky_2 \cos \psi} + C_2 e^{+iky_2 \cos \psi} \right) e^{-ikx \sin \psi} \\
P_3 &= T e^{-iky_3 \cos \psi} e^{-ikx \sin \psi}
\end{aligned}
\tag{3}
$$

Note that the incident wave's magnitude has been taken to be unity because our sole interest is the comparison of T to a simpler construction method.

Substitution of the preceding ansatz into the velocity continuity equations, Eq. (2), yields

$$\frac{1}{Z}(1 - R) = \frac{1}{Z}(B_2 - C_2) = i\omega W_U$$
$$\frac{1}{Z}\left(E_2 B_2 - \frac{C_2}{E_2}\right) = \frac{1}{Z}T = i\omega W_L \tag{4}$$

where $Z = \rho c / \sin \psi$ and $E_2 = \exp(ikH \cos \psi)$. Additional equations are obtained by substituting Eq. (3) into the panel equations of motion, Eq. (1), which become

$$i\omega Z_B W_U = 1 + R - (B_2 + C_2)$$
$$i\omega Z_B W_L = E_2 B_2 + \frac{C_2}{E_2} - T \tag{5}$$

Equations (4) and (5) constitute a set of six equations governing the six unknown coefficients R, T, B_2, C_2, W_U, and W_L. They may be solved by rewriting them in matrix form, with the coefficient matrix evaluated at each frequency in the specified interval.

The case of the double-thickness panel without an air gap fits the standard situation, so the transmission ratio in that case is described by Eq. (5.5.11). The value of Z_B is given by Eq. (5.5.31), for which the coincidence frequency is obtained by evaluating Eq. (5.5.34) at $h = 40\,\text{mm}$. In contrast, Z_R for the separated individual panels is found by setting $h = 20\,\text{mm}$. $c_{\text{bar}} = 4000\,\text{m/s}$, $\rho_{\text{pl}} = 2320\,\text{m/s}$, and $H = 0.25\,\text{m}$. The coincidence frequencies are 797 Hz for each panel in the double-wall construction and 398 Hz for a single panel of the same total thickness. The result of the computation is shown in Fig. 2.

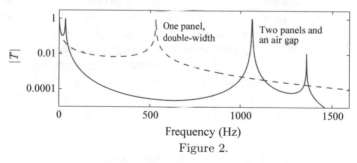

Figure 2.

At very low frequencies, each panel's impedance is essentially that of an inertial blanket. Neither configuration is effective in this frequency range. The low-frequency peak for the double-wall configuration corresponds to a system resonance, in which the combined inertial and elastic effects of the panel and cavity enclosed between the panels cancel. (This phenomenon is manifested as a degeneracy of the simultaneous equations described by Eqs. (4) and (5).) At 531 Hz, the value of c_{pl} in the double-width configuration equals the trace velocity at $\psi = 60°$, whereas matching of c_{pl} and c_{tr} for the two-panel construction occurs at 1062 Hz. The peak at 1358 Hz is another plate-cavity resonance of the two-wall configuration. Both designs generally reduce the transmission ratio. However, even though the construction that uses an air gap exhibits more peaks of $|T|$ as a function of frequency, separating the sheetrock panels by an air gap gives a much lower transmission ratio over a broad frequency range.

5.6 Closure

At this juncture, we have developed considerable expertise to investigate and understand acoustic phenomena. The concept that trace velocities must match is basic to the interpretation of experimental data. The phenomenon of supercritical incidence, which leads to evanescence in the receiving medium, also arises for waves in solid elastic media. A local impedance model of a surface is often used as a way of simplifying investigations, especially when extremely accurate predictions are not required. The developments in this chapter make it possible to address several problems of practical importance, such as how to limit the transmission of noise. However, we are limited to situations where the signal is planar. The next chapter begins to remove these limitations.

5.7 Homework Exercises

Exercise 5.1 A dispersive wave is one whose phase speed depends on its frequency; the equation describing this dependence is a dispersion equation. Consider a dispersive wave that propagates along the flat bottom of an ocean channel. The dispersion equation for this wave is $c_{\text{phase}} = 0.5 c_0 \left[1 + (\omega/\omega_{\text{ref}})^2 \right]$, where c_0 is the speed of sound in the water and $\omega_{\text{ref}} = 2000$ rad/s is a reference frequency. The bottom coincides with the xz-plane, with y defined vertically upward. The vertical displacement at the origin, $x = y = z = 0$, is $w = 3 \cos(1200t) + 0.8 \cos(3200t)$ mm. (a) Determine the pressure at the origin. (b) Determine the pressure at a height of 100 m above the origin.

Exercise 5.2 A stretched membrane is immersed in a fluid. The vertical displacement w of the membrane is governed by an inhomogeneous wave equation, specifically,

$$\Gamma \frac{\partial^2 w}{\partial x^2} - m \frac{\partial^2 w}{\partial t^2} = p_{\text{upper}} - p_{\text{lower}}$$

where p_{lower} and p_{upper} are the pressures acting on the membrane's lower and upper surfaces. The parameter m is the mass per unit surface area, and Γ is the membrane force per unit length. An excitation at a distant location has induced a wave that travels in the positive x direction along the membrane. The displacement in this wave is $w = \text{Re}\left[W \exp(i\omega(t - x/c_m)) \right]$, where ω and W are specified. Describe the phase speed c_m of this wave, which may be considered to be supersonic, and the pressure waves in the fluid above and below the membrane corresponding to specified values of ω and W.

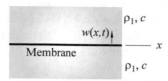

Exercise 5.2

Exercise 5.3 A wave propagates along the floor of a large tank that is filled with a liquid whose sound speed is 760 m/s. Two hydrophones located at different positions measured the waveforms depicted below. (The contribution of waves that reflect from the surface have been filtered out of these waveforms.) Hydrophone A was located at $\bar{\xi}_A = 1.8\bar{e}_y$ m, where the origin is the bottom of the tank. The elevation of hydrophone B above the floor was 0.9 m, but the horizontal position of this hydrophone was not measured, so $\bar{\xi}_B = s\bar{e}_x + 0.9\bar{e}_y$ m. Prior to these measurements, the sound speed in the liquid was measured to be 760 m/s. In addition, an accelerometer mounted on the floor indicated that the acceleration magnitude was $|\ddot{w}| = 15.46$ m/s^2. Use this data to determine the density of the liquid, the ground displacement $w(x, t)$, and the pressure field $p(x, y, t)$.

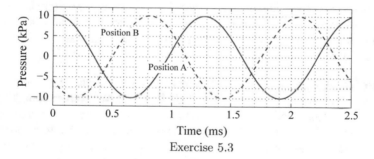

Exercise 5.3

Exercise 5.4 A plane wave in air is reflected from a locally reacting surface whose specific impedance is $\zeta = 0.3$. The given waveform is the pulse of the incident wave that is observed at point A. Determine the waveform of the total pressure (incident plus reflected) that is observed at points B and C.

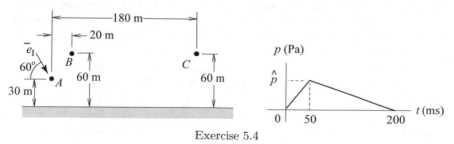

Exercise 5.4

Exercise 5.5 A plane wave in the atmosphere ($\rho_0 = 1.2\,\mathrm{kg/m^3}$, $c = 340\,\mathrm{m/s}$) is incident on a wall at $15°$ from normal. The wall is coated by a locally reacting material whose specific impedance is $\xi = 0.2-0.5i$. The RMS pressure very close to the wall is measured as $500\,\mathrm{Pa}$, and the frequency is $1.5\,\mathrm{kHz}$. (a) Determine the RMS pressure in the incident wave (in the absence of surface reflections). (b) Determine the RMS pressure at a position one meter from the wall.

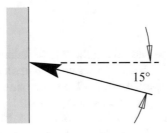

Exercise 5.5

Exercise 5.6 A $400\,\mathrm{Hz}$ plane wave in air, $\rho_0 = 1.2\,\mathrm{kg/m^3}$ and $c = 340\,\mathrm{m/s}$, is reflected from a locally reacting wall. The angle of incidence is $\psi = 60°$. When the wall is removed, the sound pressure level in the incident signal is $80\,\mathrm{dB//20}\,\mu\mathrm{Pa}$. When the wall is present, the sound pressure level very close to the wall is 74.6 dB. Also, a scan perpendicular to the surface reveals that $y = 51.5\,\mathrm{mm}$ is the nearest location to the wall at which the minimum sound pressure level occurs. (a) Determine the reflection coefficient. (b) Determine the reflection coefficient corresponding to normal incidence of the wave.

Exercise 5.7 A sound level meter is mounted on an apparatus that allows it to be placed at a variable elevation above the rigid floor of a very large room. The walls and ceiling are nonreflective. The room is used to measure the acoustic field resulting from incidence of a planar $900\,\mathrm{Hz}$ signal. Starting from a position very close to the floor, the sound level meter is moved upward slowly while being monitored. It is observed that the sound pressure level attains a minimum at a vertical distance of $0.1126\,\mathrm{m}$. A second measurement at a vertical distance of $0.18\,\mathrm{m}$ indicates that the sound pressure level is $72\,\mathrm{dB//20}\,\mu\mathrm{Pa}$. Determine the sound pressure level of the incident wave in the absence of the surface and the angle of incidence.

Exercise 5.8 The diagram depicts a conceptual view of a locally reacting boundary that is to be used to quiet a noisy restaurant. The small patches forming the upper surface move perpendicularly to the surface in a mutually independent manner. Each patch is backed by a one-degree-of-freedom system. The stiffness, mass, and damping per unit surface area are κ, μ, and β, respectively. (This is a generalization of the one-dimensional impedance model developed around Fig. 3.2.) (a) Prove that the impedance of this material model is $Z = \beta + i\,(\mu\omega - \kappa/\omega)$. (b) Suppose a plane wave having an amplitude of 4 Pa at $500\,\mathrm{Hz}$ is incident on this surface at $30°$ from normal. Determine the pressure and normal velocity amplitudes on the surface. The impedance parameters are $\beta = 300$ Rayl, $\mu = 0.2$ Rayl-s, and $\kappa = 2.4(10^6)$ Rayl/s.

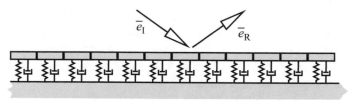

Exercise 5.8

Exercise 5.9 Consider the system in Exercise 5.8 in the case where the incident signal is a square wave whose amplitude is 4 Pa. The fundamental frequency is 250 Hz, and the angle of incidence is 30°. Plot the waveforms of the total pressure and particle velocity very close to the surface

Exercise 5.10 An incoherent acoustic plane wave in air has a sound pressure level of 80 dB//20 μPA in each octave band from 100 Hz to 6.4 kHz. It is incident at 45° on a surface whose nondimensional specific impedance varies as a function of cyclical frequency f according to $\zeta(f) = 0.3 + (0.0025f - 5.25)i$. (a) Plot the reflection factor as a function of frequency. (b) Estimate the reflection factor for each octave band as the value at the band's center frequency. Use these values to estimate the total unweighted sound pressure level for the combination of incident and reflected waves at the surface. (c) Use the reflection factors in Part (b) to estimate the total sound pressure level at a perpendicular distance of 100 mm from the wall.

Exercise 5.11 A 500-Hz plane wave in air is reflected from a horizontal surface composed of a locally reacting material. The mean-squared pressure along a line normal to the surface is measured over a range of elevations y, resulting in the below graph, but the angle of incidence ψ for the plane wave was not identified. From this graph, determine the angle of incidence, the reflection coefficient, the mean-squared pressure in the incident wave, and the specific impedance ζ. and also determine the maximum mean-squared pressure that would be observed if the angle of incidence were zero and all other parameters were unchanged.

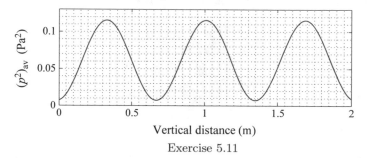

Exercise 5.11

Exercise 5.12 Two semi-infinite fluid media are separated by a planar interface, with $\rho_1 = 1000 \, \text{kg/m}^3$, $c_1 = 1500 \, \text{m/s}$, $\rho_2 = 600 \, \text{kg/m}^3$, $c_2 = 1200 \, \text{m/s}$. Plane waves in each medium are incident at the interface. The following features of the acoustic field are known: (1) The angle of incidence of the wave in medium 1 is $\psi_1 = 20°$. (2) As a result of the interaction of the two incident waves, there is no reflected/transmitted wave in medium 2. (3) The *total* pressure at the origin $x = y = 0$ is $400 \sin(\omega t)$ Pa. Determine the incident wave in medium 2 and the reflected/transmitted wave in medium 1. Describe each wave in the form $\mathrm{Re}\left[P_j \exp\left(i\bar{k}_j \cdot \bar{x} - i\omega t \right) \right]$, and state the values of P_j and $\bar{k}_j$ for each wave.

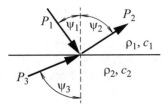

Exercise 5.12

Exercise 5.13 A 1 kHz signal in air whose sound pressure level is 100 dB is incident on the surface of a deep body of water. The media properties are $c = 340$ m/s, $\rho = 1.2$ kg/m³ for air, and $c = 1480$ m/s, $\rho = 1000$ kg/m³ for water. The angle of incidence is either $\psi_1 = 0.9\psi_{cr}$ or $\psi_1 = 1.1\psi_{cr}$, where ψ_{cr} is the critical angle for transmission into fluid 2. Determine the mean-squared pressure $\left(p^2 \right)_{av}$, mean-squared particle velocity components $\left(v_{horizontal}^2 \right)_{av}$ and $\left(v_{vertical}^2 \right)_{av}$, and average intensity $\left(\bar{I} \right)_{av}$ in the water very close to the surface.

Exercise 5.14 A plane wave having amplitude A in a semi-infinite fluid ($\rho_1 = 1200$ kg/m³, $c_1 = 2000$ m/s) half-space is obliquely incident on the interface with another semi-infinite fluid. Neither the density nor sound speed in the second fluid is known, and the angle of incidence is unspecified. (a) Derive an expression for the mean-squared pressure at an arbitrary location in fluid 1. Use the reflection coefficient to formulate this expression. (b) Based on the expression in Part (a), can a measurement of the mean-squared pressure distribution within fluid 1 solely along the line AB perpendicular to the interface be used to determine the angle of incidence? the value of c_2? the value of ρ_2? Explain each answer.

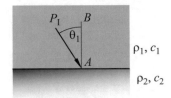

Exercise 5.14

Exercise 5.15 An acoustic wave in water is incident on a polymeric liquid that has been engineered to have a very low sound speed and density ($c = 250$ m/s, $\rho = 400$ kg/m³). The thickness of the layer of this liquid is extremely large. Determine the reflection coefficient of the wave corresponding to normal incidence and incidence at 45°.

Exercise 5.16 Solve Exercise 5.15 for the case where the polymer layer has a thickness of 100 mm, with its bottom bonded to a rigid backing. Frequencies of interest are 1.5 and 2.5 kHz.

Exercise 5.17 Solve Exercise 5.15 for the case where the polymer layer has a thickness of 100 mm, with its bottom bonded to a very thick layer of steel, for which $c = 5000$ m/s and $\rho = 7800$ kg/m³. Frequencies of interest are 1.5 and 2.5 kHz. Ignore the generation of shear waves in steel.

Exercise 5.18 Hydrazine ($\rho = 1032$ kg/m³, $c = 2092$ m/s) is a highly unstable rocket fuel. It is to be contained to a depth of 4 m in a tank whose bottom is rigid. An acoustical test to be performed will insonify it with a plane wave. Someone has suggested that for safety reasons, a mylar sheet should be placed 400 mm above the surface of the hydrazine, and the gap filled with carbon dioxide ($\rho = 1.842$ kg/m³, $c = 267$ m/s). It may be assumed that the mylar sheet is acoustically transparent. The incident plane wave has a frequency of 100 Hz, and its angle of incidence might be $\psi_1 = 0$ or 12°. For each angle, compare the reflection coefficient of the incident wave when the protective carbon dioxide layer is in place to the case where the plane wave is directly incident on the hydrazine.

Exercise 5.19 Consider a uniform normal velocity $v(x, t) = v_0 \cos(\omega t)$ at the bottom of a layer of thickness H having sound speed c_1 and density ρ_1. The top of the layer bounds an infinite fluid having sound speed c_2 and density ρ_2. Derive expressions for the insertion loss of medium 1. Do the quarter-wavelength ($H = \pi c_1/2\omega$), and half-wavelength ($H = \pi c_1/\omega$) frequencies lead to any special characteristics for these results?

Exercise 5.20 The phase speed of a wave generated in the ocean bottom by an earthquake is c_e. The vertical displacement, positive upward in this wave, is $w = \text{Re}\left[W \exp\left(i\omega\left(t - x/c_e\right)\right)\right]$. The density and sound speed of the air are ρ_1 and c_1, and these parameters for the ocean are ρ_2 and c_2. The depth of the ocean channel is H. (a) Identify the type of wave that exists in each fluid in the following situations: $c_e > c_1 > c_2$, $c_1 > c_e > c_2$, and $c_1 > c_2 > c_e$. (b) Suppose the channel depth is 50 m, and the fluid properties are $\rho_1 = 1.2$ kg/m³, $c_1 = 340$ m/s for air, and $\rho_2 = 1000$ kg/m³, $c_2 = 1480$ m/s for water. Measurements indicate that $c_e = 2000$ m/s, $\omega = 400$ rad/s, and $W = 9 \mu$m. Determine the pressure amplitude at the bottom surface and at the water-air interface.

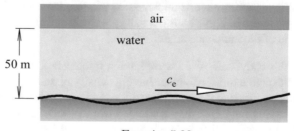

Exercise 5.20

Exercise 5.21 Repeat the analysis for Exercise 5.13 in the situation where a 0.9-m-thick layer of fluid, for which $\rho_0 = 800\,\text{kg/m}$ and $c = 1200\,\text{m/s}$, overlays the water channel.

Exercise 5.22 A plane wave in an infinite half-space 1 is obliquely incident ($\psi_1 \neq 0$) at the planar interface with layer 2, whose thickness is H. Both media are ideal fluids having densities ρ_j and sound speeds c_j. A material having local impedance Z_L is situated at the base of layer 2. Derive a set of simultaneous algebraic equations whose solution would yield the reflection coefficient at the 1–2 interface corresponding to specified angle of incidence ψ_T, frequency ω, and layer thickness L.

Exercise 5.23 An open window that is 1.4 high and 0.6 m wide is to be covered by a wool blanket whose mass is 6 kg. Consider a pressure wave from the exterior that is at normal incidence to the window. Determine the transmission coefficient into the building for frequencies ranging from 10 Hz to 1 kHz.

Exercise 5.24 A stretched membrane separates two semi-infinite domains filled with different fluids. A plane wave in fluid 1 is obliquely incident on the membrane. (a) Derive expressions for the transmission and reflection coefficients in terms of the properties of the fluids, the angle of incidence, and c_{mem}, ρ_{mem}, and h for the membrane. It may be assumed that the incidence is subcritical. (b) Consider the case where $c_1 = 800\,\text{m/s}$, $\rho_1 = 500\,\text{kg/m}^3$, $c_2 = 600\,\text{m/s}$, $\rho_2 = 1000\,\text{kg/m}^3$, $c_{\text{mem}} = 900\,\text{m/s}$, and $\rho_{\text{mem}}h = 40\,\text{kg/m}$. Determine and graph the dependence of the reflection coefficient on the angle of incidence.

Exercise 5.25 An aluminum plate, $\rho = 2700\,\text{kg/m}^3$, $E = 69\left(10^9\right)$ Pa, $\nu = 0.33$, forms a barrier in water. The thickness of the plate is 50 mm. Dissipation within the plate is negligible. (a) At what frequency will the transmission coefficient be one if the incident wave is very close to grazing incidence? (b) At what frequency will the transmission coefficient be one if the angle of incidence is $30°$? (c) If the frequency is the value identified in Part (a), what are the reflection and transmission coefficients when the angle of incidence is $30°$?

Exercise 5.26 It is desired to construct a red oak ($\rho = 770\,\text{kg/m}^3$, $c_{\text{bar}} = 3800\,\text{m/s}$, $\nu = 0$) door that will inhibit transmission of noise at 800 Hz. The thickness h has not been selected, but cost considerations dictate that $h < 200$ mm. The value of h that will be used is that which maximizes the transmission loss for a wave at an angle of incidence of $35°$. Determine this thickness by plotting the transmission loss as a function of the thickness h. Then, for this h value, evaluate and graph the transmission loss as a function of the incidence angle.

Exercise 5.27 A steady-state plane wave having frequency ω and complex amplitude P_I travels through a fluid half-space having density ρ_1 and sound speed c_1. The wave is incident on a fluid whose sound speed is c_2 and whose density is ρ_2. The latter fluid is a layer having height H that is situated over a (flat) elastic plate substrate. The pressure on the bottom of the plate is negligible. (a) What is the phase speed of the transverse wave that is induced in the plate? (b) Derive a set of solvable equations whose solution will yield the complex amplitude of the wave that is induced in the plate and the complex amplitude of the reflected wave in fluid 1. (c) Consider a system in which $\rho_1 = 800\,\text{kg/m}^3$, $c_1 = 600\,\text{m/s}$, $\rho_2 = 1200\,\text{kg/m}^3$, $c_2 = 750\,\text{m/s}$, $H = 500\,\text{mm}$, $D = 3\left(10^6\right)$ N-m, and $m_{pl} = 300\,\text{kg/m}^2$. The frequency of this system is such that $Z_B = 0$ at $\psi_1 = 50°$. Determine the reflection coefficient for $\psi_1 = 0°$, $25°$, $50°$, and $75°$.

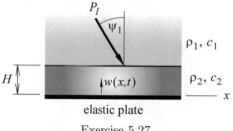

elastic plate

Exercise 5.27

Exercise 5.28 The sketch shows a sheet of flexible material that has very little rigidity. It separates a deep layer of fluid above it from a different fluid below it. The blanket is situated at height H above the rigid bottom. A plane wave P_1 in the upper fluid is incident on the sheet. Derive algebraic equations whose solution would give the reflection coefficient and the pressure at the rigid bottom.

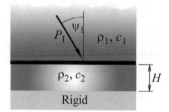

Rigid

Exercise 5.28

Exercise 5.29 Solve Exercise 5.28 for the case where the sheet separating the fluid layers is a porous blanket whose rigidity is negligible.

Chapter 6
Spherical Waves and Point Sources

Many signals have wavefronts that are not planar. The simplest to analyze are the waves generated by a vibrating spherical surface. Although a vibrating sphere is not frequently encountered in practice, the ability to generate solutions for a variety of situations makes that configuration extremely useful, both for verifying numerical techniques and for exploring fundamental issues. An important development will be the concept of a point source, which is obtained by shrinking the size of the sphere. Point sources are used as fundamental building blocks for describing the response of several general classes of problems.

6.1 Spherical Coordinates

The normal components of velocity must be continuous at the surface of a vibrating sphere. It is not likely that we could satisfy such a matching condition if we continued to use Cartesian coordinates, whereas a solution in terms of spherical coordinates is consistent with the shape of the boundary. Thus, our first task is to recast the basic acoustics equations in terms of these coordinates. Figure 6.1 defines a set of spherical coordinates that locate a point relative to an xyz reference frame. The *polar axis* is z, whose name follows from its resemblance to the axis of the Earth. The z-axis defines the *axial direction*. The *radial distance* r is measured from the origin to the point of interest. The angle from the axial direction to the radial line is the polar angle ψ. (The common notation is to use ϕ to designate the polar direction, but we have already used that symbol to denote the velocity potential.) The radial line and the polar axis define the meridional plane, and the circle in this plane, whose radius is r, is a *meridian*. Consistent with geographical terminology, xy is the equatorial plane.

The intersection of the meridional and equatorial planes is a line whose orientation in the equatorial plane is the angle θ. Figure 6.1 indicates that this angle is measured from the x-axis, but doing so is discretionary. Many individuals refer to θ

Electronic supplementary material The online version of this chapter (DOI 10.1007/978-3-319-56844-7_6) contains supplementary material, which is available to authorized users.

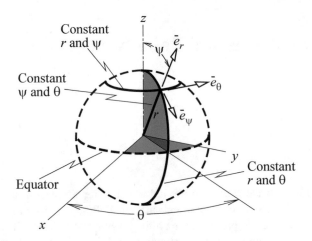

Fig. 6.1 Definition of a set spherical coordinates and associated unit vectors

as the circumferential angle. We shall reserve that name for the angle in cylindrical coordinates, which are used in the next chapter. A name consistent with the picture in Fig. 6.1 would be that θ is the meridional angle. However, we shall refer to θ as the *azimuthal angle*. The name is derived from the usage of a sextant for navigation, but we use it in the generic sense as an angle of rotation in the horizontal plane. Consider the situation depicted in Fig. 6.2, where a telescope mounted on a ship is used to observe other ships and aircraft. The upward vertical is readily identified as the z-axis, and the x-axis is the direction in which the ship is moving. The telescope is aimed at an object. The angle from the z-axis to the optical axis is the polar angle ψ. The amount that the telescope is rotated about the z-axis is the azimuthal angle θ, with $\theta = 0$ if the telescope lies in the xz-plane. The distance from the telescope to the object is the radial distance r

If the spherical coordinates (r, ψ, θ) of a point are known, its Cartesian coordinates may be found by projecting the point onto the z-axis and the xy-plane. The distance from the origin along the z-axis is $r \cos \psi$, and the distance from the origin to the

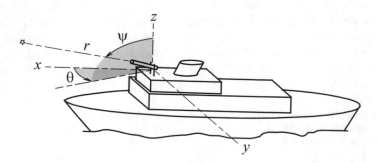

Fig. 6.2 Definition of spherical coordinates for tracking an object with a ship's telescope

projected point in the xy-plane is $R = r \sin \psi$. Thus, R would be the radial distance for polar coordinates. The resulting transformation is

$$
\begin{aligned}
x &= r \sin \psi \cos \theta \\
y &= r \sin \psi \sin \theta \\
z &= r \cos \psi
\end{aligned}
\tag{6.1.1}
$$

The inverse transformation, which gives the spherical coordinates of a point whose Cartesian coordinates are known, is found by solving the preceding relations. The result is

$$
\begin{aligned}
r &= \left(x^2 + y^2 + z^2 \right)^{1/2} \\
\psi &= \cos^{-1} \left(\frac{z}{\left(x^2 + y^2 + z^2 \right)^{1/2}} \right) \\
\theta &= \tan^{-1} \left(\frac{y}{x} \right) = \sin^{-1} \left(\frac{y}{\left(x^2 + y^2 \right)^{1/2}} \right) = \cos^{-1} \left(\frac{x}{\left(x^2 + y^2 \right)^{1/2}} \right)
\end{aligned}
\tag{6.1.2}
$$

The polar angle lies in the range $0 \le \psi \le \pi$, and the inverse cosine function is single valued in that range. In contrast, the azimuthal angle lies in a 2π range that we usually take to be $-\pi < \theta \le \pi$. Assignment of θ to the proper quadrant requires that we satisfy two of the inverse functions defining it.

A radial line is the locus of points at which the values of ψ and θ are constant, and a meridional circle is the locus of points having constant r and θ. The triad is completed with a circle of radius $r \sin \psi$, that is, the locus of points at which r and ψ are constant. This is a polar circle, analogous to the Arctic Circle. These constant coordinate lines have significance in the definition of a set of unit vectors $\bar{e}_r$, $\bar{e}_\psi$, $\bar{e}_\theta$ that provide the directions for components of a vector. The subscript of each unit vector indicates the direction in which a point would move if only that spherical coordinate were changed by an infinitesimal amount. Such a movement shifts the point tangentially along the respective curve. Thus, $\bar{e}_r$ is along the radial line, $\bar{e}_\psi$ is tangent to the meridional circle, and $\bar{e}_\theta$ is tangent to the polar circle. The representations are

$$
\begin{aligned}
\bar{e}_r &= \sin \psi \cos \theta \bar{e}_x + \sin \psi \sin \theta \bar{e}_y + \cos \psi \bar{e}_z \\
\bar{e}_\psi &= \cos \psi \cos \theta \bar{e}_x + \cos \psi \sin \theta \bar{e}_y - \sin \psi \bar{e}_z \\
\bar{e}_\theta &= - \sin \theta \bar{e}_x + \cos \theta \bar{e}_y
\end{aligned}
\tag{6.1.3}
$$

It is not difficult to show that these unit vectors form a mutually orthogonal set. This is an attribute of the general class of *orthogonal curvilinear coordinates*.

Expressions for the gradient and Laplacian when the pressure depends on each of the cylindrical coordinates are derived in Appendix A of Volume II.[1] Our initial

---

[1] J.H. Ginsberg, *Acoustics-A Textbook for Scientists and Engineers, Volume II- Applications*, Appendix A, Springer (2017).

studies shall be restricted to situations where the pressure field depends solely on the radial distance and time, so the functional dependence is merely $p(r, t)$. Such a field is said to be *radially symmetric*. The description of the pressure gradient in that situation is simpler to derive. From the chain rule, we have

$$\nabla p = \frac{\partial p}{\partial x}\bar{e}_x + \frac{\partial p}{\partial y}\bar{e}_y + \frac{\partial p}{\partial z}\bar{e}_z = \frac{\partial p}{\partial r}\left(\frac{\partial r}{\partial x}\bar{e}_x + \frac{\partial r}{\partial y}\bar{e}_y + \frac{\partial r}{\partial z}\bar{e}_z\right) \quad (6.1.4)$$

Differentiating $r^2 = x^2 + y^2 + z^2$ leads to

$$\frac{\partial r}{\partial x} = \frac{x}{r}, \quad \frac{\partial r}{\partial y} = \frac{y}{r}, \quad \frac{\partial r}{\partial z} = \frac{z}{r} \quad (6.1.5)$$

By definition the position is r in the $\bar{e}_r$ direction, so

$$\bar{e}_r = \frac{x}{r}\bar{e}_x + \frac{y}{r}\bar{e}_y + \frac{x}{r}\bar{e}_z \quad (6.1.6)$$

The result is that the gradient is

$$\boxed{\nabla p = \frac{\partial p}{\partial r}\bar{e}_r} \quad (6.1.7)$$

The Laplacian is the divergence of the gradient, $\nabla^2 p \equiv \nabla \cdot \nabla p$. Although the preceding expression for ∇p is similar to a component of the gradient in Cartesian coordinates, it contains a subtlety, in that $\bar{e}_r$ is not constant as the position changes. Thus, it is necessary to describe how $\bar{e}_r$ changes when each coordinate is changed. Doing so results in the addition of a term in the Laplacian for a radially symmetric pressure that is not analogous to the Cartesian version. To see how this term arises, we form

$$\nabla^2 p = \left(\frac{\partial}{\partial x}\bar{e}_x + \frac{\partial}{\partial y}\bar{e}_y + \frac{\partial}{\partial z}\bar{e}_z\right) \cdot \left(\frac{\partial p}{\partial r}\bar{e}_r\right)$$
$$= \frac{\partial^2 p}{\partial x \partial r}\bar{e}_x \cdot \bar{e}_r + \frac{\partial p}{\partial r}\left(\bar{e}_x \cdot \frac{\partial}{\partial x}\bar{e}_r\right) + \frac{\partial^2 p}{\partial y \partial r}\bar{e}_y \cdot \bar{e}_r$$
$$+ \frac{\partial p}{\partial r}\left(\bar{e}_y \cdot \frac{\partial}{\partial y}\bar{e}_r\right) + \frac{\partial^2 p}{\partial z \partial r}\bar{e}_z \cdot \bar{e}_r + \frac{\partial p}{\partial r}\left(\bar{e}_z \cdot \frac{\partial}{\partial z}\bar{e}_r\right) \quad (6.1.8)$$

The dot product of the unit vectors is obtained from Eq. (6.1.6). For the other terms, we differentiate the same equation, then call on Eq. (6.1.5). For example,

$$\bar{e}_x \cdot \frac{\partial}{\partial x}\bar{e}_r = \bar{e}_x \cdot \frac{\partial}{\partial x}\left(\frac{x}{r}\bar{e}_x + \frac{y}{r}\bar{e}_y + \frac{x}{r}\bar{e}_z\right) = \frac{\partial}{\partial x}\left(\frac{x}{r}\right) = \left[\frac{1}{r} - x\left(\frac{1}{r^2}\right)\left(\frac{x}{r}\right)\right]$$
$$(6.1.9)$$

Performing similar operations for the other terms ultimately leads to

$$\nabla^2 p = \frac{\partial^2 p}{\partial r^2} + \frac{2}{r}\frac{\partial p}{\partial r} \qquad (6.1.10)$$

The pressure gradient appears in Euler's equation. The linearized version of it is Eq. (4.2.6). The preceding description of the gradient leads us to conclude that the particle acceleration in a radially symmetric field is solely in the radial direction, so we may write Euler's equation as

$$\rho_0\frac{\partial v_r}{\partial t} = -\frac{\partial p}{\partial r} \qquad (6.1.11)$$

The corresponding linearized wave equation is

$$\frac{\partial^2 p}{\partial r^2} + \frac{2}{r}\frac{\partial p}{\partial r} - \frac{1}{c^2}\frac{\partial^2 p}{\partial t^2} = 0 \qquad (6.1.12)$$

It cannot be overemphasized that Eqs. (6.1.11) and (6.1.12) are restricted to radially symmetric fields.

6.2 Radially Vibrating Sphere–Time-Domain Analysis

Suppose the surface of a sphere executes a vibration in which all points synchronously move radially. Such a pattern is radially symmetric. At any instant, all points on the surface are at the same radial distance $r_0(t)$ from the center, where $r_0 = a$ is the radius of the sphere at its reference position. Because the sphere seems to expand and contract uniformly, such a vibration is occasionally called a *breathing mode*. The radial velocity is $v_0(t) = (d/dt)r_0(t) = \dot{r}_0(t)$. In the realm of linear acoustics, the displacement of the surface, which is the difference between $r_0(t)$ and a, is limited to being very small compared to an acoustic wavelength, $2\pi/k$. This enables us to say that v_0 is the radial velocity at $r = a$. The radial direction is normal to the surface. Thus, the linearized velocity continuity condition requires that the radial velocity of fluid particles that contact the sphere be v_a,

$$v_r|_{r=a} = v_a \qquad (6.2.1)$$

It follows from this relation and Euler's equation, Eq. (6.1.11), that v_r and p are radially symmetric. Thus, we seek a solution of the wave equation, Eq. (6.1.12) that satisfies Eq. (6.2.1). It is possible to perform the analysis in the time or frequency domain. We will follow the usual procedure of carrying out each analysis independently. The time-domain analysis will shed much light on underlying phenomena, but it is difficult to generalize.

6.2.1 General Solution

The first step is to derive a general solution of the wave equation. Radial symmetry allows us to use the simplified version of the wave equation in Eq. (6.1.12). It may be written as

$$\frac{1}{r}\left[\frac{\partial^2}{\partial r^2}(rp) - \frac{1}{c^2}\frac{\partial^2}{\partial t^2}(rp)\right] = 0 \tag{6.2.2}$$

The value of r is finite except at the origin, so the term inside the bracket must be zero. This is the one-dimensional wave equation for rp. Thus, the general solution for a radially symmetric, or *uniform*, spherical wave consists of simple waves that propagate toward and away from the center at phase speed c. The magnitude of the pressure is scaled by the reciprocal of the radial distance. Furthermore, the signal is launched from the sphere's surface, where $r = a$. Therefore, it makes sense to use $r - a$ as the propagation distance in both directions. Thus, we are led to recognize that the general solution for a spherical wave is

$$\boxed{p = \frac{1}{r}f\left(t - \frac{r-a}{c}\right) + \frac{1}{r}g\left(t + \frac{r-a}{c}\right)} \tag{6.2.3}$$

The functions f and g are arbitrary, as they were for the general plane wave solution. The usage of $r - a$ as the propagation distance will facilitate satisfying the continuity condition at the surface of the vibrating sphere.

Because of the presence of the $1/r$ factor, we cannot assert that pressure and particle velocity are proportional. Instead, we must use Euler's equation to determine the particle velocity. The pressure is independent of ψ and θ, so the only velocity component is radial. The arguments of the f and g functions are phase variables, so we apply the chain rule to carry out the derivative with respect to r. The derivatives of the functions are denoted with overdots because the phase variables are time quantities. Thus, Euler's equation gives

$$\rho_0 \frac{\partial v_r}{\partial t} = \frac{1}{rc}\dot{f}\left(t - \frac{r-a}{c}\right) + \frac{1}{r^2}f\left(t - \frac{r-a}{c}\right)$$

$$- \frac{1}{rc}\dot{g}\left(t + \frac{r-a}{c}\right) + \frac{1}{r^2}g\left(t + \frac{r-a}{c}\right) \tag{6.2.4}$$

If we wish to determine the particle velocity corresponding to known pressure functions, it will be necessary to integrate Eq. (6.2.4) with respect to t. The symbolic result will contain terms that are indefinite integrals of f and g. This is an awkward representation. It may be avoided by recasting the general solution such that f and g are represented as the derivatives of functions F and G. It also is convenient at this junction to separate the waves that propagate toward and away from the center, with subscripts "in" and "out," respectively. Thus, the forms we shall use are

$$p = p_{\text{out}} + p_{\text{in}}, \quad v_r = v_{\text{out}} + v_{\text{in}}$$

$$p_{\text{out}} = \frac{1}{r}\dot{F}\left(t - \frac{r-a}{c}\right), \quad v_{\text{out}} = \frac{1}{\rho_0 c}\left[\frac{1}{r}\dot{F}\left(t - \frac{r-a}{c}\right) + \frac{c}{r^2}F\left(t - \frac{r-a}{c}\right)\right]$$

$$p_{\text{in}} = \frac{1}{r}\dot{G}\left(t + \frac{r-a}{c}\right), \quad v_{\text{in}} = \frac{1}{\rho_0 c}\left[-\frac{1}{r}\dot{G}\left(t + \frac{r-a}{c}\right) + \frac{c}{r^2}G\left(t + \frac{r-a}{c}\right)\right]$$

$$(6.2.5)$$

It is problematic for some individuals that the general solution features the derivative of a function, but an equivalent result would have been obtained if we had used the velocity potential ϕ to formulate the analysis. This function, which was introduced in Sect. 4.4, also must satisfy the wave equation, but it has the advantage that pressure and particle velocity are obtained as derivatives of ϕ. Thus, the general solution for a uniform spherical wave may be written in terms of alternative arbitrary functions $\mathcal{F}$ and $\mathcal{G}$ as

$$\phi = \frac{1}{r}\mathcal{F}\left(t - \frac{r-a}{c}\right) + \frac{1}{r}\mathcal{G}\left(t + \frac{r-a}{c}\right)$$

$$p = -\rho_0\frac{\partial\phi}{\partial t} = -\frac{\rho_0}{r}\left[\dot{\mathcal{F}}\left(t - \frac{r}{c} \, \frac{a}{c}\right) + \dot{\mathcal{G}}\left(t + \frac{r-a}{c}\right)\right]$$

$$v_r = \frac{\partial\phi}{\partial r} = -\left[\frac{1}{rc}\dot{\mathcal{F}}\left(t - \frac{r-a}{c}\right) + \frac{1}{r^2}\mathcal{F}\left(t - \frac{r-a}{c}\right)\right]$$
$$+ \left[\frac{1}{rc}\dot{\mathcal{G}}\left(t + \frac{r-a}{c}\right) - \frac{1}{r^2}\mathcal{G}\left(t + \frac{r-a}{c}\right)\right]$$

$$(6.2.6)$$

Aside from a multiplicative factor of $-\rho_0$, this is the same form as that which appears in Eq. (6.2.5).

6.2.2 Radiation from a Uniformly Vibrating Sphere

The first case we consider is a sphere surrounded by a fluid whose extent is infinite. An inward propagating wave signal could not return from the infinitely distant outer boundary in a finite time, so we set the G function to zero. (This is the Sommerfeld radiation condition, which we first invoked for plane waves in a semi-infinite waveguide.) What remains is the outward propagating wave in Eq. (6.2.5),

$$p_{\text{out}} = \frac{1}{r}\dot{F}\left(t - \frac{r-a}{c}\right)$$

$$v_{\text{out}} = \frac{1}{\rho_0 c}\left[\frac{1}{r}\dot{F}\left(t - \frac{r-a}{c}\right) + \frac{c}{r^2}F\left(t - \frac{r-a}{c}\right)\right]$$

$$(6.2.7)$$

A specification of $p(t)$ at $r = a$ would give the time function $\dot{F}(t)$. Integration of this function with respect to time would give $F(t)$, from which we could evaluate the radial velocity. However, usually, it is the radial velocity $v_a(t)$ at $r = a$ that is specified. Setting $r = a$ in Eq. (6.2.7) for v_r leads to a boundary condition that is a first-order differential equation,

$$\dot{F}(t) + \frac{c}{a}F(t) = \rho_0 a c v_a(t) \tag{6.2.8}$$

There are several ways in which the solution may be obtained. An elementary one adds a particular solution $F_p(t)$ to a complementary solution $F_c(t)$, which is a real exponential whose decay constant is c/a, that is, $F_c = B \exp(-ct/a)$. The solution is finalized by making the pressure on the surface be zero for $t \leq 0$. Setting to zero, the sum of the contributions to p of the particular and complementary solutions at $t = 0$ leads to

$$-\dot{F}_c(0) \equiv \frac{c}{a}B = \dot{F}_p(0) \tag{6.2.9}$$

The resulting pressure is

$$p = \frac{1}{r}\left[\dot{F}_p(\tau) - \dot{F}_p(0)e^{-c\tau/a}\right]h(\tau)\Bigg|_{\tau = t - \frac{r-a}{c}} \tag{6.2.10}$$

where the step function is used to ensure that the function is only evaluated for $\tau \geq 0$.

The variable $\tau = t - (r - a)/c$ is sometimes called a *retarded time*. The radial distance from the surface to a point in the fluid is $r - a$, and the phase speed is c. Thus, the retarded time concisely conveys the fact that any signal that is generated on the surface will require the propagation time $(r - a)/c$ to arrive at some field point. The signal at this location will be reduced by the factor a/r relative to the pressure on the surface. We say that this reduction is *spherical spreading*. (Some individuals follow "spreading" with "loss." This terminology is not employed here because it implies a reduction in energy, which is not what happens, as will be proven when we examine spherical waves from the frequency-domain perspective.)

The particular solution in Eq. (6.2.10) represents the persistent signal induced by surface acceleration. The complementary solution represents a transition from the zero pressure state to the persistent signal. Its magnitude decreases exponentially. The time constant of this decay is a/c, which means that the transitory signal will be reduced to approximately 2% of its initial value after an interval of $4a/c$. If the induced signal is a pulse whose duration is shorter than the time constant, the homogeneous solution will persist after the pulse ends. This behavior does not arise when a piston induces a signal in a one-dimensional infinite waveguide because the particle velocity in a simple plane wave is always proportional to the pressure.

Some trends may be identified as generalities. Suppose the surface velocity is such that $\dot{v}_a(t) \gg (c/a) v_a(t)$. To quantify this condition, let us define T as the timescale for the variation of v_a. For example, T would be the duration of a pulse or the period of a repeating function. Then, the order of magnitude of $\dot{v}_a$ is $O(v_a/T)$, so rapid variation corresponds to $cT \ll a$. Furthermore, the surface velocity must be a continuous function in order that the acceleration be finite, and the surface was at rest for $t < 0$. This means that v_a must initially be zero. Because $a/c \gg T$, Eq. (6.2.8) indicates that $\dot{F}_{\mathrm{p}} \approx \rho_0 cav_a$, so that $\dot{F}_{\mathrm{p}}(0) \approx 0$. When these conditions apply, Eq. (6.2.10) indicates that the radiated pressure waveform will resemble the surface velocity, according to

$$p \approx \frac{a}{r}\rho_0 cav_a\left(t - \frac{r-a}{c}\right) \text{ if } cT \ll a \qquad (6.2.11)$$

The opposite case is that in which v_a fluctuates slowly, or a is very small. In other words, $cT \gg a$. Equation (6.2.8) tells us that $F_{\mathrm{p}} \approx \rho_0 a^2 v_a$. The initial value of $\dot{F}_{\mathrm{p}}$ is not necessarily zero in this case, but the fact that $c/a \gg 1/T$ means that the decay rate of the homogeneous solution is much greater than the rate at which the particular solution oscillates. Therefore, the homogeneous solution quickly becomes negligible. What remains is a signal that resembles the surface acceleration, specifically,

$$p \approx \frac{a}{r}\rho_0 a\dot{v}_a\left(t - \frac{r-a}{c}\right) \text{ if } T \gg a/c \text{ and } t > 4a/c \qquad (6.2.12)$$

Another trend is a general field property derived from the properties of Eq. (6.2.5). As r increases the particle velocity in an outwardly propagating wave tends to $\rho_0 cv_r = (1/r)\,\dot{F}\,(t - (r-a)/c) = p$. Locations where this relation applies are said to be in the *farfield*. Because $p = \rho_0 cv_r$ is the relation between particle velocity and pressure in a simple plane wave, the farfield behavior is sometimes said to be the *plane wave approximation*. Although this terminology is somewhat misleading because spherical spreading always occurs, it nevertheless is meaningful if we compare the signal at two close locations. Consider two radial distances r_A and r_B. A binomial series is used to describe the spherical spreading factor, such that

$$\begin{aligned}
p_A &= \frac{1}{r_A}\dot{F}\left(t - \frac{r_A - a}{c}\right)\\
p_B &= \frac{1}{r_A + (r_B - r_A)}\dot{F}\left(t - \frac{r_A - a}{c} - \frac{r_B - r_A}{c}\right)\\
&\approx \frac{1}{r_A}\left(1 - \frac{r_B - r_a}{r_A} + \cdots\right)\dot{F}\left(t - \frac{r_A - a}{c} - \frac{r_B - r_A}{c}\right)
\end{aligned} \qquad (6.2.13)$$

If the two locations are relatively close, such that $r_B - r_A \ll r_A$, then the leading term in the series suffices. In that case, p_B is well approximated as being the same as p_A, except that it is retarded additionally by the time $(r_B - r_A)/c$ required to propagate the signal from r_A to r_B. If r_A is in the farfield, then at each location $p \approx \rho_0 cv_r$. This is the behavior of a simple plane wave. In other words, if we are very far from the center from which the spherical wave originates, we may treat the propagation as though it were planar, provided that the approximation is only applied over an interval of radial distance that is much shorter than the radial distance itself. For example, if we wish to investigate how the sound wave radiated by an aircraft at high altitude is reflected at the ground, we may treat its incidence on the ground as though it were a plane wave. However, if we are interested in the interference between the incident and reflected waves well above the ground, then the spherical nature of the waves must be recognized.

EXAMPLE 6.1 A 40-mm-diameter piezoceramic sphere is made to vibrate underwater in a breathing mode. The radial displacement is $\varepsilon a \left[1 - \cos\left(\omega t\right)\right]$ starting at $t = 0$, where $\varepsilon = 2\left(10^{-6}\right)$. The water was quiescent prior to $t = 0$. Derive an expression for p as a function of r and t. Then, graph the pressure waveform at $r = 2a$ and $r = 40a$. Frequencies are 200, 20, and 2 kHz.

Significance

Determination of the radiated pressure signal corresponding to a specified surface motion is the primary focus of this problem. The significance of the rate of change of the surface velocity and the role of the complementary solution are items of special interest.

Solution

The radial velocity is the time derivative of the displacement, so we have $v_a = \varepsilon a \omega \sin\left(\omega t\right) h\left(t\right)$, which satisfies the requirement that $v_a = 0$ at $t = 0$. We shall find the particular solution of Eq. (6.2.8) by using a complex exponential representation, so we must solve

$$\dot{F} + \frac{c}{a}F = \rho_0 a c \operatorname{Re}\left(-i\varepsilon a \omega e^{i\omega t}\right)$$

The particular solution is

$$F_{\mathrm{p}} = \operatorname{Re}\left(B^{i\omega t}\right), \quad B = -i\frac{\varepsilon\rho_0 c a^2 \omega}{(i\omega + c/a)}$$

The corresponding pressure is found from Eq. (6.2.10) to be

$$p = \left(\frac{a}{r}\right)\varepsilon\rho_0 c a \operatorname{Re}\left[\frac{\omega^2}{(i\omega + c/a)}\left(e^{i\omega\tau} - e^{-c\tau/a}\right)\right]h\left(\tau\right)\Big|_{\tau = t - \frac{r-a}{c}}$$

A nondimensional representation of the solution is useful for interpretation and plotting. Introducing the wavenumber $k = \omega/c$ permits rewriting the solution as

$$\frac{p}{\varepsilon\rho_0 c^2}\left(\frac{r}{a}\right) = \operatorname{Re}\left[\frac{(ka)^2}{ika+1}\left(e^{i\omega\tau} - e^{-\omega\tau/(ka)}\right)\right]h\left(\tau\right)\Big|_{\tau = t - \frac{r-a}{c}}$$

This expression shows that graphing the nondimensional scaled pressure as a function of $\omega\tau$ only requires specification of ka. The result describes the waveform that would be seen at any value of r/a, with $\tau = 0$ being the instant when the signal arrives.

The parameter ka may be considered to be either a nondimensional frequency or radius. This parameter affects both the overall magnitude and the time dependence of the pressure. A useful way to recognize its effects is to consider the following table, which shows the contributions of the complementary and particular solutions when ka is very small and very large.

	$\left\|\dfrac{(ka)^2}{ika+1}\right\|$	$\dfrac{p}{\varepsilon\rho_0 c^2}\left(\dfrac{r}{a}\right)$	
		Particular	Complementary
$ka \ll 1$	$(ka)^2$	$(ka)^2\cos(\omega\tau)$	$(ka)^2\exp(-\omega\tau/(ka))$
$ka \gg 1$	ka/i	$ka\sin(\omega\tau)$	≈ 0

The entries for the particular and complementary solution at large ka stem from the factor being a large, negative imaginary value. Physically, this property results from the fact that the particular solution, being a sine function, starts off from zero, so the complementary solution is not needed to satisfy the initial condition that $p = 0$.

The properties in the tabulation have implications for the time response. If $ka \ll 1$, the complementary solution will be essentially zero for $\omega t > 2\pi$. The opposite case of $ka \gg 1$ leads to a complementary solution that is negligible from the outset. These observations do not tell us much about the situation where ka is an intermediate value. The actual data will clarify this question. The nondimensional frequencies are $ka = 0.17$ for $f = 2$ kHz, $ka = 1.70$ for $f = 20$ kHz, and $ka = 16.98$ for $f = 200$ kHz. The period of the oscillation is $\omega\tau = 2\pi$, which corresponds to $\omega\tau/(ka) = 0.027$, 0.27, and 2.7, respectively.

Figure 1 depicts the 2 kHz case. The dotted curve is the contribution of the particular solution, which is the steady-state signal. As we anticipated, the complementary solution for $\omega\tau$ much less than a period adds to the particular solution in order to make the pressure begin from zero. However, for $\omega\tau$ greater than 10% of a period, the complementary solution is barely evident.

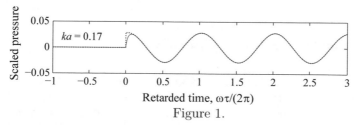

Figure 1.

Figure 2 describes the signal when the frequency is 20 kHz. This oscillation rate falls in the intermediate range. The particular solution is harmonic, but it is neither a pure sine nor pure cosine relative to $\tau = 0$. The complementary solution combines with the particular solution to start the pressure from zero. It is quite noticeable for more than half a period, with the consequence that the first pressure maximum is reduced relative to subsequent steady-state values.

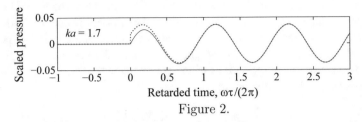

Figure 2.

The fast oscillation case is described by Fig. 3. The complementary solution is quite minor, even though it decays much more slowly than the previous cases.

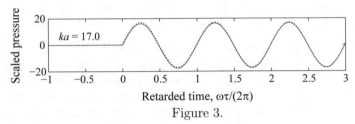

Figure 3.

These three cases are representative of general trends. Regardless of the details of the surface acceleration, it must be finite, so the surface velocity v_a must be a continuous function of time. If the acceleration is zero for $t < 0$, then it must be that $v_a = 0$ at $t = 0$. If the surface vibrates rapidly, the pressure will resemble v_a, so the complementary solution will be negligible. If the surface vibrates slowly, then the complementary solution will decay rapidly relative to the time interval over which the surface velocity oscillates. Consequently, its contribution quickly becomes negligible. In an intermediate range, the complementary solution is most noticeable. However, even then, it decays to insignificance rapidly compared to the particular solution. There are few situations in which the complementary solution for $\dot{F}$ is of much interest.

6.2.3 Acoustic Field in a Spherical Cavity

When an ideal compressible fluid is enclosed by a solid surface, the domain is said to be an *acoustic cavity*. The containing vessel examined here is a sphere that executes a uniform radial vibration $v_0(t)$. As always, the fluid's normal component of particle velocity must match this wall motion, so Eq. (6.2.1) must apply. Because the surface motion is independent of the polar and azimuthal angles, the signal that is induced within the cavity will be spherically symmetric. Thus, the general solution for a uniform spherical waves, Eq. (6.2.4), must be descriptive of the signal.

Conditions at the Center

Radiation and interior problems both require satisfaction of velocity continuity at the sphere, but the difference in the other boundary condition has a profound effect. In the exterior problem, the Sommerfeld radiation condition serves as a second boundary condition. It eliminates the inwardly propagating wave. In the case of a cavity, there must be an inwardly propagating wave because the wall movement induces a disturbance that propagates away from the wall. However, such propagation leads to a dilemma, in that inward propagation will bring the signal to $r = 0$, and the general solution features a $1/r$ factor. Thus, resolving this issue is where we begin.

The inwardly and outwardly propagating parts are

$$p_{out} = \frac{1}{r} \dot{F}\left(t - \frac{r-a}{c}\right), \quad p_{in} = \frac{1}{r} \dot{G}\left(t + \frac{r-a}{c}\right) \tag{6.2.14}$$

We cannot evaluate this expression at $r = 0$, but we can evaluate the limit. The condition that must be satisfied is that the total pressure must be finite. Thus, we require that

$$\lim_{r \to 0} (p_{out} + p_{in}) = \lim_{r \to 0} \left[\frac{\dot{F}\left(t - \frac{r-a}{c}\right) + G\left(t + \frac{r-a}{c}\right)}{r} \right] = p(0, t) \tag{6.2.15}$$

The field is a result of vibration of the sphere, so we consider the inward wave to be known. The limit as r approaches zero will be finite only if the numerator approaches zero, so we require that

$$\dot{F}\left(t + \frac{a}{c}\right) = -\dot{G}\left(t - \frac{a}{c}\right) \implies \dot{F}\left(t - \frac{r-a}{c}\right) = -\dot{G}\left(t - \frac{a}{c} - \frac{r}{c}\right) \tag{6.2.16}$$

This relation states that $\acute{F}$ at any instant is the same as $\dot{G}$ at a time that is $2a/c$ earlier. Thus, the general solution for the pressure within the cavity is

$$p = \frac{1}{r}\left[\dot{G}\left(t + \frac{r-a}{c}\right) - \dot{G}\left(t - \frac{r+a}{c}\right)\right] \tag{6.2.17}$$

This relation actually has a simple explanation. Consider Fig. 6.3, which describes the inward propagating signal that departs from diametrically opposite points A and B on the sphere. The field point $\bar{x}$ is situated on the diametral line connecting points A and B. A particular phase contemporaneously departs from both points. For the signal that departs from point A, the radial distance is $r_A = r$ and the propagation distance is $a - r_A$. Thus, the phase delay for this signal is $(a - r)/c$. The pressure in this signal therefore is $\dot{G}\left(t - (a - r_A)/c\right)/r_A$, which is the first term in the above. For the signal that departs from point B, the radial distance decreases as it propagates toward the

Fig. 6.3 Radial and
propagation distances for
inward spherical waves that
propagate along a diameter
inside a spherical cavity

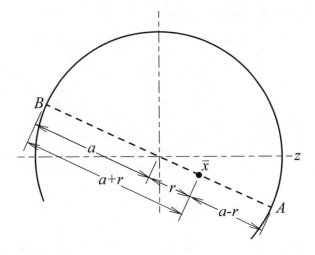

center. Beyond there, we may consider the radial distance to continue to decrease, that
is, to become increasingly large negatively. Thus, the radial distance from point B to $\bar{x}$
is $r_B = -r$, and the propagation distance is $a + r \equiv a - r_B$. The corresponding time
delay is $(a - r_B)/c$, so the pressure in this wave is $\dot{G}(t - (a - r_B)/c)/r_B$. This is the
second term in Eq. (6.2.17). Hence, a different way of viewing the signal within the
spherical cavity is that it consists of a spherical wave that propagates inward, reverses
sign as it passes through the center, then continues outward along the opposite side.

The field properties in the vicinity of the center are quite interesting. Let us expand
Eq. (6.2.17) in a Taylor series relative to $r = 0$. The resulting representation is

$$
\begin{aligned}
p &= \frac{1}{r}\left[\dot{G}\left(t - \frac{a}{c} + \frac{r}{c}\right) - \dot{G}\left(t - \frac{a}{c} - \frac{r}{c}\right)\right] \\
&= \frac{2}{c}\left[\ddot{G}\left(t - \frac{a}{c}\right) + \frac{1}{6}\left(\frac{r}{c}\right)^2 G^{iv}\left(t - \frac{a}{c}\right) + \cdots\right]
\end{aligned}
\tag{6.2.18}
$$

Application of Euler's equation to the preceding leads to an expression for the particle
acceleration near the center. Integrating the fourth derivative of G with respect to
time yields the particle velocity,

$$
v_r = -\frac{2}{3}\frac{r}{\rho_0 c^3}\dddot{G}\left(t - \frac{a}{c}\right) + \cdots \text{ as } r \to 0
\tag{6.2.19}
$$

To interpret these results, we observe that p at $r = a$ is $\dot{G}(t)/a$. Thus, the pressure
at the center at any instant t is proportional to the rate of change of the inward
pressure that departed from the surface at time $t - a/c$. This effect comes about
because the waveforms of the inward and outward waves come closer to overlapping
as r decreases, thereby canceling the singularities. The pressure varies quadratically
with small distance from the center, which means that the pressure gradient is zero at

that location. Recognition of this attribute is useful for sketching a profile of pressure along a radial line. The fact that the particle velocity is zero at the center is necessary. Otherwise, the radial velocity on one side of a diameter would be opposite the velocity on the other side, and therefore be discontinuous.

Determination of the Field

The task now is to determine the pressure field when the spherical boundary executes a specified radially symmetric oscillation $v_0(t)$. The analysis entails identifying a combination of inward and outward propagation waves that satisfy the boundedness condition at $r = 0$ and match the radial velocity of the sphere at $r = a$. We take $v_0 > 0$ to correspond to an inward velocity of the surrounding sphere. (It is the convention here to consider velocity into the fluid domain to be positive.) The fact that the radial velocity and surface pressure are related by a differential equation complicates the task of matching v_0 at $r = a$. We will circumvent this difficulty by restricting our attention to situations where $\dot{v}_0(t) \gg (c/a) v_0(t)$. This allows the general relations between p and v_r in Eq. (6.2.5) to be approximated as

$$v_{\text{out}} \approx \frac{p_{\text{out}}}{\rho_0 c} = \frac{1}{\rho_0 c r}\dot{F}\left(t - \frac{r-a}{c}\right)$$
$$v_{\text{in}} \approx -\frac{p_{\text{in}}}{\rho_0 c} = -\frac{1}{\rho_0 c r}\dot{G}\left(t + \frac{r-a}{c}\right)$$

(6.2.20)

The validity of these approximations depends in the rate at which the $\dot{F}$ and $\dot{G}$ functions change, as well as the magnitude of a. After the pressure signal has been determined, we may use the full Euler equation to evaluate the particle velocity to ascertain the error entailed in the approximation. In any event carrying out the simplified analysis will suffice to understand the basic phenomena.

The method of wave images, which was developed for plane waves, suggests how to proceed. Because $\dot{F}(t - a/c) = r p_{\text{out}}$ at $r = 0$ and $\dot{G}(t + a/c) = r p_{\text{in}}$ at $r = 0$, the singularity condition at the center, Eq. (6.2.16), is like a pressure-release boundary condition, except that it governs pressure times radial distance. Furthermore, matching $v_0(t)$ to the sum of the above expressions for inward and outward particle velocity resembles the task of satisfying a velocity boundary condition at the far end of a finite planar waveguide. This suggests that we may determine the response by considering the spherical cavity to be a one-dimensional waveguide whose end $r = 0$ is pressure-release and whose end $r = a$ has a velocity excitation. The spherical nature of the system is embedded in definition of the state variable as rp.

Physical reasoning indicates that movement of the wall generates an inward propagating wave. This is the initial contribution to the set of images, so we denote the associated wave function as G_0. By definition, v_r is positive if it is outward. Therefore, the first contribution is an inward propagating wave for which v_r at $r = a$ matches $-v_0(t) h(t)$, where the step function serves to emphasize that the movement of the spherical surface begins at $t = 0$. It follows from Eq. (6.2.20) that

$$\dot{G}_0(t) = \rho_0 c a v_0(t) h(t)$$

(6.2.21)

The value of rp at $r = 0$ due to this wave is $\dot{G}_0 (t - a/c)$. The "pressure-release" condition imposed on rp at $r = 0$ requires an outwardly propagating image F_1. The value of rp at $r = 0$ due to this wave is $\dot{F}_1 (t + a/c)$, which will cancel the pressure associated with G_0 if

$$\dot{F}_1 \left(t + \frac{a}{c} \right) = -\dot{G}_0 \left(t - \frac{a}{c} \right) \implies \dot{F}_1 (t) = -\dot{G}_0 \left(t - 2\frac{a}{c} \right) \qquad (6.2.22)$$

The F_1 wave eventually arrives at $r = a$. Because G_0 gives a signal whose particle velocity at $r = a$ matches v_0, the contribution of F_1 to the particle velocity at $r = a$ must be canceled by an inwardly propagating wave G_1. According to Eq. (6.2.20), the radial velocities at $r = a$ due to each wave are $\dot{F}_1 (t) / (\rho_0 c a)$ and $-\dot{G}_1 (t) / (\rho_0 c a)$. Their sum must be zero, so we find that

$$\dot{G}_1 (t) = \dot{F}_1 (t) = -\dot{G}_0 \left(t - \frac{2a}{c} \right) \qquad (6.2.23)$$

The contribution to rp at $r = 0$ due to this wave is $\dot{G}_1 (t - a/c)$. It is canceled by another outward wave, F_2, for which rp at $r = 0$ is $\dot{F}_2 (t + a/c)$. Setting the sum of these contributions to zero leads to

$$\dot{F}_2 \left(t + \frac{a}{c} \right) = -\dot{G}_1 \left(t - \frac{a}{c} \right) \implies \dot{F}_2 (t) = -\dot{G}_1 \left(t - 2\frac{a}{c} \right) = \dot{G}_0 \left(t - \frac{4a}{c} \right) \qquad (6.2.24)$$

The process goes on ad infinitum, with inwardly propagating images G_n generated at $r = a$ to give $v_r = 0$ and outwardly propagating images F_n set to give $rp = 0$ at $r = 0$. Thus, the next contribution to rp_{in} is

$$\dot{G}_2 (t) = \dot{F}_2 (t) = \dot{G}_0 \left(t - \frac{4a}{c} \right) \qquad (6.2.25)$$

which requires another contribution to rp_{out} given by

$$\dot{F}_3 (t) = -\dot{G}_2 \left(t - \frac{2a}{c} \right) = -\dot{G}_0 \left(t - \frac{6a}{c} \right) \qquad (6.2.26)$$

The total contribution to the functions for the inward and outward waves is the sum of the individual images, so we have established that

$$\dot{G} (t) = \sum_{n=1}^{\infty} (-1)^{n-1} \dot{G}_0 \left(t - \frac{2 (n - 1) a}{c} \right)$$

$$\dot{F} (t) = \sum_{n=1}^{\infty} (-1)^{n} \dot{G}_0 \left(t - \frac{2na}{c} \right) \qquad (6.2.27)$$

To convert these expressions for $\dot{G}$ and $\dot{F}$ to the pressure in inward and outward waves, we add $(r - a) / c$ to the argument of $\dot{G}$ for inward waves and subtract $(r - a) / c$

from the argument of $\dot{F}$ for outward waves. We also use Eq. (6.2.21) to describe $\dot{G}_0$. The resulting waves are

$$
\begin{aligned}
p_{\text{in}} &= \rho_0 c \frac{a}{r} \sum_{n=1}^{\infty} (-1)^{n-1} v_0 \left(t + \frac{r}{c} - \frac{(2n-1)a}{c} \right) h \left(t + \frac{r}{c} - \frac{(2n-1)a}{c} \right) \\
p_{\text{out}} &= \rho_0 c \frac{a}{r} \sum_{n=1}^{\infty} (-1)^{n} v_0 \left(t - \frac{r}{c} - \frac{(2n-1)a}{c} \right) h \left(t - \frac{r}{c} - \frac{(2n-1)a}{c} \right)
\end{aligned}
$$

(6.2.28)

Note that at any instant a spatial profile might feature images that lie outside of the domain of the spherical cavity; that is, they might have nonzero values for $r < 0$ or $r > a$. This is an artifice embedded in the procedure. Only the portion of any wave that lies in $0 < r < a$ is meaningful.

It is worth noting that if we desired a description that is valid when the rate of change of $v_0(t)$ is arbitrary, then it would be necessary to satisfy the full Euler equation relating pressure and particle acceleration at $r = a$, rather than the simplified version of Eq. (6.2.21). The consequence of this would be that each image would be determined as the solution of a differential equation. For G_0, the inhomogeneous term in this equation would be formed with v_{in}, while the inhomogeneous term for the subsequent G_n images would be formed from the outgoing image F_n. Such an analysis is feasible, but certainly not simple.

Several features may be deduced from Eq. (6.2.28). If we were to use the expressions to evaluate a (virtual) initial profile of $(r/a)p$, we would find that the successive images for p_{in} are stacked beyond $r = a$ with their leading wavefronts at $r = a, 3a, 5a$, etc. The initial profile of $(r/a)p_{\text{out}}$ would be stacked up in $r < 0$, with leading edges at $r = -a, -3a, -5a$, etc. Each inward image G_n that arrives at the center is matched by an inverted outward image F_{n-1}. Incidence of F_{n+1} at $r = a$ is matched by image G_{n+1}, which is upright relative to F_{n+1}, and therefore inverted relative to G_n. When G_{n+1} arrives at the center, it is matched by F_{n+2}, which is inverted relative to G_{n+1}, and therefore upright relative to G_n. Thus, G_{n+2}, which is the result of reflection of F_{n+2} at $r = a$, is the same as G_n. This is the same arrangement as that of a waveguide whose end $x = 0$ is pressure release and whose end $x = L$ is coerced to move with a specified velocity. From this, we conclude that the waveform observed at any location features a repetition of images in an interval of $4a/c$. As was shown in Sect. 2.5.3, if v_0 is periodic, then each image is endless. If the period is not commensurate with $4a/c$, then the response will be bounded. However, if the period is a rational fraction of $4a/c$, then the successive images will overlap, corresponding to a in a resonance. (This is an approximate prediction of the conditions for resonance, because it is predicated on the radius being large. The frequency-domain analysis will be applicable for any radius.)

EXAMPLE 6.2 Water ($c = 1480$ m/s, $\rho_0 = 1000$ kg/m^3) is contained within the interior of a spherical container whose radius is 600 mm. The wall of this container is a piezoceramic material that is actuated by an electrical pulse. The graph describes the wall's inward radial velocity, which is an N wave modified to give finite rise times. The duration of the pulse is $T = 0.3$ ms, and the rise times are $\Delta = T/40$. Determine the pressure waveforms at $r = 0.5a$, $0.25a$, and $0.001a$. The waveforms may be cutoff after $t = T + 4a/c$.

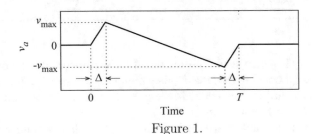

Figure 1.

Significance

This quantitative implementation of the analysis of a spherical cavity will enhance our understanding of how the inward and outward waves are generated and combined. The specific input signal and parameters will provide a basic understanding of the concept underlying an important biomedical device.

Solution

The velocity function v_0 is not too complicated to work with graphically, but great precision is needed to determine the response at very small values of r. Thus, the first step is to express v_0 in a form that is suitable for computations. The function may be represented as three linear segments, with slopes of v_{max}/Δ for the first and third, and $-2v_{max}/(T - 2\Delta)$ for the second. The straight line functions for each are initiated and terminated with step functions, so that

$$v_0 = V_{max}\frac{t}{\Delta}\left[h\left(t\right) - h\left(t - \Delta\right)\right] + V_{max}\frac{T - 2t}{T - 2\Delta}\left[h\left(t - \Delta\right) - h\left(t - T + \Delta\right)\right]$$
$$+ V_{max}\frac{t - T}{\Delta}\left[h\left(t - T + \Delta\right) - h\left(t - T\right)\right]$$

According to Eq. (6.2.21), the function $\dot{G}_0\left(t\right)$ is $\rho_0 c a v_0\left(t - a/c\right)$. The inward propagating wave images arrive at $r = 0$ when $t = a/c$, $3a/c$, and $5a/c$, whereas the outward images arrive at $r = a$ when $t = 2a/c$, $4a/c$. The duration of the pulse is $T = 0.3$ ms and $a/c = 0.4054$ ms. Because $T < a/c$, two inward and two outward images will be seen at any location within the specified interval. Thus, we shall halt the summations in Eq. (6.2.28) at $n = 2$.

The programming effort begins with construction of a function or subroutine that evaluates $v_0\left(t\right)$ for an array of t values. We define a column vector $\{t\}$ of time instants in the interval $0 \leq t \leq 4a/c + T$, which are the time values for the waveforms. Initialization of the data ends by defining zero column vectors of the same size to hold $\{p_{in}\}$ and $\{p_{out}\}$. The wave images are evaluated in a loop running from $n = 1$ to $n = 2$. Within this loop, retarded times for the inward waves are placed in vectors $\{\tau_{in}\} = \{t\} + (r - (2n - 1)a)/c$ and $\{\tau_{out}\} = \{t\} - (r + (2n - 1)a)/c$. The pressure in the nth inward wave is $(-1)^{n-1}(a/r)\rho_0 v_0(\{\tau_{in}\})$ and that in the nth outward wave is $(-1)^{n-1}(a/r)\rho_0 v_0(\{\tau_{in}\})$. These vectors are added to the prior values of $\{p_{in}\}$ and $\{p_{out}\}$.

Figure 2 describes the waveform at $r = 0.5a$. The duration of the pulse is small, so each image is seen individually. To assist interpretation of the waveform, the instant at which each arrives is marked with a dot. The first arrival, at $t = 0.5a/c = 0.203$ ms, is the G_0 image. At $t = 1.5a/c$ the F_1 image that departed from the origin at $t = a/c$ arrives. The pressure-release condition on rp inverts $\dot{F}_1$. It reflects in-phase at $r = a$ to generate $\dot{G}_1$, which arrives at $r = a/2$ when $t = 2.5a/c$, and so on.

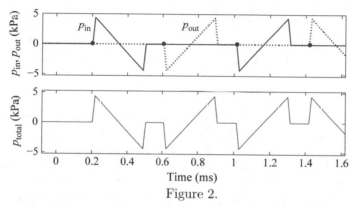

Figure 2.

As the radial distance is reduced, the value of r/a might be sufficiently small that an outward image arrives prior to the termination of the inward image that generated it. This behavior is displayed in Fig. 3 for $r = a/4$. In the region of overlap, the total pressure is constant within the interval where the images have rates of change that are equal in magnitude but oppositely signed. Furthermore, because the magnitude of each wave is inversely proportional to the radial distance, the magnitude of each waveform is twice the value at $r = a/2$.

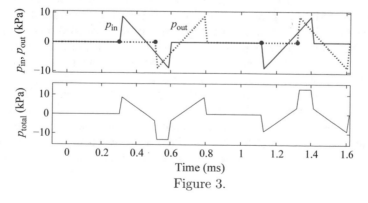

Figure 3.

Figure 4, for $r = 0.001a$, shows an enormous increase in the magnitudes of p_{in} and p_{out} as a result of the inverse dependence on r. It can be seen in the first graph that the individual waveforms are like those at $r = a/2$, except that p_{out} arrives soon after p_{in}. The time difference between their arrivals is $2r/c = 40.54\ \mu s$, which is so small that p_{out} seems to be an inverted version of p_{in}. However, the arrival time of p_{out} is not exactly the same as for p_{in}, so the total pressure is not zero.

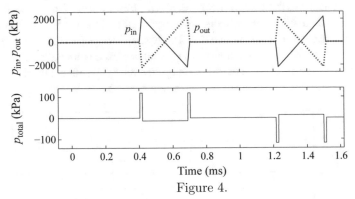

Figure 4.

The pressure at $r \ll a$ is described by Eq. (6.2.18). Until the G_1 image enters the picture, Eq. (6.2.21) indicates that $\dot{G}(t) = \rho_0 c a v_0 (t - a/c)$. Hence, the pressure at $r \approx 0$ should be $p = (2/c) \rho_0 c a \dot{v}_0 (t - a/c)$. The derivative of v_0 is v_{max}/Δ in the first and third intervals, where v_0 increases, and $-2 v_{max}/(T - 2\Delta)$ in the middle interval. Thus, after the signal arrives at $t = a/c$, it should have a constant value of $(2/c) \rho_0 c a (v_{max}/\Delta) = 118.4$ kPa over an interval Δt and then fall to a value of $(2/c) \rho_0 c a (-2 v_{max}/(T - 2\Delta)) = -13.2$ kPa for an interval $T - 2\Delta$, then rise to $(2/c) \rho_0 c a (v_{max}/\Delta)$ for another Δt interval. After that, the pressure should be zero until the $\dot{G}_1$ image arrives. These values are the same as the plotted values of p_1 and p_2. Subsequent contributions to the waveform replicate this sequence of constant values, except that each is inverted relative to the preceding one.

The discussion shows that the peak pressure at the center is proportional to v_{max}/Δ, so increasing the peak velocity and decreasing Δ can lead to a large pressure at $r = 0$. This behavior has been exploited in a device called a *lithotripter*, which is used to break up kidney stones without performing surgery. The actual device uses an elliptical cavity filled with water. An ellipse has two foci, which allows the patient to be positioned such that the kidney stone is at one focus and a transducer that generates the initial wave is at the other. Nevertheless, the fundamental phenomena are like those for a spherical cavity. Because living tissue is mostly water, a pressure wave generated by the transducer is transmitted into the patient with little modification. The idea is that reducing the rise time for the transducer's motion will raise the pressure level. This, coupled with the rapid fluctuation, often fractures the stone into smaller pieces that may pass through the urethra without excessive pain. (The pressure levels are sufficiently high that accurate models must account for nonlinear effects.)

6.3 Radially Vibrating Sphere–Frequency-Domain Analysis

The differential nature of the relation between pressure and particle velocity inhibits extensive exploration of radially symmetric spherical waves in the time domain. Handling situations where a boundary is locally reacting is another complication. These difficulties are circumvented in a frequency-domain formulation because a time derivative becomes an $i\omega$ factor. Another benefit of analyzing spherical waves in the frequency domain is that such an analysis is required in the next chapter, where the restriction to radially symmetric waves will be removed.

6.3.1 General Solution

The frequency-domain ansatz of a radially symmetric pressure wave is

$$p = \operatorname{Re}\left(P\left(r\right)e^{i\omega t}\right) \tag{6.3.1}$$

Substitution of this representation into Eq. (6.2.2) leads to the radially symmetric version of the Helmholtz equation,

$$\frac{1}{r}\left[\frac{\partial^2}{\partial r^2}\left(rP\right) + k^2\left(rP\right)\right] = 0 \tag{6.3.2}$$

The bracketed term, which is the one-dimensional Helmholtz equation for rP, must vanish because r is finite except at the center. Thus, the general solution for rP is like the forward and backward waves in the planar case. Here, the directions are said to be outward and inward, so we have

$$\boxed{\begin{aligned} P &= P_{\text{out}} + P_{\text{in}} \\ P_{\text{out}} &= B_1\frac{e^{-ikr}}{r}, \quad P_{\text{in}} = B_2\frac{e^{ikr}}{r} \end{aligned}} \tag{6.3.3}$$

The coefficients B_1 and B_2 may be complex; usually, their value will be obtained by satisfying boundary conditions.

An expression for the particle velocity is found from the frequency-domain Euler's equation,

$$V_r = -\frac{1}{i\omega\rho_0}\frac{dP}{dr} = \frac{1}{\rho_0 c}\left[\left(1 - \frac{i}{kr}\right)B_1\frac{e^{-ikr}}{r} - \left(1 + \frac{i}{kr}\right)B_2\frac{e^{ikr}}{r}\right] \tag{6.3.4}$$

Equations (6.3.3) and (6.3.4) are what would be obtained from the time-domain analysis if the wave functions were

$$\dot{F} = \text{Re}\left(B_1 e^{i\omega(t-r/c)}\right), \quad \dot{G} = \text{Re}\left(B_2 e^{i\omega(t+r/c)}\right) \tag{6.3.5}$$

The determination of the radial velocity in the time domain entails integration of the arbitrary $\dot{F}$ and $\dot{G}$ functions. In the frequency domain, these functions are harmonic, so the integration is replaced by an algebraic equation.

Equation (6.3.3) constitutes the general solution. That means that it describes any radially symmetric field, regardless of the nature of the boundaries. Fulfillment of those details sets the B_1 and B_2 coefficients.

6.3.2 Radiation from a Radially Vibrating Sphere

The concern here is with a sphere having radius a that is surrounded by a fluid in an infinite domain. The surface of the sphere executes the same harmonically varying radial displacement at all locations, thereby generating a radially symmetric outwardly propagating wave. The Sommerfeld radiation condition states that a signal cannot arrive from infinity in a finite amount of time. We therefore discard the part of the general solution of the Helmholtz equation that is the inwardly propagating wave. The particle velocity in the outgoing wave is given by Eq. (6.3.4). Matching it to the sphere's velocity, whose complex amplitude is denoted as V_0, gives

$$\frac{1}{\rho_0 \omega}\left(k - \frac{i}{a}\right) B_1 \frac{e^{-ika}}{a} = V_0 \implies B_1 = \rho_0 c V_0 a \frac{ka}{ka-i} e^{ika} \tag{6.3.6}$$

The corresponding pressure and particle velocity fields are

$$\boxed{\begin{aligned} P &= \rho_0 c V_0 \left(\frac{ka}{ka-i}\right)\left(\frac{a}{r}\right) e^{-ik(r-a)} \\ V_r &= V_0 \left(\frac{ka}{ka-i}\right)\left(1 - \frac{i}{kr}\right)\left(\frac{a}{r}\right) e^{-ik(r-a)} \end{aligned}} \tag{6.3.7}$$

Embedded in these expressions is the relation of P and V_r at any field point,

$$\boxed{V_r = \left(1 - \frac{i}{kr}\right)\frac{P}{\rho_0 c}} \tag{6.3.8}$$

Whereas the discussion of trends in the time-domain solution qualitatively alluded to rates of change, we now have frequency as a measure of rapidity. Thus, a rapidly changing $v_0(t)$ now translates to $ka \gg 1$, which alternatively may viewed as a statement that the sphere's radius is large compared to an acoustic wavelength $2\pi/k$. The farfield limit is one where $kr \gg 1$, that is, r is much greater than an acoustic wavelength. This region will be of much interest for a variety of situations. Taking $1/(kr)$ to be negligible in Eq. (6.3.8) leads to

$$V_r \approx \frac{P}{\rho_0 c}; \quad kr \gg 1 \qquad (6.3.9)$$

This is the property encountered in a simple plane wave, but unlike the plane wave case, the pressure amplitude depends inversely on the radial distance.

The high-frequency limit, at which $ka \gg 1$, is also of interest. Because $kr \geq ka$, the high-frequency limit is characterized by

$$V_r \approx \frac{P}{\rho_0 c} \approx V_0 \left(\frac{a}{r}\right) e^{-ik(r-a)}; \quad ka \gg 1 \qquad (6.3.10)$$

This behavior is observed everywhere. In it, the signal replicates the velocity on the surface, with an a/r scale factor as a result of spherical spreading and a phase lag due solely to the time to propagate the signal from the surface. The difference between the farfield and high-frequency trend is subtle. The former describes the relationship between the particle velocity and pressure that are observed at a specific large radial distance. The latter states that the same relation between pressure and particle velocity is observed everywhere, but it goes further by relating the signal to the velocity on the surface.

The low-frequency/small sphere limit, in which $ka \ll 1$, also is important. The trend in this case is

$$\left.\begin{array}{l} P \approx \rho_0 c V_0 ka \left(\dfrac{a}{r}\right) i e^{-ik(r-a)} \\[2ex] V_r = V_0 ka \left(\dfrac{a}{r}\right) \left(i + \dfrac{1}{ka}\dfrac{a}{r}\right) e^{-ik(r-a)} \end{array}\right\} ka \ll 1 \qquad (6.3.11)$$

The factor ka reduces the pressure amplitude relative to the value $\rho_0 V_0 (a/r)$ it would have in the high-frequency limit. In addition, the i factor represents a 90° phase lead. Close to the sphere, where r is only slightly greater than a, the particle velocity amplitude is essentially comparable to V_0 and the pressure is proportional to $i\omega V_0$. In other words, the pressure in the vicinity of a small sphere replicates the surface acceleration, whereas the particle velocity replicates the surface velocity. The corollary is that V_r close to the surface is much larger in magnitude than $|P| / (\rho_0 c)$ and it lags the pressure by 90°. Increasing the radial distance merely modifies the proportionality of pressure to surface acceleration by introducing the spherical spreading factor a/r and the propagation phase shift $e^{-ik(r-a)}$. The farfield approximation applies at large $kr \gg 1$ for any frequency. Thus, as the radial distance increases, the particle velocity transitions from being proportional to the surface velocity to a proportionality to acceleration.

The fact that the relation between particle velocity and pressure depends so strongly on the values of ka and kr has significant implications for the radiated acoustic power. It is best to begin an examination by considering the instantaneous intensity. Only the radial component is nonzero, such that

$$
\begin{aligned}
I_r &= \frac{1}{4} \left(P e^{i\omega t} + \text{c.c.} \right) \left(V_r e^{i\omega t} + V_r^* e^{-i\omega t} \right) \\
&= \frac{1}{2} \,\text{Re} \left(P V_r^* + P V_r e^{2i\omega t} \right)
\end{aligned}
\tag{6.3.12}
$$

Substitution of Eq. (6.3.7) yields

$$
\begin{aligned}
I_r = \frac{1}{2} \rho_0 c \left(\frac{a}{r} \right)^2 \text{Re} &\left[|V_0|^2 \frac{(ka)^2}{(ka)^2 + 1} \left(1 + \frac{i}{kr} \right) \right. \\
&\left. + V_0^2 \left(\frac{ka}{ka - i} \right)^2 \left(1 - \frac{i}{kr} \right) e^{i2(\omega t - k(r-a))} \right]
\end{aligned}
\tag{6.3.13}
$$

The nonoscillatory term is the time-averaged intensity,

$$
\boxed{(I_r)_{\text{av}} = \frac{1}{2} \rho_0 c \, |V_0|^2 \frac{(ka)^2}{(ka)^2 + 1} \left(\frac{a}{r} \right)^2 = \frac{|P|^2}{2\rho_0 c}}
\tag{6.3.14}
$$

Thus, the ratio of the amplitude of the oscillating part to the mean value is

$$
\frac{|I_r|_{\text{oscillating}}}{(I_r)_{\text{av}}} = \left[1 + \frac{1}{(kr)^2} \right]^{1/2}
\tag{6.3.15}
$$

From this, we see that in the farfield, the fluctuating and average parts are nearly equal in magnitude. Because $kr \equiv ka(r/a)$ and $kr \geq ka$, the same behavior is observed everywhere if $ka \gg 1$, whereas the fluctuating part near the sphere is much greater than the mean value if $ka \ll 1$.

The time-averaged radiated power may be obtained by integration over a sphere centered at the origin. The surface area is $4\pi r^2$, and $(I_r)_{\text{av}}$ is constant along the surface. It follows that

$$
\boxed{\mathcal{P} = 4\pi r^2 \, (I_r)_{\text{av}} = 2\pi \rho_0 c a^2 \, |V_0|^2 \frac{(ka)^2}{(ka)^2 + 1}}
\tag{6.3.16}
$$

Suppose a very large planar surface executes a normal vibration whose amplitude is V_0. The power radiated by a patch having area $4\pi a^2$ would be $4\pi a^2 \left(|P|^2 / (2\rho_0 c) \right)$. Thus, the factor $(ka)^2 / \left[(ka)^2 + 1 \right]$ in Eq. (6.3.16) may be considered to describe the efficiency with which the sphere radiates power. The low-frequency/small radius

case, $ka \ll 1$, represents very low efficiency. In contrast, if $ka \gg 1$, the sphere radiates almost as efficiently as the vibrating plane.

This trend is quite similar to the behavior that arises when a surface wave is
induced on the planar boundary of a half-space. The half wavelength of the surface
motion along the plane is like a, in the sense that the velocity at opposite ends of a
radial line in the sphere are 180° are in the opposite sense. Hence, small ka is like the
situation where the wavelength on the planar surface is much smaller than an acoustic
wavelength. As was noted in the surface wave analysis of the previous chapter, it
usually is the case that little power will be radiated if the acoustic wavelength is large
compared to the distance over which a surface motion varies significantly.

The energy densities (energy per unit volume) provide a different way of viewing
the field. The time-averaged kinetic energy per unit volume is

$$
\begin{aligned}
T_{\text{av}} &= \frac{1}{2}\rho_0 |V_r|^2 = \frac{1}{2}\rho_0 |V_0|^2 \frac{(ka)^2}{(ka)^2 + 1} \left(\frac{a}{r}\right)^2 \left[1 + \left(\frac{1}{kr}\right)^2\right] \\
&= \frac{(I_r)_{\text{av}}}{c}\left[1 + \left(\frac{1}{kr}\right)^2\right]
\end{aligned}
\tag{6.3.17}
$$

while the average acoustic potential energy per unit volume is

$$
U_{\text{av}} = \frac{|P|^2}{2\rho_0 c} = \frac{1}{2}\rho_0 |V_0|^2 \frac{(ka)^2}{(ka)^2 + 1}\left(\frac{a}{r}\right)^2 = \frac{(I_r)_{\text{av}}}{c}
\tag{6.3.18}
$$

Thus, in the entire field radiated by a large sphere, $ka \gg 1$, as well as the farfield for
any case, there is an equipartition of energy, $T_{\text{av}} \approx U_{\text{av}}$. In contrast, in the nearfield
of a small source, which corresponds to $kr \ll 1$, the kinetic energy is dominant.

The power radiated by multiple sources was explored in Sect. 4.5.3, where it was
proven that the time-averaged power flowing out of any surface that encloses the
sources will equal the total power radiated by the sources. Suppose there are two
spherical sources, each of which executes a specified harmonic motion in the radial
direction. The total pressure is the scalar sum of the individual contributions, while
the particle velocity is a vector sum. We wish to evaluate the power they radiate. A
first thought is that we can apply the relation between pressure and particle velocity to
describe the field from each sphere and then superpose the field of each. The radiated
power would then be obtained by adding the output from each source. Whether this
reasoning is faulty depends on several features.

A primary question is whether the spheres are vibrating coherently. In the context
of harmonic motion, coherence reduces to a question of whether the spheres vibrate
at the same frequency. Suppose the spheres vibrate at different frequencies. Then,
evaluation of the mean-squared pressure and average intensity can ignore any sum
and difference frequencies when the signals are multiplied (see Sect. 1.2). Thus, a
simple summation of individual time averaged quantities is appropriate when the
spheres do not vibrate coherently. It follows that the average radiated power in this
case is the sum of the individual outputs.

The situation is more complicated if the spheres vibrate at the same frequency. A fundamental feature is the fact that each sphere is an obstacle for the field of the other. The signal from one sphere reflects off the other, thereby generating additional signals that only exist when both spheres are present. This is called *mutual scattering*. Analysis of this phenomenon can be quite difficult, because it requires simultaneous consideration of the field radiated by both spheres with the surface velocity on each sphere being dependent on the field properties of the other sphere.

An analysis of mutual scattering might need to contend with another issue. We began with the assumption that the radial velocity of each sphere is known. However, the sphere is some sort of transducer, which means that the radial velocity is actuated by an electrical signal that drives a mechanical element, such as a piezoceramic. The stresses induced by the electrical signal move the element in opposition to the pressure field acting on the sphere. If that pressure field is strongly modified by the field from the other sphere, the radial velocity will be altered.

Mutual scattering may be ignored if the distance between the spheres is sufficiently large compared to both radii. Let d be the distance between centers, a_n be the radius of source n, and V_n be the surface velocity. An upper bound, independent of frequency, for the pressure from sphere 1 at the center of sphere 2 is $\rho_0 c V_1 (a_1/d)$. If $V_1 a_1/d$ is much less than V_2, then we may ignore the presence of the field from sphere 1 when we analyze the field from sphere 2. A similar statement obviously applies when we consider the influence of sphere 2 on sphere 1. We may generalize this observation to state that mutual scattering of coherently vibrating transducers may be ignored if their surface velocities have comparable magnitudes and the separation distances are large compared to the size of the transducers. An analysis that ignores mutual scattering often is justifiable, but it is questionable for a tightly grouped array of transducers, such as some types of loudspeaker systems.

Even if mutual scattering is unimportant, evaluation of the radiated power is not a trivial task. If we try to integrate the radial component of the time-averaged intensity on each sphere, we need to include the other sphere's pressure and particle velocity in the evaluation of the time-averaged intensity, which means that we must describe the dependence of the other field on the spherical coordinates of the first sphere.

These complications may be avoided if the radius of the surrounding spherical surface is very large. Doing so would allow us to invoke the farfield approximation for the field radiated by each vibrating sphere. The particle velocity in the signal radiated by each sphere then is proportional to the pressure in that signal, and both pressure and particle velocity are a superposition of the contribution of each radiator. Thus, the time-averaged intensity in the farfield is representable as

$$\left(\bar{I}_{\text{farfield}} \right)_{\text{av}} = \frac{1}{2\rho_0 c} \, \text{Re} \left[(P_1 + P_2) \left(P_1^* \, (\bar{e}_r)_1 + P_2^* \, (\bar{e}_r)_2 \right) \right] \tag{6.3.19}$$

where $(\bar{e}_r)_1$ and $(\bar{e}_r)_2$ are the radial directions from the center of each spherical source to a field point on the enclosing sphere. The radiated power is the integral over the surface of the surrounding sphere of the normal component of this intensity vector. Evaluation of that integral would require that we describe how P_1 and P_2, as well as $(\bar{e}_r)_1$ and $(\bar{e}_r)_2$, vary along the surface of the enclosing sphere. Such a description requires a geometrical analysis that accounts for the fact that the spherical sources are not concentric. The farfield description of radial distance for arbitrary points, which is part of the development in Sect. 6.4.3, would considerably expedite this analysis.

EXAMPLE 6.3 Two 50-mm-radius spherical sources in the atmosphere are arranged as shown in the sketch. When both A and B are active, the sound pressure level at position C is 86 dB//20 μPa. Turning off source B reduces the sound pressure level at position C to 80 dB. (a) Suppose that source A has a frequency of 800 Hz and source B has a frequency of 900 Hz. Determine the sound pressure level at position C when only B is active, and the time-averaged intensity at point C when both sources are active. (b) Repeat the analysis of case (a) in the situation where both sources oscillate in-phase at a frequency of 800 Hz.

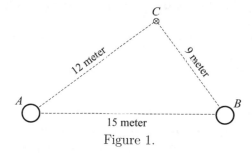

Figure 1.

Significance

The vectorial nature of intensity and the difference between incoherent and coherent signals are especially important when one deals with multiple sources.

Solution

The relationship between the given sound pressure levels and the corresponding mean-squared pressure does not depend on the frequency. Thus, we begin with

$$\left(p_A^2\right)_{av} = p_{ref}^2\left(10^{80/10}\right) = 0.040 \text{ Pa}^2$$
$$\left(p_{A+B}^2\right)_{av} = p_{ref}^2\left(10^{86/10}\right) = 0.15924$$

How to proceed depends strongly on whether the sources radiate at different frequencies or the same frequency.

Case a—Incoherent Sources

The relative phase of incoherent signals is irrelevant to an evaluation of their combined effect. Therefore, the mean-squared pressure due to both spheres is the sum of the individual contributions in the spheres. We are given the combined value and the value due to sphere A, so the contribution of sphere B is the difference between the two given values. In other words,

$$\left(p_B^2\right)_{\text{av}} = \left(p_{A+B}^2\right)_{\text{av}} - \left(p_A^2\right)_{\text{av}} = 0.11924$$

$$\mathcal{L}_B = 10 \log\left(\frac{\left(p_B^2\right)_{\text{av}}}{p_{\text{ref}}^2}\right) = 84.7 \text{ dB}$$

The radial component of intensity for a spherical source is proportional to the mean-squared pressure at that location, so

$$(I_r)_{\text{av,A}} = \frac{\left(p_A^2\right)_{\text{av}}}{\rho_0 c} = 0.0980, \quad (I_r)_{\text{av,B}} = \frac{\left(p_B^2\right)_{\text{av}}}{\rho_0 c} = 0.2926 \text{ mW/m}^2$$

We use a Cartesian coordinate system whose x-axis is oriented from A to B to describe each radial direction from a source to point C. Triangle ABC is similar to a "3-4-5" right triangle, so

$$(\bar{e}_r)_A = 0.8\bar{e}_x + 0.6\bar{e}_y, \quad (\bar{e}_r)_B = -0.6\bar{e}_x + 0.8\bar{e}_y$$

The corresponding vector sum of the intensity from each source is

$$\bar{I} = (I_r)_{\text{av,A}} (\bar{e}_r)_A + (I_r)_{\text{av,A}} (\bar{e}_r)_B = -0.0960\bar{e}_x + 0.2926\bar{e}_y \text{ mW/m}^2$$

Case b—Coherent Sources

When both spheres oscillate at the same frequency, addition of their signals must account for their relative phase. The phase at the field point depends on the phase at each source and the phase lag due to propagation to the field point. It is given that the spheres vibrate in-phase, so we may take both surface velocities to be real values V_A and V_B. The first of Eq. (6.3.7) tells us that the complex pressure amplitude at location C due to each source is proportional to the corresponding surface velocity, according to

$$P_A = \rho_0 c V_A \left(\frac{ka}{ka - i}\right)\left(\frac{a}{r_A}\right) e^{-ik(r_A - a)} = Z_A V_A, \quad Z_A = 0.9914 + 0.1959i$$

$$P_B = \rho_0 c V_B \left(\frac{ka}{ka - i}\right)\left(\frac{a}{r_B}\right) e^{-ik(r_B - a)} = Z_B V_B, \quad Z_B = 1.1382 + 0.7210i$$

$$\tag{1}$$

where the parameters used for this calculation are $\rho_0 = 1.2 \text{ kg/m}^3$, $c = 340 \text{ m/s}$, $a = 0.05 \text{ m}$, $\omega = 1600\pi \text{ rad/s}$, $r_A = 12 \text{ m}$, and $r_B = 9 \text{ m}$.

The (real) value of V_A may be found from the given sound pressure level when only A is active,

$$\left(p_A^2\right)_{\text{av}} = 0.040 \text{ Pa}^2 = \frac{1}{2} |P_A|^2 = \frac{1}{2} |Z_A| V_A^2$$

$$V_A = 0.2799 \text{ m/s}$$

$$\tag{2}$$

According to the first of Eq. (1) the corresponding pressure at point C due to source A is

$$P_A = 0.2775 + 0.0548i \text{ Pa}$$

The contribution of sphere B to the pressure at point C may be found by equating the value of $\left(p^2_{A+B}\right)_{av}$ established at the outset to the coherent sum of the contribution of each sphere. For this operation, we use the second of Eq. (1) to describe P_B and recognize that V_B is real. Thus, we write

$$\left(p^2_{A+B}\right)_{av} = \frac{1}{2}|P_A + P_B|^2 = \frac{1}{2}\left[|P_A|^2 + |Z_B|^2\, V_B^2 + 2\,\text{Re}\left(P_A Z_B^*\right) V_B\right] \quad (3)$$
$$= p^2_{\text{ref}}\left(10^{86/10}\right)$$

All quantities in this expression other than V_B are known, so we have derived a quadratic equation for that velocity,

$$C_2 V_B^2 + C_1 V_B + C_0 = 0$$
$$C_2 = \frac{1}{2}|Z_B|^2 = 0.90770, \quad C_1 = \text{Re}\left(P_A Z_B^*\right) = 0.35536,$$
$$C_0 = \frac{1}{2}|P_A|^2 - \left(p^2_{A+B}\right)_{av} = -0.11924$$

We select the positive root because we have taken $V_A > 0$ and V_B is in-phase with V_A,

$$V_B = 0.21618 \text{ m/s}$$

Then, Eq. (1) gives

$$P_B = 0.24606 + 0.15587i \text{ Pa}$$

The sound pressure level due to source B is

$$\mathcal{L}_B = 20 \log\left(\frac{|P_B|}{\sqrt{2}\, p_{\text{ref}}}\right) = 80.3 \text{ dB}$$

Evaluation of the intensity requires that the complex amplitude vector of each contribution be evaluated. We already have determined the individual pressures. The corresponding radial velocities are found from Eq. (6.3.8) to be

$$(V_r)_A = \frac{P_A}{\rho_0 c\left(1 - \dfrac{i}{kr_A}\right)} = 0.6793 + 0.1382i \text{ mm/s}$$

$$(V_r)_B = \frac{P_B}{\rho_0 c\left(1 - \dfrac{i}{kr_B}\right)} = 0.60019 + 0.38654i \text{ mm/s}$$

Each radial velocity is in the direction of the unit vector from the center of the sphere to the field point. The resultant velocity is the vector sum of each contributions

$$\bar{V} = (V_r)_A \, (\bar{e}_r)_A + (V_r)_B \, (\bar{e}_r)_B = (18334 - 0.12137i) \, \bar{e}_x$$
$$+ \, (0.88774 + 0.39214i) \, \bar{e}_y \text{ mm/s}$$

From this value, we find the time-averaged intensity to be

$$\bar{I}_{av} = \frac{1}{2} \text{Re} \left((P_A + P_B) \, \bar{V}^* \right) = 0.03521 \bar{e}_x + 0.2737 \bar{e}_y \text{ mW/m}^2$$

There is little doubt that the coherent case requires substantially more attention to details. Evaluation of intensity in either case is a common source of error for students who forget that it is a vector.

6.3.3 Standing Waves in a Spherical Cavity

The time-domain investigation of spherical waves in a cavity was limited to large spheres. Without that restriction, determination of the reflected inward wave would require solution of a linear differential equation. The algebraic nature of the frequency-domain relationship between pressure and particle velocity enables us to consider any size sphere, so ka is arbitrary.

We begin with the problem that initiated the time-domain investigation: a spherical cavity whose wall executes a radial vibration. The radial velocity at $r = a$ is set at $v_r = \text{Re}\left(V_0 e^{i\omega t}\right)$, with v_r taken to be positive for outward motion in order to reduce the opportunity for sign errors. Inward and outward waves are both permitted here, so the general solution of the Helmholtz equation, which was stated in Eq. (6.3.3), applies here

$$P = \frac{1}{r} \left(B_1 e^{-ikr} + B_2 e^{ikr} \right) \tag{6.3.20}$$

The velocity boundary condition at $r = a$ provides one equation for the arbitrary coefficients B_1 and B_2. The other equation is the requirement that the pressure at the center must be finite. Because r occurs in the denominator, a finite limit requires that the quantity in the parentheses vanish as $r \to 0$, which leads to

$$\lim_{r \to 0} P \text{ is finite} \implies B_1 = -B_2 \tag{6.3.21}$$

When this relation applies, the expression for P may be rewritten as

$$P = C \frac{\sin (kr)}{kr} \tag{6.3.22}$$

where the coefficient B_2 has been replaced by $C/(2ki)$. The corresponding expression for particle velocity is found from Euler's equation to be

$$V_r = -\frac{C}{i\rho_0 c} \left(\frac{kr \cos (kr) - \sin (kr)}{k^2 r^2} \right) \tag{6.3.23}$$

Both P and V_r have finite limits as $r \to 0$. Their series expansions for small r are

$$P \approx C \left[1 - \frac{1}{6} (kr)^2 + \cdots \right]$$
$$V_r \approx \frac{C}{i \rho_0 c} \left(\frac{kr}{3} + \cdots \right) \tag{6.3.24}$$

Thus, the coefficient C is the pressure at the center and the particle velocity is zero there. These properties are consistent with the conclusions of the time-domain analysis in the case where the function for the inward wave is $\dot{G} = \text{Re} \left((B_2/r) e^{i\omega(t-r/c)} \right)$.

The radial dependence of P on r is $\sin(kr)/(kr)$. In the next chapter, we will see that it is one of a class of solutions of Bessel's differential equation. This one is the spherical Bessel's function of order zero, whose notation is

$$j_0(kr) \equiv \frac{\sin(kr)}{kr} \tag{6.3.25}$$

Nothing is gained by using this notation in the current context.

To determine the coefficient C, we match V_r at $r = a$, as given by Eq. (6.3.23), to the radial velocity V_0 of the wall

$$-\frac{C}{i\rho_0 c} \left[\frac{ka \cos(ka) - \sin(ka)}{(ka)^2} \right] = V_0 \tag{6.3.26}$$

From this, we find that

$$\boxed{\begin{aligned} P &= -i \rho_0 c V_0 \left(\frac{a}{r} \right) \frac{ka \sin(kr)}{ka \cos(ka) - \sin(ka)} \\ V_r &= V_0 \left(\frac{a}{r} \right)^2 \frac{kr \cos(kr) - \sin(kr)}{ka \cos(ka) - \sin(ka)} \end{aligned}} \tag{6.3.27}$$

The ratio of P/V_0 is imaginary, whereas V_r/V_0 is real. This means that the signal is a standing wave, in which the radial profiles of pressure and particle velocity maintain a fixed shape. The pressure is $90°$ out-of-phase from the velocity. Otherwise, the time-averaged power input to the fluid at $r = a$ would be nonzero, which is not possible because the fluid is ideal and the domain is closed. If we rewrite kr as $ka\,(r/a)$, we may consider $P/(\rho_0 c V_0)$ to be a function of ka and r/a.

Figure 6.4 displays typical profiles of $\text{Im}(p)$ along a radial line. At any frequency, the pressure amplitude decreases in an oscillatory manner as r increases. The zeros, at which $\sin(ka\,(r/a)) = 0$, are radii of pressure nodal spheres. The radii at which $dP/dr = 0$ are nodal spheres on which the radial particle velocity is zero. These values occur at $\cot(ka\,(r/a)) = 1/(ka\,(r/a))$. Only for reasonably large values of ka do the velocity nodal spheres occur midway between the pressure nodes.

It will be noted that the pressure in Fig. 6.4 indicates that $|P|$ at $r = 0$ is larger for $ka = 8$ than it is for the other frequencies. A frequency plot of $|P|$ as a function of ka explains why this is so. Figure 6.5 is such a plot for two radial positions.

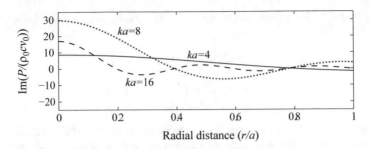

Fig. 6.4 Nondimensional pressure, Im $(P) / (\rho_0 c V_0)$, as a function of radial distance for several nondimensional frequencies

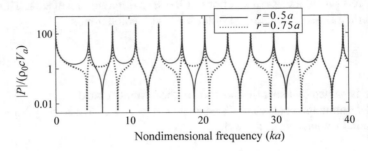

Fig. 6.5 Frequency response of a spherical cavity excited by a radially symmetric vibration V_0 at the outer wall

At frequencies that appear to be spaced regularly, the pressure at both locations rises to a peak value. In fact, a continuous frequency sweep rather than a fine sampling would show infinite values. According to Eq. (6.3.27), the pressure everywhere is singular when ka is a root of $\tan(ka) = ka$. These are resonances, and the corresponding values of ω are natural frequencies of the cavity. As will be shown shortly, they are the frequencies at which an initial field can be sustained without any movement of the wall. A motion of the wall at any of these frequencies is an excitation when no excitation is needed. As we saw with the method of images for a planar waveguide, the result is continual addition of energy to the closed field, leading to unbounded growth of the pressure. Ultimately, nonlinear and/or dissipative effects would limit this growth.

The other prominent feature of Fig. 6.5 is the very low pressures at frequencies between some resonances. Equation (6.3.27) states that $P = 0$ wherever $\sin(ka(r/a)) = 0$. Thus, when the range of ka is swept through at a specific r, the value of P will be zero whenever $ka = n\pi a/r$. At those frequencies, the selected value of r constitutes a nodal sphere. As evidenced by Fig. 6.4, more than one location at a specific frequency might be a nodal spheres.

In general, the modes of a closed domain are obtained by seeking conditions under which the general solution of the Helmholtz equation is a nontrivial function that satisfies the homogeneous version of the boundary conditions. (Homogeneous boundary conditions are those for which neither the pressure nor particle velocity are imposed by external agencies.) The identification of cavity modes is important for its own sake. In addition, a modal description provides the foundation for a time-domain analysis of closed domains. This topic will be explored in Chap. 10.

For the radially symmetric field in a spherical cavity, the conditions satisfied by a mode function are finiteness at $r = 0$, and $V_r = 0$ at $r = a$ in Eq. (6.3.26). These conditions are satisfied by the expression in Eq. (6.3.22) if either $C = 0$, which must be discarded because it leads to $P \equiv 0$, or else

$$\boxed{ka \cos (ka) - \sin (ka) = 0} \qquad (6.3.28)$$

This is the *characteristic equation* for the radially symmetric modes of a spherical cavity. The roots $ka = \kappa_n$ are the eigenvalues of the system. After these values have been determined, the natural frequencies are $\omega_n = \kappa_n c/a$, and the modal pressure functions obtained from Eq. (6.3.22) are

$$\boxed{\psi_n = \frac{\sin (\kappa_n (r/a))}{\kappa_n (r/a)}} \qquad (6.3.29)$$

When ka is very large, the first term in Eq. (6.3.28) is dominant, so the higher eigenvalues are well approximated by setting $\cos (\kappa_n) = 0$. This is the same characteristic equation as that for the modes of a planar waveguide that has one open end and one rigid end, as described by Eq. (3.2.29). This resemblance was observed in the time-domain analysis, where the method of images was used. The development suggested that rP behaves like P for a pressure-release/rigid plane waveguide. Given this observation, it is useful to compare the zeros of the characteristic equation to those obtained from the high-frequency approximation. The lowest few values appear in the following tabulation.

Mode #	κ_1	κ_2	κ_3	κ_4	κ_5
Root of Eq. (6.3.28)	0	4.493	7.725	10.904	14.066
$\cos(\kappa_n) = 0$		1.571	4.712	7.854	10.996	14.137

The third and higher eigenvalues are very well predicted by the approximation. This observation suggests that it is reasonable to employ a plane wave approximation to handle internal reflections in the time domain if the FFT of $v_0 (t)$ indicates that the most dominant part of the spectrum is in the range where $ka \geq \kappa_3$.

Figure 6.6 displays the modes corresponding to the tabulated eigenvalues. It can be seen that the they are finite at $r = 0$ with a zero slope and that the slope of each mode at $r = a$ also is zero.

At the low end of the frequency spectrum, the first eigenvalue gives a zero natural frequency. The Helmholtz equation at zero frequency reduces to $d^2 (RP) /dr^2 = 0$, whose solutions are $P = b_1 + b_2/r$. Finite P at the origin sets $b_2 = 0$ and $v_r = 0$ at $r = a$ is satisfied by any value of b_1. Thus, the zero-frequency mode is one in which

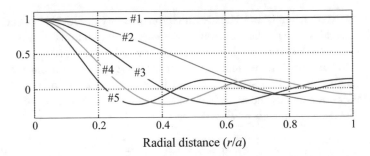

Fig. 6.6 Radially symmetric mode functions for a spherical cavity with an immobile wall

the pressure is the same at all locations. This mode is a specific manifestation of the large cavity lumped element of a planar waveguide. As was stated in Sect. 3.5.1, the pressure in a large volume at very low frequencies is approximately distributed uniformly.

EXAMPLE 6.4 Suppose the wall of a spherical cavity is a balloon that is impermeable, but lightly tensioned. (a) Determine the natural frequencies and mode functions if the pressure outside the balloon is atmospheric. Graph the mode functions as a function of r/a for the three lowest natural frequencies. (b) Consider the situation wherein the pressure outside the balloon pulsates as $\mathrm{Re}\left(P_0 e^{i\omega t}\right)$. Derive an expression for the complex pressure amplitude $P\left(r\right)$, and evaluate the frequency dependence of the pressure at $r = 0$ and $r = a/2$.

Significance

This example will enhance recognition of the close relationship between forced response and modal properties. It also will show that changing the type of boundary conditions for a cavity typically alters the details of a response, but not the general phenomenological features.

Solution

The general solution that is consistent with the finiteness condition at $r = 0$ is given by Eq. (6.3.22),

$$P = C\frac{\sin\left(kr\right)}{kr} \tag{1}$$

The field corresponding to setting $P = P_0$ at $r = a$ is

$$P = P_0\frac{a}{r}\frac{\sin\left(kr\right)}{\sin\left(ka\right)} \tag{2a}$$

The value at $r = a/2$ is readily obtained by direct evaluation of Eq. (2a), whereas P at $r = 0$ is found as a limit. For small kr, we know that $\sin\left(kr\right) \approx kr$. Correspondingly, we find that

$$P|_{r=0} = P_0 \frac{ka}{\sin(ka)} \tag{2b}$$

To find the modal properties, we seek conditions for which the pressure is not identically zero when $P_0 = 0$. Evaluating Eq. (1) at $r = a$ in this case gives

$$0 = C \frac{\sin(ka)}{ka} \tag{3}$$

If $C = 0$, then the field is zero everywhere, so it must be that $\sin(ka) = 0$. The roots of this equation are the eigenvalues κ_n, so the natural frequencies are defined by

$$\kappa_n \equiv \frac{\omega_n a}{c} = n\pi, \quad n = 1, 2, \dots \tag{4}$$

The boundary conditions are satisfied for arbitrary C at these frequencies, so we may set C as we wish. Our choice is that the modal pressure should be one at the center, which leads to the mode functions being defined as

$$\psi_n = \frac{\sin(\kappa_n r/a)}{\kappa_n r/a} \tag{5}$$

Note that there is no zero-frequency mode. This is because the associated mode function must be a uniform distribution, and $\psi_0 = 0$ is the only uniform value that is consistent with the zero pressure condition at $r = a$.

Figure 1 shows the mode functions ψ_n for $n = 1, 2,$ and 3. It can be seen that $d\psi_n/dr = 0$ at $r = 0$ as required by the finiteness conditions at the center and that $\psi_n = 0$ at $r = a$.

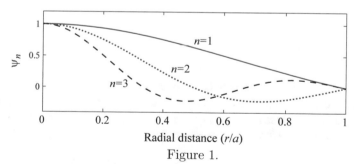

Figure 1.

Figure 2 displays the frequency response at $r = a/2$ and $r = 0$, which are obtained by evaluating Eq. (2) as functions of ka. That there are resonant frequencies is evident from the fact that there are very narrow spikes. However, the response at the center displays twice as many resonances as the response at $r = 0.5a$. This seems to violate the stated property that the pressure everywhere is infinite at a resonance. Furthermore, there is no frequency at which $|P| = 0$ at either radial position. The explanation of both features is that the pressure at $r = 0.5a$ is indicated by Eq. (2a) to be zero if

$ka/2 = m\pi$, which is satisfied if ka is an even multiple of π. However, those also are the eigenvalue for the even numbered modes. Evaluation of $\sin(n\pi) / \sin(2n\pi)$ via l'Hopitals's rule shows that the pressure at $r = 0.5a$ in this case is $(-1)^n P_0$. Thus, P at $r = 0.5a$ is finite whenever $ka = 2n\pi$, whereas P at $r = 0$ is unbounded at those frequencies.

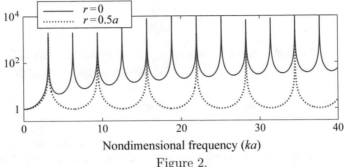

Figure 2.

6.4 Point Sources

According to Eq. (6.3.7), the pressure output by a sphere radiating to an open domain is proportional to the product of the vibration amplitude V_0 and the square of the sphere's radius a. As $a \to 0$, the output also will vanish unless $a^2 V_0$ remains finite. A *point source* is this idealization. Many techniques for analyzing and diagnosing complicated sound fields are founded on a superposition of point sources.

6.4.1 Single Source

The perspective of the time domain is where we start. As a is progressively decreased to a small value, the time derivative term in the differential equation for the wave function, Eq. (6.2.8), loses significance. In the limit $a \to 0$, that equation indicates that $F \to \rho_0 a^2 v_0(t)$. Hence, the pressure field radiated by a very small vibrating sphere is

$$p(r, t) = \frac{\dot{F}\left(t - \dfrac{r}{c}\right)}{r} = \frac{\rho_0 a^2 \dot{v}_0\left(t - \dfrac{r}{c}\right)}{r} \tag{6.4.1}$$

This is the same as the low-frequency/small radius approximation in Eq. (6.2.12), except that the extreme smallness of a allows the radial distance from the sphere to be represented as r.

If $p(r, t)$ is to remain finite as $a \to 0$, the acceleration of the surface must increase without bound. The approach by which this is done originates from the definition of the sphere's instantaneous volume. Let r_s denote its instantaneous radius, with $r_s = a$ being the reference state. At any instant, the sphere's volume is $V_s = (4/3)\pi r_s^3$. The derivative of this quantity at any instant is $dV_s/dt = 4\pi r_s^2 \dot{r}_s$, but $\dot{r}_s = v_0$. Thus, $4\pi a^2 v_0$ is the rate of change of the displaced fluid volume if the radius is taken to differ little from the reference value a. This rate of change is denoted as the *volume velocity*,

$$Q_s(t) = 4\pi a^2 v_0(t) \tag{6.4.2}$$

The fluid mass displaced by the sphere is $\rho_0 V_s$, so $\rho_0 Q_s$ is the rate at which fluid mass is displaced by vibration of the sphere. The pressure amplitude in Eq. (6.4.1) is proportional to $\rho_0 \dot{Q}_s$, which is referred to as the *mass acceleration*.

A different perspective is to consider a fixed sphere of radius a. The mass flowing across any differential patch of a stationary surface is $\rho_0 \bar{v}_s \cdot \bar{n} dS$. For a radially symmetric field $\bar{v}_s \cdot \bar{n} = v_r$, which is v_0 everywhere on the sphere. The surface area of the sphere is $4\pi a^2$, so the mass flow rate is $4\pi \rho_0 a^2 v_0 \equiv \rho_0 Q_s$ and $4\pi \rho_0 a^2 \dot{v}_0$ is the rate of change of the mass flow rate. If $v_0 > 0$, mass is flowing out of the sphere, which is equivalent to saying that it is being added to the fluid.

A point source is obtained when Q_s is used to describe the field, and the result is taken to apply for any nonzero r. The corresponding expression for particle velocity is found from Euler's equation. Thus, the time-domain point source field is

$$
\boxed{
\begin{aligned}
p(r, t) &= \frac{\rho_0 \dot{Q}_s\left(t - \dfrac{r}{c}\right)}{4\pi r}, \quad r > 0 \\
v_r(r, t) &= \frac{1}{4\pi r c}\dot{Q}_s\left(t - \frac{r}{c}\right) + \frac{1}{4\pi r^2}Q_s\left(t - \frac{r}{c}\right)
\end{aligned}
}
\tag{6.4.3}
$$

Note that injecting mass into the fluid at a steady rate does not generate sound, even though doing so moves the fluid particles.

A similar analysis may be applied to the frequency-domain description of a radially symmetric field. When ka is very small, the pressure in Eq. (6.3.7) becomes

$$P = i\frac{\rho_0 \omega a^2 V_0}{r}e^{-ikr} \tag{6.4.4}$$

Because the surface velocity is harmonic, we can write the volume velocity as $dV_s/dt = \text{Re}\left(\hat{Q}e^{i\omega t}\right)$, where

$$\hat{Q} = 4\pi a^2 V_0 \tag{6.4.5}$$

Correspondingly, the field of a point source in the frequency domain is described by

$$P = \frac{i\omega\rho_0\hat{Q}}{4\pi r}e^{-ikr}$$

$$V_r = \frac{\hat{Q}}{4\pi r}\left(ik + \frac{1}{r}\right)e^{-ikr}$$

(6.4.6)

This is fully equivalent to the complex amplitude of time-domain representation in the case of a harmonic volume velocity, for which $Q_s\,(t - r/c) = \mathrm{Re}\left(\hat{Q}e^{i\omega(t-r/c)}\right)$.

6.4.2   Green's Function

A question whose answer has far-reaching ramifications is, what equation is satisfied by the field radiated by a point source? A useful tool for this analysis is the Dirac delta function, which was introduced in Sect. 1.4.1 to describe an impulsive function of time. We begin by generalizing the definition of a Dirac delta in order to use it to describe spatial distributions.

Dirac Delta Functions

A Dirac delta function is defined to be zero everywhere except one point where it is singular. To compensate for this gap in its definition, the function is further defined in terms of a property of its integral. A delta function may be considered to be the limit of a tall narrow function as its width goes to zero. Two such functions are depicted in Fig. 6.7. Each function is defined such that its integral over any x interval that includes $x_0 < x < x_0 + \varepsilon$ is one. As $\varepsilon \to 0$, the maximum value of each therefore must grow without bound.

Fig. 6.7 Two continuous function that approach a Dirac delta function as their width $\varepsilon \to 0$

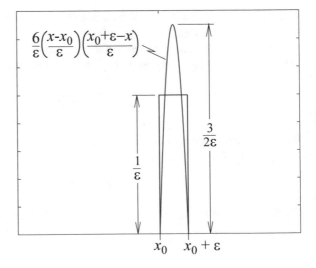

Ultimately, if we actually bring ε to zero, both functions become spikes of infinite height situated at x_0. The details of the shape between x_0 and $x_0 + \varepsilon$ become irrelevant, so both become the delta function $\delta(x - x_0)$. This function is not defined at $x = x_0$, but it is zero everywhere else. The usage of $x - x_0$ as the argument reminds us that the function is nonzero only at the distance x for which its argument is zero.

These two properties are generalized to define a Dirac delta function in any domain $\mathcal{D}$. This domain may be two- or three-dimensional, so we let $\bar{x}$ denote an arbitrary field point, and $\bar{x}_0$ denotes the point at which the function is singular. The position of the field point relative to the singularity is $\bar{x} - \bar{x}_0$, so a delta function whose singularity is situated at $\bar{x}_0$ is denoted as $\delta(\bar{x} - \bar{x}_0)$. The basic properties

$$\boxed{\begin{aligned} &\delta(\bar{x} - \bar{x}_0) = 0 \text{ if } \bar{x} \neq \bar{x}_0 \\ &\iiint_{\mathcal{D}} \delta(\bar{x} - \bar{x}_0)\, d\mathcal{D} = 1 \text{ if } \bar{x}_0 \in \mathcal{D} \end{aligned}} \qquad (6.4.7)$$

An important property follows as a corollary of the preceding definition. Suppose we wish to integrate the product of an analytical function $f(\bar{x})$ and a delta function, with the singularity being situated somewhere in $\mathcal{D}$. To understand this, let $\chi(\bar{x} - \bar{x}_0, \varepsilon)$ denote a continuous function that is zero for points that are not close to $\bar{x}_0$, $|\bar{x} - \bar{x}_0| > \varepsilon$, but whose integral over $\mathcal{D}$ is unity. Thus, we can say that $\chi(\bar{x} - \bar{x}_0, \varepsilon) \to \delta(\bar{x} - \bar{x}_0)$ as $\varepsilon \to 0$. Rather than evaluating the integral of $f(\bar{x})\delta(\bar{x} - \bar{x}_0)$, we will evaluate the integral of $f(\bar{x})\chi(\bar{x} - \bar{x}_0, \varepsilon)$ and then take evaluate the limit. In other words,

$$\iiint_{\mathcal{D}} f(x)\,\delta(x - \bar{x}_0)\, d\mathcal{D} = \lim_{\varepsilon \to 0} \iiint_{|\bar{x} - \bar{x}_0| < \varepsilon} f(\bar{x})\,\chi(\bar{x} - \bar{x}_0, \varepsilon)\, d\mathcal{D} \qquad (6.4.8)$$

Now, consider decreasing ε to zero. Because $f(\bar{x})$ is analytical, the central limit theorem tells us that $f(\bar{x}) \to f(\bar{x}_0)$ as $\bar{x} \to \bar{x}_0$. Thus, if ε is sufficiently small, we may factor the analytical function out of the integral

$$\iiint_{\mathcal{D}} f(\bar{x})\,\chi(\bar{x} - \bar{x}_0, \varepsilon)\, d\mathcal{D} \approx f(\bar{x}_0) \iiint_{|\bar{x} - \bar{x}_0| < \varepsilon} \chi(\bar{x} - \bar{x}_0, \varepsilon)\, d\mathcal{D} \qquad (6.4.9)$$

This approximation is increasingly accurate as ε decreases. Furthermore, the definition of $\delta(\bar{x} - \bar{x}_0, \varepsilon)$ is that it has a unit integral. Thus, the limit gives

$$\boxed{\iiint_{\mathcal{D}} f(\bar{x})\,\chi(\bar{x} - \bar{x}_0)\, d\mathcal{D} = f(\bar{x}_0)} \qquad (6.4.10)$$

One way of viewing this property is that integrating the product of an analytical function $f(\bar{x})$ and a Dirac delta function is a process that filters out all values of $f(\bar{x})$ other than its value at the singularity.

A subtlety is embedded in these properties. The definition of the differential element of a domain depends not only on the spatial dimensionality, but also on the coordinate system that is in use. To explore this issue, first consider the one-dimensional delta function $\delta(x - x_0)$ obtained as the limit of the functions in Fig. 6.7 as $\varepsilon \to 0$. For this domain, $d\mathcal{D} = dx$, so that

$$\boxed{\int_a^b \delta(x - x_0)\, dx = 1 \text{ if } a \leq x_0 \leq b} \qquad (6.4.11)$$

The multidimensional Dirac delta function has been written generically as $\delta(\bar{x} - \bar{x}_0)$, but how is it defined if we describe $\bar{x}_0$ in terms of a specific coordinate system? If we are working in rectangular Cartesian coordinates, then x_0, y_0, z_0 would be the coordinates of the singularity point $\bar{x}_0$ and the domain element would be $d\mathcal{D} = dx\, dy\, dz$. The Dirac delta function for that geometry would be

$$\delta(\bar{x} - \bar{x}_0) = \delta(x - x_0)\, \delta(y - y_0)\, \delta(z - z_0) \qquad (6.4.12)$$

By virtue of Eq. (6.4.11), integration over x, y, and z would each give a unit value if each integration range includes the respective coordinate of $\bar{x}_0$.

Now, suppose that we are interested in a radially symmetric situation with the singularity at the origin, $\bar{x}_0 = \bar{0}$. Because of radial symmetry, we may use a spherical shell volume element $d\mathcal{D} = 4\pi r^2\, dr$. Simply setting $\delta(\bar{x}) = \delta(r)$ will not give the desired integral property in this geometry because it would lead to $\delta(\bar{x} - \bar{x}_0)\, d\mathcal{D}$ being represented as $\delta(r)\left(4\pi r^2\, dr\right)$. Hence, according to Eq. (6.4.10), integration over any domain including the origin will give zero. A factor of $1/\left(4\pi r^2\right)$ must be inserted into the definition of the Dirac delta function, in order to cancel the $4\pi r^2$ factor in $d\mathcal{D}$. Thus,

$$\delta(\bar{x}) = \frac{\delta(r)}{4\pi r^2} \qquad (6.4.13)$$

Now, consider a planar situation in which the singularity is at the origin, $\bar{x}_0 = \bar{0}$, and polar coordinates (R, θ) is to be used. Then, the differential area element for a radially symmetric field would be $d\mathcal{D} = 2\pi R\, dR$. The factor $2\pi R$ must be canceled by the delta function, so that

$$\delta(\bar{x}) = \frac{\delta(R)}{2\pi R} \qquad (6.4.14)$$

Field Equations for a Point Source

To see why a Dirac delta function is relevant to our study of point sources, consider substituting the frequency-domain solution for a point source, Eq. (6.4.6), into the Helmholtz equation. At any finite radial distance r from the source, evaluation of

$\nabla^2 P + k^2 P$ will give zero. However, $r = 0$ is a singularity. Let us designate the source's location as the origin. Then, we may say that P satisfies an inhomogeneous version of the Helmholtz equation

$$\nabla^2 P + k^2 P = \hat{C}\delta(\bar{x}) \tag{6.4.15}$$

But what is the constant $\hat{C}$? To answer this, we invoke the properties of a Dirac delta function by integrating this inhomogeneous Helmholtz equation over a spherical domain having finite radius r_0 centered on the origin. Application of Eq. (6.4.7) to the right side gives

$$\iiint_{\mathcal{D}} \left(\nabla^2 P + k^2 P\right) d\mathcal{D} = \hat{C} \tag{6.4.16}$$

The first term on the left side may be evaluated with the aid of Green's theorem

$$\iiint_{\mathcal{D}} \nabla^2 P \, d\mathcal{D} = \iint_{S} \bar{n}(x_s) \cdot \nabla P \, dS = \left. \frac{dP}{dr} \right|_{r=r_0} \left(4\pi r_0^2\right)$$

$$= \left[\frac{i\omega\rho_0 \hat{Q}}{4\pi} \left(-\frac{ik}{r_0} - \frac{1}{r_0^2} \right) e^{-ikr_0} \right] \left(4\pi r_0^2\right) \tag{6.4.17}$$

The value of r_0 must be greater than zero, but it may be arbitrarily small. Progressively reducing r_0 causes the second term within the parentheses to become dominant, so that

$$\iiint_{\mathcal{D}} \nabla^2 P \, d\mathcal{D} \rightarrow -i\omega\rho_0 \hat{Q} \tag{6.4.18}$$

The other term on the left side is

$$\iiint_{\mathcal{D}} k^2 P \, d\mathcal{D} = \frac{i\omega\rho_0 \hat{Q}}{4\pi} \int_0^{r_0} k^2 \left(\frac{1}{r}\right) e^{-ikr} \left(4\pi r^2 dr\right) \tag{6.4.19}$$

Because kr_0 is restricted to being very small, the complex exponential is essentially one over the range of integration. This approximation allows the integral to be evaluated, with the result that

$$\iiint_{\mathcal{D}} k^2 P \, d\mathcal{D} = \frac{1}{2} i\omega\rho_0 \hat{Q} k^2 r_0^2 \tag{6.4.20}$$

Decreasing kr_0 causes this integral to become negligibly small.

When the results of analyzing each term are substituted back into Eq. (6.4.16), we find that

$$\hat{C} = -i\omega\rho_0 \hat{Q} \tag{6.4.21}$$

Thus, we have established that the field of a point source is characterized by

$$
\boxed{
\begin{aligned}
\nabla^2 P + k^2 P &= -i\omega\rho_0 \hat{Q}\delta(\bar{x}) \\
P &= \frac{i\omega\rho_0\hat{Q}}{4\pi r}e^{-ikr}
\end{aligned}
}
\tag{6.4.22}
$$

Recall that $\rho_0\hat{Q}$ is the complex amplitude of the rate at which mass flows out of the source, so $\hat{C}$ is the negative of the mass acceleration.

In the time domain, the point source field satisfies an inhomogeneous version of the wave equation. The steps leading to this equation are like those followed thus far. The pressure in Eq. (6.4.3) satisfies the wave equation everywhere except the origin. At that location, the wave equation will be singular with a coefficient C that might be time-dependent, so we begin with

$$
\nabla^2 p - \frac{1}{c^2}\frac{\partial^2 p}{\partial t^2} = C(t)\,\delta(\bar{x})
\tag{6.4.23}
$$

As we did for the frequency-domain analysis, we surround the origin with a sphere of radius r_0 and integrate this inhomogeneous wave equation over the sphere. Application of Green's theorem to the Laplacian term gives

$$
\iiint_{\mathcal{D}} \nabla^2 p\, d\mathcal{D} = \iint_{\mathcal{S}} \bar{n}(x_s)\cdot\nabla p\, d\mathcal{S} = \left.\frac{dp}{dr}\right|_{r=r_0}(4\pi r_0^2)
\tag{6.4.24}
$$

Substitution of the time-domain description of the field radiated by a point source solution, Eq. (6.4.3), gives

$$
\iiint_{\mathcal{D}} \nabla^2 p\, d\mathcal{D} = \left[\frac{\rho_0\ddot{Q}_s\left(t-\dfrac{r_0}{c}\right)}{4\pi r_0}\left(-\frac{1}{c}\right) - \frac{\rho_0\dot{Q}_s\left(t-\dfrac{r_0}{c}\right)}{4\pi r_0^2}\right](4\pi r_0^2)
\tag{6.4.25}
$$

At this juncture, we introduce the fact that r_0 is an arbitrarily small nonzero value, so that

$$
\iiint_{\mathcal{D}} \nabla^2 p\, d\mathcal{D} \rightarrow -\rho_0\dot{Q}_s(t)
\tag{6.4.26}
$$

When p is the point source field, evaluation of the second integral gives

$$
\iiint_{\mathcal{D}} \frac{\partial^2 p}{\partial t^2}\, d\mathcal{D} = -\frac{\rho_0}{4\pi}\int_0^{r_0} \left(\frac{1}{r}\right)\ddot{Q}\left(t-\frac{r}{c}\right)(4\pi r^2 dr)
\tag{6.4.27}
$$

The r^2 factor in the differential element cancels the $1/r$ spreading factor. The largest value of r in the integrand is r_0, which is very small, so this integral is negligible. Thus, the result of integrating Eq. (6.4.23) over the small sphere is that

$$C = -\rho_0 \dot{Q}(t) \tag{6.4.28}$$

Once again, the coefficient is the negative of the mass acceleration.

Free-space Green's Function

A Green's function is the pressure field that is obtained when a point source having a certain mass acceleration is placed at location $\bar{x}_0$, not necessarily the origin. For a specified spatial region, the frequency-domain field of a point source depends only on the values of $\bar{x}_0$ and $\bar{x}$, so we shall denote it as $G(\bar{x}, \bar{x}_0)$. (The fact that this function also depends on ω is evident, so it is not displayed in the notation.) A problematic aspect is that there is no universal agreement regarding the volume velocity to use in this definition. Some individuals select it such that $i\omega\rho_0 \hat{Q} = 4\pi$, and some insert a minus sign to this definition. The corresponding pressure field for a source at the origin is $\pm e^{-ikr}/r$, which is simple to write. The difficulty is that we will develop general theorems that employ a Green's function. If we use this simple definition of the Green's function, the consequence will be that it is necessary to introduce a factor to adjust the theorem according to the spatial dimensionality of the system.

This adjustment is not required if we **define a Green's function to be the point source solution for a unit mass acceleration,** regardless of the spatial dimensionality. Then, the equation it satisfies always is

$$\nabla^2 G + k^2 G = -\delta(\bar{x} - \bar{x}_0) \tag{6.4.29}$$

Our interest here is with a fluid medium that has no outer boundary. Correspondingly, we say that the function that satisfies this field equation is the *free-space Green's function* in the frequency domain.

In the case of an unbounded three-dimensional domain, setting $\omega\rho_0\hat{Q} = 1$ in Eq. (6.4.6) gives the Green's function for a source at the origin. To generalize this definition to situations where the point source is not at the origin, it is sufficient to note that r represents the distance between the source and field point. If the source is located arbitrarily, this position variables may be found by taking the magnitude of the difference between the positions, that is

$$r \equiv |\bar{x} - \bar{x}_0|, \quad \bar{e}_r = \frac{\bar{x} - \bar{x}_0}{r} \tag{6.4.30}$$

The resulting free-space Green's function is

$$G(\bar{x}, \bar{x}_0) = \frac{e^{-ik|\bar{x}-\bar{x}_0|}}{4\pi |\bar{x} - \bar{x}_0|} \tag{6.4.31}$$

The time-domain Green's function for any region is the pressure that is generated at an arbitrary field point $\bar{x}$ at time t when a point source is situated $\bar{x}_0$. The mass acceleration is defined to be a unit impulse that occurs at time t_0, so that $\rho_0 \dot{Q}_s = \delta(t - t_0)$. According to Eq. (6.4.28), the inhomogeneous term in the wave equation is defined by $C(t) = -\delta(t - t_0)$. Thus, the time-domain Green's function $g(\bar{x}, t, \bar{x}_0, t_0)$ satisfies

$$\nabla^2 p - \frac{1}{c^2}\frac{\partial^2 p}{\partial t^2} = -\delta(\bar{x} - \bar{x}_0)\,\delta(t - t_0) \tag{6.4.32}$$

The corresponding Green's function is obtained by setting $\rho_0 \dot{Q}_s = \delta(t - t_0)$ in Eq. (6.4.3) and also replacing the radial distance r with the source-field point distance. The result is that

$$g(\bar{x}, t, \bar{x}_0, t_0) = \frac{\delta(t - t_0 - |\bar{x} - \bar{x}_0|/c)}{4\pi |\bar{x} - \bar{x}_0|} \tag{6.4.33}$$

An approximation of this signal may be obtained by setting off a strong electrical spark between two closely spaced electrodes. The location of these electrodes marks $\bar{x}_0$ and the instant when the spark is initiated is t_0. According to this relation, the initiation of the spark at the origin at t_0 will be observed at a field point at time $t_0 + |\bar{x} - \bar{x}_0|/c$ as a very short burst whose strength is reduced by spherical spreading.

EXAMPLE 6.5 The volume velocity per unit length of a line source changes instantaneously at all locations. Until $t = 0$, it is zero, at which instant it jumps to a constant value q_0. This corresponds to a mass acceleration per unit length that everywhere is $\rho_0 q_0 \delta(t)$. The length of the line source is $2L$. Determine the pressure signal received at a field point on the plane that bisects the line source.

Significance

This example will enhance our understanding of the Dirac delta function. The result will disclose some fundamental differences in the nature of two-dimensional models relative to three-dimensional systems.

Solution

A sketch is helpful for defining the position parameters. The z-axis is defined to coincide with the line source, and the origin of xyz is placed at the midpoint of the source. Every plane that contains the line source presents the same picture. Accordingly, we may place the field point on the x-axis. This is the arrangement described in Fig. 1, where a segment of the line source of length dz_0 is situated at distance z_0 from the origin. This segment is a point source whose mass acceleration is the infinitesimal amount $[\rho_0 q_0 \delta(t)]\, dz_0$. The distance from this source to the field point is $|\bar{x} - \bar{x}_0| = [(z_0)^2 + x^2]^{1/2}$.

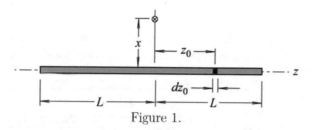

Figure 1.

The expression in Eq. (6.4.33) is the pressure from a mass acceleration that is a unit Dirac delta. Hence, the contribution of this infinitesimal source is obtained by multiplying that expression by $\rho_0 q_0 dz_0$. The total pressure received at the field point is the superposition of the contribution of each infinitesimal source, which leads to an integration over all z_0. The instant when the impulse is initiated is $t_0 = 0$, so we have

$$p = \int_{-L}^{L} \frac{(\rho_0 q_0 dz_0)}{4\pi} \frac{\delta\left(t - \left((z_0)^2 + x^2\right)^{1/2}/c\right)}{\left((z_0)^2 + x^2\right)^{1/2}} \tag{1}$$

This expression may be simplified slightly by observing that the integrand is an even function of z_0, so the integration range may be reduced to $0 < z_0 < L$, with the result doubled. The more difficult aspect stems from the fact that the Dirac delta function in Eqs. (6.4.7) and (6.4.10) features an argument in which the integration variable occurs linearly. For this reason, we introduce into Eq. (1) a change of variables $u = \left((z_0)^2 + x^2\right)^{1/2}/c$, whose inverse is $z_0 = \left(c^2 u^2 - x^2\right)^{1/2}$. The field point is fixed, so $dz_0 = c^2 u\, du / \left(c^2 u^2 - x^2\right)^{1/2}$. The result is that Eq. (1) is transformed to

$$p = \frac{\rho_0 q_0}{2\pi} \int_{x/c}^{(L^2+x^2)^{1/2}/c} \left(\frac{\delta(t-u)}{cu}\right)\left(\frac{c^2 u\, du}{\left(c^2 u^2 - x^2\right)^{1/2}}\right)$$

The integral may be found by invoking Eq. (6.4.10). It gives a nonzero value if the delta function occurs in the range of integration, which means here that $x/c < t < \left(L^2 + x^2\right)^{1/2}/c$. If t is in this interval, then the integral is the factor of $\delta(t - u)$ evaluated at $u = t$. Outside the interval the result is zeroed by a pair of step functions. The result is

$$p = \frac{\rho_0 c q_0}{2\pi} \frac{1}{\left[(ct)^2 - x^2\right]^{1/2}} \left[h(ct - x) - h\left(ct - \left(L^2 + x^2\right)^{1/2}\right)\right] \tag{2}$$

To plot this expression, we introduce a nondimensional time $t' = ct/L$, which leads to

$$\frac{pL}{\rho_0 c q_0} = \frac{1}{2\pi} \frac{1}{\left[(t')^2 - \left(\frac{x}{L}\right)^2 \right]^{1/2}} \left[h\left(t' - \frac{x}{L}\right) - h\left(t' - \left(1 + \frac{x^2}{L^2}\right)^{1/2}\right) \right]$$

The only parameter affecting the dependence of the nondimensional pressure $pL/(\rho_0 c q_0)$ on the nondimensional time ct/L is the distance ratio x/L. This function is plotted in Fig. 2.

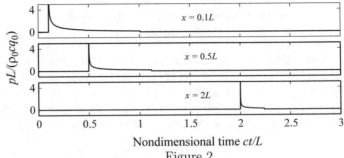

Nondimensional time ct/L

Figure 2.

The minimum distance from the field point to the line source is x, and the maximum distance is $\left(x^2 + L^2\right)^{1/2}$. Division of these distances by c gives the time when the signal begins and ends, respectively. According to Eq. (2), the effect of increasing L with x fixed is to increase the duration of the signal. The concept of a two-dimensional line source is that L is infinite. Increasing x with L fixed has the opposite effect, in that the duration decreases. If $L \ll x$, the signal has the appearance of the delta function for a point source.

To understand the time signature, let τ be the elapsed time following arrival of the signal at a field point, $\tau = t - x/c$. In the early portion of the signal, $\tau \ll x/c$, so writing $c^2 t^2 - x^2$ as $c\tau(c\tau + 2x)$ leads to an approximation of Eq. (2) as

$$p \approx \frac{\rho_0 c q_0}{2\pi} \frac{1}{(2c\tau x)^{1/2}}, \quad 0 \le ct - x \ll x$$

This shows that the early signal decays as the reciprocal of the square root of time. Equally important is the proportional dependence on $1/x^{1/2}$ at any instant. This is a characteristic of *cylindrical spreading*, which is a term that follows from the fact that the integral of p^2 over a cylinder of radius x at fixed τ will be constant.

Another important limit of Eq. (2) is the case where the line source is extremely long. In the limit of infinite L, the value of ct is always less than $\left(x^2 + L^2\right)^{1/2}$, so Eq. (2) reduces to

$$\lim_{L \to \infty} p = \frac{\rho_0 c q_0}{2\pi} \frac{h(ct - x)}{\left[(ct)^2 - x^2\right]^{1/2}} \tag{6.4.34}$$

This is the pressure field for an *impulsive line source*. Compare this expression to Eq. (6.4.3) in the case of an *impulsive point source*, for which $\dot{Q}_s = Q_0 \delta(t)$. The pressure in this case is

$$p(r, t) = \frac{\rho_0 Q_0 \delta\left(t - \dfrac{r}{c}\right)}{4\pi r} \tag{6.4.35}$$

The fields of the two types of impulsive sources are fundamentally different. The nonzero pressure for a point source occurs solely on a sphere whose radius is ct. It is an impulse function whose strength decreases reciprocally with increasing distance, or equivalently elapsed time. The line source is singular when ct equals the radial distance from the field point to the line, but it persists after the singularity passes. The pressure approaches zero asymptotically at very large values of ct/x. Some individuals refer to this persistent part as the "tail."

Green's Functions for Finite Domains

More than one Green's function may be defined for a specific system. A free-space Green's function is the one that is most commonly employed in a variety of applications because it may be identified without being concerned with boundary conditions. This advantage also might be found to be a disadvantage if the system has finite extent, because the free-space Green's function does not satisfy the boundary conditions of that system. One way of constructing a Green's function for such a system is to use superposition by supplementing the free-space function with a term to be determined. The ansatz in the frequency domain is

$$G(\bar{x}, \bar{x}_0) = \frac{e^{-ikr}}{4\pi r} + F(\bar{x}, \bar{x}_0) \tag{6.4.36}$$

Because the free-space Green's function satisfies the inhomogeneous Helmholtz equation, the function $F(x, x_0)$ must be a solution of the homogeneous equation,

$$\nabla^2 F + k^2 F = 0 \tag{6.4.37}$$

The aspect that inhibits wide use of this concept is the necessity to satisfy boundary conditions. When the superposition form is substituted, the free-space term will have some value on the boundary that F must offset. For example, if a boundary is pressure-release, then it will be necessary that

$$F(\bar{x}_s, \bar{x}_0) = -\frac{e^{-ikr}}{4\pi r}, \quad r = |\bar{x}_s - \bar{x}_0|, \quad \bar{x}_s \in \mathcal{S} \tag{6.4.38}$$

In general, the superposition transfers the inhomogeneity from the Helmholtz equation to the boundary condition. It might not be possible to solve for $F(\bar{x}, \bar{x}_0)$, and even if one could, the effort is likely to be formidable. However, in certain circumstances, it will be possible to construct the function with a minimal effort.

This is the case for systems for which the method of images, developed later in this chapter, is suitable.

Reciprocity

In the development, $G\left(\bar{x},\bar{x}_0\right)$ has been defined to be the pressure at field point $\bar{x}$ radiated by a source having unit mass acceleration at $\bar{x}_0$. *Reciprocity* is the property that the source and field points may be swapped in this definition, in other words

$$\boxed{G\left(\bar{x}_0,\bar{x}\right)=G\left(\bar{x},\bar{x}_0\right)} \qquad (6.4.39)$$

That the free-space Green's function is reciprocal is evident from the fact that it only depends on the distance between the source and field points. However, the reciprocity principle also is satisfied by Green's functions that satisfy boundary conditions. The proof of this property serves as a precursor for the derivation of the Kirchhoff–Helmholtz integral theorem in the next chapter, so pursuing it is doubly useful.

In the scenario under consideration, the Green's function satisfies the boundary condition on all portions of the boundary. These are taken to be those for a locally reacting surface having impedance Z_s, which includes the limiting cases of rigid and pressure-release surfaces. The surface normal is defined to point into the fluid, and the particle velocity associated with the Green's function is $-\nabla G\left(\bar{x},\bar{x}_0\right)/(i\omega\rho_0)$. Thus, the Green's function we are considering here is one that satisfies

$$\nabla^2 G\left(\bar{x},\bar{x}_0\right)+k^2 G\left(\bar{x},\bar{x}_0\right)=-\delta\left(\bar{x}-\bar{x}_0\right)$$
$$G\left(\bar{x},\bar{x}_0\right)=\frac{Z_s}{i\omega\rho_0}\bar{n}\left(\bar{x}\right)\cdot\nabla G\left(\bar{x},\bar{x}_0\right),\quad \bar{x}\in\mathcal{S} \qquad (6.4.40)$$

A subtlety of these equations is that $\bar{x}_0$ is fixed; all gradients are evaluated at the field point $\bar{x}$.

To prove that a Green's function is reciprocal, we also consider the function for another source point $\bar{y}_0$, which satisfies Eq. (6.4.40) with $\bar{y}_0$ instead of $\bar{x}_0$. The first step is to multiply the field equation for $G\left(\bar{x},\bar{x}_0\right)$ by $G\left(\bar{x},\bar{y}_0\right)$ and also to multiply the field equation for $G\left(\bar{x},\bar{y}_0\right)$ by $G\left(\bar{x},\bar{x}_0\right)$. We take the difference of these equations, which cancels the terms whose coefficient is k^2. The remaining terms are

$$G\left(\bar{x},\bar{y}_0\right)\nabla^2 G\left(\bar{x},\bar{x}_0\right)-G\left(\bar{x},\bar{x}_0\right)\nabla^2 G\left(\bar{x},\bar{y}_0\right)$$
$$=-G\left(\bar{x},\bar{y}_0\right)\delta\left(\bar{x}-\bar{x}_0\right)+G\left(\bar{x},\bar{x}_0\right)\delta\left(\bar{x}-\bar{y}_0\right) \qquad (6.4.41)$$

The next step is to integrate this equation over the entire acoustic domain. The filtering property in Eq. (6.4.10) is used to evaluate the integral on the right side, which leads to

$$\iiint\limits_{\mathcal{D}} [-G(\bar{x},\bar{y}_0)\delta(\bar{x}-\bar{x}_0)+G(\bar{x},\bar{x}_0)\delta(\bar{x}-\bar{y}_0)]d\mathcal{D}=-G(\bar{x}_0,\bar{y}_0)+G(\bar{y}_0,\bar{x}_0)$$
$$(6.4.42)$$

Application of Green's integral theorem to the left side of Eq. (6.4.41) gives

$$\iiint\limits_{\mathcal{D}} \left[G\left(\bar{x}, \bar{y}_0\right) \nabla^2 G\left(\bar{x}, \bar{x}_0\right) - G\left(\bar{x}, \bar{x}_0\right) \nabla^2 G\left(\bar{x}, \bar{y}_0\right) \right] d\mathcal{D}$$

$$= \iiint\limits_{\mathcal{D}} \nabla \cdot \left[G\left(\bar{x}, \bar{y}_0\right) \nabla G\left(\bar{x}, \bar{x}_0\right) - G\left(\bar{x}, \bar{x}_0\right) \nabla G\left(\bar{x}, \bar{y}_0\right) \right] d\mathcal{D}$$

$$- \iiint\limits_{\mathcal{D}} \left[\nabla G\left(\bar{x}, \bar{y}_0\right) \cdot \nabla G\left(\bar{x}, \bar{x}_0\right) - \nabla G\left(\bar{x}, \bar{x}_0\right) \cdot \nabla G\left(\bar{x}, \bar{y}_0\right) \right] d\mathcal{D}$$

$$= - \iint\limits_{\mathcal{S}} \left[G\left(\bar{x}, \bar{y}_0\right) n\left(\bar{x}\right) \cdot \nabla G\left(\bar{x}, \bar{x}_0\right) - G\left(\bar{x}, \bar{x}_0\right) n\left(\bar{x}\right) \cdot \nabla G\left(\bar{x}, \bar{y}_0\right) \right] d\mathcal{S}$$

$$(6.4.43)$$

where the minus sign in the preceding the integral stems from defining $\bar{n}$ to point into $\mathcal{D}$, that is into the fluid.

If the surface is rigid, then $n\left(\bar{x}_s\right) \cdot \nabla G\left(\bar{x}_s, \bar{x}_0\right) = 0$ for any point $\bar{x}_s$ on the surface. A pressure-release surface condition, $Z_s = 0$, requires that $G\left(\bar{x}_s, \bar{x}_0\right) = 0$. If Z_s is nonzero, the boundary condition in Eq. (6.4.40) is used to eliminate the gradient of the Green's function. Doing so leads to recognition that the two terms in the surface integral are equal. The result is that the left side of Eq. (6.4.42) is zero if the surface is locally reacting. Thus, the right side of Eq. (6.4.42) must vanish, which leads to

$$G\left(\bar{y}_0, \bar{x}_0\right) = G\left(\bar{x}_0, \bar{y}_0\right) \qquad (6.4.44)$$

The location of $\bar{y}_0$ is arbitrary. Replacing it with the symbol $\bar{x}$ leads to Eq. (6.4.39).[2]

EXAMPLE 6.6 Consider plane waves in a waveguide whose cross-sectional area is $\mathcal{A}$. Determine the frequency-domain free-space Green's functions for this system. Prove that the result is consistent with the reciprocity principle.

Significance

This example affords the opportunity to interpret the general concepts of Green's function from the familiar perspective of plane waves. Doing so will clarify the meaning of a Green's function and its physical implications.

Solution

The free-space Green's function is the solution for an unbounded domain, which here is $-\infty < x < \infty$. The source location is x_0, so the inhomogeneous Helmholtz equation for the Green's function is

---

[2]A proof of the reciprocity principle in the case where the surface is the boundary of an elastic structure was performed by M. Lyamshev, "A question in connection with the principle of reciprocity in acoustics," Sov. Phys. Dokl., Vol. 5, (1959) 405–409.

$$\frac{d^2 G}{dx^2} + k^2 G = -\delta \left(x - x_0 \right)$$

The analysis begins by integrating the inhomogeneous Helmholtz equation over some interval that contains x_0, so that $a < x, x_0 < b$. A differential element of the domain is $\mathcal{A} dx$, but the area factor appears in all terms. Thus,

$$\int_a^b \frac{\partial^2 G}{\partial x^2} dx + k^2 \int_a^b G \, dx = -\int_a^b \delta \left(x - x_0 \right) dx = -1$$

The first integral may be evaluated directly. (Writing $\partial^2 G / \partial^2 x^2$ as $\nabla \cdot \nabla G$ shows that this integration actually is the one-dimensional version of Green's theorem.) Doing so in combination with the fundamental definition of a Dirac delta function gives

$$\left. \frac{\partial G}{\partial x} \right|_{x=b} - \left. \frac{\partial G}{\partial x} \right|_{x=a} + k^2 \int_a^b G \, dx = -1$$

The next step shrinks the integration interval to zero by letting $a = x_0^-$ and $b = x_0^+$, which denotes that they are infinitesimally smaller and larger than x_0. A crucial property now comes into play. We know that in any region where there is no source, the pressure can only consist of plane waves. If the amplitude of the pressure anywhere is infinite, then the pressure will be infinite everywhere. Consequently, G must be finite everywhere, including x_0. The integrand in the preceding relation therefore is finite, but the integral extends over an infinitesimal interval. It follows that the integral is zero. What remains states that the gradient of G must change discontinuously. This condition must be supplemented by an explicit statement that G is continuous at x_0. Thus, we have

$$G|_{x=x_0^+} = G|_{x=x_0^-}$$
$$\left. \frac{\partial G}{\partial x} \right|_{x=x_0^+} = \left. \frac{\partial G}{\partial x} \right|_{x=x_0^-} - 1$$

These relations are said to be *jump conditions*

In addition to the jump conditions, the Green's function must satisfy the radiation condition for an infinite domain, which requires that signals can only propagate away from a source toward the distant boundary. The source associated with the Dirac delta function essentially splits the waveguide into two. In the portion where $x_0 < x < \infty$, waves may only propagate in the positive x direction, whereas in $-\infty < x < x_0$ only waves that propagate in the negative x direction are admissible. Thus, the Green's function must be such that

$$G \left(x, x_0 \right) = \begin{cases} B_1 e^{-ik(x-x_0)}; & x > x_0 \\ B_2 e^{ik(x-x_0)}; & x < x_0 \end{cases}$$

The coefficients are determined by satisfying the jump condition, which gives

$$B_1 = B_2 \text{ and } B_2 = -B_1 - \frac{i}{k}$$

The result is that

$$G(x, x_0) = \begin{cases} -\dfrac{i}{2k} e^{-ik(x-x_0)}; & x > x_0 \\ -\dfrac{i}{2k} e^{ik(x-x_0)}; & x < x_0 \end{cases}$$

The result is plotted in Fig. 1. It will be noted that $\text{Re}(G(x, x_0))$ has a slope discontinuity at $x = x_0$, whereas $\text{Im}(G(x, x_0))$ is smooth. It also can be seen that $G(x, x_0)$ is an even function relative to x_0.

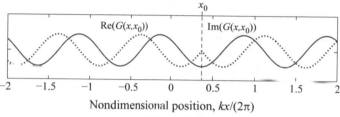

Figure 1.

The most direct way to prove that the solution for $G(x, x_0)$ is consistent with the reciprocity principle is to observe that if $x > x_0$, then $|x - x_0| = x - x_0$, whereas if $x < x_0$, then $|x - x_0| = -(x - x_0)$. Thus, the solution may be written as a single form

$$G(x, x_0) = -\frac{i}{2k} e^{-ik|x-x_0|}$$

This shows that the one-dimensional free-space Green's function is like the three-dimensional version, in that both only depend on the distance between the field point and source point.

6.4.3 Point Source Arrays

The principle of superposition tells us that the field radiated by a collection of sources at a variety of locations may be evaluated as a simple addition of the individual contributions. Using trigonometry to determine the distance from each source to a field point would be cumbersome, so we will develop an alternative approach that exploits vector algebra. An important development will be identification of a simple representation of the farfield of a set of sources.

Evaluation of the Field

We consider a collection of N sources at frequency ω whose individual volume velocity amplitudes are $\hat{Q}_n$. The location of the nth source is given by the position vector $\bar{x}_n$. The mass acceleration of each source gives rise to an inhomogeneous term $-i\omega\rho_0\hat{Q}_n\delta\left(\bar{x}-\bar{x}_n\right)$ in the Helmholtz equation, so the pressure field is the solution of

$$\nabla^2 P + k^2 P = -\sum_{n=1}^{N} i\omega\rho_0\hat{Q}_n\delta\left(\bar{x}-\bar{x}_n\right) \tag{6.4.45}$$

The free-space Green's function is the field for a source having unit mass acceleration. Hence, the total field is

$$P = \sum_{n=1}^{N} \frac{i\omega\rho_0\hat{Q}_n}{4\pi}\left(\frac{e^{-ikr_n}}{r_n}\right) \tag{6.4.46}$$

In many situations, the specific values of the volume velocities are not of interest. If that is the case, the constant parameters for a source may be grouped as a single coefficient $\hat{A}_n$ called the *monopole amplitude*, whose definition is

$$\hat{A}_n = \frac{i\omega\rho_0\hat{Q}_n}{4\pi} \tag{6.4.47}$$

which leads to the pressure being

$$\boxed{P = \sum_{n=1}^{N} \hat{A}_n \frac{e^{-ikr_n}}{r_n}, \quad r_n = \left|\bar{x}-\bar{x}_n\right|} \tag{6.4.48}$$

The magnitude of A_n divided by one length unit is the pressure amplitude radiated by the nth source at a distance of one length unit from it. In SI units, A_n is given in Pa-m.

Although the preceding does meet the objective of displaying the spatial dependence, it is obscure. One thing that can be deduced is that if the field point is very far from all of the sources, then the value of $1/\left|\bar{x}-\bar{x}_n\right|$ will not differ much between all of the sources. Recognition of this feature leads to the farfield approximation for a set of sources.

Consider the three sources in Fig. 6.8a. The geometrical definition of the farfield is that the distance from the field point $\bar{x}$ to each source is sufficiently large to consider the radial lines from $\bar{x}$ to each $\bar{x}_n$ to be parallel. This is the configuration depicted in the figure. Such a condition requires that each of these radial distances be much greater than the maximum distance between a pair of sources, which may be stated as

$$\left|\bar{x}-\bar{x}_n\right| \gg \max\left(\left|\bar{x}_n-\bar{x}_j\right|\right) \tag{6.4.49}$$

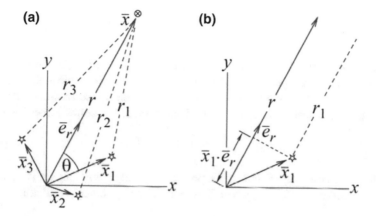

Fig. 6.8 Geometrical parameters for the analysis of the field radiated by a set of point sources, **a** nearfield description, **b** farfield limit

As is done in Fig. 6.8a, placing the origin within the set of sources expedites the analysis. A condition that assures this placement is

$$|\bar{x}_n| \leq \max \left(|\bar{x}_n - \bar{x}_j| \right) \qquad (6.4.50)$$

which merely states that the distance from the origin to each source should not exceed the largest distance between a pair of sources.

The distance from the origin to the field point is $|\bar{x}| = r$, and the unit vector from the origin to $\bar{x}$ is denoted as $\bar{e}_r$. These parameters may be found by evaluating

$$r = |\bar{x}|, \quad \bar{e}_r = \frac{\bar{x}}{r} \qquad (6.4.51)$$

Because all lines in Fig. 6.8b are essentially parallel and end at the same field point, the difference between r and r_n is the projection of $\bar{x}_n$ onto $\bar{e}_r$. Although the figure is two-dimensional, the sources may be arranged arbitrarily. Hence, it is useful to have a vectorial description of this evaluation of the source/field point distances, which is

$$\boxed{r_n = r - \bar{x}_n \cdot \bar{e}_r} \qquad (6.4.52)$$

This validity of this approximation may be proven by applying the law of cosines to the triangle in Fig. 6.8a formed from $\bar{x}_1$, $\bar{x}$, and $\bar{x} - \bar{x}_1$. The known sides are the first two, and the angle θ in Fig. 6.8a between these sides also is known. In the case of the first source, the law of cosines states that

$$r_1 = \left(r^2 + |\bar{x}_1|^2 - 2r |\bar{x}_1| \cos \theta \right)^{1/2} \qquad (6.4.53)$$

No approximation is contained in this relation. However, if $r \gg |\bar{x}_1|$, a power series expansion in $|\bar{x}_1| / r$ will converge rapidly. The result is

$$r_1 = r \left[1 - \frac{|\bar{x}_1|}{r} \cos \theta + \frac{1}{2} \frac{|\bar{x}_1|^2}{r^2} (\sin \theta)^2 + \cdots \right] \tag{6.4.54}$$

As r increases, terms whose order is quadratic or higher in $|\bar{x}_1|/r$ become increasingly insignificant compared to the linear term. (It is not acceptable to drop the linear term because doing so would give $r_1 \approx r$, which would lead to all source signals having the same phase at the field point.) Truncation of this series expansion leads to

$$r_1 = r - |\bar{x}_1| \cos \theta \tag{6.4.55}$$

The angle between r_1 and $\bar{e}_r$ is θ, so the preceding is fully consistent with Eq. (6.4.52). Now consider Eq. (6.4.48) when $\tilde{r}_n$ is represented by this approximation,

$$P = \sum_{n=1}^{N} \hat{A}_n \left(\frac{1}{r - \bar{x}_n \cdot \bar{e}_r} \right) e^{-ikr} e^{ik\bar{x}_n \cdot \bar{e}_r} \tag{6.4.56}$$

The factor e^{-ikr} is the phase lag associated with propagation from the origin, and $e^{ik\bar{x}_n \cdot \bar{e}_r}$ are compensations of this phase lag that account for the fact that the propagation distance from a source is different from r. Smallness of $|\bar{x}_n|/r$ makes it permissible to approximate the fraction as $1/r$. However, the phase shift associated with the complex exponential containing $\bar{x}_n$ cannot be ignored, because k is not necessarily small. Factoring out all terms that are independent of the source locations leads to the *farfield approximation* of the pressure radiated by a collection of sources

$$P_{\mathrm{ff}} = \frac{e^{-ikr}}{r} \sum_{n=1}^{N} \hat{A}_n e^{ik\bar{x}_n \cdot \bar{e}_r} \tag{6.4.57}$$

To obtain a slightly more definitive form, let us locate the field point with spherical coordinates centered at the origin. The radial distance is r, while the polar and azimuthal angles are ψ and θ, respectively. Let us define the polar axis to be z, so that $\bar{x} = r\bar{e}_r = r \sin \psi \cos \theta \bar{e}_x + r \sin \psi \sin \theta \bar{e}_y + r \cos \psi \bar{e}_z$. The location of each source is defined by its coordinates x_n, y_n, z_n, so the preceding description gives

$$P_{\mathrm{ff}} = \frac{e^{-ikr}}{r} \sum_{n=1}^{N} \hat{A}_n e^{ik(x_n \sin \psi \cos \theta + y_n \sin \psi \sin \theta + z_n \cos \psi)} \tag{6.4.58}$$

Suppose we consider field points on a sphere of radius r. At one of these points, we will find that $|P_{ff}|$ is maximum. Let us denote the angles locating this point as ψ_0 and θ_0. We may define a reference monopole amplitude, A_0, to be that of a point source at the origin whose radiated pressure equals the farfield pressure along the radial line defined by ψ_0 and θ_0. This allows us to describe the farfield pressure as

$$r\,|P_{\text{ff}}| = \max\left(\left|\sum_{n=1}^{N}\hat{A}_n e^{ik\bar{x}_n\cdot\bar{e}_r}\right|\right) \equiv A_0 \text{ at } \psi = \psi_0 \text{ and } \theta = \theta_0$$

$$P_{\text{ff}} = \frac{A_0 e^{-ikr}}{r}\mathcal{D}_{\text{ff}}(\psi,\theta)$$

$$\mathcal{D}_{\text{ff}}(\psi,\theta) = \frac{1}{A_0}\sum_{n=1}^{N}\hat{A}_n e^{ik(x_n\sin\psi\cos\theta + y_n\sin\psi\sin\theta + z_n\cos\psi)}$$

(6.4.59)

The quantity $\mathcal{D}_{\text{ff}}(\psi,\theta)$ is the *farfield directivity*. Because of the way it is defined, it lies in the range $0 \le |\mathcal{D}_{\text{ff}}| \le 1$. It is typically displayed in a polar plot showing $|\mathcal{D}_{\text{ff}}|$ as a function of the polar angle ψ at various θ, but it also might be displayed as a spherical plot with $|\mathcal{D}_{\text{ff}}|$ being the radial coordinate. In many situations, the field is axisymmetric, in which case a single plot of $|\mathcal{D}_{\text{ff}}|$ versus ψ will suffice. When the range of $|\mathcal{D}_{\text{ff}}|$ is very large, it is quite common to use a decibel scale to plot $|\mathcal{D}_{\text{ff}}|$.

We encountered a similar form for the farfield pressure in Sect. 4.5.3 when we considered power flow from arbitrary sources. The representation in Eq. (4.5.43) is very much like the preceding expression for P_{ff}, except that the previous expression was scaled by the radiated power. The directivity function $f(\psi,\theta)$ could not be characterized there because it depends on the nature of the sources.

We cannot use superposition to add the radiated power of each source because they might be spaced too closely to justify the assumption that mutual scattering is negligible. Instead, we follow the approach in Chap. 4 where the sources were surrounded by a very large sphere. The fluid inside this sphere is ideal so the power radiated by the sources into this interior domain averaged over one period must equal that which flows across the sphere. Euler's equation gives the particle velocity, which has a component in each direction

$$\bar{V} = \frac{A_0}{i\omega\rho_0}\frac{e^{-ikr}}{r}\left[\left(ik+\frac{1}{r}\right)\mathcal{D}_{\text{ff}}\bar{e}_r + \frac{1}{r}\frac{\partial\mathcal{D}_{\text{ff}}}{\partial\psi}\bar{e}_\psi + \frac{1}{r\sin\psi}\frac{\partial\mathcal{D}_{\text{ff}}}{\partial\theta}\bar{e}_\theta\right]$$

(6.4.60)

As r increases, the terms whose order is $1/r^2$ become insignificant, which leads to the farfield velocity being essentially radial in a ratio to pressure that is defined by the characteristic impedance

$$\bar{V}_{\text{ff}} = \frac{P_{\text{ff}}}{\rho_0 c}\bar{e}_r$$

(6.4.61)

In the present situation, the field depends on the polar angles, so it is necessary to find the total radiated power by actually evaluating an integral

$$\mathcal{P} = \int_0^\pi\int_{-\pi}^\pi \frac{P_{\text{ff}}P_{\text{ff}}^*}{2\rho_0 c}r^2\sin\psi\,d\theta d\psi$$

$$\hat{P} = \frac{1}{2\rho_0 c}\sum_{n=1}^{N}\sum_{j=1}^{N}\hat{A}_j^*\hat{A}_n \int_0^{\pi}\int_{-\pi}^{\pi} e^{ik(x_j-x_n)\sin\psi\cos\theta}\, e^{ik(yj-y_n)\sin\psi\sin\theta}\, e^{ik(zj-z_n)\cos\psi}$$

$$\sin\psi d\theta d\psi \tag{6.4.62}$$

The primary reason for the explicit description in the last form of $\mathcal{P}$ is that it highlights the interaction of sources. If those effects were not important, then those terms of the double summation for which $j \neq n$ would not be present

We have referred to the set of sources as a collection or set. When the sources are arranged in an orderly manner, they usually are said to be an *array*. One that is commonly encountered is a *planar array*, in which many sources are arranged on a regular grid in a plane. The sources may be driven in phase with the objective of simulating a plane wave over an area that is so large that it would not be feasible to use a single transducer. Another common arrangement is a *line array*, which has the sources situated on a straight line. The idea here is that by adjusting the relative phase of the sources, the farfield directivity may be maximized in a desired direction, thereby "aiming" the sound.

EXAMPLE 6.7 A set of N point sources are equally spaced at interval Δ along a straight line. The monopole amplitudes are equal in magnitude but have different phase lags, so that $\hat{A}_n = \hat{A}\exp(i\phi_n)$. This system is symmetric relative to the line on which the sources lie, so it follows that defining this line to be the polar axis z for a set of spherical coordinates leads to a field that depends only on the radial distance r and polar angle ψ. As shown in the sketch, the origin is placed at the center of the array. For the case where N is an even number, derive an expression for the farfield pressure as a function of ψ. Then, specialize the result to the case of a *phased line array*, which is obtained if the electrical signal driving each source is delayed by an amount that is proportional to the relative position of that source. Specifically, if z_n is the coordinate of source n, then let $\phi_n = \gamma k z_n$, where γ is a constant parameter. Plot the directivity for $N = 6$ and $k\Delta = \pi/2$ for cases in which $\gamma = 0, 1/2, \sqrt{3}/2$, and 1.

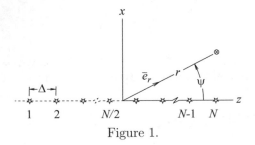

Figure 1.

Significance

A phased line array is an exceptional application of the theory of point sources. It illustrates the fact that interference phenomena can be useful.

Solution

The first step is to describe the farfield for an arbitrary set of monopole amplitudes, after which we will introduce the specified phase angles. It is useful to number the sources such that $n = 1$ marks the source that is farthest behind the origin, which corresponds to the source positions being

$$\bar{x}_n = z_n \bar{e}_z, \quad z_n = \left(n - \frac{1}{2} - \frac{N}{2} \right) \Delta, \quad n = 1, 2, ..., N \tag{1}$$

Because of the axisymmetry of the system, the field point may be placed in the xz plane, so that

$$\bar{e}_r = \cos \psi \bar{e}_z + \sin \psi \bar{e}_x \tag{2}$$

In this arrangement, we have $\bar{x}_n \cdot \bar{e}_r = z_n \cos \psi$, so Eq. (6.4.57) gives

$$P_{\text{ff}} = \hat{A} \frac{e^{-ikr}}{r} \sum_{n=1}^{N} \exp^{i(\phi_n - kz_n \cos \psi)} \tag{3}$$

When the phase angles are the specified ones, the preceding becomes

$$P_{\text{ff}} = \hat{A} \frac{e^{-ikr}}{r} \sum_{n=1}^{N} e^{i(\gamma - \cos \psi)k(n - N/2 - 1/2)\Delta} \tag{4}$$

This expression can be simplified. Toward that end, we observe that every source at a positive z_n is matched by one at a negative coordinate, with a opposite signed phase angle. Sources $N/2 + 1, N/2 + 2, ...$ respectively match sources $N/2, N/2 - 1, ...$ To account for this, we split the summation and change the summation index such that the source number increases with increasing distance from the origin. The first part accounts for $1 < n \le N/2$, so the summation index is changed to $j = N/2 - n + 1$. The second part accounts for $N/2 + 1 \le n < N$, so we set $j = n - N/2$. This converts Eq. (4) to

$$P_{\text{ff}} = \hat{A} \frac{e^{-ikr}}{r} \sum_{j=1}^{N/2} \left[e^{i(\gamma - \cos \psi)k(1/2 - j)\Delta} + e^{i(\gamma - \cos \psi)k(j - 1/2)\Delta} \right]$$

$$= 2\hat{A} \frac{e^{-ikr}}{r} \sum_{j=1}^{N/2} \cos \left[(\cos \psi - \gamma) k \left(j - \frac{1}{2} \right) \Delta \right] \tag{5}$$

The expanded form of this description is

$$
P_{\text{ff}} = 2\hat{A}\frac{e^{-ikr}}{r} \left\{ \cos\left[\frac{1}{2}\left(\cos\psi - \gamma\right)k\Delta\right] + \cos\left[\frac{3}{2}\left(\cos\psi - \gamma\right)k\Delta\right] \right.
$$
$$
\left. + \cdots + \cos\left[\frac{N-1}{2}\left(\cos\psi - \gamma\right)k\Delta\right] \right\}
$$
(6)

An identity[3] states that

$$
\cos(\alpha) + \cos(3\alpha) + \cdots \cos((2N-1)\alpha) = \frac{\sin(2N\alpha)}{2\sin(\alpha)}
$$
(7)

This matches the terms in Eq. (6), with α and $2N$ in the identity replaced by $(\cos\psi - \gamma)k\Delta/2$ and N, respectively. Thus, the farfield of a phased line array is

$$
P_{\text{ff}} = \hat{A}\left(\frac{e^{-ikr}}{r}\right)\frac{\sin\left(\frac{N}{2}\left(\cos\psi - \gamma\right)k\Delta\right)}{\sin\left(\frac{1}{2}\left(\cos\psi - \gamma\right)k\Delta\right)}
$$
(8)

A general property is that $r\,|P_{\text{ff}}|$ has a maximum at $\psi_0 = \cos^{-1}(\gamma)$. This attribute is most readily recognized by the fact that each cosine term in Eq. (6) equals one at that angle. There are $N/2$ terms in the series, so setting all cosine terms to one leads to $\max\left(r\,|P|_{\text{ff}}\right) = \hat{A}N$. In this case, the time delay associated with ϕ_n compensates for differences in the propagation time from each source to the field point at angle ψ_0. The significant feature is that γ is a selectable parameter, which allows the radiated signal to be maximized in a desired direction. A broadside array corresponds to $\psi_0 = \pi/2$, which requires $\gamma = 0$, whereas an end-fire array corresponds to $\psi_0 = 0$, for which $\gamma = 1$. (The case $\gamma = -1$ is the same as reversing the z-axis and setting $\gamma = 1$.)

The equivalent monopole amplitude $\hat{A}_0$ defined in Eq. (6.4.59) is that of a single source whose farfield pressure would equal $\hat{A}_0/r$ along ψ_0. Thus, $\hat{A}_0 = N\hat{A}$. The corresponding farfield directivity is

$$
\mathcal{D}_{\text{ff}}(\psi) = \frac{\sin\left(\frac{N}{2}\left(\cos\psi - \gamma\right)k\Delta\right)}{N\sin\left(\frac{1}{2}\left(\cos\psi - \gamma\right)k\Delta\right)}
$$
(9)

---

[3] Junger and Feit, *Sound, Structures, and Their Interactions*, ASA Press, 1993 reprint, p. 56.

The polar plots of $|\mathcal{D}_{ff}|$ as a function of ψ for the specified four values of γ and $\kappa\Delta = \pi/2$ appear in Fig. 2. The arrow in the plot marks $\psi = 0$. The field is independent of the azimuthal angle, which means that the same polar plot would be obtained for any plane containing the z-axis. Thus, a radial line in the plot, such as $\psi = \psi_0$ for the maximum $|\mathcal{D}_{ff}|$, represents a cone on which $\mathcal{D}_{ff}$ is constant.

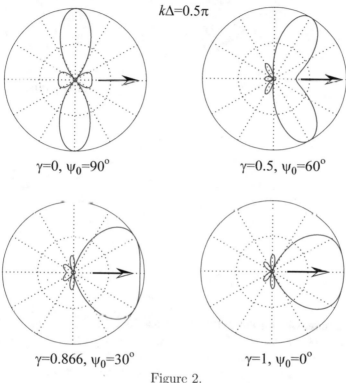

$k\Delta=0.5\pi$

$\gamma=0,\ \psi_0=90°$

$\gamma=0.5,\ \psi_0=60°$

$\gamma=0.866,\ \psi_0=30°$

$\gamma=1,\ \psi_0=0°$

Figure 2.

The plots confirm that the directivity is one at $\psi_0 = \cos^{-1}(\gamma)$, but some other features also are evident. In each case, local maxima occur at other angles, although these maxima correspond to $|\mathcal{D}_{ff}| < 1$. These local maxima are said to be *side lobes*. Each of the side lobes in the graphs are at least 10 dB lower than the main lobe at ψ_0, so one could say that the sound is aimed in the direction of ψ_0. However, if the intent is to aim sound the broadness of the main lobe is problematic because a projected signal would be at nearly the same level over a range of angles.

To attain a narrower main lobe, we may alter the spacing Δ between sources. Reducing $k\Delta$ below $\pi/2$ would yield main lobes that are wider than those in the plots. Thus, let us consider raising $k\Delta$. The directivity patterns in Fig. 3 describe the behavior for $k\Delta = 2.5\pi$, again with $N = 6$. The increased spacing between sources does give a much narrower main lobe, but some of the side lobes now correspond to $|\mathcal{D}_{ff}| = 1$. This is counter to the objective of limiting the range of angles over which the signal is audible.

$k\Delta=2.5\pi$

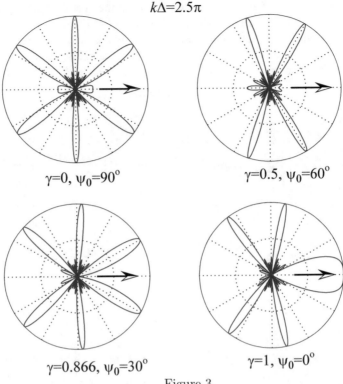

$\gamma=0,\ \psi_0=90^\circ$ $\gamma=0.5,\ \psi_0=60^\circ$

$\gamma=0.866,\ \psi_0=30^\circ$ $\gamma=1,\ \psi_0=0^\circ$

Figure 3.

To understand why the side lobes have grown, consider Eq. (9) for the directivity. The argument of the sine term in the numerator is N times the argument of the sine term in the denominator, which is $\chi = (\cos\psi - \gamma)\,k\Delta/2$. If χ is any multiple of π, both terms will be zero, and a limit analysis would show that the directivity evaluates to one in that case. When viewed as a function of ψ for fixed $k\Delta$, χ is a cosine curve whose amplitude is $k\Delta/2$ from which is subtracted a mean value of $\gamma k\Delta/2$. At $\psi = 0$, we have $\chi = (1 - \gamma)\,k\Delta/2$, while at $\psi = \pi$ it is $\chi = (-1 - \gamma)\,k\Delta/2$. Although only positive γ was considered here, it might be desirable that ψ_0 be greater than 90°, so it is possible that $-1 \le \gamma \le 1$. Avoiding a side lobe for which $|\mathcal{D}_{\mathrm{ff}}| = 1$ in any case requires that $k\Delta < \pi$. In other words, the source spacing should be less than half the acoustic wavelength.

Another design parameter is the total array length, which is $N\Delta$. Holding the length fixed and increasing the number of sources, which reduces the spacing, generally will reduce the sidelobes. In addition, it is possible to introduce *shading*, which is a process of adjusting the monopole amplitudes to attain desirable properties. The length itself can be a design issue, particularly for underwater sonar applications. A low-frequency array must be very large. For example, setting $\Delta = \pi/k$ at 50 Hz gives $\Delta \approx 29$ m.

An interesting variant is based on the fact that transducers can also be used to receive signals. Suppose a set of microphones or hydrophones is arranged as a line array. A constant relative time delay between adjacent signals may be applied electronically to the output from each transducer before the received signals are added. The time delay corresponds to a constant relative delay factor γ at a specified frequency. Different values of γ may be applied to the individual signals. The value for which the combined signal is maximum would mark the direction from which the sound is arriving, according to $\psi_0 = \cos^{-1}(\gamma)$. A phased array radar is the electromagnetic implementation of this concept.

6.4.4 Method of Images

Superposition of point source signals has many usages beyond analyzing arrays. One application adds to a real set image sources that do not exist in order to satisfy boundary conditions. We begin by developing the concept in the frequency domain.

The Basic Concept

In Fig. 6.9, the field point $\bar{x}$ is situated arbitrarily on the plane that bisects the line connecting two sources at $\bar{x}_1$ and $\bar{x}_2$. The intersection of this bisecting plane and the plane of the figure is the bisecting line. The unit vectors are $\bar{e}_n$ parallel to the line from $\bar{x}_2$ to $\bar{x}_1$, and therefore perpendicular to the bisecting plane, and $\bar{e}_b$ parallel to the bisecting line. These direction are used to describe the radial unit vectors from each source to the field point.

Fig. 6.9 Analysis of the signal on the plane bisecting the line connecting two point sources

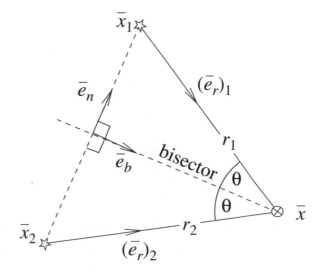

The volume velocity of each source is $\hat{Q}_n$, so the pressure at the field point is

$$P = \left(i\omega\rho_0\hat{Q}_1\right) G\left(\bar{x}, \bar{x}_1\right) + \left(i\omega\rho_0\hat{Q}_2\right) G\left(\bar{x}, \bar{x}_2\right) \qquad (6.4.63)$$

We previously derived a farfield description of the particle velocity for a point source, but here we need an expression that is valid everywhere. Euler's equation serves this purpose, but there are several ways to implement it. One that will prove to be particularly useful as a general approach exploits the definitions of a gradient and a Green's function.

The only position-dependent quantity in the free-space Green's function $G\left(\bar{x}, \bar{x}_0\right)$ is the radial distance. Thus, the gradient of this function may be found by the chain rule

$$\nabla G\left(\bar{x}, \bar{x}_0\right) = \nabla r \frac{\partial G}{\partial r} \qquad (6.4.64)$$

The description of ∇r may be found either heuristically or mathematically, but either approach requires that we recognize that the gradient is at the field point. Thus, ∇r describes the change in r when the source location is fixed. A gradient indicates the direction in which a scalar changes most rapidly, and the magnitude of the gradient is the maximum increment in this scalar for a differential increment of the position. When $\bar{x}_0$ is fixed, a constant value of r defines a sphere centered on $\bar{x}_0$. The direction of most rapid change is normal to this sphere, so it is the unit vector extending from $\bar{x}_0$ to $\bar{x}$; that is, it is $\bar{e}_r$. A differential position shift in this direction is dr, which is also the increment of the scalar r. Thus, we find that

$$\boxed{\nabla r = \bar{e}_r = \frac{\bar{x} - \bar{x}_0}{r}} \qquad (6.4.65)$$

To prove this formally, the field point and source point are represented in terms of their coordinates

$$r = \left[(x - x_0)^2 + (y - y_0)^2 + (z - z_0)^2\right]^{1/2} \qquad (6.4.66)$$

The gradient at the field point is

$$\begin{aligned}
\nabla r &= \frac{\partial r}{\partial x}\bar{e}_x + \frac{\partial r}{\partial y}\bar{e}_y + \frac{\partial r}{\partial z}\bar{e}_z \\
&= \frac{(x - x_0)\,\bar{e}_x + (y - y_0)\,\bar{e}_y + (z - z_0)\,\bar{e}_z}{\left[(x - x_0)^2 + (y - y_0)^2 + (z - z_0)^2\right]^{1/2}}
\end{aligned} \qquad (6.4.67)$$

The numerator is $\bar{x} - \bar{x}_0$ and the denominator is r, so this expression and Eq. (6.4.65) are equivalent.

When this identity is used to evaluate Euler's equation for the pressure in Eq. (6.4.63), the particle velocity is found to be

$$\bar{V} = -\hat{Q}_1 \frac{\partial}{\partial r_1} G(\bar{x}, \bar{x}_1)(\bar{e}_r)_1 - \hat{Q}_2 \frac{\partial}{\partial r_2} G(\bar{x}, \bar{x}_2)(\bar{e}_r)_2$$
$$= \hat{Q}_1 \left(ik + \frac{1}{r_1}\right) G(\bar{x}, \bar{x}_1)(\bar{e}_r)_1 + \hat{Q}_2 \left(ik + \frac{1}{r_2}\right) G(\bar{x}, \bar{x}_2)(\bar{e}_r)_2 \tag{6.4.68}$$

These expressions are valid for any field point. For a field point on the bisecting plane, $r_2 = r_1$, so that $G(\bar{x}, \bar{x}_2) = G(\bar{x}, \bar{x}_1)$. In addition, θ is the angle from $\bar{e}_b$ to both radial unit vectors. The corresponding forms of Eqs. (6.4.63) and (6.4.68) are

$$p = i\omega\rho_0 \left(\hat{Q}_1 + \hat{Q}_2\right) G(\bar{x}, \bar{x}_1)$$
$$\bar{V} = \left[\left(-\hat{Q}_1 + \hat{Q}_2\right) \sin\theta \bar{e}_n + \left(\hat{Q}_1 + \hat{Q}_2\right) \cos\theta \bar{e}_b\right] \left(ik + \frac{1}{r_1}\right) G(\bar{x}, \bar{x}_1)$$
$$\tag{6.4.69}$$

The appearance of Q_1 and Q_2 only as a sum or difference is the central feature. If $\hat{Q}_2 = \hat{Q}_1$, then the particle velocity in the $\bar{e}_n$ direction is zero. The location of $\bar{x}$ on the bisecting plane is arbitrary, so it follows that

> The particle velocity normal to the bisecting plane between two identical sources is zero everywhere on that plane.

Note that the pressure on the bisecting plane is doubled. These are the conditions for a rigid boundary. Now, consider the case where the volume velocities are equal in magnitude, but opposite in sign; that is, they are 180° out-of-phase. In that case, $\hat{Q}_2 = -\hat{Q}_1$, so the pressure on the bisecting plane is zero. This is true for any point on the bisector, so we have found that

> The pressure on the bisecting plane between two sources having equal strength, but opposite sign, is zero everywhere on that plane.

This is the condition that occurs on a pressure-release surface.

The importance of these two cases is that they enable us to add *virtual sources* to those that actually exist to attain either a rigid or pressure-release condition on a planar surface. To do so, we consider the surface to be either a positive mirror for a rigid surface or a negative mirror for a pressure-release surface. The image of the actual source on the physical side is placed behind the mirroring surface in a region that is not part of the physical domain, hence the term "virtual." The sign of the image source is assigned according to whether the surface is rigid or pressure-release.

This construction is depicted in Fig. 6.10. The source is at distance H above the surface, so the image is at distance H behind the surface. The place where a line from the image to the field point intersects the surface is the location where a ray from the actual source reflects from the surface. The angle of reflection equals the angle of incidence, and the distance from the actual source along the reflected path equals the distance from the image to the field point. For a rigid surface, the image source is the

Fig. 6.10 Construction of
the image of a point source
that is reflected from a planar
surface

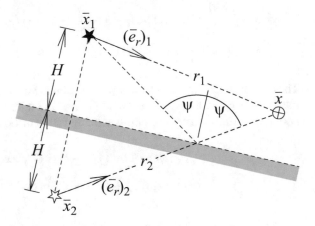

same as the actual source, so the reflection coefficient is one. In contrast, the image
source is the negative of the actual source for a pressure-release surface, which
corresponds to the reflection coefficient being minus one. Thus, the source/image
construction is an extension of the reflection law for plane waves. Each ray from
the actual source that is incident on the surface is reflected like a ray from a plane
wave. However, there is spherical spreading, so the pressure decreases reciprocally
with distance along a ray. If we were to consider several field points, the reflected
ray arriving at each would correspond to a different angle of incidence.

Given this similarity in the reflection law, it might seem as though a point source
above a planar surface with arbitrary local impedance could be treated in a similar
manner. To do so would be erroneous. To see why recall that unless Z is zero or
infinite, the reflection coefficient depends on the angle of incidence. Hence, the
monopole amplitude of each ray emanating from the image would have a value that
depends on the ray's angle of incidence on the surface. In other words, the image
would be required to have a directivity. The dilemma is that the signal radiated by
such a source would not satisfy the Helmholtz equation. There is no simple way to
set up a solution for the reflection of a point source from a planar boundary unless
that boundary's impedance is zero or infinite.[4]

In the special case where the mass acceleration of the source is unity, Eq. (6.4.69)
becomes the Green's function for half-space bounded by a rigid or pressure-release
plane. Thus,

$$G\left(\bar{x}, \bar{x}_1\right) = \frac{e^{-ikr_1}}{4\pi r_1} \pm \frac{e^{-ikr_2}}{4\pi r_2} \tag{6.4.70}$$

where the alternative sign is plus for a rigid boundary and minus for the pressure-
release case.

---

[4]The theory of geometric acoustics, which is the topic of Chap. 11, applies at very high frequencies. According to it, using images to describe reflection from a locally reacting is a reasonable
approximation.

In some situations, it is useful to describe the radial distance in terms of the coordinates of the source and field point. Figure 6.10 leads to these relations. The xz-plane is defined to contain the source at $\bar{x}_1$ and the field point $\bar{x}$, with the z-axis normal to the surface. The xz coordinates of the field point are designated as $x = \xi$ and $z = \eta$, and the source coordinates are $x_1 = 0$, $z_1 = H$. The image is at $x_2 = 0$, $z_2 = -H$. Then, the distances along the ray paths are

$$r_1 = |\bar{x} - \bar{x}_1| = \left[\xi^2 + (\eta - H)^2\right]^{1/2}$$
$$r_2 = |\bar{x} - \bar{x}_2| = \left[\xi^2 + (\eta + H)^2\right]^{1/2} \tag{6.4.71}$$

Other situations, such as creation of a computer code, call for a vectorial description of the radial distances. Presumably, the normal $\bar{e}_n$ to the reflecting plane, the source position $\bar{x}_1$, and field point position $\bar{x}$ are specified in terms of a convenient xyz coordinate system. The height of the source above the plane is $H = \bar{x}_1 \cdot \bar{e}_n$. The image source is at H behind the surface and therefore $2H$ from the actual source in the direction of $-\bar{e}_n$. Therefore, $\bar{x}_2 = \bar{x}_1 - 2(\bar{x}_1 \cdot \bar{e}_n)\bar{e}_n$.

We have proven that any Green's function satisfies the reciprocity property, but let us verify that this is so for the specific function derived here. Thus, we let $\bar{x}_1$ be the field point, and set $\bar{x}$ as the source location. We us a prime to denote the radial distances associated with this switched arrangement. We have $\bar{x} = \xi\bar{e}_x + \eta\bar{e}_z$, so the image location is $\bar{x}_2' = \xi\bar{e}_x - \eta\bar{e}_z$. The reciprocal Green's function is

$$G(\bar{x}_1, \bar{x}) = \frac{e^{-ikr_1'}}{4\pi r_1'} \pm \frac{e^{-ikr_2'}}{4\pi r_2'} \tag{6.4.72}$$

The line connecting the field point $\bar{x}_1$ and source at $\bar{x}_0$ is the same as it is in Fig. 6.10, so $r_1' \equiv r_1$. The distance along the reflected path in this switched arrangement is

$$r_2' = |\bar{x}_1 - \bar{x}_2'| = \left[\xi^2 + (H + \eta)^2\right]^{1/2} \equiv r_2 \tag{6.4.73}$$

The respective distances are the same, so we have proven that $G(\bar{x}_1, \bar{x}) = G(\bar{x}, \bar{x}_1)$ for the Green's function obtained from the method of images.

The image concept is equally applicable in the time domain. If a point source has volume velocity $Q_s(t)$, then the pressure observed at a field point will be

$$p(\bar{x}, t) = \frac{\rho_0 \dot{Q}_s(t - r_1/c)}{4\pi r_1} \pm \frac{\rho_0 \dot{Q}_s(t - r_2/c)}{4\pi r_2} \tag{6.4.74}$$

where r_1 and r_2 are, respectively, the distances to the field point from the source and its image. Setting the mass acceleration to a Dirac delta function at time t_0, $\rho_0 \dot{Q}_s(t) = \delta(t - t_0)$, converts the preceding to the time-domain Green's function for the half-space

$$g\left(\bar{x}, t, \bar{x}_0, t_0\right) = \frac{\delta\left(t - t_0 - r_1/c\right)}{4\pi r_1} \pm \frac{\delta\left(t - t_0 - r_2/c\right)}{4\pi r_2} \tag{6.4.75}$$

There is much similarity between Eq. (6.4.74) and the results of our earlier analysis of plane wave reflection. The primary differences are that in the previous study, the ray that is incident on the surface is parallel to the direct ray (the ray of the incident signal that intersects the field point), and the signal in the earlier analysis does not spread spherically. If we place the source point $\bar{x}_0$ far above the surface and the field point $\bar{x}$ close to the surface, then the direct ray and the ray that is incident on the surface will be nearly parallel. Furthermore, as was demonstrated in Sect. 6.2.2, the value of r_2 will differ little from r_1 so both signals will have essentially the same amplitude at the field point.

Generalization

A Green's function for a half-space bounded by a rigid plane is useful for its own sake, and it will be crucial later to our analysis of radiation from a vibrating plane. However, it is equally valuable as a basic concept that can be extended to systems that have several rigid or pressure-release surfaces. In such cases, we situate images of the actual source such that the combination gives zero normal velocity or zero pressure on each surface. In many situations, the added images might require additional images to cancel their contribution on other surfaces.

To see how the concept is applied, consider Fig. 6.11 where the acoustic domain is bounded by two perpendicularly intersecting planes. The plane of the figure is the one that intersects the source and is perpendicular to both surfaces. Each reflecting surface extends infinitely. The volume velocity of the source is Q_1. Suppose the reflecting surface in the xz plane is rigid, while the one in the yz-plane is pressure-release. The image of Q_1 with respect to the xz plane is Q_2, while the image with respect to the yz plane is Q_3. The normal particle velocity of Q_2 on the xz surface must cancel the contribution of Q_1 on that surface, so $Q_2 = Q_1$. The pressure of Q_3 must cancel the pressure of Q_1 on the yz plane, so $Q_3 = -Q_1$. However, these images are not sufficient. Incidence on the yz surface of the Q_2 image gives a pressure that must be canceled. Furthermore, Q_3 gives a normal velocity on the xz surface. Both effects are resolved by adding image Q_4. To cancel the Q_2 pressure along yz, we set $Q_4 = -Q_2$, while cancelation of the normal velocity on xy requires $Q_4 = Q_3$. Both conditions reduce to $Q_4 = -Q_1$.

After the set of images has been identified, the pressure and particle velocity at a field may be determined by adding the contributions of the images to that of the actual source. The only limitation is that the field point must be situated in the actual domain of the fluid.

The method of images is not limited to planar arrangements. In Fig. 6.12, a point source Q_1 is in the corner formed by three perpendicularly intersecting planes. The required set of images is identified by extending each plane infinitely as a mirror. Any source, real or virtual, in front of a mirror must have an image at an equal distance behind the mirror. The construction in the figure shows that the actual source has seven

Fig. 6.11 A point source in a corner of two long walls and the associated image sources

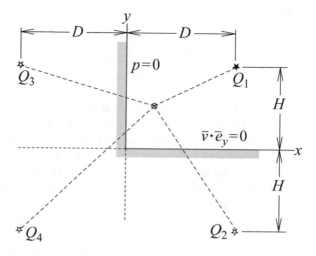

Fig. 6.12 Source Q_1 in a corner and the associated virtual images according to the method of images

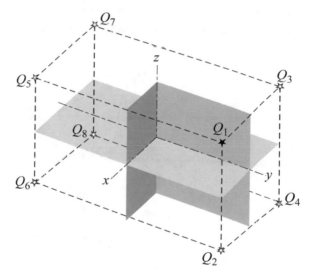

images, one in each octant of an xyz coordinate system defined by the intersections of the reflecting planes.

The identification of the images is a geometric construction. Whether a reflecting surface is rigid or pressure-release affects the sign of the volume velocity assigned to each image. For example, if each surface in Fig. 6.12 is rigid, then it would be necessary that $Q_n = Q_1$ for $n = 2$ to 7. Now, suppose that xz and yz are pressure-release and xy is rigid. Then, images behind the xy-plane would have the same volume velocity as those in front, that is, $Q_2 = Q_1$, $Q_4 = Q_3$, $Q_8 = Q_7$, and $Q_6 = Q_5$. Images behind the yz plane would reverse the sign of the matching one in front, which gives $Q_3 = -Q_1$, $Q_4 = -Q_2$, $Q_8 = -Q_6$, and $Q_7 = -Q_5$.

Sign reversal for image pairs relative to the pressure-release xz plane requires that $Q_5 = -Q_1, Q_6 = -Q_2, Q_7 = -Q_3$, and $Q_8 = -Q_4$. All of these requirements are met by $Q_2 = Q_7 = Q_8 = Q_1, Q_3 = Q_4 = Q_5 = Q_6 = -Q_1$.

The necessary condition to use the method of images is that the reflecting surfaces are rigid or pressure-release planes, but it is not sufficient. In some cases, generating images leads to an infinite set of images, all at a finite distance from the source. Addition of the pressure from each such image would not be a convergent process. This occurs if the walls in Fig. 6.11 or Fig. 6.12 intersect at an arbitrary angle. (In the next example, there also will be an infinite number of images. However, those images occur at increasing distance from the actual source, which leads to a convergent sum for the total pressure.) Another reason that the image process might fail is that it might not be possible to find a set of volume velocities for the images that consistently satisfy the rigid and/or pressure-release requirements. Figure 6.13 depicts a system where this conflict occurs. The angle between the surfaces is $60° = \pi/3$.

Figure 6.13 indicates that a source and its images are at the same distance from the corner. Each image is defined by the angle of the radial line to it from the corner. A source and its image are at equal distances on opposite sides of their reflecting plane, which means that the radial angles measured from that plane must be equal. The sense

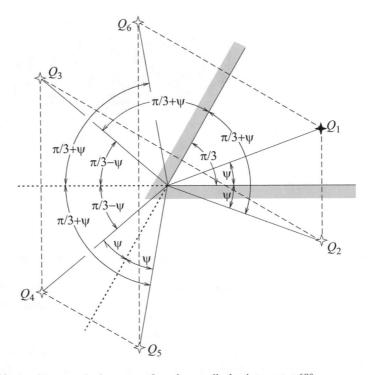

Fig. 6.13 A point source in the corner of two long walls that intersect at $60°$

of the angle will be denoted as cw for clockwise and ccw for counterclockwise. One has discretion as to the sequence to follow for identifying images.

The source is Q_1 at angle ψ ccw from the horizontal surface. Its image relative to the horizontal surface is Q_2 at ψ cw from that surface. Because Q_2 is at angle $\pi/3 + \psi$ cw from the angled wall, image Q_3 is at angle $\pi/3 + \psi$ ccw from that wall. The angle from the extension of the horizontal surface to Q_3 is $\pi/3 - \psi$ cw, so Q_4 is at $\pi/3 - \psi$ ccw from the horizontal extension. This places Q_4 at angle ψ cw from the extension of the angled wall, so Q_5 is at ψ ccw from that surface. Then, Q_5 is at $\pi/3 + \psi$ ccw from the extended horizontal surface, so Q_6 is at $\pi/3 + \psi$ cw from that surface. The radial line to Q_6 is at $2\pi/3 - (\pi/3 + \psi)$ ccw from the angled wall, so its image is at $\pi/3 - \psi$ cw from that wall. However, this image coincides with the actual source Q_1, so no further images can be constructed. Thus, the source has five images. Each lies in a $\pi/3$ sector in a mirrored placement.

The condition that a surface is rigid or pressure-release sets the strength of each image. If both surfaces are rigid, then $Q_n = Q_1, n = 2, ..., 6$. To identify the strengths when both surfaces are pressure-release, let us proceed in the sequence in which the images are numbered. Doing so shows that $Q_2 = -Q_1$, $Q_3 = -Q_2 = Q_1$, $Q_4 = -Q_3 = -Q_1$, $Q_5 = -Q_4 = Q_1$, and $Q_6 = -Q_5 = -Q_1$. Note that Q_6 also is an image of Q_1, which is satisfied by the result that $Q_6 = -Q_1$.

Now, consider the case where the horizontal surface is rigid and the angled one is pressure-release. We proceed in the image numbering sequence. Doing so shows that $Q_2 = Q_1$, $Q_3 = -Q_2 = -Q_1$, $Q_4 = Q_3 = -Q_1$, $Q_5 = -Q_4 = Q_1$, and $Q_6 = Q_5 = Q_1$. However, Q_6 also is an image of Q_1 with respect to the angled (pressure-release) surface, so it must be that $Q_6 = -Q_1$. This condition conflicts with the requirement that Q_6 be the positive image of Q_5. Thus, even though we can perform the geometrical construction with which images are placed, no set of images is consistent with the reflection coefficients. From this, we conclude that the method of images requires a suitable geometry and a suitable set of reflection coefficients.

EXAMPLE 6.8 Consider the waveguide contained between two parallel rigid planes whose extent is infinite. The distance between the planes is H. A point source is situated at distance $s \le H/2$ from one surface, so it is $H - s$ from the other surface. Derive an expression for the pressure at an arbitrary field point in the channel. Then, for the case where the source is symmetrically located, $s = H/2$, and $H = 2\pi/k$, evaluate $|P|$ as function of distance from the source along a line parallel to the planes.

Significance

This waveguide is a model configuration that arises in several contexts, including the ocean and HVAC systems. The set of images in this system is somewhat different because they extend infinitely, which makes the evaluation of the field more challenging.

Solution

We begin by following the basic principles for identifying a suitable set of images. Let the surface plane that is closer to the source be A and the other be B. We define an xyz coordinate system such that xy coincides with plane A and the z-axis intersects the source. We let the volume velocity of the source be Q_1. Its coordinates are $x_1 = y_1 = 0$, $z_1 = s$. The z-axis is normal to both planes, so all virtual images are situated on the z-axis. Consequently, the field is axisymmetric about the z-axis. In a strict sense, we should use cylindrical coordinates defined relative to this axis to describe the pressure field. However, because the dependence is the same in any direction perpendicular to the z-axis, we can consider the dependence in the xz plane.

It is helpful to view Fig. 1 as we proceed to construct the images. The source is at distance s in front of plane A, so the first image is identified as Q_2 behind plane A at $z_2 = -s$. Sources Q_1 and Q_2 constitute a pair of images at distance H in the negative z direction in front of the plane. Their reflections from plane B is described by a pair of sources Q_3 and Q_4 centered at distance H behind plane B. This places

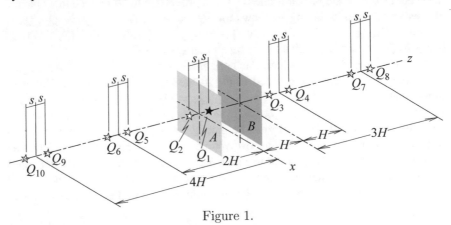

Figure 1.

the center of Q_3 and Q_4 at $z = 2H$. In turn, plane A sees Q_3 and Q_4 as a pair of sources centered at distance $2H$ ahead of it. This requires another pair of images centered at distance $2H$ behind it, which places the center of that pair at $z = -2H$. The center of this pair is at distance $3H$ in front of mirror B, so their images are Q_7 and Q_8 centered at distance $3H$ behind B, at $Z = 4H$. The process goes on ad infinitum. The result is that the images occur in pairs whose center is spaced at distance $2H$ extending to infinity in both directions along the z-axis. Each image must have the same volume velocity as its generator because both surfaces are rigid, which leads to $Q_n = Q_1, n = 2, 3, \dots$.

There are several algorithms with which the calculation may be implemented. Regardless of the approach, it will be necessary to truncate the number of sources for the calculation. Let us consider N pairs of sources behind plane A (negative z) and N pairs behind plane B (positive x). In combination with the actual source and its image relative to plane A, this gives $2N + 1$ pairs of sources. The spacing between the centers of a pair of sources is $2H$. If we designate as $n = 1$ the pair that is farthest in the negative z direction, then the centers are at $z = \zeta_n = (n - N - 1)(2H)$, $n = 1, 2, ..., 2N + 1$. The source locations for each pair are $z = Z_{2n-1} = \zeta_n - s$ and $z = Z_{2n} = \zeta_n + s$, so there at $2(2N + 1)$ individual sources for the computation. The computation will require consideration of many field points, so the z_j values are saved for reuse at each field point.

An arbitrary field point is situated at x, z, so the radial distance from source j to a field point is

$$r_j = \left[x^2 + (z - z_j)^2\right]^{1/2}, \quad j = 1, ..., 2(2N + 1) \tag{1}$$

All sources have volume velocity Q_1, so a column vector of the pressures at each field point due to source j is computed according to

$$P_j = i\omega\rho_0\hat{Q}_1 \frac{e^{-ikr_j}}{4\pi r_j}$$

It is useful to nondimensionalize the distance as $\tilde{r}_j = kr_j$. The total pressure is obtained by summing the contribution of each pair

$$\frac{P}{\rho_0\omega k\hat{Q}_1} = \frac{i}{4\pi} \sum_{j=1}^{2(2N+1)} \frac{e^{-ik\tilde{r}_j}}{\tilde{r}_j} \tag{2}$$

The operations in Eqs. (1) and (2) may be carried out in a vectorized manner using the column vector $\{z\}$ of source locations to create a column vector $\{r\}$ for each field point.

The summation of source pressures at a specific $\bar{x}$ will converge because the pressure radiated by a point source decays as $1/r_j$ and the distances increase as the sources become more distant from the reflecting planes. However, it might be necessary that the number of sources be large. The largest contributions come from the closest sources, which are the actual source and its immediate images. A reasonable truncation criterion is to neglect images whose distance is greater than a large multiple of the smallest distance. Consider the pressure at an arbitrary location (x, z). The source closest to this point is #1. The distance from this point is $r_1 \geq x$ because $0 \leq z \leq H$. The distance from the origin to source N is $\xi_N = N(2H) + s$. Thus, the distance from the farthest source to the field point is $r_N \geq 2NH$. Let us set as the truncation criterion that sources may be ignored if they are farther from the field point than $20r_1$. If we use for this criterion the smallest possible distances, we are led to

$$N \geq \text{ceil}\left(\frac{20x}{2H}\right) \tag{3}$$

It is requested that the pressure be evaluated along a line running outward from the z-axis, but it is not specified how far. Let us take the range as 50 wavelengths, so that $0 < kx < 50(2\pi)$. It might seem that the fineness with which the x coordinate should be sampled should be set by the wavelength. That would be true if we wished to construct an instantaneous pressure profile. Rather, fluctuations of the pressure magnitude are set by interference between the sources. We have no a priori knowledge of this process. Therefore, let us anticipate that using twenty points per wavelength along the axis is adequate. (It is!) This leads to 1000 points along the axis. Equation (3) for the largest x and $kH = 2\pi$ gives $N = 501$. The computation at each field point entails adding the contribution of 1003 pairs of sources. Vectorized computational software requires a relatively brief CPU time.

The result for $s = H/2$ appears in Fig. 2. Superficially, it appears acceptable, but the result actually is quite wrong. One clue has to do with the dashed smooth line. Recall the discussion in Sect. 4.5.3 of the power radiated by sources. In the present situation, no energy is transferred across the rigid walls, so power flows outward from the z-axis within the channel formed by the walls. Let us surround the z-axis with a cylinder of radius x and height H. The power radiated by the source must equal the power that flows out across this cylinder. The surface area of the surrounding cylinder increases in proportion to x. Because the power flow across the cylinder is constant, the intensity at large x must decrease as $1/x$. This leads to the expectation that $|P|$ should decrease as $1/x^{1/2}$. The dashed curve shows this dependence based on the pressure at the farthest x, that is, $p(x_{max})(x_{max}/x)^{1/2}$. Notwithstanding the fluctuations in the computed pressure, there seems to be a substantial disagreement between the results and the expectation.

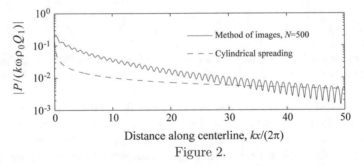

Figure 2.

The fluctuations themselves are a reason to be suspicious. They might be interference resulting from positive and negative reinforcement of individual sources, which are uniformly spaced at two wavelengths when $kH = 2\pi$ and $s = H/2$. However, if this were so, slightly changing s or H should make a significant difference, but additional computations would show that this does not occur.

The error in our computations is not unusual when infinite series must be evaluated. One begins with an estimate of the number of terms that are required, but all too often the fact that this is an estimate is forgotten. That is what happened here. The criterion for N in Eq. (3) is grossly inadequate. Increasing N by a factor of 10 is not adequate. A factor of 100, for which $N = 50,000$, isresults in Fig. 3 indicate that there is no interference effect, and the data fits a $1/x^{1/2}$ dependence quite well. The overall result is to substantially increase the pressure at all locations relative to the results for $N = 500$.

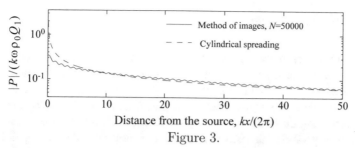

Figure 3.

This problem can also be analyzed in terms of waveguide modes, which is the topic of Chap. 9. The modal solution confirms that $|P|$ decays smoothly as $1/x^{1/2}$. Such a solution consists of a sum of mode functions. That sum also is infinite, but convergence at large kx is attained with a relatively small number of modes. The convergence properties are switched if the field point is situated close to the source. The contributions of images relative to that of the source fall off rapidly for small kx, so N may be very small. In contrast, the number of modes required for convergence does not change drastically with distance. Thus, the method of images is more efficient for field points that are close to the source. This duality of solutions is illustrative of many acoustical systems. Specifically, several analytical approaches might be available, and each might be the most suitable or efficient for a specific purpose or parameter range. Knowing which approach to use is part of what one must learn, but even experienced researchers sometimes find that their initial approach was not the best.

6.5 Dipoles, Quadrupoles, and Multipoles

Even though a point source does not actually exist, the concept led us to some important insights, as well as analyses of a variety of systems. Thus, it should not be surprising that much can be gained by considering what happens when two or more point sources are brought together. The development begins with a concept that has a familiar analog in mechanics. When two forces are equal in magnitude, opposite in direction, and have different lines of action, their sole effect on a rigid body is to cause it to rotate. The measure of this rotational effect is the moment whose magnitude is the product of the magnitude of either force and the perpendicular distance between the lines along which the forces act. The magnitude of this moment will decrease as this distance decreases, unless the magnitude of the force increases. A point couple is the limit as the distance decreases to zero with the force increased to maintain the moment at a constant value. Similarly, an *acoustic dipole* is the limit as two point source whose volume velocities are equal in magnitude, but oppositely signed, are brought together such that the product of the volume velocity and the separation distance is constant. After we evaluate the dipole signal, we will use the development to look at other ways that point sources may be brought together without annihilating their combined field.

6.5.1 The Dipole Field

The oppositely phased sources in Fig. 6.14 are separated by distance d, which ultimately will be shrunk to zero. The reference point $\bar{x}_0$ is midway between the sources, and the z-axis intersects both sources. Our convention will be to define the sense of the z-axis to be from the negative source to the positive one. The field is axisymmetric, so considering the plane containing the field point and the z-axis gives a result that is generally valid. The picture in the figure resembles that in Fig. 6.8a we are interested in the behavior when $kd \ll 1$, but it is not adequate to employ the farfield description developed from that figure. This is so because our objective is to obtain a description of the pressure field that is valid everywhere, including locations where $kr \ll 1$. Thus, we will carry out a formal series expansion in which all terms are truncated at the same order.

Fig. 6.14 Arrangement of two sources of opposite phase but equal strength. A dipole is obtained by letting $d \to 0$

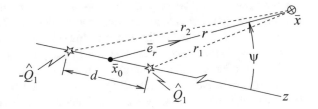

The law of cosines relates r_1 and r_2 to r, d, and the polar angle ψ. Then application of the binomial series for a square root based on $d/r \ll 1$ leads to

$$r_1 = \left[r^2 + \left(\frac{d}{2}\right)^2 - 2r\left(\frac{d}{2}\right)\cos\psi\right]^{1/2} = r\left[1 - \left(\frac{d}{2r}\right)\cos\psi + \cdots\right]$$

$$r_2 = \left[r^2 + \left(\frac{d}{2}\right)^2 - 2r\left(\frac{d}{2}\right)\cos(\pi-\psi)\right]^{1/2} = r\left[1 + \left(\frac{d}{2r}\right)\cos\psi + \cdots\right]$$

$$(6.5.1)$$

These expressions are used to synthesize the combined pressure field, with terms that are of the order of $(d/r)^2$ and smaller ignored. This gives

$$P = \frac{i\omega\rho_0\hat{Q}_1}{4\pi}\frac{e^{-ik\left(r-\frac{d}{2}\cos\psi\right)}}{r - \frac{d}{2}\cos\psi} + \frac{i\omega\rho_0\left(-\hat{Q}_1\right)}{4\pi}\frac{e^{-ik\left(r+\frac{d}{2}\cos\psi\right)}}{r + \frac{d}{2}\cos\psi}$$

$$= \frac{i\omega\rho_0\hat{Q}_1}{4\pi}\left(\frac{e^{-ikr}}{r}\right)\left[\left(1+\frac{d}{2r}\cos\psi\right)e^{i\frac{kd}{2}\cos\psi} - \left(1-\frac{d}{2r}\cos\psi\right)e^{-i\frac{kd}{2}\cos\psi}\right]$$

$$= \frac{i\omega\rho_0\hat{Q}_1}{4\pi}\left(\frac{e^{-ikr}}{r}\right)\left[2i\sin\left(\frac{kd}{2}\cos\psi\right) + \frac{d}{r}(\cos\psi)\cos\left(\frac{kd}{2}\cos\psi\right)\right]$$

$$(6.5.2)$$

The dipole field is obtained by letting $kd \to 0$. Doing so will lead to $P \to 0$ unless $\hat{Q}_1 d$ is held constant. The monopole amplitude of either source is $\hat{A} = \omega \rho_0 \hat{Q}_1 / (4\pi)$. The *dipole moment* D is defined to be $\hat{A}d$, that is,

$$D = \hat{A}d = \frac{i\omega\rho_0 \hat{Q}_1 d}{4\pi} \qquad (6.5.3)$$

We use D to eliminate $\hat{Q}_1$ from Eq. (6.5.2). In addition, the smallness of kd leads to $\sin((kd/2)\cos\psi) \approx (kd/2)\cos\psi$ and $\cos((kd/2)\cos\psi) \approx 1$. The resulting limit is

$$\boxed{P_{\text{dipole}} = D(\cos\psi)\left(\frac{ik}{r} + \frac{1}{r^2}\right)e^{-ikr}} \qquad (6.5.4)$$

It is noteworthy that Eq. (6.5.1) is the same as the farfield description of radial distance. If we had used $1/r_1 \approx 1/r_2 \approx 1/r$ to describe the spherical spreading factors, we would have lost the $1/r^2$ term in the preceding. This term is the dominant effect when $kr \ll 1$, so the result would not have been descriptive of the field in the vicinity of the dipole.

Soon we will examine what happens when dipoles are combined. Replicating the line of analysis followed here will be even more challenging. Fortunately, there is an alternative approach that is not geometrically based. It employs a Taylor series expansion to describe a function that depends on more than one spatial coordinate. Let us review the concept in two dimensions. Let $F(\bar{x})$ denote a function whose value depends on coordinates x and y. Suppose we know the value of the function and all of its derivatives with respect to x and y at some reference position $\bar{x}_0$, whose coordinates are x_0, y_0. This knowledge allows us to determine the function's value at another point $\bar{x}_0 + \bar{\xi}$, where $\bar{\xi} = \xi_1 \bar{e}_x + \xi_2 \bar{e}_y$. The series expansion is

$$\begin{aligned}
F(x_0 + \xi_1, y_0 + \xi_2) &= F(x_0, y_0) + \xi_1 \frac{\partial}{\partial x_0} F(x_0, y_0) + \xi_2 \frac{\partial}{\partial y_0} F(x_0, y_0) \\
&+ \frac{1}{2}\xi_1^2 \frac{\partial^2}{\partial x_0^2} F(x_0, y_0) + \frac{1}{2}\xi_2^2 \frac{\partial^2}{\partial y_0^2} F(x_0, y_0) \\
&+ \xi_1 \xi_2 \frac{\partial^2}{\partial x_0 \partial y_0} F(x_0, y_0) + \cdots
\end{aligned} \qquad (6.5.5)$$

It is useful to have a representation that does not depend explicitly on position coordinates. This is achieved by using the gradient to define a differential operator

$$\bar{\zeta} \cdot \nabla_0 \equiv (\xi_1 \bar{e}_x + \xi_2 \bar{e}_y) \cdot \left(\frac{\partial}{\partial x_0}\bar{e}_x + \frac{\partial}{\partial y_0}\bar{e}_y\right) = \xi_1 \frac{\partial}{\partial x_0} + \xi_2 \frac{\partial}{\partial y_0} \qquad (6.5.6)$$

This enables us to rewrite the Taylor series as

$$F(\bar{x}_0 + \bar{\zeta}) = F(\bar{x}_0) + \bar{\xi} \cdot \nabla_0 F + \frac{1}{2}(\bar{\xi} \cdot \nabla_0)^2 F + \cdots \qquad (6.5.7)$$

The dipole field may be derived by expanding the free-space Green's function for each source in a Taylor series. That quantity depends on the position of two points, the field point $\bar{x}$ and source location $\bar{x}_0$. For this reason, it is imperative to recognize that ∇_0 is the gradient at $\bar{x}_0$ with $\bar{x}$ fixed. The positive source forming the dipole is located at $\bar{x}_0 + \bar{d}/2$, and the negative source is at $\bar{x}_0 - \bar{d}/2$, where $\bar{d}$ is the position vector that extends from the negative to the positive source,

$$\bar{d} = \bar{x}_1 - \bar{x}_2 \tag{6.5.8}$$

(In Fig. 6.14 $\bar{d} = d\bar{e}_z$.) The pressure field radiated by the two sources is

$$P = i\omega\rho_0 \hat{Q}_1 \left[G\left(\bar{x}, \bar{x}_0 + \frac{\bar{d}}{2}\right) - G\left(\bar{x}, \bar{x}_0 - \frac{\bar{d}}{2}\right) \right] \tag{6.5.9}$$

The Taylor series expansion relative to the central point $\bar{x}_0$ (with $\bar{x}$ fixed) is

$$P = i\omega\rho_0 \hat{Q}_1 \left[G\left(\bar{x}, \bar{x}_0\right) + \frac{\bar{d}}{2} \cdot \nabla_0 G\left(\bar{x}, \bar{x}_0\right) + \frac{1}{2}\left(\frac{\bar{d}}{2} \cdot \nabla_0\right)^2 G\left(\bar{x}, \bar{x}_0\right) + \cdots \right]$$

$$- i\omega\rho_0 \hat{Q}_1 \left[G\left(\bar{x}, \bar{x}_0\right) - \frac{\bar{d}}{2} \cdot \nabla_0 G\left(\bar{x}, \bar{x}_0\right) + \frac{1}{2}\left(\frac{\bar{d}}{2} \cdot \nabla_0\right)^2 G\left(\bar{x}, \bar{x}_0\right) + \cdots \right]$$

$$= i\omega\rho_0 \hat{Q}_1 \bar{d} \cdot \nabla_0 G\left(\bar{x}, \bar{x}_0\right) + \cdots \tag{6.5.10}$$

The omitted terms are of the order of $\hat{Q}_1 d^3$ and smaller, so their significance as $d \to 0$ decreases in comparison with the term that is retained.

The free-space Green's function only depends on the separation distance r, so we will use the chain rule to evaluate its gradient. This gives rise to the need to evaluate $\nabla_0 r$. Equation (6.4.65) describes ∇r. Because $r = |\bar{x} - \bar{x}_0|$, it must be that $\nabla_0 r$, which is obtained by displacing $\bar{x}_0$ with $\bar{x}$ fixed, is the negative of ∇r, so that

$$\boxed{\nabla_0 r = -\nabla r = -\frac{\bar{x} - \bar{x}_0}{r} = -\bar{e}_r} \tag{6.5.11}$$

This property allows us to transfer the gradient of the Green's function from the source location to the field point, so that

$$\boxed{\nabla_0 G\left(\bar{x}, \bar{x}_0\right) = (\nabla_0 r) \frac{d}{dr} G\left(\bar{x}, \bar{x}_0\right) = -(\nabla r) \frac{d}{dr} G\left(\bar{x}, \bar{x}_0\right) = -\nabla G\left(\bar{x}, \bar{x}_0\right)} \tag{6.5.12}$$

The pressure field that results is

$$P = \frac{i\omega\rho_0 \hat{Q}_1}{4\pi} \bar{d} \cdot \bar{e}_r \left(\frac{ik}{r} + \frac{1}{r^2}\right) e^{-ikr} \tag{6.5.13}$$

The *dipole moment vector* $\bar{D}$ is defined to be the limiting value of the coefficient as $\bar{d} \to 0$ with $\hat{Q}_1 d$ held constant at D,

$$\bar{D} = \lim_{d \to 0} \left(\left. \frac{i\omega\rho_0 \hat{Q}_1}{4\pi} \bar{d} \right|_{\hat{Q}_1 = D/d} \right) \qquad (6.5.14)$$

Thus, the dipole pressure field is given by

$$\boxed{P_{\text{dipole}} = \bar{D} \cdot \bar{e}_r \left(\frac{ik}{r} + \frac{1}{r^2} \right) e^{-ikr} \equiv -4\pi \bar{D} \cdot \nabla G\left(\bar{x}, \bar{x}_0 \right)} \qquad (6.5.15)$$

If the gradient form in the preceding is used, it is important to recognize that $\nabla G(\bar{x}, \bar{x}_0)$ is evaluated at the field point, with the coordinates of $\bar{x}_0$ fixed. Also, it should be noted that a dipole is defined uniquely by $\bar{D}$. This is analogous to a point source, which is described by the monopole amplitude $\hat{A}$. When $\bar{D}$ is parallel to the z-axis, as in Fig. 6.14, then $\bar{D} \cdot \bar{e}_r = D \cos \psi$, which shows that this description is equivalent to Eq. (6.5.4).

Some applications require a description of the particle velocity in a dipole field. Euler's equation in conjunction with $\nabla r = \bar{e}_r$ gives

$$\boxed{\bar{V}_{\text{dipole}} = -\frac{1}{i\omega\rho_0} \frac{d}{dr} \left(P_{\text{dipole}} \right) \bar{e}_r = \frac{\bar{D} \cdot \bar{e}_r}{\rho_0 \omega} \left(\frac{ik^2}{r} + 2\frac{k}{r^2} - \frac{2i}{r^3} \right) \left(e^{-ikr} \right) \bar{e}_r}$$

$$(6.5.16)$$

The farfield of the dipole corresponds to radial distances that are much greater than a wavelength, $kr \gg 1$, so that $1/(kr)$ is the only important term. Therefore, the farfield behavior is

$$\left(P_{\text{dipole}} \right)_{\text{ff}} = \bar{D} \cdot \bar{e}_r \left(\frac{ik}{r} \right) e^{-ikr}, \quad \left(\bar{V}_{\text{dipole}} \right)_{\text{ff}} = \frac{\bar{D} \cdot \bar{e}_r}{\rho_0 c} \left(\frac{ik}{r} \right) e^{-ikr} \bar{e}_r = \frac{\left(P_{\text{dipole}} \right)_{\text{ff}}}{\rho_0 c} \bar{e}_r$$
$$(6.5.17)$$

The maximum positive pressure amplitude is observed in the direction of $\bar{D}$, where it is $D(ik/r)e^{-kr}$. Setting $\bar{D} \cdot \bar{e}_r = D \cos \psi$ shows that the farfield directivity is

$$\mathcal{D}_{\text{ff}}(\psi) = |\cos \psi| \qquad (6.5.18)$$

This quantity is plotted in Fig. 6.15. It is two tangent circles whose centers are situated on the axis in which the dipole moment is oriented. That no signal is received along $\psi = 90°$ could have been anticipated from the method of images for a pressure-release surface. What could not be anticipated is the broad range of polar angles over which the pressure amplitude is relatively undiminished. For example, it is not until $\psi = 60°$ that the directivity is -6 dB.

Fig. 6.15 Farfield directivity
of a dipole as a function of
the polar angle ψ measured
relative to the dipole
moment $\bar{D}$

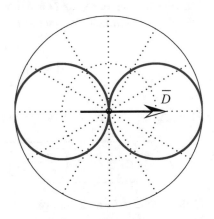

The Taylor series approach is equally applicable to describe a dipole in the time
domain. The mass accelerations of the sources are $\pm \rho_0 \dot{Q}(t)$, so the combined field is

$$
\begin{aligned}
p &= \frac{\rho_0}{4\pi} \left[\frac{1}{|\bar{x} - (\bar{x}_0 + \bar{d}/2)|} \dot{Q}\left(t - \frac{|\bar{x} - (\bar{x}_0 + \bar{d}/2)|}{c}\right) \right. \\
&\qquad \left. - \frac{1}{|\bar{x} - (\bar{x}_0 - \bar{d}/2)|} \dot{Q}\left(t - \frac{|\bar{x} - (\bar{x}_0 - \bar{d}/2)|}{c}\right) \right] \\
&= \frac{\rho_0}{4\pi} \left\{ \frac{1}{|\bar{x} - \bar{x}_0|} \dot{Q}\left(t - \frac{|\bar{x} - \bar{x}_0|}{c}\right) \right. \\
&\qquad \left. + \frac{\bar{d}}{2} \cdot \nabla_0 \left[\frac{1}{|\bar{x} - \bar{x}_0|} \dot{Q}\left(t - \frac{|\bar{x} - \bar{x}_0|}{c}\right) \right] + \cdots \right\} \\
&\quad - \frac{\rho_0}{4\pi} \left\{ \frac{1}{|\bar{x} - \bar{x}_0|} \dot{Q}\left(t - \frac{|\bar{x} - \bar{x}_0|}{c}\right) \right. \\
&\qquad \left. - \frac{\bar{d}}{2} \cdot \nabla_0 \left[\frac{1}{|\bar{x} - \bar{x}_0|} \dot{Q}\left(t - \frac{|\bar{x} - \bar{x}_0|}{c}\right) \right] + \cdots \right\} \\
&= \frac{\rho_0}{4\pi} \bar{d} \cdot \nabla_0 \left[\frac{1}{|\bar{x} - \bar{x}_0|} \dot{Q}\left(t - \frac{|\bar{x} - \bar{x}_0|}{c}\right) \right]
\end{aligned}
\tag{6.5.19}
$$

Because $r \equiv |\bar{x} - \bar{x}_0|$, the chain rule for differentiation leads to

$$
p = -\frac{\rho_0}{4\pi} \bar{d} \cdot \bar{e}_r \frac{\partial}{\partial r} \left[\frac{1}{r} \dot{Q}\left(t - \frac{r}{c}\right) \right]
\tag{6.5.20}
$$

The time-domain dipole moment $\bar{D}(t)$ is defined to be

$$
\bar{D}(t) = \frac{\rho_0}{4\pi} \bar{d} \, \dot{Q}(t)
\tag{6.5.21}
$$

so the dipole field in the time domain reduces to

$$p = \bar{e}_r \cdot \left[\frac{1}{rc} \dot{\bar{D}} \left(t - \frac{r}{c} \right) + \frac{1}{r^2} \bar{D} \left(t - \frac{r}{c} \right) \right] \qquad (6.5.22)$$

EXAMPLE 6.9 Two oppositely phased harmonic sources $\pm \hat{Q}$ are situated at $\pm (d/2) \bar{e}_z$. Evaluate the pressure magnitude along the radial lines at $\psi = 0$ and $\psi = 45°$ and compare each distribution to the dipole field. Also, compare the farfield directivity to the dipole's property. Consider cases where $kd = 2\pi$, π, $\pi/2$, and 0.1π.

Significance

The dipole field is often used to discuss trends observed in the acoustic radiation from objects. With this qualitative evaluation, we will get some insight into the significance of the formulas, and also gain experience in how to recognize such trends.

Solution

The evaluation of pressure dependence on radial distance must consider all values of r, so the farfield approximation is not appropriate to the evaluation of P as a function of r. Instead, the radial distances r_1 and r_2 from each source to a field point will be evaluated according to the law of cosines, Eq. (6.5.1), with ψ held constant. The pressure then is given by

$$P = \frac{i\omega\rho_0 \hat{Q}}{4\pi} \left[\frac{e^{-ikr_1}}{r_1} - \frac{e^{-ikr_2}}{r_2} \right] \equiv \frac{i\omega k \rho_0 \hat{Q}}{4\pi} \left[\frac{e^{-ikr_1}}{kr_1} - \frac{e^{-ikr_2}}{kr_2} \right]$$

To compare this field to that of a dipole, we use Eq. (6.5.3) to represent the dipole moment, so Eq. (6.5.4) becomes

$$P_{\text{dipole}} = \frac{i\omega k \rho_0 \hat{Q}}{4\pi} (kd) (\cos \psi) \left(\frac{i}{kr} + \frac{1}{(kr)^2} \right) e^{-ikr}$$

The factor $\omega k \rho_0 \hat{Q}$ will be used to nondimensionalize all pressures.

Evaluation of pressure along various radial lines has an important purpose. At large distances from any set of sources the value of $r |P|$ depends only on the spherical angles relative to a central reference point. The radial distance beyond which $r |P|$ is constant along any radial line marks the beginning of the farfield. Thus, the first part of the present analysis will enable us to identify where the farfield of a pair of sources begins.

The first graph in Fig. 1 indicates that if two sources are separated by a wavelength, $kd = 2\pi$, the value of $r |P|$ has not quite reached a constant value at $kr = 8\pi$. This means that the farfield behavior is not attained until kr is beyond this distance. Furthermore, the farfield signal for this separation is much weaker than the equivalent dipole's. The magnitude of $r\,|P|$ for the dipole is proportional to $[1 + 1/\,(kr)^2]^{1/2}$, so at a radial distance of one-half of a wavelength, $kr = \pi$, this factor is less than 5% larger than its asymptotic value of one. The graphs demonstrate that the two-source combination approaches a dipole as kd decreases. Note that the spike in the radial plots for $\psi = 0$ is a consequence of the fact that $r = 0$ is the midpoint between the sources, so $r = d/2$ along the radial line at $\psi = 0$ places the field point on the positive source.

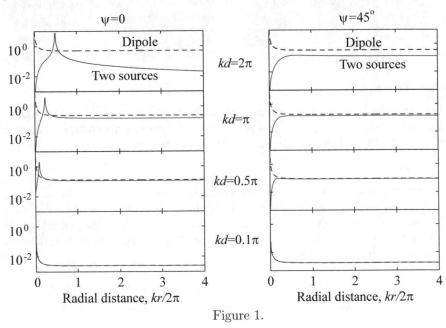

Figure 1.

To compare directivities, we employ Eq. (6.4.57) for the pair of sources. This entails setting $\bar{x}_1 = -\bar{x}_2 = (d/2)\,\bar{e}_z$, $\bar{e}_r = \bar{e}_z \cos\psi + \bar{e}_x \sin\psi$, and using Eq. (6.4.47) to replace the monopole amplitudes with volume velocities. The result is

$$P_{\text{ff}} = \left(\frac{i\omega k \rho_0 \hat{Q}}{4\pi}\right) \frac{e^{-ikr}}{kr} \left[e^{ik(d/2)\cos\psi} - e^{-ik(d/2)\cos\psi} \right]$$

$$\equiv \left(\frac{i\omega k \rho_0 \hat{Q}}{4\pi}\right) \frac{e^{-ikr}}{kr} (2i) \sin\left(\frac{kd}{2}\cos\psi\right)$$

To compare the magnitude of the pressure, as well as its dependence on ψ, to the dipole field, we divide P_{ff} by the maximum value of P_{dipole} at large kr, which occurs at $\psi = 0$. This value is

$$\max\left(\left|P_{\text{dipole}}\right|\right) = \left(\frac{i\omega k\rho_0\hat{Q}}{4\pi}\right)\frac{kd}{kr}$$

Thus, rather than plotting the true directivity of the set of sources, we will compare $\left|P_{\text{ff}}/\max\left(P_{\text{dipole}}\right)\right|$ as a function of ψ to $\cos\psi$, where

$$\frac{|P_{\text{ff}}|}{\max\left(P_{\text{dipole}}\right)} = \frac{2\sin\left(\dfrac{kd}{2}\cos\psi\right)}{kd}$$

The results are shown in Fig. 2. It can be seen that the farfield directivity for $kd = 2\pi$ does not resemble that of the dipole, but the cases of smaller kd do. The farfield pressure of separate sources is lower in all directions than that of the dipole. Nevertheless, if one were to see a measured directivity for radiation from any object that consisted of two rounded lobes, it would be appropriate to say that the source behaves like a dipole, at least to a first approximation.

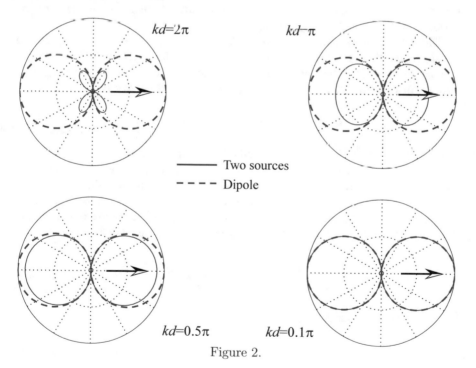

Figure 2.

6.5.2 Radiation from a Translating Rigid Sphere

As noted in the previous example, vibrating bodies sometimes have a directivity pattern that suggests dipole-like behavior. This tendency results from imparting to

Fig. 6.16 Velocity on the
surface of a transversely
oscillating rigid sphere

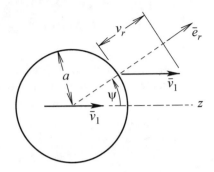

the fluid a particle velocity that is 180° out-of-phase on opposite sides of the body.
The truest incarnation of this behavior is a sphere that translates back and forth
without changing its shape. The field it generates fits the dipole solution. To see why
this is so, consider Fig. 6.16 where a sphere of radius a is oscillating as a rigid body at
velocity $v_1 = \mathrm{Re}\,(V_1 \exp{(i\omega t)})$ horizontally. The definition of rigid body translation
is that all points in the body experience the same displacement, so all points on the
surface have this translational velocity.

The radial line to a surface point is oriented at angle ψ relative to $\bar{v}_1$. In the
inviscid approximation, only the radial component of the surface velocity generates
an acoustic signal. Continuity of the motion of the sphere and the surrounding fluid
requires that this velocity component match that of the fluid, that is

$$\bar{v} \cdot \bar{e}_r = v_1 \bar{e}_z \cdot \bar{e}_r \implies V_r = V_1 \cos \psi \tag{6.5.23}$$

The crucial feature is that the particle velocity in a dipole field, which is described in
Eq. (6.5.16), also has a radial velocity that is proportional to $\cos \psi$. The polar angle
there is measured from the direction of the dipole moment. It follows that if we
align $\bar{D}$ with the z-axis in Fig. 6.16 and select $|\bar{D}|$ such that $\bar{V} \cdot \bar{e}_r$ in the dipole field
evaluated at $r = a$ matches the radial velocity of the sphere, then we will have found
the field radiated by the sphere. Thus, we seek the dipole moment $\bar{D}$ for which

$$V_r = \frac{\bar{D} \cdot \bar{e}_r}{\rho_0 \omega} \left(\frac{ik^2}{a} + 2\frac{k}{a^2} - \frac{2i}{a^3} \right) e^{-ika} = V_1 \cos \psi \tag{6.5.24}$$

Setting $\bar{D} = D\bar{e}_z$ leads to

$$D = \frac{i\omega \rho_0 a^3 V_1}{2 + 2ika - (ka)^2} e^{ika} \tag{6.5.25}$$

According to Eq. (6.5.4), the corresponding pressure is

$$P_{\text{dipole}} = D \,(\cos \psi) \left(\frac{ik}{r} + \frac{1}{r^2} \right) e^{-ikr} \tag{6.5.26}$$

Substitution of D from above gives

$$P = \rho_0 c V_1 \frac{(ka)^2}{2 - (ka)^2 + 2ika} \left(-1 + \frac{i}{kr}\right) \frac{a}{r} e^{-ik(r-a)} (\cos \psi) \qquad (6.5.27)$$

It is instructive to compare this pressure to that generated by a radially vibrating sphere, which is the first of Eq. (6.3.7). Because $r \geq a$, kr and ka have comparable orders of magnitude close to the sphere, whereas the farfield is such that $kr \gg ka$. The high-frequency limit is defined as $ka \gg 1$, in which case $kr \gg 1$ everywhere. This means that the farfield approximation is applicable at all locations. In contrast, the low-frequency limit, $ka \ll 1$, leads to $kr \ll 1$ in the vicinity of the sphere, whereas $kr \gg 1$ in the farfield. The consequence is that the relative magnitudes of the two fields in the low-frequency limit is range-dependent. The asymptotic forms are tabulated below (Table 6.1).

According to this table, if the vibration amplitudes in each type of motion are equal, then the pressure near the sphere at very low frequencies will have the same order of magnitude, but the farfield pressure for the translating sphere will be smaller by a factor of ka. In contrast, in the high-frequency limit, the pressure fields generated by both motions will have similar orders of magnitude everywhere.

In some applications, it is necessary to know the resultant force applied to the sphere by the fluid. Because the pressure distribution is axisymmetric, this resultant must act parallel to the z-axis. The normal to the surface of the sphere is $\bar{e}_r$, so the axial force acting on a patch of surface area is $(-pdS\bar{e}_r) \cdot \bar{e}_z$. The differential area element appropriate to this axisymmetric situation is a ring whose radius is $a \sin \psi$ and whose width is $ad\psi$. Consequently, the force resultant is

$$F_z = -\int_0^\pi p|_{r=a} (\bar{e}_r \cdot \bar{e}_z) (2\pi a \sin \psi) (ad\psi) \qquad (6.5.28)$$

Substitution of Eq. (6.5.27), combined with setting $\bar{e}_r \cdot \bar{e}_z = \cos \psi$ yields

$$F_z = \frac{4}{3}\pi a^2 \rho_0 c V_1 \frac{ka (ka - i)}{2 - (ka)^2 + 2ika} \qquad (6.5.29)$$

Table 6.1 Comparison of the pressure radiated form a sphere that is vibrating radially or transversely

Frequency range	Region	$Pe^{ik(r-a)}/(\rho_0 c V_1)$ Translational vibration	$Pe^{ik(r-a)}/(\rho_0 c V_0)$ Radial vibration
$ka \ll 1$	Close to the sphere	$\frac{i}{2}ka \left(\frac{a}{r}\right)^2 \cos \psi$	$ika\frac{a}{r}$
$ka \ll 1$	Far from the sphere	$-\frac{(ka)^2}{2}\frac{a}{r} \cos \psi$	$ika\frac{a}{r}$
$ka \gg 1$	Everywhere	$\frac{a}{r} \cos \psi$	$\frac{a}{r}$

The force exerted by the sphere on the surrounding fluid is $-F_z$. The low and high-frequency limits of this force are

$$ka \ll 1 \implies -F_z \approx \frac{4}{3}\pi a^3 \rho_0 \left(\frac{i}{2}\omega V_1\right)$$

$$ka \gg 1 \implies -F_z \approx \frac{4}{3}\pi a^2 \rho_0 c V_1 \tag{6.5.30}$$

Some interesting interpretations emerge when we recognize that $i\omega V_1$ is the acceleration of the sphere. Thus, at low frequencies, $-F_z$ is proportional to the sphere's acceleration. The mass of the fluid displaced by the sphere is $(4/3)\pi a^3 \rho_0$, so $-F_z$ is half the value required to accelerate the displaced body of fluid. At high frequency, the value of $-F_z$ is proportional to the velocity of the sphere. In this case, we observe that πa^2 is the maximum cross-sectional area of the sphere, and $\rho_0 c V_1$ is the pressure in a plane wave corresponding to a particle velocity V_1. Thus, $-F_z$ at high frequencies is $4/3$ of the force that would result if the plane wave pressure were applied to a circle of radius a.

Radiated power is another useful metric for comparing the effects of radial and transverse oscillation of a sphere. The power radiated by the radially oscillating sphere is given by Eq. (6.3.16). We have two ways in which to evaluate the power radiated by a translating sphere. One exploits the availability of an expression for F_z. The force exerted by the sphere is $\mathrm{Re}\,(-F_z \exp(i\omega t))\,\bar{e}_z$, and the velocity of all points on the surface is $\mathrm{Re}\,(V_1 \exp(i\omega t))\,\bar{e}_z$. The power input to the fluid is the product of these quantities. The alternative derivation surrounds the sphere with a large virtual sphere whose radius is r_0. This radius is taken to be sufficiently large to permit setting the radial velocity to $p/(\rho_0 c)$. The time-averaged intensity along this sphere is $\mathrm{Re}\,(P_{\mathrm{ff}} P_{\mathrm{ff}}^*)/(2\rho_0 c)\,\bar{e}_r$. The normal to the surrounding sphere is $\bar{e}_r$. Hence, the time-averaged power is the integral of the intensity over the surface of the surrounding sphere. Thus, the ways in which the power may be evaluated are

$$\mathcal{P}_{\mathrm{av}} = \frac{1}{2}\,\mathrm{Re}\left[(-F_z V_1^*)\right] = \int_0^\pi \left.\frac{P_{\mathrm{ff}} P_{\mathrm{ff}}^*}{2\rho_0 c}\right|_{r=r_0} (2\pi r_0 \sin\psi)\, r_0 d\psi \tag{6.5.31}$$

Both evaluations must yield the same result because there are no sources between the surface of the oscillating sphere and the surrounding sphere. Obtaining the final form requires some elaborate manipulations. The result is

$$\boxed{\mathcal{P} = \frac{2}{3}\pi \rho_0 c a^2 |V_1|^2 \frac{(ka)^4}{(ka)^4 + 4}} \tag{6.5.32}$$

A comparison of this expression to Eq. (6.3.16) for power radiated by a radially vibrating sphere is instructive. Their ratio is

$$\frac{\mathcal{P}_{\text{translation}}}{\mathcal{P}_{\text{radial}}} = \frac{1}{3} \left| \frac{V_1}{V_0} \right|^2 \frac{(ka)^2 \left[(ka)^2 + 1 \right]}{(ka)^4 + 4} \tag{6.5.33}$$

If the velocities are equal, the translating sphere at low ka radiates much less power that the radially oscillating sphere, which itself was found to radiate inefficiently at low frequencies. In contrast at high ka, the ratio is $1/3$, so that both motions feature comparable radiated powers.

There are situations where it would be more useful to have descriptions of the time-averaged intensity and radiated power in terms of the dipole moment, rather than the translational velocity. Such expressions may be obtained by using Eq. (6.5.25) to eliminate V_1 from the preceding expressions. The results are

$$\boxed{\begin{aligned} (I_r)_{\text{av}} &= \frac{k^2 |D|^2 (\cos \psi)^2}{2\rho_0 c r^2} \\ \mathcal{P} &= \frac{2\pi}{3} \frac{k^2 |D|^2}{\rho_0} \end{aligned}} \tag{6.5.34}$$

Usually, it is sufficient to view the field radiated by a transversely oscillating sphere from the perspective of the frequency domain, but Eq. (6.5.22) may be adapted if we seek a time-domain description. The radial particle acceleration on the sphere's surface is found by substituting that equation into Euler's equation, which gives

$$\begin{aligned} \dot{v}_r &= -\frac{1}{\rho_0} \frac{\partial p}{\partial r} \\ &= \frac{1}{\rho_0} \bar{e}_r \cdot \left[\frac{1}{rc^2} \ddot{\bar{D}} \left(t - \frac{r}{c} \right) + \frac{2}{r^2 c} \dot{\bar{D}} \left(t - \frac{r}{c} \right) + \frac{2}{r^3} \bar{D} \left(t - \frac{r}{c} \right) \right] \end{aligned}$$

The spheres's translational velocity is an arbitrary function $v_1(t)$. The radial particle acceleration at the sphere's surface must be same as the radial acceleration of the sphere. We take both the motion of the sphere and the dipole moment to be parallel to the z-axis, so that $\bar{v}_1 = v_1 \bar{e}_z$ and $\bar{D} = D \bar{e}_z$. Because $\bar{e}_r \cdot \bar{e}_z = \cos \psi$, continuity of the radial acceleration at the sphere's surface is enforced by matching $\dot{v}_r$ at $r = a$ to the radial component of $\dot{v}_1 \bar{e}_z$. This gives

$$\ddot{D} \left(t - \frac{a}{c} \right) + 2\frac{c}{a} \dot{D} \left(t - \frac{a}{c} \right) + 2\frac{c^2}{a^2} D \left(t - \frac{a}{c} \right) = \rho_0 c^2 a \dot{v}_1(t) \tag{6.5.35}$$

Rather than dealing with a time delay, it is easier to introduce the retarded time $\tau = t - a/c$, and observe that $d/d\tau \equiv d/dt$. Then, the velocity continuity condition at the spheres's surface requires that

$$\ddot{D}(\tau) + 2\frac{c}{a} D(\tau) + 2\frac{c^2}{a^2} D(\tau) = \rho_0 c^2 a \dot{v}_1 \left(\tau + \frac{a}{c} \right) \tag{6.5.36}$$

This is a second-order linear differential equation. It arises in the theory of vibrations as the equation of motion for a critically damped one-degree-of-freedom oscillator. After it is solved for zero initial conditions, the field at any location may be found from Eq. (6.5.22).

EXAMPLE 6.10 A set of hydrophones have been arranged to measure the pressure radiated by a submerged sphere. It is known that the vibration of the sphere is a combination of a breathing mode and translation, but the amplitude and phase of each are not known. Furthermore, the direction in which the translational displacement occurs also is not known. It is desired to identify these vibrational properties by processing the pressure signal received by the microphones. The sphere has a radius of 5 m, and the fluid is seawater at $10\,^\circ$C, for which $\rho_0 = 1027$ kg/m^3 and $c = 1490$ m/s. Measurements are taken at a frequency of 100 Hz, and four hydrophones are used. The coordinates of the hydrophones relative to an xyz coordinate system centered on the sphere with z-oriented northward and x vertically upward are tabulated below, accompanied by the amplitude and phase angle of the complex pressure amplitude measured at each hydrophone. From this information, determine the amplitude and phase angle for the breathing mode's velocity V_0, and the amplitude, phase angle, and direction of the translational velocity amplitude $V_1\bar{e}_v$.

Hydrophone	Coordinates(m)			$P = \|P\|\exp^{i\Delta}$	
	x	y	z	$\|P\|$ (kPa)	Δ (deg)
1	0	0	10	8.421	-128.235
2	0	0	5	16.575	-12.789
3	5	0	5	5.677	3.567
4	0	5	5	13.472	-63.452

Significance

This is an inverse problem in which measured data is used to determine system properties. It is sometimes referred to as a problem in target identification. The procedure followed here is not used as a standard procedure. Nevertheless, the formulation does demonstrate that acoustical signals carry the basic properties of the systems that generate them, which is the core notion underlying many experimental procedures. Along the route to solving this problem, we will become intimately acquainted with the basic equations for sound radiation from a sphere.

Solution

Our approach is to form a model equation that describes the data being processed. Matching this equation to the measurements will lead to equations for the unknown parameters in the model equation. The model equation is the superposition of the pressures for radial and transverse vibration of a sphere. The given position of each of the hydrophones will be denoted as $\bar{\xi}_j$ in order to avoid confusion with the general use of $\bar{x}_n$ to denote source positions. Thus, the position, radial distance, and radial direction to a hydrophone are

$$\bar{\xi}_j = x_j\bar{e}_x + y_j\bar{e}_y + z_j\bar{e}_z, \quad j = 1, 2, 3, 4$$
$$r_j = \left(x_j^2 + y_j^2 + z_j^2\right)^{1/2}, \quad (\bar{e}_r)_j = \frac{x_j}{r_j}\bar{e}_x + \frac{y_j}{r_j}\bar{e}_y + \frac{y_j}{r_j}\bar{e}_z \tag{1}$$

The coefficients in Eqs. (6.3.7) and (6.5.27) that depend exclusively on the system parameters are grouped as B_0 and B_1, respectively. This leads to a representation of the pressure at each sensor as

$$P_j = \left[V_0 B_0 \alpha_j + V_1 B_1 \beta_j \cos\psi_j\right] e^{-ik(r_j-a)} \tag{2}$$

where

$$B_0 = \rho_0 c \frac{ka}{ka-i} = (1.2492 + 0.5925i)10^6 \text{ Rayl}$$

$$B_1 = \rho_0 c \frac{ika}{(ka)^2 + 2ika - 2} = (0.5726 + 0.3321) 10^6 \text{ Rayl}$$

(The nondimensional frequency corresponding to the given parameters is $ka = 2.1085$.) The coefficients α_j and β_j depend on the distance from the origin to each hydrophone, specifically,

$$\alpha_j = \frac{a}{r_j}, \quad \beta_j = ika\frac{a}{r_j} + \frac{a^2}{r_j^2} \tag{3}$$

The proper interpretation of $\cos\psi$ is crucial to the application of Eq. (2). The general definition for a dipole is that ψ is the angle between the radial line and the dipole moment vector. For a translating sphere, the latter direction is the translational velocity $V_1\bar{e}_v$. Thus, ψ_j is the angle between $\bar{e}_v$ and $(\bar{e}_r)_j$. At this stage, neither the complex amplitude V_1 nor the direction $\bar{e}_v$ is known. We describe $\bar{e}_v$ in terms of a polar angle ψ_v and an azimuthal angle θ_v, both of which must be determined. Thus, we set

$$\bar{e}_v = \sin\psi_v \cos\theta_v\bar{e}_x + \sin\psi_v \sin\theta_v\bar{e}_y + \cos\psi_v\bar{e}_z \tag{4}$$

A dot product of $\bar{e}_v$ with the $(\bar{e}_r)_j$ gives the value of $\cos\psi_j$ to be used in Eq. (2). This operation gives

$$\cos\psi_j = (\bar{e}_r)_j \cdot \bar{e}_v = \frac{x_j}{r_j}\sin\psi_v \cos\theta_v + \frac{y_j}{r_j}\sin\psi_v \sin\theta_v + \frac{z_j}{r_j}\cos\psi_v \tag{5}$$

It is evident that at a specified position, the value of $\cos\psi_j$ depends only on the two unknown angles describing $\bar{e}_v$. Because the pressure at each sensor has been measured, Eq. (2) contains four unknowns: V_0, V_1, ψ_v, and θ_v. There are four measurement locations, so it appears that evaluating Eq. (2) at each of the four hydrophones should lead to as solvable set of equations.

The question then is how to carry out the solution in light of the fairly complicated dependence on the angles? The answer lies in the observation that $x_j = y_j = 0$ at points 1 and 2, so that $r_j = z_j$ and $\cos \psi_j = \cos \psi_v$. Thus Eq. (2) at these points reduces to

$$P_j = \left[B_0 \alpha_j V_0 + B_1 \beta_j V_1 \cos \psi_v \right] e^{-ik(z_j - a)}, \quad j = 1, 2 \tag{6}$$

The two equations represented by Eq. (6) may be solved for the values of V_0 and $V_1 \cos \psi_v$. Doing so gives

$$V_0 \equiv \gamma_1 = \frac{\beta_2 P_1 e^{ik(z_1-a)} - \beta_1 P_2 e^{ik(z_2-a)}}{B_0 (\alpha_1 \beta_2 - \alpha_2 \beta_1)} \tag{7}$$
$$= 7.6427 - 2.3642i = 8e^{-0.3i} \text{ mm/s}$$

$$V_1 \cos \psi_v \equiv \gamma_2 = \frac{\alpha_2 P_1 e^{ik(r_1-a)} - \alpha_1 P_2 e^{ik(r_2-a)}}{B_1 (\alpha_2 \beta_1 - \alpha_1 \beta_2)}$$
$$= -3.6623 - 3.0847i = 4.7883 e^{-2.4416i} \text{ mm/s}$$

The values of V_0 and $V_1 \cos \psi$ are denoted as γ_1 and γ_2 for later use.

Next, we turn to points 3 and 4, for which $y_3 = 0$ and $x_4 = 0$. Correspondingly, Eqs. (2) and (5) for these points give

$$\left[V_0 B_0 \alpha_3 + V_1 B_1 \beta_3 \left(\frac{x_3}{r_3} \sin \psi_v \cos \theta_v + \frac{z_3}{r_3} \cos \psi_v \right) \right] e^{-ik(r_3-a)} = P_3$$
$$V_0 B_0 \alpha_4 + V_1 B_1 \beta_4 \left(\frac{y_4}{r_4} \sin \psi_v \sin \theta_v + \frac{z_4}{r_3} \cos \psi_v \right) e^{-ik(r_4-a)} = P_4$$

The values of V_0 and $V_1 \cos \psi_v$ are γ_1 and γ_2, respectively, in Eq. (7). We move these quantities and the known coefficients to the right side. This manipulation leads to

$$V_1 \sin \psi_v \cos \theta_v \equiv \gamma_3 = \frac{r_3}{B_1 \beta_3 x_3} \left[P_3 e^{ik(r_3-a)} - B_0 \alpha_3 \gamma_1 - B_1 \beta_3 \frac{z_3}{r_3} \gamma_2 \right]$$
$$V_1 \sin \psi_v \sin \theta_v \equiv \gamma_4 = \frac{r_4}{B_1 \beta_4 y_4} \left[P_4 e^{ik(r_4-a)} - B_0 \alpha_4 \gamma_1 - B_1 \beta_4 \frac{z_4}{r_4} \gamma_2 \right] \tag{8}$$

For the given set of parameters, these values are

$$\gamma_3 = 9.4552 + 7.9640i = 12.362 e^{0.7i} \text{ mm/s}$$
$$\gamma_4 = -3.4414 - 2.8987i = 4.4994 e^{-2.4416i} \text{ mm/s}$$

The crucial observation at this stage comes from returning to Eq. (4). According to it, γ_2, γ_3, and γ_4 are the projections of the complex velocity vector $V_1 \bar{e}_v$ onto the z-, x-, and y-axis, respectively. Because $\bar{e}_v$ is constant for translation in a fixed direction, the components of all γ_j values must have the same phase angle. In other words, we know that

$$V_1 \bar{e}_v = U_1 \exp(i\delta)\bar{e}_v = \gamma_3 \bar{e}_x + \gamma_4 \bar{e}_y + \gamma_2 \bar{e}_z \tag{9}$$

where U_1 and $\bar{e}_v$ are real. (This property is evident in the listed values of γ_2, γ_3, and γ_4, which indicate that $\text{angle}(\gamma_2) = \text{angle}(\gamma_4) = \text{angle}(\gamma_3) - \pi$. Because $\text{angle}(\gamma_3)$ value is 180° out-of-phase from the others, $\bar{e}_v \cdot \bar{e}_y$ will be negative.) To find V_1, ψ_v, and θ_v from the preceding relation, we divide it by $\exp(i\delta_2)$, where $\delta_2 = \text{angle}(\gamma_2)$, and then take the magnitude of the result, which is used to construct the unit vector, specifically

$$U_1 \bar{e}_v = \exp(-i\delta_2)\left(\gamma_2 \bar{e}_x + \gamma_3 \bar{e}_y + \gamma_4 \bar{e}_z\right)$$
$$U_1 = \left(\gamma_2^2 + \gamma_3^2 + \gamma_4^2\right)^{1/2} = 0.014 \text{ m/s}$$
$$\bar{e}_v = \frac{(U_1 \bar{e}_v)}{U_1} = -0.88302\bar{e}_x + 0.32139\bar{e}_y + 0.34202\bar{e}_z$$

A comparison of this result for $\bar{e}_v$ to Eq. (4) leads to

$$\cos\psi_v = \bar{e}_v \cdot \bar{e}_z, \quad \cos\theta_v = \frac{\bar{e}_v \cdot \bar{e}_x}{\sin\psi_v}, \quad \sin\theta_v = \frac{\bar{e}_v \cdot \bar{e}_y}{\sin\psi_v} \tag{10}$$

Both $\cos\theta_v$ and $\sin\theta_v$ are required to place θ_v in the proper quadrant. The result is

$$\psi_v = 1.2217 = 70°, \quad \theta_v = 2.7925 = 160°$$

The complex velocity vector for the translation is

$$V_1 = U_1 e^{i\delta} = -10.708 - 9.019i \text{ mm/s}$$

The parameters used to generate the given hydrophone readings were $V_0 = 7.6427 - 2.3642i$ mm/s, $V_1 = 10.708 + 9.019i$ mm/s, $\psi_v = 110°$, $\theta_v = -20°$. The value of V_0 in the first of Eq. (7) is the same as the true value. Viewed as a vector, the identified translational velocity also matches the actual motion. This is so because the identified direction angles yield a direction $\bar{e}_v$ that is opposite the one used to generate the data, and the identified value of V_1 is the negative of the generating value.

6.5.3 The Quadrupole Field

Given that bringing two equal strength point sources together might not lead to their mutual annihilation, it is reasonable to conjecture whether bringing two opposite dipoles together might lead to another kind of field. Because the dipole moment

(a)   **(b)**   **(c)**

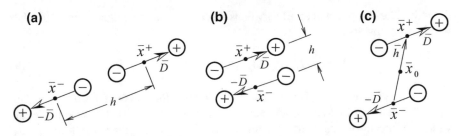

Fig. 6.17 Some possible arrangements of a pair of opposite dipoles

is a vector quantity, the combination of two dipoles only offers the possibility of complete cancelation if the vectors are equal in magnitude but opposite in direction. If that is the case, raising the dipole moment as the distance between the dipoles decreases might negate the cancelation effect. Some possible arrangements appear in Fig. 6.17.

Decreasing the distance h between the dipole centers in case (a), where the line between the dipole centers is parallel to the dipole moment, is one possibility. Is case (b) different? There the line between the centers is perpendicular to the centers. If the distance h between the centers in each case decreases and the dipole moments increase such that Dh is constant, will the same field result? What if the line between the dipole centers is oriented arbitrarily relative to the direction of the moment, as in case (c)? The value of the Taylor series expansion is that it will identify all possible dipole combinations without any presuppositions on our part.

In the general case described by Fig. 6.17c, the positive dipole is situated at $\bar{x}^+$ with dipole moment $\bar{D}$, while the opposite dipole is situated at $\bar{x}^-$. Their relative position is $\bar{h} = \bar{x}^+ - \bar{x}^-$, and the midpoint $\bar{x}_0$ between these centers will be the reference point. We use Eq. (6.5.15) to describe the pressure field of each dipole, so that

$$P = -4\pi \bar{D} \cdot \nabla G\left(\bar{x}, \bar{x}^+\right) + 4\pi \bar{D} \cdot \nabla G\left(\bar{x}, \bar{x}^-\right) \tag{6.5.37}$$

By definition, $\bar{x}^+ = \bar{x}_0 + \bar{h}/2$ and $\bar{x}^- = \bar{x}_0 - \bar{h}/2$. We use these representations to expand the pressure in a Taylor series about $\bar{x}_0$. Terms whose order of magnitude is $\left|\bar{h}\right|^2$ or smaller will not be needed, so the expansion steps are

$$\begin{aligned}
P &= -4\pi \bar{D} \cdot \nabla G\left(\bar{x}, \bar{x}_0 + \bar{h}/2\right) + 4\pi \bar{D} \cdot \nabla G\left(\bar{x}, \bar{x}_0 - \bar{h}/2\right) \\
&= \left(\frac{\bar{h}}{2} \cdot \nabla_0\right)\left[-4\pi \bar{D} \cdot \nabla G\left(\bar{x}, \bar{x}_0\right)\right] + \left(-\frac{\bar{h}}{2} \cdot \nabla_0\right)\left[4\pi \bar{D} \cdot \nabla G\left(\bar{x}, \bar{x}_0\right)\right] + \cdots \\
&= -4\pi\left(\bar{h} \cdot \nabla_0\right)\left[\bar{D} \cdot \nabla G\left(\bar{x}, \bar{x}_0\right)\right]
\end{aligned} \tag{6.5.38}$$

The sequence in which the gradients are applied is unimportant. Swapping the orders, so that the gradient at $\bar{x}_0$ is taken first allows us to employ Eq. (6.5.12), according to which $\nabla_0 G\left(\bar{x}, \bar{x}_0\right) = -\nabla G\left(\bar{x}, \bar{x}_0\right)$, so that

$$\boxed{P = 4\pi \left(\bar{D} \cdot \nabla\right) \left(\bar{h} \cdot \nabla\right) G\left(\bar{x}, \bar{x}_0\right)} \tag{6.5.39}$$

To understand how to carry out the gradients, let us examine the terms in component form. For brevity, we will consider a two-dimensional operation, but the result will be generalized to a three-dimensional description. Thus, we write

$$
\begin{aligned}
\left(\bar{D} \cdot \nabla\right)\left(\bar{h} \cdot \nabla\right) &= \left[\left(D_x \bar{e}_x + D_y \bar{e}_y\right) \cdot \left(\frac{\partial}{\partial x}\bar{e}_x + \frac{\partial}{\partial y}\bar{e}_y\right)\right]\left[\left(h_x \bar{e}_x\right.\right. \\
&\quad \left.\left. + h_y \bar{e}_y\right) \cdot \left(\frac{\partial}{\partial x}\bar{e}_x + \frac{\partial}{\partial y}\bar{e}_y\right)\right] \\
&= D_x h_x \frac{\partial^2}{\partial x^2} + D_y h_y \frac{\partial^2}{\partial y^2} + \left(D_x h_y + D_y h_x\right)\frac{\partial^2}{\partial x \partial y} \tag{6.5.40}
\end{aligned}
$$

It should be obvious from the pattern what the representation for a three-dimensional gradients would be, but writing it out would be tedious. Fortunately, there is a shorthand notation. Rather than denoting the Cartesian coordinates as x, y, z, we write them as x_1, x_2, x_3. In this notation, vector components are denoted by the coordinate indices. The preceding differential operator becomes a double sum over these indices. The corresponding representation of the pressure in Eq. (6.5.39) is

$$P = 4\pi \left(\sum_{j=1}^{3} D_j \frac{\partial}{\partial x_j}\right)\left(\sum_{n=1}^{3} h_n \frac{\partial}{\partial x_n}\right) G\left(\bar{x}, \bar{x}_0\right) \tag{6.5.41}$$

A *quadrupole* is obtained in the limit as $\left|\bar{h}\right| \to 0$ with $\left|\bar{h}\right|\left|\bar{D}\right|$ held at a finite value. This means that we should replace the combination $D_j h_n$ with a single constant coefficient, according to

$$\hat{Q}_{j,n} = \lim_{\left|\bar{h}\right| \to 0} D_j h_n \neq 0 \tag{6.5.42}$$

Thus, a general quadrupole field is described by

$$P = 4\pi \sum_{j=1}^{3}\sum_{n=1}^{3} \hat{Q}_{j,n} \frac{\partial^2}{\partial x_j \partial x_n} G\left(\bar{x}, \bar{x}_0\right) \tag{6.5.43}$$

This representation contains a subtlety stemming from the fact that the order in which the derivatives are taken does not affect the mixed terms. For example, the term for $j = 1$ and $n = 2$ has the same spatial dependence as the term for $j = 2$ and $n = 1$. The strength of their combined contribution to the field is $\hat{Q}_{1,2} + \hat{Q}_{2,1}$. Suppose this quantity were known, for example, by extracting the value from measured data. It would be futile to attempt to identify the individual contributions. For this reason, we shall combine the two contributions into a single parameter according to

$$Q_{j,j} = \hat{Q}_{j,j}$$
$$Q_{j,n} = \hat{Q}_{j,n} + \hat{Q}_{n,j} \text{ if } j \neq n$$
(6.5.44)

A corollary of combining the coefficients is that the double sum in Eq. (6.5.41) contains only six independent terms, three corresponding to $j = n$, and three terms for which $j \neq n$. This combined form may be represented in summation form by removing index combinations that overlap, so that

$$P = 4\pi \sum_{j=1}^{3} \sum_{n=j}^{3} Q_{j,n} \frac{\partial^2}{\partial x_j \partial x_n} G(\bar{x}, \bar{x}_0)$$
(6.5.45)

The free-space Green's function is $\exp(-ikr)/(4\pi r)$, so the expanded representation of a general quadrupole field is

$$P = \left(Q_{1,1} \frac{\partial^2}{\partial x_1^2} + Q_{2,2} \frac{\partial^2}{\partial x_2^2} + Q_{3,3} \frac{\partial^2}{\partial x_3^2} + Q_{1,2} \frac{\partial^2}{\partial x_1 \partial x_2} \right.$$
$$\left. + Q_{1,3} \frac{\partial^2}{\partial x_1 \partial x_3} + Q_{2,3} \frac{\partial^2}{\partial x_2 \partial x_3} \right) \left(\frac{e^{-ikr}}{r} \right)$$
(6.5.46)

Strength factors that have a repeated subscript are *longitudinal quadrupoles*, and terms whose strength factor contains different subscripts are *lateral quadrupoles*. For brevity, we will refer to them by their index; for example, a 2-2 quadrupole is longitudinal, whereas a 1-3 quadrupole is lateral. To understand the reason for this terminology, consider the arrangement of dipoles prior to letting $\bar{h}$ shrink to zero. The 3-3 quadrupole corresponds to $\bar{D} = D_3 \bar{e}_3$ and $\bar{h} = h_3 \bar{e}_3$. This arrangement is depicted in Fig. 6.18a. The line connecting the positive and negative dipoles is parallel to the dipole moment. Note that although the figure seems to indicate a "plus-minus-minus-plus" sequence of sources, that is merely for pictorial purposes. The 3-3 quadrupole reduces h_3 to zero such that the "plus" source of one is brought to the "minus" source of the other. Figure 6.18b and c depicts the 1-1 and 2-2 longitudinal quadrupoles.

Whereas the picture for a longitudinal quadrupole has the sources arranged along a line, the lateral quadrupole arrangement shows the sources at the corners of a rectangle. The lateral 1-2 quadrupole $Q_{1,2}$ is depicted in Fig. 6.18d. This figure makes it evident why the two contributions, $\hat{Q}_{1,2}$ and $\hat{Q}_{2,1}$, are indistinguishable. The definition of $\hat{Q}_{1,2}$ is that it is the limit of two dipoles $\pm D_1 \bar{e}_1$ separated by $\bar{h} = h_2 \bar{e}_2$. This corresponds in Fig. 6.18d to the dipoles being formed from a pair of point sources along the x direction. Similarly, $\hat{Q}_{2,1}$ corresponds to the limit of a pair of dipoles $\pm D_2 \bar{e}_2$ separated by $\bar{h} = h_1 \bar{e}_1$. In the figure this corresponds to the dipoles being formed from a pair of point sources along the y direction. This is the same arrangement of point sources as that which leads to the $\hat{Q}_{1,2}$ quadrupole. The situations for $\hat{Q}_{2,3}$ and $\hat{Q}_{1,3}$ in Fig. 6.18e and f, respectively, are similar. We shall refer to any of the configurations described by this figure as an *elemental quadrupole*.

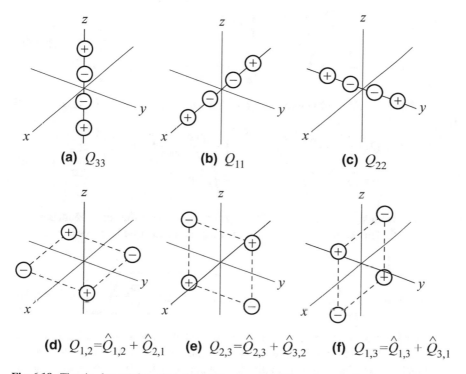

Fig. 6.18 The six elemental quadrupoles relative to a set of Cartesian coordinates

To describe the pressure field explicitly in terms of the position coordinates of a field point, it is convenient to let the quadrupole's position $\bar{x}_0$ be the origin. The free-space Green's function is $e^{-ikr}/(4\pi r)$ and $r = \left(x_1^2 + x_2^2 + x_3^2\right)^{1/2}$, so the chain rule allows us to evaluate the derivatives in Eq. (6.5.41). A few identities are useful for this purpose. In the following, j and n are any two indices, and δ_{jn} is the Kronecker delta, which is one if $j = n$ and zero otherwise.

$$\frac{\partial r}{\partial x_n} = \frac{x_n}{r}, \quad \frac{\partial^2 r}{\partial x_j \partial x_n} = \frac{r^2 \delta_{jn} - x_j x_n}{r^3}$$

$$\frac{\partial}{\partial x_n}\left(\frac{e^{-ikr}}{r}\right) = \frac{\partial r}{\partial x_n}\frac{d}{dr}\left(\frac{e^{-ikr}}{r}\right) = \frac{x_n}{r}\frac{d}{dr}\left(\frac{e^{-ikr}}{r}\right)$$

$$\frac{\partial^2}{\partial x_j \partial x_n}\left(\frac{e^{-ikr}}{r}\right) = \left(\frac{r^2\delta_{jn} - x_j x_n}{r^3}\right)\frac{d}{dr}\left(\frac{e^{-ikr}}{r}\right) + \frac{x_j x_n}{r^2}\frac{d^2}{dr^2}\left(\frac{e^{-ikr}}{r}\right)$$

$$(6.5.47)$$

The field of a longitudinal quadrupole is axisymmetric about its axis. It is convenient to align z with this axis. The preceding identities with $j = n = 3$ in combination with Eq. (6.5.46) give the field of the (3-3) dipole,

$$P = Q_{,33}\left[\left(\frac{r^2 - x_3^2}{r^3}\right)\frac{\partial}{\partial r}\left(\frac{e^{-ikr}}{r}\right) + \frac{x_3^2}{r^2}\frac{\partial^2}{\partial r^2}\left(\frac{e^{-ikr}}{r}\right)\right]$$

$$= -Q_{3,3}\left[\left(\frac{x_1^2 + x_2^2}{r^3}\right)\left(\frac{ik}{r} + \frac{1}{r^2}\right) + \frac{x_3^2}{r^2}\left(\frac{k^2}{r} + \frac{2ik}{r^2} - \frac{2}{r^3}\right)\right]e^{-ikr} \quad (6.5.48)$$

A lateral quadrupole is not symmetric about any axis. The (1-2) dipole is described by

$$P = Q_{1,2}\left[\left(-\frac{x_1 x_2}{r^3}\right)\frac{\partial}{\partial r}\left(\frac{e^{-ikr}}{r}\right) + \frac{x_1 x_2}{r^2}\frac{\partial^2}{\partial r^2}\left(\frac{e^{-ikr}}{r}\right)\right]$$

$$= -Q_{1,2}\left(\frac{x_1 x_2}{r^2}\right)\left(\frac{k^2}{r} - \frac{3ik}{r^2} - \frac{3}{r^3}\right)e^{-ikr} \quad (6.5.49)$$

Farfield representations are obtained by dropping terms that are proportional to $1/r^2$ or $1/r^3$. The result is that the farfield of the elemental quadrupoles is described by

$$(P_{\text{ff}})_{\text{long}} = -k^2 Q_{j,j}\frac{x_j^2}{r^2}\frac{e^{-ikr}}{r}, \quad (P_{\text{ff}})_{\text{lat}} = -k^2 Q_{j,n}\frac{x_j x_n}{r^2}\frac{e^{-ikr}}{r} \quad (6.5.50)$$

A directivity plot describes $r|P|$ as a function of the direction to a field point in the farfield. This direction is defined by the polar angle ψ from the z-axis and the azimuthal angle θ. The function to plot is obtained by using the coordinate transformation in Eq. (6.1.1) to eliminate the Cartesian coordinates from the preceding. This yields an expression for $r P_{\text{ff}}$ as a function of ψ and θ. Typical are the expressions for the 3-3 and 1-2 quadrupoles, which are

$$(P_{\text{ff}})_{\text{long}} = -k^2 Q_{3,3}(\cos\psi)^2\left(\frac{e^{-ikr}}{r}\right)$$

$$(P_{\text{ff}})_{\text{lat}} = -Q_{1,2}k^2(\sin\psi)^2\sin\theta\cos\theta\left(\frac{e^{-ikr}}{r}\right) \quad (6.5.51)$$

The radial distance in each of Fig. 6.19 is the scaled value of $|r P|$ and values of ψ and θ define the direction of $\bar{e}_r$. These plots are the three-dimensional analog of a polar directivity plot. The longitudinal field exhibits two lobes, which is reminiscent of the dipole case, but the lobes are narrower. The lateral quadrupole has four identical lobes whose maxima occur at $\psi = 90°$, $\theta = \pm 45°, \pm 135°$.

A different visualization show slices through the three-dimensional data. Such a slice for the longitudinal 3-3 quadrupole appears first in Fig. 6.20. It is evident here that the lobes are narrower than those of the dipole. Two views are used for the lateral 1-2 quadrupole. The plot in the xy-plane clearly shows the four lobes and their maxima. The other view shows a cutting plane that bisects two opposite lobes. This plane contains the z-axis and is at 45° from the xz and yz planes. The lobes are seen to be broader in this plane than in the xy-plane.

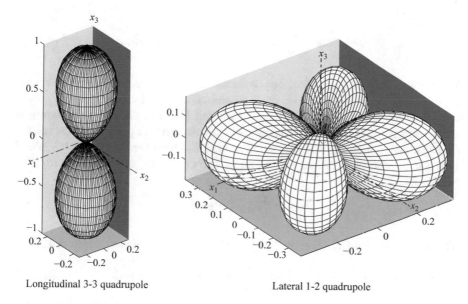

Longitudinal 3-3 quadrupole

Lateral 1-2 quadrupole

Fig. 6.19 Farfield directivity of longitudinal and lateral dipoles

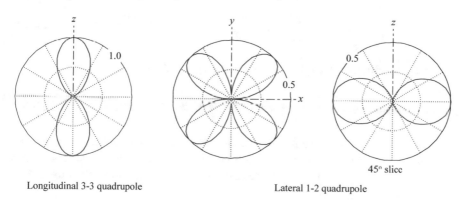

Longitudinal 3-3 quadrupole

Lateral 1-2 quadrupole

Fig. 6.20 Slices through the directivity patterns of longitudinal and lateral dipoles

An arbitrary quadrupole consists of a sum of the elemental fields. It follows that only the combination of elemental quadrupoles is meaningful. However, such a combination generally will not resemble any of the elemental types, and it might not actually be a quadrupole. For example, consider the field radiated by longitudinal quadrupoles aligned with each coordinate axis, each of whose strength is Q. Based on Fig. 6.19, one might think that the combined directivity has lobes aligned along each positive and negative coordinate axis. However, the farfield pressure obtained from Eq. (6.5.49) in this case is

$$P_{\text{ff}} = -k^2 \left(\frac{Qx_1^2 + Qx_2^2 + Qx_3^2}{r^2} \right) \frac{e^{-ikr}}{r} \equiv -k^2 Q \frac{e^{-ikr}}{r} \qquad (6.5.52)$$

In other words, this combination is equivalent to a point source, with no directionality.

In general, the field resulting from a combination of elemental quadrupoles is critically dependent on the relative strength and phase of each. A demonstration of this aspect is found by considering a coordinate system x_1', x_2', x_3' that is rotated by angle ϕ about the x_3 axis relative to $x_1 x_2 x_3$. Suppose the signal is a 1-2 lateral quadrupole relative to this new set of axes. The farfield description in Eq. (6.5.50) gives

$$P = -k^2 Q_{1,2}' \left(\frac{x_1' x_2'}{r^2} \right) \frac{e^{-ikr}}{r} \qquad (6.5.53)$$

The values of x_1' and x_2' are related to x_1 and x_2 through a rotation transformation,

$$x_1' = x_1 \cos\phi + x_2' \sin\phi, \quad x_2' = -x_1 \sin\phi + x_2 \cos\phi \qquad (6.5.54)$$

Substituting these relations into the lateral quadrupole pressure leads to

$$P_{\text{ff}} = -k^2 \left(Q_{1,1} \frac{x_1^2}{r^2} + Q_{2,2} \frac{x_2^2}{r^2} + 2 Q_{1,2} \frac{x_1 x_2}{r^2} \right) \frac{e^{-ikr}}{r} \qquad (6.5.55)$$

where the quadrupole strengths relative to the $x_1 x_2 x_3$ coordinate system are

$$Q_{11} = -Q_{22} = -\frac{1}{2} \hat{Q}_{1,2}' \sin 2\phi, \quad Q_{1,2} = \frac{1}{2} Q_{1,2}' \cos 2\phi \qquad (6.5.56)$$

Thus, although this field actually is a single lateral quadrupole relative to the rotated coordinate system, its mathematical representation relative to the original coordinate system makes it appear to be a combination of one lateral and two longitudinal quadrupoles.

Any alteration of these strengths fundamentally changes the field. Figure 6.21 displays the result of modifying the strength factors listed above. All views are down the x_3-axis. Case (a) depicts the result of setting the $Q_{j,n}$ values in accord with the rotation transformation. The result is a rotated lateral quadrupole, as expected. In case (b), $Q_{1,1}$ has been set to zero. This directivity is not axisymmetric. Rather, the small lobes that are orthogonal to the two main lobes are aimed in the $x_1 x_2$ plane. Case (c) is what is obtained if the sign of $Q_{1,1}$ is reversed, while case (d) reverses the sign of $Q_{1,2}$. The latter rotates the dipole by $\pi - \phi$, while the former shows little evidence of the presence of a lateral quadrupole.

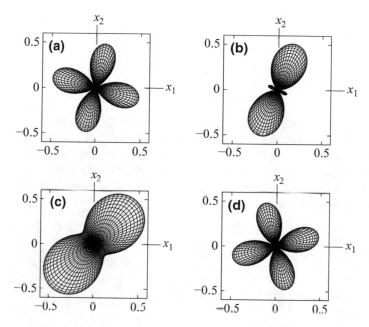

Fig. 6.21 Quadrupoles based on rotating a lateral quadrupole that intially was aligned along the xy-axis. **a** The correct rotation, **b** effect of removal of the rotated 1-1 longitudinal quadrupole, **c** effect of reversing the sign of the rotated 1-1 longitudinal quadrupole, **d** effect of reversing the sign of the rotated 1-2 lateral quadrupole

A corollary of this discussion is that one must consider the combination of the quadrupoles to understand the field properties. What appears to be a complicated set of quadrupoles when viewed from the perspective of one coordinate system might actually be quite simple when viewed relative to another coordinate system.

In many situations, the directional dependence of intensity, rather than the pressure directivity, is of interest. As always, the farfield is such that the velocity is essentially radial and that component is proportional to the pressure through the characteristic impedance. Thus, these quantities are

$$
\begin{aligned}
\left(\bar{I}_{\text{long}}\right)_{\text{ff,av}} &= \frac{|Q_{3,3}|^2}{2\rho_0 c} \frac{k^4}{r^2} \left(\cos\psi\right)^4 \bar{e}_r \\
\left(\bar{I}_{\text{lat}}\right)_{\text{ff,av}} &= \frac{|Q_{12}|^2}{2\rho_0 c} \frac{k^4}{r^2} \left(\sin\psi\right)^4 \left(\sin\theta\right)^2 \left(\cos\theta\right)^2 \bar{e}_r
\end{aligned}
\tag{6.5.57}
$$

The radiated power is evaluated as an integral of the radial component of intensity over a sphere. The radius r_0 of this sphere is taken to be sufficiently large that usage of the farfield forms is justified. The lateral quadrupole field is not axisymmetric, so the differential area element is formed from differential increments of $d\psi$ and $d\theta$, that is, $dS = (r\sin\psi d\theta)(r d\psi)$. The ranges of integration are $0 < \psi < \pi$, $-\pi < \theta < \pi$, with the result that

$$\boxed{\begin{aligned} \dot{\mathcal{P}}_{\text{long}} &= \frac{2\pi}{5} \frac{k^4}{\rho_0 c} \left| Q_{3,3} \right|^2 \\[2mm] \mathcal{P}_{\text{lat}} &= \frac{2\pi}{15} \frac{k^4}{\rho_0 c} \left| \hat{Q}_{1,2} \right|^2 \end{aligned}} \qquad (6.5.58)$$

These relations apply equally to quadrupoles having other polarities. The radiated power for a quadrupole that is not one of the fundamental types is examined in the following example. An expression for the power radiated by an arbitrary combination of elemental quadrupoles is presented there. The result will show that the total power might not equal the sum of the power radiated by each elemental quadrupole in isolation from the others.

EXAMPLE 6.11   Consider a quadrupole that is obtained from a pair of dipoles in the limit as $h \to 0$ with Dh constant, where $\bar{D}$ is situated in the $x_1 x_3$ plane at angle χ from the x_3-axis, and $\bar{h}$ is situated in the $x_2 x_3$ plane at angle η from the z-axis. What are the elemental strengths Q_{jn} relative to the $x_1 x_2 x_3$ coordinates system, and what is the radiated power?

Significance

The field properties of a combination of fundamental quadrupoles seldom is examined in texts. The result is not intuitive.

Solution

Identification of the values of Q_{jn} is not difficult because it is covered by Eq. (6.5.42). It is given that $\bar{D} = D (\sin \chi \bar{e}_1 + \cos \chi \bar{e}_3)$ and $\bar{h} = h (\sin \eta \bar{e}_2 + \cos \eta \bar{e}_3)$. Longitudinal quadrupoles are obtained as a product of components of $\bar{D}$ and $\bar{h}$ along the same axis, while laterals are a product of components along different axes. Thus, the given arrangement is a combination of one longitudinal and three lateral quadrupoles whose strengths are

$$Q_{12} = D_1 h_2 + D_2 h_1 = Dh \sin \chi \sin \eta, \quad Q_{13} = D_1 h_3 + D_3 h_1 = Dh \sin \chi \cos \eta$$

$$Q_{23} = D_2 h_3 + D_3 h_2 = Dh \cos \chi \sin \eta, \quad Q_{33} = D_3 h_3 = Dh \cos \chi \cos \eta$$

Figure 1 is a typical spherical plot of $r \, |P|$ corresponding to $k^2 h D = 1$. The appearance of this plot could not have been anticipated from the given $\bar{D}$ and $\bar{h}$ values.

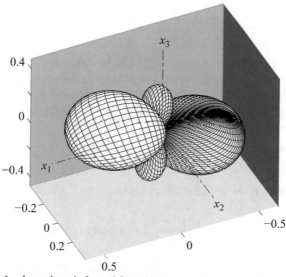

Figure 1. A quadrupole formed from dipole moments at $\chi = 30°$, $\eta = 150°$.

The radiated power may be evaluated by integrating the radial component of the farfield intensity over a large sphere, as was done to derive Eq. (6.5.57). The time-averaged farfield intensity is radial, so that $(I_r)_{\text{av}} = |P|^2 / (2\rho_0 c)$. It is necessary to account for interactions between different elemental quadrupoles. To do so, we let P_{jn} denote the field associated with Q_{jn} and use this decomposition to describe $|P|^2 = PP^*$. This leads to

$$(I_r)_{\text{av}} = \frac{1}{2\rho_0 c} (P_{12} + P_{13} + P_{23} + P_{33}) \left(P_{12}^* + P_{13}^* + P_{23}^* + P_{33}^* \right)$$

$$= \frac{1}{2\rho_0 c} \left(|P_{12}|^2 + |P_{13}|^2 + |P_{23}|^2 + |P_{33}|^2 \right) + \frac{1}{\rho_0 c} \text{Re}(P_{12} P_{13}^*$$

$$+ P_{12} P_{23}^* + P_{12} P_{33}^* + P_{13} P_{23}^* + P_{13} P_{33}^* + P_{23} P_{33}^*)$$

As a preliminary to evaluating the surface integral of $(I_r)_{\text{av}}$, it is helpful to return to the farfield representations in terms of the Cartesian coordinates, Eq. (6.5.50). The dependence of the term $P_{33} P_{12}^*$ is $x_1 x_2 x_3^2$. This is an odd function with respect to x_1 and x_2, that is, $f(-x_1, x_2, x_3) = f(x_1, -x_2, x_3) = -f(x_1, x_2, x_3)$. The significance of this observation is that the sphere over which we integrate the radial intensity is situated symmetrically relative to the coordinate planes. If we divide the sphere into two parts to the left and right of the $x_1 x_3$ plane, the integral over the hemispheres will cancel because each point at a certain x_2 on one hemisphere is matched by a mirror image point whose coordinate is $-x_2$. A similar argument applies for each coordinate plane. Consequently, any product of quadrupole fields that contains an odd power of either x_1, x_2, or x_3 will not contribute to an integral over the surface of the sphere. What remains are squares and products of longitudinal dipoles and squares of the lateral quadrupoles. In the present situation, the result is that

$$\mathcal{P} = \frac{1}{2\rho_0 c} \iint\limits_{S} \left(|P_{12}|^2 + |P_{13}|^2 + |P_{23}|^2 + |P_{33}|^2\right) dS$$

The three lateral quadrupoles only differ in how they are distributed relative to the coordinate axes. Consequently, the integral of one will be equally descriptive of the other three. Equation (6.5.58) describes the contribution of the P_{12} field, as well as the P_{33} field. Applying the former to the P_{13} and P_{23} contributions leads to

$$\mathcal{P} = \frac{2\pi}{15} \frac{k^4}{\rho_0 c} \left(|Q_{12}|^2 + |Q_{13}|^2 + |Q_{23}|^2 + 3|Q_{33}|^2\right)$$

Substitution of the expressions for the dipole strengths in terms of Dh reduces this to

$$\mathcal{P} = \frac{\pi}{30} \frac{k^4 D^2 h^2}{\rho_0 c} \left[1 + 3\left(\cos\chi\cos\eta\right)^2\right]$$

The result that the radiated power is the same as the sum of the output from each elemental quadrupole acting in isolation is not general. Because of the behavior as odd functions with respect to two coordinates, the power radiated by the elemental lateral quadrupoles always is a simple sum, with no mutual interaction, nor any interaction with the longitudinal quadrupoles. In contrast, the longitudinal quadrupoles have a mutual interactive effect on the radiated power. In part, this is associated with the fact that some features of a simple point source are embedded in their sum. For an arbitrary quadrupole, the radiated power is

$$\boxed{\begin{aligned}\mathcal{P} = \frac{2\pi k^4}{15\rho_0 c} &\left[3\left(|Q_{11}|^2 + |Q_{22}|^2 + |Q_{33}|^2\right) + \left(|Q_{12}|^2 + |Q_{23}|^2 + |Q_{13}|^2\right)\right. \\ &\left. + 2\,\mathrm{Re}(Q_{11}Q_{22}^* + Q_{22}Q_{33}^* + Q_{11}^*Q_{33})\right]\end{aligned}}$$

$$(6.5.59)$$

An interesting observation pertains to the situation where longitudinal quadrupoles have equal strength and the lateral quadrupoles are zero. We previously saw that this combination forms a point source. The radiated power in that case is $2\pi k^4 / (\rho_0 c)$, which is five times the isolated output from any one of these quadrupoles.

6.5.4 Multipole Expansion

The logical extension of the developments thus far is treatment of an arbitrary collection of point sources. The pressure field is the sum of the individual contributions, as described by Eq. (6.4.46). We will apply the Taylor series expansion to this expression. To do so, we define the reference point $\bar{x}_0$ to be situated somewhere within the set, as prescribed by Eq. (6.4.50). The relative position of the source is

$$\bar{\xi}_n = \bar{x}_n - \bar{x}_0 \qquad (6.5.60)$$

It is convenient to use the monopole amplitudes $\hat{A}_n = i\omega\rho_0 \hat{Q}_n / (4\pi)$ to describe the strength of each point source, so the expansion is

$$P = \sum_{m=1}^{N} 4\pi\hat{A}_m \left[G\,(\bar{x}, \bar{x}_0) + \bar{\xi}_m \cdot \nabla_0 G\,(\bar{x}, \bar{x}_0) + \frac{1}{2}\left(\bar{\xi}_m \cdot \nabla_0\right)^2 G\,(\bar{x}, \bar{x}_0) + \cdots \right]$$

$$(6.5.61)$$

The preceding is valid regardless of how $G\,(\bar{x}, x_0)$ is defined. To advance the analysis, we require that $G\,(\bar{x}, \bar{x}_0)$ be the free-space Green's function, which depends solely on r. This property allows us to invoke Eq. (6.5.12) to replace the gradient at $\bar{x}_0$ with the gradient at $\bar{x}$. In addition, we replace the operator $\bar{\xi}_n \cdot \nabla$ by its equivalent representation in terms of the position components $\xi_{m,j}$. The series expansion that results is

$$P = \sum_{m=1}^{N} 4\pi \hat{A}_m G\,(\bar{x}, \bar{x}_0) - \sum_{m=1}^{N} 4\pi \hat{A}_m \sum_{j=1}^{3} \xi_{m,j} \frac{\partial}{\partial x_j} G\,(\bar{x}, \bar{x}_0)$$

$$+ \frac{1}{2}\sum_{m=1}^{N} 4\pi \hat{A}_m \sum_{j=1}^{3}\sum_{n=1}^{3} \xi_{m,j}\xi_{m,n} \frac{\partial^2}{\partial x_j \partial x_n} G\,(\bar{x}, \bar{x}_0) + \cdots$$

$$(6.5.62)$$

It is useful to view this expression as a set of differentiations applied to a sum of contributions from all sources. Such a form results when the sum over m is brought inside the other sums, so that

$$P = 4\pi\sum_{m=1}^{N} \hat{A}_m G\,(\bar{x}, \bar{x}_0) - 4\pi\sum_{j=1}^{3}\sum_{m=1}^{N} \hat{A}_m \xi_{m,j} \frac{\partial}{\partial x_j} G\,(\bar{x}, \bar{x}_0)$$

$$+ 4\pi\sum_{j=1}^{3}\sum_{n=1}^{3}\sum_{m=1}^{N} \frac{\hat{A}_m}{2} \xi_{m,j}\xi_{m,n} \frac{\partial^2}{\partial x_j \partial x_n} G\,(\bar{x}, \bar{x}_0) + \cdots$$

$$(6.5.63)$$

We have found that the field of a point source in free space is proportional to $G\,(\bar{x}, \bar{x}_0)$, so the first term represents a sum of N monopole contributions to the overall field. Reference to Eq. (6.5.15) leads to interpretation of the second term as a sum of N dipoles whose moment in direction j is $\hat{A}_n \xi_{n,j}$. The third term appeared in Eq. (6.5.41) with a different coefficient of the second derivative. That equation was an intermediate step in the derivation of the quadrupole field. As was done there, the terms for $j \neq m$ are combined in order that the field be represented as a set of independent contributions. The specific form is

$$P = 4\pi \hat{A} G (\bar{x}, \bar{x}_0) - 4\pi \sum_{j=1}^{3} D_j \frac{\partial}{\partial x_j} G (\bar{x}, \bar{x}_0)$$
$$+ 4\pi \sum_{j=1}^{3} \sum_{n=j}^{3} Q_{j,n} \frac{\partial^2}{\partial x_j \partial x_n} G (\bar{x}, \bar{x}_0) + \cdots$$

(6.5.64)

The derivatives of the Greens' function are described by Eq. (6.5.47). The strength factors are

$$\hat{A} = \sum_{m=1}^{N} \hat{A}_m, \quad D_j = \sum_{m=1}^{N} \hat{A}_m \xi_{m,j}$$
$$Q_{j,j} = \frac{1}{2} \sum_{m=1}^{N} \hat{A}_m \left(\xi_{m,j} \right)^2, \quad Q_{j,n} = \sum_{m=1}^{N} \hat{A}_m \xi_{m,j} \xi_{m,n} \text{ if } n \neq j$$

(6.5.65)

The derivatives of the Greens' function are described by Eq. (6.5.47).

Equation (6.5.64) is said to be a *multipole expansion*. The equivalent monopole amplitude is $\hat{A}$; the components of the equivalent dipole moment vector are D_j, and $Q_{j,m}$ are the quadrupole strengths. The terms that are omitted by truncation of the series begin with octupoles, which are obtained by pairing opposite quadrupoles.

The farfield approximation applies if r is much greater than the largest distance to a source, which is satisfied if $r \gg \max(|\zeta_m|)$. In that region, we may approximate the derivatives in Eq. (6.5.47) by dropping terms that are proportional to $1/r^2$ and smaller. The resulting farfield version of the multipole expansion is

$$P_{\text{ff}} = \left[\hat{A} + ik \sum_{j} = 1^3 D_j \frac{x_j}{r} \right.$$
$$\left. - k^2 \sum_{j} = 1^3 \sum_{m} = j^3 Q_j m \frac{x_j x_m}{r^2} + \cdots \right] \left(\frac{e^{-ikr}}{r} \right)$$

(6.5.66)

where the x_i values are the coordinates of the field point at which P is observed.

A possible cause for confusion is the fact that the field of a collection of sources at an arbitrary field point is described by both Eqs. (6.4.48) and (6.5.64). The farfield counterparts of these relations are Eqs. (6.4.57) and (6.5.66), respectively. The first set of relations place no limitation on the relative position of the sources. The descriptions derived here are Taylor series expansions of $1/r_n$ and $\exp(-ikr_n)$ based on distance from the central reference point $\bar{x}_0$. It follows that the multipole expansion will converge slowly if the distances $|\bar{\xi}_n|$ from the origin to any of the sources are large fraction of the wavelength.

For this reason, the multipole expansion is most useful if the collection of point sources is a *compact set*. This term means that all sources are situated in a region whose size is much smaller than a wavelength, so that all $\left|\bar{\xi}_n\right|$ are small compared to $2\pi/k$. Consider substitution of Eq. (6.5.65) into (6.5.66). The result would show that each multipole features factors whose order of magnitude is $k\left|\bar{\xi}_n\right|$ raised to the power that is the order of the multipole. That is, the monopole does not contain that factor, the dipoles depend on $k\left|\bar{\xi}_n\right|$ to the first power, and the quadrupoles' order of magnitude is $k\left|\bar{\xi}_n\right|$ squared. Thus, smallness of $k\left|\bar{\xi}_n\right|$ means that the contribution of each type becomes progressively smaller with increasing order. A consequence of this property is that a good approximation for the farfield of a compact set of sources is obtained by judicious truncation of the multipole series. If $\hat{A}$ is not zero, then the set of sources is essentially a monopole, while $\hat{A} = 0$ and one or more D_j values being nonzero mark the behavior as dipole-like. Quadrupole behavior requires that $\hat{A}$ and all of the D_j values be zero. It is important to remember the compactness requirement when one decides to use a multipole expansion. For example, it cannot be used for a phased array because the spacing between adjacent sources is a large fraction of a wavelength, and the size of the array is much greater than a wavelength.

EXAMPLE 6.12 Three points sources have been micromachined on an acoustically transparent substrate. Their signal is transmitted through the air in order to actuate micromotors. The sources are arranged equidistantly along a circle whose radius is $\varepsilon = 0.12$ mm, and the monopole amplitude $\hat{A}_j$ of each source is adjustable. The frequency is 300 kHz. (a) Determine the $\hat{A}_j$ values that will yield a sound pressure level of 10 dB at 50 mm from the origin at any angular position. (b) Determine the $\hat{A}_j$ values that will produce a sound pressure level of 10 dB at 50 mm from the origin on the x_2-axis, and no signal at 50 mm from the origin on the x_1-axis. (c) Determine the $\hat{A}_j$ values that will produce a sound pressure level of 10 dB at 50 mm from the origin on the x_1- and x_2-axis, and no signal at 50 mm from the origin on the x_3-axis.

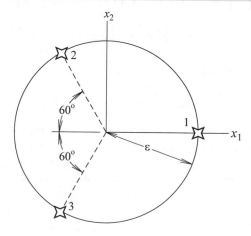

Figure 1.

Significance

This problem will enhance our understanding of how point source fields combine. The specific configuration, which features three sources, is intended to address a question that often is raised, specifically, why is the multipole expansion of a triad of sources not called a "tripole"?

Solution

A multipole expansion is an accurate description if sources form a compact set. The frequency is 300 kHz, and the sound speed is 340 m/s, so the wavelength is $c/f = 1.1333$ mm. This is more than five times greater than the distance between sources, so the condition is met. Furthermore, the field point is at $r = 50$ mm, which is much larger than the distance between sources. Thus, usage of the farfield description in Eq. (6.5.66) is justified. The pressure amplitude corresponding to 10 dB//20 μPa is 89.44 μPa.

The center of the circle locating the sources is the origin in the problem statement. Because that location is situated in the midst of the sources, it is suitable for the analysis. The source positions are

$$\bar{\xi}_1 = \varepsilon \bar{e}_1, \quad \bar{\xi}_2 = \varepsilon\left(-0.5\bar{e}_1 + 0.866\bar{e}_2\right), \quad \bar{\xi}_3 = \varepsilon\left(-0.5\bar{e}_1 - 0.866\bar{e}_2\right)$$

Case (a) calls for an omnidirectional field, which will be obtained if the monopole amplitude is nonzero. This condition places no restriction on the individual source strengths, but we can assure that it is an accurate description if we also set the dipole moment to zero. The equivalent multipole strengths are described by Eq. (6.5.65), so we require that

$$\frac{1}{0.05}\left|\hat{A}_1 + \hat{A}_2 + \hat{A}_3\right| = 89.44\left(10^{-6}\right)$$

$$\left(\hat{A}_1\bar{\xi}_1 + \hat{A}_2\bar{\xi}_2 + \hat{A}_3\bar{\xi}_3\right)\cdot\bar{e}_1 = \varepsilon\left(\hat{A}_1 - 0.5\hat{A}_2 - 0.5\hat{A}_3\right) = 0$$

$$\left(\hat{A}_1\bar{\xi}_1 + \hat{A}_2\bar{\xi}_2 + \hat{A}_3\bar{\xi}_3\right)\cdot\bar{e}_2 = \varepsilon\left(0.866\hat{A}_2 - 0.866\hat{A}_3\right) = 0$$

The solution of these equations is

$$\hat{A}_1 = \hat{A}_2 = \hat{A}_3 = 1.49\left(10\right)^{-6} \text{ Pa-m}$$

If the monopole amplitude is nonzero, there is no direction in which the pressure is zero. Thus, to obtain the field properties in case (b), we set the monopole strength to zero. The pressure in any direction perpendicular to the dipole's moment vector is zero, so the specified properties could be obtained with a dipole oriented along the x_2-axis. However, a 2-2 longitudinal quadrupole would also give the desired property. Let us consider the dipole first. We wish that $\hat{A} = 0$ and $\bar{D} = D\bar{e}_2$ for the set of sources. Thus, we set the monopole amplitude to zero, and the dipole moment components are matched to $D\bar{e}_2$. In addition, we use Eq. (6.5.66) to set the pressure at $x_2 = 0.05$ m, $x_1 = x_3 = 0$. Because $r = x_2 = 0.05$ m at this location, we have

$$\left| \hat{A}_1 + \hat{A}_2 + \hat{A}_3 \right| = 0$$

$$\left(\hat{A}_1 \bar{\xi}_1 + \hat{A}_2 \bar{\xi}_2 + \hat{A}_3 \bar{\xi}_3 \right) \cdot \bar{e}_1 = \varepsilon \left(\hat{A}_1 - 0.5\hat{A}_2 - 0.5\hat{A}_3 \right) = 0$$

$$\left(\hat{A}_1 \bar{\xi}_1 + \hat{A}_2 \bar{\xi}_2 + \hat{A}_3 \bar{\xi}_3 \right) \cdot \bar{e}_2 = \varepsilon \left(0.866\hat{A}_2 - 0.866\hat{A}_3 \right) = D$$

$$\left(kD\frac{x_2}{r} \right) \frac{1}{r} = \frac{5544D}{0.05} = 89.44 \left(10^{-6} \right)$$

The solution of these equations is

$$\hat{A}_1 = 0, \quad \hat{A}_2 = -\hat{A}_3 = 8.066 \left(10^{-10} \right) \text{ Pa-m}$$

The other alternative for case (b) is to combine the sources into a quadrupole. To do so, the equivalent monopole amplitude and each component of the dipole moment must be zero. Thus, it is necessary that

$$\left| \hat{A}_1 + \hat{A}_2 + \hat{A}_3 \right| = 0$$

$$\left(\hat{A}_1 \bar{\xi}_1 + \hat{A}_2 \xi_2 + \hat{A}_3 \xi_3 \right) \cdot \bar{e}_1 = \varepsilon \left(\hat{A}_1 - 0.5\hat{A}_2 - 0.5\hat{A}_3 \right) = 0$$

$$\left(\hat{A}_1 \bar{\xi}_1 + \hat{A}_2 \bar{\xi}_2 + \hat{A}_3 \bar{\xi}_3 \right) \cdot \bar{e}_2 = \varepsilon \left(0.866\hat{A}_2 - 0.866\hat{A}_3 \right) = 0$$

This is a set of three homogeneous equations for the three source amplitudes. The only solution is that $\hat{A}_1 = \hat{A}_2 = \hat{A}_3 = 0$, in which case there is no signal. Thus, an equivalent dipole is the combination that will produce the condition in case (b).

The field in case (c) has zero pressure in the x_3 direction and equal pressures in the x_1 and x_2 directions. One possibility is a dipole in the $x_1 x_2$-plane at 45° from the x_1-axis. This corresponds to setting $\bar{D} = D (\cos 45° \bar{e}_1 + \sin 45° \bar{e}_2)$. (Orientations at 45° in the opposite sense or at ±135° yield equivalent results because we are only interested in the pressure amplitude.) The desired properties could also be obtained from a pair of longitudinal quadrupoles aligned at 45° in the $x_1 x_2$ plane, or from a lateral quadrupole rotated by 45° in the manner of Fig. 6.21. However, canceling the monopole amplitude and dipole moments in order that the three sources combine to be any type of quadrupoles would require that all $\hat{A}_j = 0$, as it did in case (b). Thus, the only possibility is the dipole combination. The equations in this case are

$$\left| \hat{A}_1 + \hat{A}_2 + \hat{A}_3 \right| = 0$$

$$\left(\hat{A}_1 \bar{\xi}_1 + \hat{A}_2 \bar{\xi}_2 + \hat{A}_3 \bar{\xi}_3 \right) \cdot \bar{e}_1 = \varepsilon \left(\hat{A}_1 - 0.5\hat{A}_2 - 0.5\hat{A}_3 \right) = 0.7071D$$

$$\left(\hat{A}_1 \bar{\xi}_1 + \hat{A}_2 \bar{\xi}_2 + \hat{A}_3 \bar{\xi}_3 \right) \cdot \bar{e}_2 = \varepsilon \left(0.866\hat{A}_2 - 0.866\hat{A}_3 \right) = 0.7071D$$

$$k \sum_{j=1}^{3} D_j \frac{x_j}{r} \left(\frac{1}{r} \right) = k \, (0.7071D) \left(\frac{x_1 + x_2}{r} \right) \left(\frac{1}{r} \right) = 89.44 \left(10^{-6} \right)$$

The points of interest are on the x_1- or x_2-axis at $r = 0.05$ m from the origin. The solution is

$$\hat{A}_1 = 5.378 \left(10^{-10}\right), \quad \hat{A}_2 = 1.968 \left(10^{-10}\right), \quad \hat{A}_3 = -7.346 \left(10^{-10}\right) \text{ Pa-m}$$

To close this example, it is useful to take an overview of the result. We were unable to generate a quadrupole by manipulating the three source strengths. This is because a multipole expansion will begin at the quadrupole terms only if the equivalent monopole amplitude and three dipole moment are zero. Such a condition requires that the individual source strengths satisfy four equations, which cannot be done with only three sources. The underlying explanation may be found in Fig. 6.18, which shows that any type of quadrupole consists is the limit of four sources that approach each other. Three sources cannot replicate this behavior.

6.6 Doppler Effect

The Doppler effect is the shift in frequency when the source generating a signal moves. For many individuals, their first encounter with the Doppler effect is the change of pitch they hear when a passing locomotive emits a horn burst. This is the situation we use to introduce the phenomenon.

6.6.1 Introduction

Consider the following scenario. A locomotive moves along a straight track at constant speed v. It is a very foggy day. To warn people who might be crossing the railroad tracks, the locomotive operator periodically blows the horn. The period is T, and the first burst occurs at $t = 0$. Figure 6.22 depicts the bursts as a sequence of impulses. At time $t = T$, the locomotive has moved distance vT, and the first burst has propagated by distance cT down the track. The second burst is emitted at this instant. At time $t = 2T$, both bursts have propagated distance cT relative to their positions at $t = T$. The locomotive, which now is at $2vT$, emits the third burst.

Inspection of Fig. 6.22 and consideration of the situation for subsequent bursts leads to recognition that the spatial separation of bursts is $cT - vT$. Thus, if t_n is the instant when the nth burst is heard at some location ahead of the locomotive, then burst $n + 1$ will be heard at time $t_n + (cT - vT)/c$. In other words, an observer in front of the locomotive will perceive that the period of the bursts is $(1 - v/c)T$. Periodic signals are describable by Fourier series. The fundamental frequency of the signal emitted by the locomotive is denoted as ω_0, and the fundamental frequency heard by the observer is ω. We have established that

Fig. 6.22 Propagation ahead
of a moving source that
emits a periodic sequence of
impulses

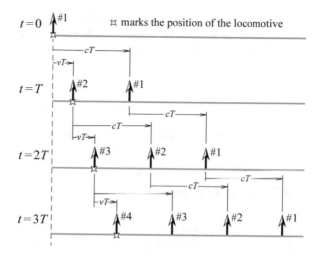

$$\omega_0 = \frac{2\pi}{T}, \quad \omega = \frac{2\pi}{(1 - v/c)\,T} = \frac{\omega_0}{1 - v/c} \qquad (6.6.1)$$

In other words, forward movement of the source of a sound is perceived by observers ahead of the source as an increased frequency. This is referred to as an upward *Doppler shift*. It should be noted that the analysis requires $v < c$; that is, the source is moving subsonically. In the supersonic case, $v > c$, the signal that is emitted will arrive after the source arrives at the observer.

What about observers who are behind the moving source? Figure 6.23 describes this situation. The analogous construction shows that the impulses behind the source

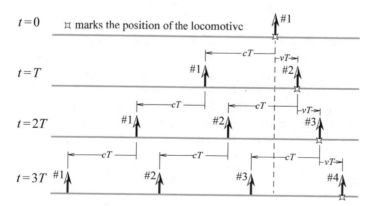

Fig. 6.23 Propagation behind a moving source that emits a periodic sequence of impulses

are separated by distance $cT + vT$, which corresponds to periodic arrival at such observers at intervals of $(1 + v/c)T$. The consequence is that the frequency that is heard is lowered, being given by

$$\boxed{\omega = \frac{\omega_0}{1 + v/c}} \qquad (6.6.2)$$

This is a downward Doppler shift. It is the same as the upward shift for negative v.

6.6.2 Moving Fluid

When a wave propagates through a steadily moving fluid, a fixed observer hears a different frequency from what would be heard if the observer moved in unison with the fluid. This situation arises when there is a steady wind, as well as in some parts of an HVAC systems. Our derivation of the wave equation assumed that the particle velocity is due solely to the acoustic signal. Correspondingly, we used an earth-fixed coordinate system to formulate analyses. Suppose that in the absence of an acoustic disturbance, the fluid flows at a constant velocity $\bar{v}_f$ in an arbitrary, but constant, direction. A fundamental aspect of nonrelativistic mechanics is that equations of motion may be formulated with respect to any inertial reference frame without altering its form. A coordinate system $x'y'z'$ that moves in unison with the particles of fluid translates at the constant flow velocity; it constitutes an inertial reference frame. The corollary of this observation is that the wave equation formulated with respect to $x'y'z'$ governs the propagation of acoustic signals through the flowing fluid, that is,

$$\frac{\partial^2 p}{\partial (x')^2} + \frac{\partial^2 p}{\partial (y')^2} + \frac{\partial^2 p}{\partial (z')^2} - \frac{1}{c^2}\frac{\partial^2 p}{\partial t^2} = 0 \qquad (6.6.3)$$

We consider a plane harmonic wave from the perspective of an observer who is moving in unison with $x'y'z'$. From that observer's perspective, the frequency is ω_0, the phase speed is c, and the propagation direction is $\bar{e}'$. Because the wave equation applies relative to $x'y'z'$, the signal is described relative to this coordinate system in the same way as it is for propagation relative to a fixed coordinate system. The origin of time is unimportant, so we represent the signal as a pure sine function. The following are equivalent representations,

$$p = A \sin\left[\omega_0\left(t - \frac{\bar{x}' \cdot \bar{e}'}{c}\right)\right] \equiv A \sin\left(\omega_0 t - \bar{k}' \cdot \bar{x}'\right), \quad \bar{k}' = \frac{\omega_0}{c}\bar{e}' \qquad (6.6.4)$$

What is the position dependence with respect to the earth-fixed xyz coordinates? The coordinate transformation will answer this question. We consider $x'y'z'$ at time $t = 0$ to have coincided with the stationary xyz. At time $t > 0$, the origin of $x'y'z'$ will be situated relative to xyz at $\bar{v}_f t$. The position $\bar{x}$ with respect to the fixed coordinates system is related to the position $\bar{x}'$ with respect to the moving coordinate system by a Galilean (that is, translational) transformation, which states that

$$\bar{x}' = \bar{x} - \bar{v}_f t \qquad (6.6.5)$$

Thus, the dependence of the plane wave pressure from the perspective of an observer on the fixed xyz coordinate system is

$$p = \mathrm{Re}\left\{ A \sin\left[\omega_0 t - \bar{k}' \cdot \left(\bar{x} - \bar{v}_f t \right) \right] \right\} \qquad (6.6.6)$$

The phase variable is $\left(\omega_0 + \bar{k}' \cdot \bar{v}_f \right) t - \bar{k}' \cdot \bar{x}$. The coefficient of t is the frequency ω heard by a fixed observer. Thus, we have established that this frequency and the wavenumber vector $\bar{k}$ measured relative to xyz are

$$\boxed{\omega = \omega_0 \left(1 + \frac{v_f}{c} \bar{e}_x \cdot \bar{e}' \right), \quad \bar{k} = \bar{k}'} \qquad (6.6.7)$$

The magnitude of the wavenumber for a plane wave is frequency divided by sound speed, according to which $|\bar{k}| = \omega/c_{xyz}$, where c_{xyz} is the phase speed relative to xyz. Equality of $\bar{k}$ and $\bar{k}'$ leads to the conclusion that the wavenumbers point in the same direction and that they have equal magnitude. It follows that $\omega/c_{xyz} = \omega_0/c$, so that

$$\boxed{\bar{e} = \bar{e}', \quad c_{xyz} = c + \bar{v}_f \cdot \bar{e}'} \qquad (6.6.8)$$

That the phase speed relative to the fixed reference frame is increased by the component of the flow velocity in the propagation direction is not surprising. (It is like our velocity when we walk on a moving sidewalk.) The frequency shows an upward Doppler shift if the component of flow velocity in the propagation direction is positive, which is like the property we established for the train signal. However, it is not obvious that a fixed observer perceives the same propagation direction as an observer who moves in unison with the flow.

6.6.3 Subsonic Point Source

Our initial exploration of the Doppler effect was limited to what is heard directly ahead of and behind the moving source of the sound. The more profound question is what is heard by an observer at an arbitrary location? This question is relevant to any radiating object whose speed is a substantial fraction of the speed of sound, such as an automobile, a machine part, or an aircraft, because a radiating object looks like a point if it is viewed at a sufficiently long range. To avoid complications, we will consider the case where the radiated field is radially symmetric when the source is stationary. In other words, we wish to establish the frequency, wavenumber vector, and waveform when a point source moves along a straight line at constant velocity.

We define the path of the source to be the x-axis, so the velocity of the source is $v\bar{e}_x$. The development begins with the subsonic case, in which the source moves slower than the speed of sound. The analytical solution also is valid for the supersonic case. However, interpretation and evaluation of that solution require special consideration. We will address those issues after the subsonic case has been fully explored.

To gain insight into what happens, let us construct the analog of Fig. 6.22 for the case of a point source. Its mass acceleration is proportional to $\cos(\omega_0 t)$, starting at $t = 0$. We define the x-axis to coincide with the source's path and place the origin at the source's location at $t = 0$. The development will track the constant phase wavefronts that are emitted by the source at $t = 0$, $2\pi/\omega_0$, etc. The value of rp is a maximum at each of these instants.

The fluid is stationary. Consequently, the signal that is emitted by the source at a specific instant propagates spherically from the instantaneous location of the source. At $t = 0$, the first maximum departs from the source and spreads spherically from the origin. At $t = T$, the radius of this first sphere of maximum pressure is cT, still centered at the origin. This is the first instant depicted in Fig. 6.24. At this time, the source, which is at $x = vT$, emits the second maximum sphere of maximum pressure. At $t = 2T$, the first pressure maximum is situated on a sphere of radius $2cT$ centered at the origin. The elapsed time for the second maximum at this instant is T, so the radius of the second sphere of maximum pressure is cT. This sphere is centered at the location from which it was emitted, which is at distance vT. At this instant, the third maximum is emitted. This sequence continues as long as the source is active. In Fig. 6.24, the value of τ for a sphere is the time at which the sphere was emitted by the source.

Certain features are evident in the figure. The spatial separation between spheres of maximum pressure is not constant, so there is a Doppler shift everywhere. An observer who is situated on the path of the source would see the same separation as the observer in the earlier scenario of impulses from a locomotive. However, there does not seem to be an obvious rule for evaluating this separation for any other observer.

Let us track the propagation of the signal emitted by the source at an arbitrary time τ. For a harmonic source at frequency ω_0, the phase variable is $\omega_0 \tau$, so the pressure emitted by the source at this instant is

$$\boxed{rp = \mathrm{Re}\left(Ae^{i\theta}\right), \quad \theta = \omega_0 \tau} \qquad (6.6.9)$$

At any instant $t > \tau$, this signal has propagated spherically over a time interval of $t - \tau$ at the speed of sound, so the radius of the sphere on which this phase may be observed is $c(t - \tau)$. Another viewpoint is that this radius is the distance from the source's location when this phase was emitted, which is $v\tau$ along the x-axis, to a field point on the sphere. Thus, it must be that

$$\boxed{r = \left[(x - v\tau)^2 + y^2 + z^2\right]^{1/2} = c(t - \tau)} \qquad (6.6.10)$$

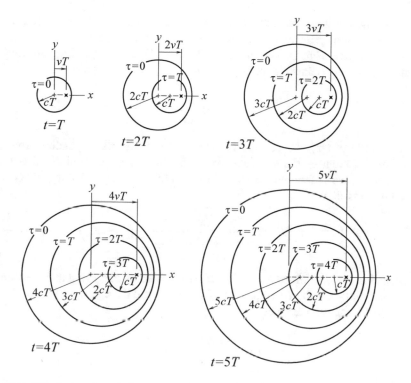

Fig. 6.24 Wavefronts of maximum pressure for the signal radiated by a harmonic point source that moves along a straight path at constant speed v

This equation defines the locus of points at which a specific phase $\omega_0 \tau$ is situated at a specific instant $t > \tau$.

An observer stationed at an arbitrary field point will see a sequence of instants at which one of the spheres depicted in Fig. 6.24 passes by. Each such instant corresponds to a maximum value of rp, so the interval between successive passages marks the perceived period. If this observer looks outward to the adjacent sphere on which rp equals the current value, then the distance would be interpreted as a wavelength. However, it is difficult to convert this perception to an quantitative description, so we shall take a different approach.

For a plane wave, the phase variable of a harmonic wave may be written as $\theta = \omega t - k\bar{x} \cdot \bar{e}$. If we know θ, we may evaluate the frequency as $\partial\theta/\partial t$. Furthermore, $\nabla\theta = -k\bar{e} = -\bar{k}$. In the case of a spherical wave, the phase variable is $\theta = \omega t - kr$, where $r = |\bar{x} - \bar{x}_0|$. Thus, $\omega = \partial\theta/\partial t$ in this case also. Furthermore, we proved that $\nabla r = \bar{e}_r$, so $\nabla\theta = -k\bar{e}_r = -\bar{k}$ for spherical waves. These properties are general. That is, the frequency and wavenumber vector may be determined according to

$$\boxed{\omega = \frac{\partial\theta}{\partial t}, \quad \bar{k} = -\nabla\theta}$$

(6.6.11)

The derivatives will not necessarily lead to constant values of ω and/or $\bar{k}$, in which case the definitions provide local descriptions at a specific instant.

Let us examine what the definitions in Eq. (6.6.11) tell us regarding the Doppler effect for a moving point source. Differentiation of Eq. (6.6.9) gives

$$\omega = \omega_0 \frac{\partial \tau}{\partial t} \tag{6.6.12}$$

The derivative of $\partial \tau / \partial t$ measures the rate at which signals originating from the successive locations of the source arrive at the measurement point. To determine this derivative, we differentiate both descriptions of r in Eq. (6.6.10) implicitly with respect to t as follows

$$\frac{\partial}{\partial t}\left[(x - v\tau)^2 + y^2 + z^2\right]^{1/2} \equiv \frac{(x - v\tau)}{\left[(x - c\tau)^2 + y^2 + z^2\right]^{1/2}}\left(-v\frac{\partial \tau}{\partial t}\right) = c\left(1 - \frac{\partial \tau}{\partial t}\right) \tag{6.6.13}$$

This equation may be solved for $\partial \tau / \partial t$. It is simplified by observing that the numerator of the fraction is the x component of $\bar{x} - v\tau \bar{e}_x$ and the denominator is r. Thus, the preceding equation reduces to

$$\left[c - v\frac{(\bar{x} - v\tau \bar{e}_x)\cdot \bar{e}_x}{r}\right]\frac{\partial \tau}{\partial t} = c \tag{6.6.14}$$

The radial unit vector relative for the sphere corresponding to $\theta = \omega_0 \tau$ is

$$\bar{e}_r = \frac{(\bar{x} - v\tau \bar{e}_x)}{r} \tag{6.6.15}$$

Hence, the preceding relation leads to

$$\frac{\partial \tau}{\partial t} = \frac{1}{1 - \dfrac{\bar{v}\cdot \bar{e}_r}{c}} \tag{6.6.16}$$

Correspondingly, we find from Eq. (6.6.12) that the perceived frequency is

$$\omega = \frac{\omega_0}{1 - \dfrac{v}{c}\left(\dfrac{x - v\tau}{r}\right)} \equiv \frac{\omega_0}{1 - \dfrac{\bar{v}\cdot \bar{e}_r}{c}} \tag{6.6.17}$$

Now that we have a description of the frequency shift, let us determine the wavenumber. Because the phase is $\theta = \omega_0 \tau$, Eq. (6.6.11) gives $\bar{k} = -\omega_0 \nabla \tau$. To evaluate $\nabla \tau$, we take the gradient of both sides of Eq. (6.6.10),

$$\nabla r = \frac{\partial r}{\partial x}\bar{e}_x + \frac{\partial r}{\partial y}\bar{e}_y + \frac{\partial r}{\partial z}\bar{e}_z + \frac{\partial r}{\partial \tau}$$

$$\nabla \tau \equiv \bar{e}_r + \frac{(x - c\tau)(-v)}{\left[(x - c\tau)^2 + y^2 + z^2\right]^{1/2}}\nabla \tau = -c\nabla \tau \qquad (6.6.18)$$

This relation is simplified by Eq. (6.6.15), with the result that the gradient of τ is

$$\nabla \tau = -\frac{\bar{e}_r}{c - \bar{v} \cdot \bar{e}_r} \qquad (6.6.19)$$

In view of Eq. (6.6.17), the wavenumber vector obtained from this gradient is

$$\boxed{\bar{k} = -\omega_0 \nabla \tau = \frac{\omega}{c}\bar{e}_r} \qquad (6.6.20)$$

In other words, the wavenumber vector at a field point is oriented radially from the location at which the instantaneous signal originated. Its magnitude is the same at that of a spherical wave emitted by a stationary source at frequency ω. This result could have been anticipated. As shown in Fig. 6.24, each phase propagates as an ordinary spherical wave—movement of the source merely leads to each phase being a spherical wave having a different location for its center.

Equation (6.6.17) states that the frequency heard by a stationary observer is increased if the radial component of the source's velocity is positive, and reduced if it is negative. If the observer is directly ahead of the source, then $\bar{e}_r = \bar{e}_x$ and $\bar{v} \cdot \bar{e}_r = v$. The Doppler shift in this case is the same as Eq. (6.6.1) for the one-dimensional case. A stationary observer at an arbitrary location will hear a time-dependent frequency as a result of the dependence of $\bar{e}_r$ on τ.

An interesting observation is that the derivation considered the source to be following a straight path at a constant velocity, but doing so was merely a way of specifying the position of the source as a function of time. The development is equally valid if $v\tau\bar{e}_x$ is replaced by the a different specification of the source's motion.

How one may evaluate the waveform at a specific location is addressed in the next example. In the course of the analysis, it will be evident that the entire development does not require that $|\bar{v}| < c$. Indeed, the formulation is equally applicable for a supersonically moving source. However, evaluation and interpretation of the signal properties in that case requires special consideration. The example introduces these issues, and they are resolved in the following section.

EXAMPLE 6.13 Abby is standing at sea level. She hears a distant aircraft that is in level flight at an altitude of 4 km at constant speed v. Its straight path will eventually bring it directly over Abby. The single turbine engine radiates a broad spectrum of frequencies, but one at 400 Hz is dominant. Determine as a function of elapsed time the frequency and pressure waveform heard by

Abby if the aircraft is moving subsonically at $v = 0.6c$. Then, explain how
these observations are altered when the aircraft is supersonic with $v = 1.5c$.

Significance

Examining what a stationary observer hears will greatly increase understanding of
the significance of the phase variable's dependence on time and position, and thereby
enhance our understanding of the formulation.

Solution

In order to emphasize the Doppler shift effect, we shall ignore the role of ground
reflections. Analysis of the direct wave begins with the observation that Eq. (6.6.9)
with $\omega_0 = 2\pi (400)$ rad/s defines the value of the pressure corresponding to given
values of τ and r. The value of r is the radial distance to an observer whose coordinates
are x, y, z from the location where a signal was emitted, τ is the time that the signal
was emitted, and t is the time that the signal arrives at the observer. Equating both
sides of Eq. (6.6.10) yields an equation relating t and τ. We wish to evaluate p as
a function of t at the observer's location. The most direct approach is to let τ have
a range of values beginning with the initial time, at which the source emitted an
audible signal and ending when it is out of the range of our interest. Let τ_n denote
one of these values. Then, we may find the corresponding value of t_n as a simple
manipulation of the second part of Eq. (6.6.10). With that value known, the first part
of that equation gives the value of r_n, which is the distance to the observation point
from the location where the phase $\omega_0 \tau$ left the source. The relations are

$$t_n = \tau_n + \frac{1}{c}\left[(x - v\tau_n)^2 + y^2 + z^2\right]^{1/2}, \quad r_n = c(t_n - \tau_n) \tag{1}$$

The monopole amplitude A is not specified, and no statement gives the pressure
observed at any location. Consequently, we can only evaluate the pressure relative to
its amplitude at a reference location. For this purpose, we observe that the minimum
radial distance from the source to the observer, and therefore the maximum pressure
amplitude, corresponds to the aircraft being directly overhead. The radial distance
when the aircraft is directly overhead is H, so we take the reference pressure to be
$|A|/H$.

Let us place the origin of xyz at the location of the aircraft when it is overhead,
and let $\tau = 0$ mark this instant. The phase lag of the source's signal is unimportant,
so we consider the source to be a pure sine. Correspondingly, the relative pressure
heard by the observer at instant t_n is

$$\frac{p_n}{p_{max}} = \frac{H}{r_n}\sin(\omega_0 \tau_n) \tag{2}$$

The frequency heard by the observer at each instant is given by Eq. (6.6.17),

$$\omega_n = \frac{\omega_0}{1 - \dfrac{v}{c}\left(\dfrac{x - v\tau_n}{r_n}\right)} \tag{3}$$

Our algorithm creates a column array of τ_n values, evaluates corresponding arrays of t_n and r_n values according to Eq. (1), and then evaluates arrays of p_n/p_{max} and ω_n/ω_0 values according to Eqs. (2) and (3). The only question is what range of τ values should we use. Abby's coordinates are $x = 0$, $y = -H$, $z = 0$. Because $\tau = 0$ marks the instant when the source is at the origin, the distance from the origin to the source is $v\tau$. The aircraft is initially heard when it is far away, so the lowest value of τ should be a large negative value. Similarly, when it is out-of-range, τ will be a large positive value. We begin with a guess that a maximum range of 40 km, which corresponds to $-40,000/v \le \tau_n \le 40,000/v$, is sufficient.

We take $c = 340$ m/s and set $v = 144$ m/s. The first graph in Fig. 1 displays $v\tau$ as a function of t. This information tells us the horizontal location of the aircraft when it emitted the signal that is heard at time t. The frequency in the second graph describes ω as a function of t. The frequency changes rapidly when the aircraft is nearly overhead. When the aircraft is far away, it is essentially moving directly toward or away from the observer, like the one-dimensional situation. The Doppler shift in these cases is $\omega = \omega_0 / (1 \mp v/c)$.

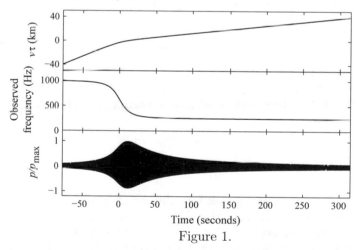

Figure 1.

The waveform heard by the observer is described in the third graph. The individual curve is not discernible because many cycles occur over a one second interval, so the width of a plotted line exceeds the period of the oscillation. If we were to zoom in on a very short interval, we would see a harmonic oscillation whose frequency is the current value of ω. The envelope of the signal indicates that the amplitude changes slowly relative to the period of the oscillation. Beyond the 40-km range used for the plot, the pressure amplitude essentially decays inversely with $v\tau$, which is a close approximation of the range.

Now, let us consider the supersonic case. Equations (1)–(3) were derived without restriction on the value of c, so we may implement the computational algorithm here also. We set $v = 510$ m/s, for which the values of $v\tau$ and ω as functions of elapsed time appear in Fig. 2. These graphs are unlike those for the subsonic case. The two most notable features are that there is no signal prior to some instant and that there are two values of τ for t afterward. The interpretation is that there is a specific instant at which the sound generated by the source first arrives at a field point, and that subsequent to that instant, two signals arrive at that point. One was generated at a location $v\tau^{(1)}$ whose value is on the lower branch of the graph of $v\tau$ as a function of t. Thus, this location is increasingly far back on the x-axis as t increases. The other source is emitted at $v\tau^{(2)}$ on the upper branch. This location moves ahead of the observer as t increases.

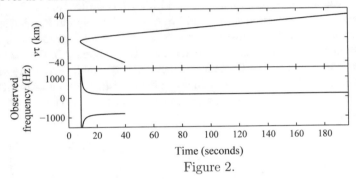

Figure 2.

The supersonic phenomena are too interesting and profound for us to limit our attention to a specific situation. Thus, we shall investigate the supersonic case in greater detail and then return to this example to consider what is heard by the observer.

6.6.4 Supersonic Point Source

The analysis in the previous example found that two signals are heard at a field point when a source moves supersonically. That observation is the general behavior. It has a simple mathematical explanation. The computational algorithm in the example considered Eq. (6.6.10) to provide the t value corresponding to a specific τ. Only one value of t is obtained in this viewpoint. However, it would be equally correct to consider that expression to be a quadratic equation whose solution gives the value of τ at a specified t. In the subsonic casem, both roots of this quadratic equation are real, but one corresponds to $\tau > t$, which gives $r < 0$ and violates the principal of causality. In the supersonic case, both roots for τ are imaginary until t attains a value t_{arr} that is the arrival time at the observer. After this time, both roots of the quadratic equation are real and satisfy $\tau < t$, as required by causality. From this, we conclude that it is generally true that an observer at any instant $t > t_{arr}$ hears two signals that were emitted by the source when it was at each of the $v\tau$ values.

Fig. 6.25 Development of the Mach cone that results when a point source has a constant velocity that exceeds the speed of sound

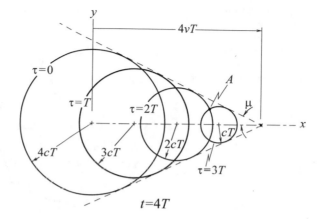

The preceding is the mathematical view, but what is the physical explanation for this behavior? To answer this question, we need to reconsider Fig. 6.24. Spheres of constant phase are depicted in Fig. 6.25, wherein $\tau = 0$ has been defined to correspond to the source being at the origin. Anyone who has studied compressible fluid dynamics will recognize the figure as depicting a *Mach cone*. (It is a cone because the field is symmetric about the x-axis.) It occurs because the source is moving faster than the acoustic signal that departs from it, so that spheres of constant phase have the appearance of being left behind. Ahead of the cone, there is no signal. The angle μ of the cone is the *Mach angle*, which is found from the figure to be

$$\mu = \sin^{-1}\left(\frac{c}{v}\right) \equiv \sin^{-1}\left(\frac{1}{M}\right) \qquad (6.6.21)$$

where $M = v/c$ is the *Mach number*.

The picture of constant phase spheres in Fig. 6.25 helps to explain why an observer receives signals that left the source when it was at two different locations. The instant that is depicted there is $t = 4T$. Two spheres corresponding to different τ values intersect at any location $\bar{x}$ inside the Mach cone. For example, point A in the figure lies on the spheres whose radii are cT and and $2cT$. The latter left the source at $\tau = 2T$ and the former departed at $\tau = 3T$. These features tell us that the sound heard by an observer at point A when $t = 4T$ came from the supersonically moving source at two instants, $\tau = 2T$ and $\tau = 3T$. At a later time, the apex of the Mach cone will have moved to the right and the radii of the spheres depicted in the figure will be larger. Two different spheres will intersect point A at that instant. Both spheres will be centered on the x-axis, with the center of the larger sphere situated father back and the smaller sphere centered farther ahead.

The movement of the Mach cone is different from the direction in which any signal propagates, which is radially outward from the location at which it originated. Thus, at the instant the Mach cone arrives at an observer, which is denoted as t_{arr}, the

Fig. 6.26 Arrival of the
Mach cone at field point A,
and the associated positions
of the source at the current
instant and the instant it
emitted the arriving signal

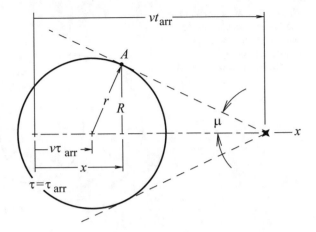

signal is propagating perpendicularly to the surface of the cone. We may determine
t_{arr} by examining the geometry of the Mach cone. Figure 6.26 shows the constant
phase sphere at t_{arr} for an arbitrary point A. The definition of the Mach cone is that
point A is the location where the sphere is tangent to the Mach cone. The constant
phase sphere containing this point was emitted at time τ_{arr}, so the center of this sphere
is situated at $v\tau_{arr}$ on the x-axis.

We wish to determine t_{arr} and τ_{arr} in terms of the x, y, z coordinates of point A.
We may do so by constructing the projection of point A onto the x-axis. The distance
along the x-axis from the projection to the source is $v t_{arr} - x$, while the distance from
the projection point to the point where the arriving signal was emitted is $x - v\tau_{arr}$.
The transverse distance is $R = (y^2 + z^2)^{1/2}$. A trigonometric evaluation of Fig. 6.26
reveals that

$$v t_{arr} - x = R \cot \mu, \quad x - v\tau_{arr} = R \tan \mu \tag{6.6.22}$$

Elimination of μ with the aid of Eq. (6.6.21) leads to

$$t_{arr} = \frac{1}{v} \left[x + \left(y^2 + z^2 \right)^{1/2} \left(\frac{v^2}{c^2} - 1 \right)^{1/2} \right]$$

$$\tau_{arr} = \frac{1}{v} \left[x - \left(\frac{y^2 + z^2}{\frac{v^2}{c^2} - 1} \right)^{1/2} \right] \tag{6.6.23}$$

The last graphs of Example 6.13 indicated that the plot of $v\tau$ as a function of t has an infinite slope at $t = t_{\text{arr}}$. This property is general. Equation (6.6.17) indicates that $\partial\tau/\partial t$ is infinite if

$$\frac{x - vt}{r} = \frac{c}{v} \tag{6.6.24}$$

In general, $r = c\,(t - \tau)$, which is the second description in Eq. (6.6.10). Thus, the preceding equation indicates that the arrival of Mach cone at an observer is characterized by

$$x - vt_{\text{arr}} = \frac{c^2}{v}\,(t_{\text{arr}} - \tau_{\text{arr}}) \tag{6.6.25}$$

Substitution of Eq. (6.6.23) would lead to identical satisfaction of this equation. Hence, we have verified that an observer hears no signal prior to arrival of the Mach cone and that its arrival features an infinite frequency at that instant.

We saw in the previous example that the plot of $v\tau$ versus t tends to straight lines on the upper branch, for which $\tau > \tau_{\text{arr}}$, and on the lower branch on which $\tau < \tau_{\text{arr}}$. This behavior arises because at large τ the source location for the instantaneous signal is on the x-axis far ahead or far behind the field point. There is little difference between the radial distance and the distance along the x axis in that case. The corollary of this property is that the Doppler shift for both branches tend to the constant values $\omega = \omega_0 / (1 \mp v/c)$ observed in the one-dimensional case. The slope on the lower branch of the τ versus t plot is negative, which means that the frequency is negative. There is nothing anomalous about this feature, because the phase is $\theta = \omega_0 \tau$, so decreasing τ merely means that the phase of that signal decreases with increasing t.

The next example returns to Example 6.13 to develop the manner in which the waveform at an arbitrary field point may be evaluated when $v > c$. Before we address that task, it is appropriate to note that our concern is with the acoustical signal that is generated by a harmonically varying source. An object that moves supersonically generates an aerodynamic shock wave. This shock wave is the "sonic boom" that can be quite surprising if one is not expecting its arrival. The sudden pressure rise of the shock wave occurs when the Mach cone passes, while the oscillating signal radiated by a supersonically moving source would be heard after the sonic boom passes. Thus, in the scenario of the example, the sound radiated by the engine would be heard after the sonic boom dies out.

EXAMPLE 6.14 Consider the situation addressed in Example 6.13, in which an aircraft in level flight along a straight path at an altitude of 4 km passes directly above an observer on the ground. The signal of interest is a steady 400 Hz tone. Determine the frequency and waveform heard by the observer when the speed is Mach 1.5.

Significance

This example will provide a detailed view of the behavior of the Mach cone, and how the received signal is affected by the simultaneous arrival of signals from two locations of the source. The results will show some unexpected features.

Solution

How we shall evaluate the derived formulas depends on the computational resources that are available. Because $\partial \tau / \partial t$ is infinite when the Mach cone arrives, the initial signal contains very high frequencies. A very small value of Δt is needed to capture the features of the initial waveform, but the time interval for a maximum range of 40 km is of the order of 80 s. Sampling this interval with a small Δt would require many computations. We could employ a vectorized computational algorithm, which is much more efficient. However, because the data is processed in core memory, memory management would be a problem if the amount of data to be stored exceeds the working space.

These difficulties are addressed by using a variable time step scheme. Let Δt be a representative time step for late times. If this increment were used uniformly, the time instants would be $n\Delta t$. To shorten the sampling interval for early time and eventually approach Δt at later time, the time instants will be set by multiplying Δt by a function of n that is very small when t_n is close to t_{arr}, but rises to a value close to one when $t_n \gg t_{arr}$. A quadratic dependence on n works very well, specifically

$$ t_n = t_{arr} + n\Delta t \, (n/N)^2 \,, \quad n = 1, ..., N $$

We select Δt based on the long range properties. As $|r| \to \infty$, both sources seem to be moving directly toward or away from the field point. The Doppler-shifted frequencies then approach $f = f_0 / |1 \pm v/c|$, which gives 227 and 800 Hz for $v = 1.5c$. The higher frequency requires the smaller Δt, so we select Δt to give eight sampling instants for one period of an 800 Hz signal, which gives $\Delta t = (1/800)/8 = 0.156$ ms. The value of N is set to give the desired time interval. At $n = N$, the formula for t_n gives $t_N = t_{arr} + N\Delta t$. Equation (6.6.23) gives $t_{arr} = 8.769$ s. Setting $t_N = 80$ s requires that $N = 4.57 \left(10^5\right)$, which we round up to $N = 5 \left(10^5\right)$.

The roots $\tau_n^{(1)}$ and $\tau_n^{(2)}$ of Eq. (6.6.10) at each instant are obtained by evaluating the coefficients of the quadratic equation,

$$ A = v^2 - c^2, \quad B = c^2 t_n - vx, \quad C = x^2 + y^2 + z^2 - (ct)^2, \quad D = B^2 - AC $$

$$ \tau_n^{(1)} = \frac{-B - D^{1/2}}{A}, \quad \tau_n^{(2)} = \frac{-B + D^{1/2}}{A} $$

The frequency corresponding to each value of τ is found from Eq. (6.6.17). Evaluation of the pressure requires values of the radial distance from each source location to the observer, which is $r_n^{(j)} = c \left(t_n - \tau_n^{(j)} \right)$. The superposition of the signals received from the two originating source locations is

$$\frac{p}{p_{\text{ref}}} = \frac{4000}{r_n^{(1)}} A e^{i\omega_0 \tau_n^{(1)}} + \frac{4000}{r_n^{(2)}} A e^{i\omega_0 \tau_n^{(2)}}$$

The reference pressure used here is the pressure at the observer's location when a stationary unit source is placed directly overhead, where the radius is 4000 m.

The first graph in Fig. 1 shows the time dependence of $v\tau$. The first of Eq. (6.6.23) was already stated to give $t_{\text{arr}} = 8.769$ s, and the second equation tells us that the initial signal was emitted by the source when $\tau_{\text{arr}} = -7.015$ s. As time goes by, the distance $v\tau^{(2)}$ to the forward source location increases, while the distance $v\tau^{(1)}$ to the rear source location becomes larger negatively. The rear location moves back more quickly than the forward one moves ahead, which means that the distance to the rear source location grows more quickly. The consequence is that the signal coming from the rear location becomes less important as time elapses. The second graph shows that the frequency of the signal coming from each source location starts by being very large (in fact, singular at t_{arr}), The asymptotic values are $f_0/(1 + v/c) = 267$ Hz for the forward source and $f_0/(1 - v/c) = 800$ Hz for the rear location.

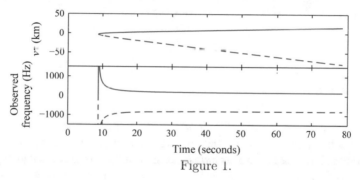

Figure 1.

Figure 2 shows the waveform heard by the observer on the ground. It is not possible to discern the individual fluctuations in the first graph, which spans the full-time range. However, it does show that the amplitude decreases, with a $1/(vt)$ behavior being a good fit after $t = 20$ s. The second graph zooms in on an initial 40-ms interval during which the Mach cone arrives. We see a series of beats, which is a consequence of the frequencies corresponding to the two-source locations being close and their amplitudes being nearly equal. The period of the beat depends on the difference of the frequencies, while the average of the two frequencies sets the oscillation rate within each beat. The third graph describes the waveform in an interval well after t_{arr}. There is no indication of beating. Rather, two distinct frequencies are evident. The contribution of the forward source is dominant because it comes from a closer location. Its frequency is lower, so the contribution of the backward location appears as a high-frequency fluctuation superposed on a low-frequency carrier tone. This term becomes less dominant as t increases, with the ultimate result that the received signal eventually becomes nearly harmonic.

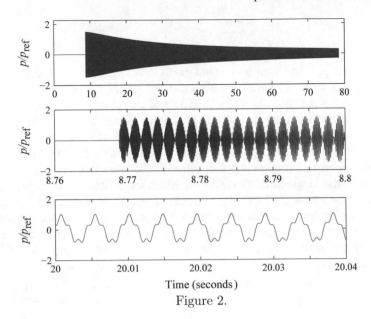

Figure 2.

6.7   Closure

We have devoted a great deal of attention to the study of spherical waves, first as individual entities and then in a variety of combinations, such as arrays, dipoles, and quadrupoles. These studies are important for their own sake. But it is equally important that spherical waves, in combination with plane waves, constitute fundamental building blocks for the analysis and interpretation of many complicated acoustical systems. Indeed, our studies thus far have led to identification of core concepts and analytical tools. The chapters that follow apply this knowledge base to a diverse range of topics that cover a broad range of subjects in which acousticians are active.

6.8   Homework Exercises

Exercise 6.1 The surface of a sphere was at rest until $t = 0$. At that instant, it begins a constant acceleration $\dot{v}_0$. The acceleration continues until $t = T$, when it ceases. Derive an expression for the pressure as a function of r and t. Examine this result to identify the set of nondimensional parametric combinations that allow the waveform at any radial distance to be described by a single plotted curve. Use these combinations to determine the conditions for which the pressure waveform at any location is nearly proportional to the velocity on the surface.

Exercise 6.2 The graph is the waveform measured at an unspecified radial distance ξ from a vibrating sphere at a great depth in a lake. The radius of the sphere is 200 mm, and $t = 0$ is the instant when the vibration was initiated. Use this graph to determine (a) the distance ξ, (b) the pressure waveform at the surface of the sphere, and (c) the dependence of the pressure on the radial distance from the center of the sphere when $t = 0.46$ ms.

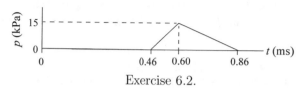

Exercise 6.2.

Exercise 6.3 A sphere having radius a displaces radially such that the field in the open space of a fluid is spherically symmetric. Before the sphere began to move the fluid was quiescent at homogeneous ambient conditions. The pressure at distance $r = \xi$ from the center of a sphere is observed to be

$$p = \frac{\rho_0 c^3}{a} Mt \exp(-\beta t) h(t)$$

where M is a small dimensionless constant and $h(t)$ is the step function. (a) At what time did the sphere begin to move? (b) Derive an expression for the pressure on the surface of the sphere as a function of time. (c) In the case where $\beta = 0.2c/a$, what is the largest magnitude of the sphere's radial velocity, and when does it occur?

Exercise 6.4 A sphere whose radius is a undergoes a spherically symmetric radial vibration at its interface with an ideal fluid. Before the sphere began to move the fluid was quiescent at homogeneous ambient conditions. The pressure at distance $r = \xi$ from the center of the sphere is observed to be

$$p = \rho_0 c^2 M \frac{\Delta}{t + \Delta} h(t)$$

where M is a small dimensionless constant, Δ is a positive constant whose units are seconds, and $h(t)$ is the step function. (a) Derive an expression for the radial velocity at $r = \xi$. (b) Derive an expression for the radial velocity of the surface of the sphere as a function of time.

Exercise 6.5 The wall enclosing a spherical cavity executes a vibration that is a two cycles of a sine function, that is, $v_r(0, t) = v_a \sin(2\pi t/T)$ if $0 < t < 2T$ and $v_r = 0$ outside this interval. In the situation of interest, $cT/a = 0.5$. This value is sufficiently small to consider $\dot{v}_0(t) \gg (c/a) v_0(t)$, which is a requirement for the use of Eq. (6.2.20). Determine the pressure $p/(\rho_0 c v_0)$ as a function of ct/a at $r = 0$, $a/2$, and a. The interval for the evaluations may be limited to $t < 12T$.

Exercise 6.6 An old research paper described the result of experimental measurement of the steady-state pressure radiated by a harmonically vibrating spherical source. The medium was air, $c = 340$ m/s, $\rho = 1.2$ kg/m^3. The result was described as $p = \text{Re}\left[P\left(r\right)\exp\left(i\omega t\right)\right]$, with $P\left(r\right) = 40\exp\left(-i2\pi/3\right)$ Pa at $r_1 = 2$ m, and $P\left(r\right) = 8i$ Pa at $r_2 = 10$ m. Unfortunately, the value of ω was not stated in the paper. (a) Determine the possible values of the frequency ω corresponding to this data. (b) Assume that the frequency is the lowest value obtained in Part (a), and the sphere's radius is $a = 100$ mm. What is the complex amplitude of the radial velocity on the surface of the source? (c) What is the time-averaged power radiated by the sphere?

Exercise 6.7 The steady-state acoustic signal measured at 25 m from the center of a 400 mm spherical source in air is observed to be $p = 0.10\cos\left(20\pi t\right)$ Pa. (a) Determine the pressure and radial particle velocity on the surface of the source. Give the amplitude and phase angle relative to a pure cosine for each quantity. (b) Determine the time-averaged acoustic power radiated by the source based on the farfield behavior. (c) Determine the instantaneous radial intensity observed on the surface of the source. What are the maximum and minimum values of this intensity? (d) Use the intensity in Part (c) to evaluate the time-averaged radiated power, and compare the result to the value obtained by considering the farfield behavior.

Exercise 6.8 The surface of a sphere whose radius is a undergoes a beating vibration, $v_r = V_0[\sin\left(\Omega - \sigma/2\right)t) + \sin\left(\Omega + \sigma/2\right)t)]$, $\sigma \ll \Omega$. Does the radiated pressure signal at an arbitrary location $r > a$ also constitute a beat? If so what is the amplitude and frequency of the envelope function?

Exercise 6.9 The sketch shows a virtual circle whose radius is R_0. It is situated at distance H from a sphere. The surface of the sphere executes a radially symmetric vibration, thereby inducing a spherical wave that propagates across the plane. The task is to determine the time-averaged radiated power that flows across the plane. The configuration is axisymmetric, so all quantities required to perform this evaluation can be described solely in terms of the polar angle ψ measured from the normal to the plane that intersects the center of the sphere. Also, it is permissible to consider kH to be large, so that all points on the plane are in the farfield. Plot the power flowing across this plane ratioed to the total radiated power as a function of R_0/H. Explain the limit as $R_0/H \rightarrow \infty$.

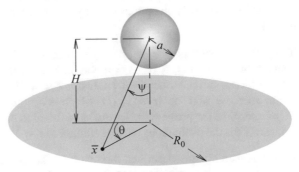

Exercise 6.9

Exercise 6.10 Two spherically symmetric sources, both of whose radius is a, vibrate radially, such that their surface velocities are

$$(v_r)_1 = \varepsilon c \cos(\omega_1 t) \quad (v_r)_2 = \varepsilon c \sin(\omega_2 t)$$

where c is the sound speed and ε is a small positive number. The separation distance L is sufficiently large to permit each sphere to radiate independently of the other. Derive expression expressions for $(p^2)_{\text{av}}$ and $\bar{I}_{\text{av}}$ at the position that is midway between the sources if ω_1 differs significantly from ω_2.

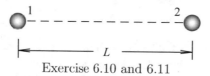

Exercise 6.10 and 6.11

Exercise 6.11 Two spherically symmetric sources, both of whose radius is a, vibrate radially at the same frequency, but 90° out-of-phase, such that their surface velocities are $(v_r)_1 = \varepsilon c \cos(\omega t)$, $(v_r)_2 = \varepsilon c \sin(\omega t)$ m/s. The separation distance L is sufficiently large to permit each sphere to radiate independently of the other. The field point of interest is midway between the sources. For this location, derive expressions for the complex pressure amplitude, the mean-squared pressure, and the time-averaged intensity.

Exercise 6.12 The two spheres in the sketch have the same radius a. They are vibrating harmonically, such that their radial surface velocities are

$$(v_r)_A = \text{Re}(A \exp(i\omega t)), \quad (v_r)_B = \text{Re}(B \exp(i\omega t))$$

where A and B are arbitrary complex coefficients. The separation L of the sources is sufficiently large to neglect their interaction, and $kL \gg 1$, so farfield approximations for the field at location C are valid. Derive an expression for $\bar{I}_{\text{av}}$ at $\bar{x}_C$. Are there any set of values of A, B, L, and s for which $\bar{I}_{\text{av}}(\bar{x}_C) = 0$?

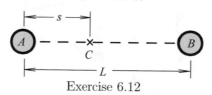

Exercise 6.12

Exercise 6.13 Three spherically symmetric underwater projectors are equally spaced along a circle whose radius is $L = 1.5$ m. The diameter of each source is 300 mm. The sources oscillate in-phase at 5 kHz. The power radiated by each source is 5 W, and the spacing radius L is sufficiently large that this power is the same as what it would radiate if the others were not present. Determine the time-averaged intensity at a field point $\bar{x}$ that is situated at $r = 9$ m, $\theta = 45°$.

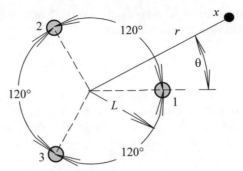

Exercise 6.13 and 6.14.

Exercise 6.14 Consider the field radiated by the three underwater projectors in Exercise 6.13. Determine the θ-dependence of the mean-squared pressure along the circle $r = 9$ m in the plane of the diagram. Perform this evaluation for frequencies of 50, 500, and 5000 Hz.

Exercise 6.15 The sketch depicts a 5×5 planar square array of point sources. All sources are in- phase, and their monopole amplitudes are equal. Evaluate $|P|$ as a function of x along three lines that are parallel to the x-axis: $z = D/2$, $y = 0$; $z = 2D$, $y = 0$, and $z = 8D$, $y = 0$. Frequencies for this evaluation are $kD = 0.5$, 2, and 8.

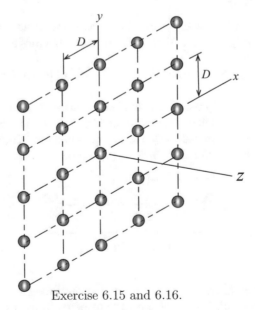

Exercise 6.15 and 6.16.

Exercise 6.16 The sketch depicts a 5×5 planar square array of point sources. All sources are in-phase, and their monopole amplitudes are equal. Evaluate and graph the angular dependence of the farfield mean-squared pressure at fixed r for field points in the xz-plane. Frequencies for this evaluation are $kD = 0.5$, 2, and 8.

Exercise 6.17 An acoustic field consists of the superposition of a plane wave propagating in the x direction and a uniform spherical wave generated by a point source
at $x = y = 0$. The frequency for both signals is ω. The monopole amplitude A is
real, whereas the complex amplitude B of the plane wave has arbitrary magnitude
and phase. (a) Derive an expression for the time-averaged intensity as a function of
the radial distance r and polar angle ψ. (b) Determine all possible values of B/A for
which the intensity at $x = a$, $y = 0$ is identically zero at all times. (c) Evaluate the
time-averaged power radiated by the point source by surrounding it with a sphere.
Compare the result to the power that would be radiated if the plane wave were not
present.

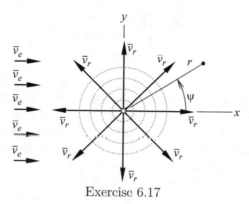

Exercise 6.17

Exercise 6.18 A point source is situated at distance H above a rigid surface. The
volume velocity of the source is $Q = Q_0 \exp(-ct/H) h(t)$. Draw the pressure wave
forms measured at points $\bar{x}_1$, $\bar{x}_2$, and $\bar{x}_3$.

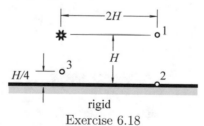

Exercise 6.18

Exercise 6.19 Point source A is situated near the corner where a pressure-release
floor meets a rigid wall. The wall and floor extend to sufficiently long distances
to permit considering them to be infinitely long, and no other boundaries are significant. The source begins to emit a harmonic signal at $t = 0$, such that $Q(t) =
Q_0 \sin(\pi t/T) h(t)$, where $T = L/c$. It is desired to examine the pressure field
received at field point $\bar{x}$ as a consequence of the presence of the source. (a) What
is the earliest value of t at which the sound field measured at $\bar{x}$ is nonzero? (b) At
what time t does the sound field at $\bar{x}$ attain a steady-state condition? (c) Derive an

expression for the amplitude of the pressure measured at $\bar{x}$ after the steady-state condition has been attained. (d) Sketch a graph of the waveform, p versus t, that would be measured at $\bar{x}$ in the interval between the instants in Parts (a) and (b).

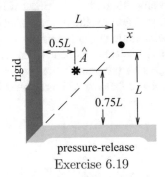

pressure-release

Exercise 6.19

Exercise 6.20 A source whose volume velocity is $\mathrm{Re}(Q \exp(i\omega t))$ is situated at equal distances from the floor and the wall, and all other boundaries are very distant. The floor may be taken to be rigid, and the wall is pressure-release. Derive an expression for the time-averaged intensity at an arbitrary point in the farfield. Describe the position of this point in terms of spherical coordinates whose axial direction $\bar{e}_z$ coincides with the corner, so that the plane of the diagram corresponds to the polar angle $\psi = 90°$ and the projected radial line in the sketch is at azimuthal angle θ. (b) Graph the angular dependence of $r^2 (I_r)_{av}$ in the xy, xz, and yz planes. Frequencies of interest are $kd = 2$ and 10.

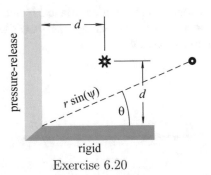

rigid

Exercise 6.20

Exercise 6.21 A dipole $D\bar{e}_y$ and a monopole A radiate at 2 kHz. They are separated by 4 m, as described in the sketch. The medium is air: $c = 340$ m/s, $\rho_0 = 1.2$ kg/m³. If only the monopole is active, the acoustic pressure at point B is $2\cos(\omega t)$ Pa. If only the dipole is active, the acoustic pressure at point B is $3\sin(\omega t - 2\pi/3)$ Pa. (a) Determine the values of A and D. (b) Determine the mean-squared pressure at point B when both sources are active. (c) Determine the complex amplitude of the particle velocity (vector) at point B when both sources are active. (d) Determine the time-averaged intensity at point B when both sources are active.

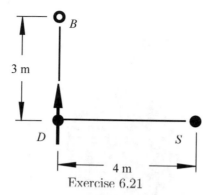

Exercise 6.21

Exercise 6.22 A sphere of radius a oscillates back and forth with maximum velocity v_1, simultaneously with undergoing a radial oscillation in which the maximum radial velocity is v_0. The oscillations occur in-phase, so the radial velocity on the surface is $v_r = [v_0 + v_1 \cos(\psi)] \cos(\omega t)$. Derive expressions for the pressure at an arbitrary field point, and for the time-averaged radiated power. Then, determine the ratio v_1/v_0 for which the pressure amplitude at a field point situated at $r \gg 2\pi/k$, $\psi = 0$, is twice the value that would be obtained if v_1 were zero.

Exercise 6.23 A sphere whose mass is M is suspended from the ceiling by a spring whose stiffness is K. An oscillatory force $F = \mathrm{Re}\left(\hat{F} \exp(i\omega t)\right)$, positive downward, induces a steady-state vibration. In addition, the acoustic force acts on opposition to movement of sphere. The displacement u is measured from the static position in which the spring force balances the gravitational force. Consequently, the equation of motion is $M\ddot{u} + Ku = \mathrm{Re}\left(\left(\hat{F} - F_z\right)\exp(i\omega t)\right)$, where F_z is the force resultant in Eq. (6.5.29). The steady-state displacement has the form $u = \mathrm{Re}\left(U \exp(i\omega t)\right)$. The natural frequency of this system if it were in a vacuum is $\omega_{\mathrm{nat}} = (K/M)^{1/2}$. (a) Derive an algebraic expression for U. System parameters appearing in this equation should be $\sigma = \omega_{\mathrm{nat}}a/c$ and $\mu = (4/3)\rho_0 a^3/M$, which is the ratio of the mass of the fluid displaced by the sphere to the sphere's mass. (b) Evaluate and plot the nondimensional frequency response $|U| M\omega_{\mathrm{nat}}^2/\hat{F}$ as a function of ka in the range from zero to 2σ. Cases of interest $\sigma = 1$ and $\sigma = 10$ when $\mu = 0.1$ and 1. (c) When damping from a dashpot is incorporated in the in-vacuo system, the maximum value of $|U| M\omega_{\mathrm{nat}}^2/\hat{F} = 0.5/\zeta$, where ζ is the ratio of critical damping. The frequency at which that maximum occurs is very close to $ka = \sigma$. For each of the cases in Part (b), identify the value of ka at which the maximum value $|U|$ occurs, and use that maximum to determine the equivalent ratio of critical damping resulting from the fluid loading. Do the results suggest any general trends?.

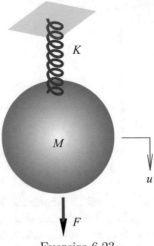

Exercise 6.23

Exercise 6.24 When an object executes a rigid body rotation about a fixed axis, the velocity of any point in the body is $\bar{v} = \bar{\Omega} \times \bar{x}$, where $\bar{\Omega}$ is the angular velocity, parallel to the rotation axis, and $\bar{x}$ is the position from any point on the rotation axis to the point of interest. The sphere in the figure executes such a rotation about the vertical shaft. The rotation angle about this axis oscillates with a very small amplitude, that is, $\Theta = \varepsilon \sin \omega t$, with $|\varepsilon| \ll 1$ and $\Omega = d\Theta/dt$. Describe the acoustic pressure field induced by this rotation. Identify any conditions the radius a, offset distance D, and oscillation rate ω must satisfy in order to permit a linearized analysis. *Hint*: The motion of any rigid body may be represented as a superposition of the movement of its center and a rotation about its center. In the case of a sphere, the velocity of any point on the surface in the rotational part is tangential to the surface. Therefore, the rotational part of the motion does not radiate sound. The consequence is that the sphere may be considered to translate at the velocity of its center.

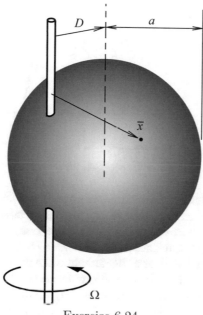

Exercise 6.24

Exercise 6.25 Suppose that instead of the rotation angle in Exercise 6.24 undergoing a small oscillation, the rotation rate is constant at $d\Theta/dt = \Omega_0$. Identify any limitations on a, D, and Ω_0 that must be satisfied in order that a linearized analysis be valid. Assume that these conditions are satisfied to analyze the pressure radiated by the sphere. *Hint:* The velocity normal to the surface is zero in any rotation about the sphere's center. Represent the center's velocity in terms of components relative to a fixed coordinate system whose z-axis coincides with the shaft, and then, apply the principle of superposition.

Exercise 6.26 A pair of oppositely directed dipoles $D\bar{e}_z$ and $-D\bar{e}_z$ are situated on the z-axis at $\varepsilon\bar{e}_z$ and $-\varepsilon\bar{e}_z$, respectively. (a) Derive an expression for the farfield directivity in terms of the polar angle θ and the value of $k\varepsilon$. (b) Derive a multipole expansion for the acoustic field. The series should include terms up to, and including, quadrupoles, but none higher. (c) Compare the results of Parts (a) and (b) for $k\varepsilon = 0.25$, 1, and 2.

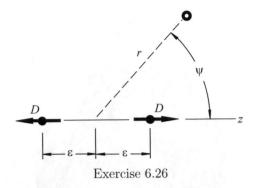

Exercise 6.26

Exercise 6.27 Two identical point sources whose volume velocity is Q_0 are situated at points $x = \pm L/2$, $z = H$ above a pressure-release surface at $z = 0$. (a) Describe the limiting behavior of the farfield when $L \ll 2\pi/k$ and H is not small. (b) Describe the limiting behavior of the farfield when $H \ll 2\pi/k$ and L is not small. (c) Describe the limiting behavior of the farfield when $H \ll 2\pi/k$ and $L \ll 2\pi/k$. (d) Consider situations in which the sources oscillate in air at 500 Hz. Cases of interest are the four combinations of $L = 0.1\pi/k$ and $2\pi/k$, with $H = 0.1\pi/k$ and $2\pi/k$. For each combination of L and H, plot $|P|r/(\omega\rho_0 Q_0)$ as a function of the polar angle ψ to field point $\bar{x}$, which is situated in the farfield on the xz plane. Compare each plot to the limits in Parts (a)–(c).

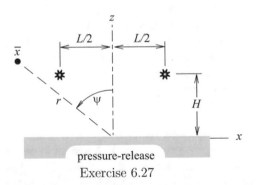

Exercise 6.27

Exercise 6.28 Consider two monopoles having amplitude $\hat{A}_1$ at position $\bar{x}_0 + (d/2)\bar{e}$ and amplitude $\hat{A}_2$ at position $\bar{x}_0 - (d/2)\bar{e}$. The magnitude and phase angle of each monopole amplitude are arbitrary, and the value of d is very small compared to $2\pi/k$, but finite. (a) Express the combined acoustic pressure due to these sources in terms of monopoles, dipoles, quadrupoles, etc. centered at $\bar{x}_0$. (b) Derive an expression for the power radiated by this pair of monopoles.

Exercise 6.29 Four equal sources having monopole amplitudes A_0 are situated at $\bar{\xi}_1 = \bar{0}$, $\bar{\xi}_2 = h\bar{e}_x$, $\bar{\xi}_3 = h\bar{e}_y$, $\bar{\xi}_4 = h\bar{e}_z$. (a) Derive a multipole expansion for the farfield pressure P_{ff} of this collection. Include the quadrupole contributions, but nothing of higher order. (b) Consider cases where $kh = 0.1\pi$ and $kh = \pi$. For each case, use the multipole expansion to evaluate $|P_{\mathrm{ff}}|\,r/A_0$ along the x-axis along the line $x = y$, $z = 0$, and along the line $x = y = z$. Compare each value to the one obtained from a direct summation of the individual sources.

Exercise 6.30 The sketch shows six-point sources located at equal distance s along the positive and negative x, y, and z axes. (a) Derive a multipole expansion describing this set of six sources in the case where the source strengths have arbitrary values A_m, $m = 1, 2, ..., 6$. Distributions higher than quadrupole may be ommitted from the expansion. Then, assume that the sources constitute a compact set. Based on that assumption: (b) Find the combination of source strengths for which the pressure field is a dipole whose moment is dirccted in the $\bar{e}_y$ direction. (c) Find the combination of source strengths for which the pressure field is a longidudinal quadrupolc that is aligned with the y-axis? (d) Find the combination of source strengths for which the pressure field is a lateral dipole in the xy plane? (e) Find the combination of source strengths for which the pressure field is a dipole whose moment is in direction $\bar{e}_D = 0.7071\left(\bar{e}_y + \bar{e}_z\right)$?

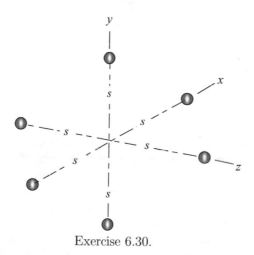

Exercise 6.30.

Exercise 6.31 The analysis in Sect. 6.4 found that the acoustic field generated by a point mass source situated at position $\bar{x}_0$ is governed by the inhomogeneous wave equation

$$\nabla^2 p - \frac{1}{c^2}\frac{\partial^2 p}{\partial t^2} = -\rho_0 \dot{Q}\,(t)\,\delta\,(\bar{x} - \bar{x}_0)$$

where $\bar{x}$ is the field point at which p is observed, and Q is the volume velocity. (a) Explain why this inhomogeneous equation is valid in the situation where $\bar{x}_0$ is a function of time. (b) In the situation depicted below, a monopole at position $\bar{x}_0$ is suspended from a cable that swings back and forth like a pendulum. The source strength varies harmonically at frequency ω, such that $Q = Q_0 \sin(\omega t)$. The supporting cable is very long, so the location of the monopole may be considered to be a vibration in the horizontal direction: $\bar{x}_0 = \varepsilon \cos(\Omega t)\,\bar{e}_x$, with $\Omega \ll \omega$. Identify the conditions that ε, ω, and Ω for which a multipole expansion relative to the stationary position $\bar{x}_0$ based on a compact set of sources is valid. Subject to the assumption that these conditions are met, derive an expression for the pressure at field point $\bar{x}$ in terms of the radial distance r, and polar angle θ. Only the two largest order terms in a multipole expansion are required. What frequencies would be heard at $\bar{x}$ in the farfield?

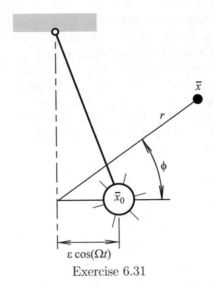

$\varepsilon \cos(\Omega t)$

Exercise 6.31

Exercise 6.32 It was established in Exercise 6.8 that the inhomogeneous wave equation applies when the source location $\bar{x}_0$ is time-dependent. Consider the situation where a pair of oppositely phased monopoles A and $-A$ are mounted at the ends of a short rod that spins about the vertical axis at angular speed Ω, which is much lower than the frequency ω of the monopoles. The bar's length is $d = \pi/20k$. Begin by deriving an expression for the farfield pressure at a point on the x-axis corresponding to an arbitrary θ that is constant. Substitution of $\theta = \Omega t$ yields the description of the field for the rotating pair of monopoles. From that result, identify the frequencies and RMS pressure that would be heard by an observer in the farfield aalong the x-axis.

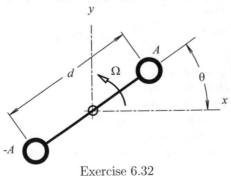

Exercise 6.32

Exercise 6.33 Beth is at the Bonneville Salt Flats, where she is observing the test of a turbine powered race car. She is positioned 100 m from the straight path the race car follows. She records the sound emitted by the vehicle from the instant when it is 2 km from her until it has traveled 2 km beyond her location. In this interval, the vehicle's speed 320 m/s. The primary sound emitted by the race car is a steady 5 kHz tone. At 50 m from the vehicle, the sound pressure level is 140 dB//20 μPa, and the signal may be assumed to spread spherically. Determine the frequency and amplitude of the signal measured by Beth as a function of time.

Exercise 6.34 Consider the scenario described in Fig. 6.22, in which a locomotive emits an impulse signal at a constant interval T. The difference here is that the locomotive has a constant acceleration $\dot{v}_0$, starting from rest at $x = 0$ when $t = 0$. The first impulse is emitted at $t = 0$, so the jth impulse is emitted by the locomotive from its current position at time $(j - 1)\,t$. (a) Derive an expression for the distance x_j from the origin of impulse j at an arbitrary time $t > (j - 1)\,T$. (b) Derive an expression for the distance x_{j+1} from the origin to impulse $j + 1$ at an arbitrary time t. (c) Use the results of Parts (a) and (b) to deduce the time interval Δ between the instant when a stationary listener at $x = \xi$ hears impulse j until she hears impulse $j + 1$. (d) As in the prior analysis, $2\pi/T$ is the frequency relative to the locomotive, whereas $2\pi/\Delta$ represents an instantaneous fundamental frequency heard by a stationary observer ahead of the locomotive. Does the expression for $2\pi/\Delta$ indicate that the Doppler-shifted frequency is constant. (e) It is possible that $x_{j+1} = x_j$ at some instant $\tilde{t}$. What is the significance of such an event?

	Layout: **T1 Standard**	Book ID: **439427_1_En**	Book ISBN: **978-3-319-56843-0**
	Chapter No.: **7**	Date: **3-5-2021** Time: **2:25 pm**	Page: **1/1**

Correction to: Acoustics—A Textbook for Engineers and Physicists

Correction to:
J. H. Ginsberg, *Acoustics—A Textbook for Engineers and Physicists, Volume I,* **https://doi.org/10.1007/978-3-319-56844-7**

In the original version of this book, the following belated corrections have been updated:

The Electronic Supplementary materials have been included in Chaps. 1–6.
The book and the chapters has been updated with the change.

The updated version of the book can be found at
https://doi.org/10.1007/978-3-319-56844-7

© Springer International Publishing AG 2021   C1
J. H. Ginsberg, *Acoustics—A Textbook for Engineers and Physicists, Volume I,*
https://doi.org/10.1007/978-3-319-56844-7_7

Index

A

Absolute pressure, 97, 324
Absorption coefficient
    at an interface, 370
    definition, 229
    relation to dissipative processes, 230,
        233, 237, 245
Acceleration, 299, 437
Acoustical energy, 329, 331
Acoustical Society of America, 2
Acoustical tile, 370, 419
Acoustical transmission line, 248
Acoustic impedance, 209, 259, 265, 273
Acoustic perturbation, 97
Acoustic source, *see* Source
Acoustic-structure interaction, 3
Adiabatic assumption, 103
Adiabatic compressibility, 105
Air, properties of, 106, 108, 239
Aliasing, 57
Ambient conditions, 58, 102, 239
Amplitude, 5, 122
Analogy
    electical circuit, 201
    for an elastic cable, 181
    for a stretched cable, 172
Angle of incidence, 358, 381
Angle of reflection, 359, 381
Angle of transmission, 373, 383
Arrival time, 314
Atmosphere, properties of, 108, 239
Attenuation, 228
Attenuation factor, 155
Averaging, 18, 74, 275
Axisymmetric field, 346, 487

Azimuthal angle, 434
Azimuthal function, 436

B

Backward characteristic, 115, 194
Backward propagating wave, 115, 118, 198
B/A coefficient, 105
Baffle, 275
Barrier, 395, 408
Beating signal, 22
Bessel function
    cylindrical, 88, 276
    spherical, 462
Boundary condition
    for a cable, 173, 181
    for a plane wave, 132, 134
    for a waveguide, 159, 197
    linearization of, 132, 348
    on a vibrating sphere, 439, 440
Boundary layer, 243
Branch cut, 353, 385
Breathing mode, 437, 442
Broadband signal, 74
Broadside array, 490
Bulk modulus, 104, 106

C

Cable, stretched, 170
Causality, principle of, 72, 548
Cavitation, 375
Cavity, acoustic, 273, 444
Cavity modes, 465
C.c., definition, 14

© Springer International Publishing AG 2018
J.H. Ginsberg, *Acoustics—A Textbook for Engineers and Physicists, Volume I*,
DOI 10.1007/978-3-319-56844-7